PLATE I (*Frontispiece*)

Shining Cliff,   Middleton Dale: Upper Monsal Dale Beds, forming cliff, overlain by Eyam Limestones   (L 187)

NATURAL ENVIRONMENT RESEARCH COUNCIL

INSTITUTE OF GEOLOGICAL SCIENCES

MEMOIRS OF THE GEOLOGICAL SURVEY OF GREAT BRITAIN

ENGLAND AND WALES

# Geology of the Country around Chapel en le Frith

(*Explanation of One-inch Geological Sheet 99, New Series*)

BY

I. P. STEVENSON, M.Sc. and G. D. GAUNT, B.Sc.

*Palaeontology by*

M. Mitchell, M.A., W. H. C. Ramsbottom, Ph.D. and M. A. Calver, M.A.

*Petrography by*

R. K. Harrison, M.Sc.

*LONDON*
HER MAJESTY'S STATIONERY OFFICE
*1971*

*The Institute of Geological Sciences
was formed by the
incorporation of the Geological Survey of Great Britain
and the Museum of Practical Geology
with Overseas Geological Surveys
and is a constituent body of the Natural Environment Research Council*

SBN 11 880115 5 *

Printed in England for Her Majesty's Stationery Office
by Hull Printers Ltd., Willerby, Hull HU10 6DH

# PREFACE

THE DISTRICT represented on the New Series One-inch Geological Sheet 99 (Chapel en le Frith) was first geologically surveyed by E. Hull, John Phillips and W. W. Smyth. Old Series maps were published in 1852 (81 NE and 81 SE) and 1864 (81 NW and 81 SW). Subsequent additions by A. H. Green, C. le Neve Foster, J. R. Dakyns and E. Hull were published in 1866, 1867 and 1874. The western part of the district was described in a memoir by Hull and Green published in 1866 and the remainder in 1887 in a memoir by A. H. Green, C. le Neve Foster and J. R. Dakyns.

Small marginal areas of the Chapel en le Frith sheet were surveyed during work on adjoining sheets between 1927 and 1950. Systematic fieldwork was started in 1951 by Mr. I. P. Stevenson, later joined by Mr. R. A. Eden for a few seasons, and then by Mr. G. D. Gaunt, with whose help the sheet was completed in 1961. The six-inch maps covered by the various surveyors are listed on p. xii. Solid and drift editions of the one-inch map were published in 1967. Part of the Chapel en le Frith district is covered by a special 1 : 25,000 sheet (Edale and Castleton) published in 1969.

Mr. Stevenson has written most of the present memoir and was responsible for its compilation. Mr. Gaunt wrote parts of the Introduction and chapters on Millstone Grit Series, Pleistocene and Recent Deposits, Mineral Products and Water Supply. The memoir was edited by Mr. D. R. A. Ponsford and Dr. A. W. Woodland. Mr. M. Mitchell identified fossils from the Carboniferous Limestone and contributed a chapter on the Stratigraphical Palaeontology of the series. Dr. W. H. C. Ramsbottom and Mr. M. A. Calver identified fossils from the Millstone Grit and Coal Measures respectively and contributed to the relevant stratigraphical chapters. Petrography of the sedimentary and igneous rocks is by Mr. R. K. Harrison.

Thanks are given to Dr. D. E. Owen (Director) and Mr. D. Rushton, of the Manchester Museum, and to Mr. I. E. Burton, Curator and Librarian, Buxton Library and Museum, for allowing us to borrow bones of Pleistocene mammals from Doveholes; also to Dr. A. Sutcliffe, British Museum (Natural History) and Mr. H. E. P. Spencer, late of the Ipswich Museum, for examining them on our behalf.

Acknowledgement is here made to the Derwent Valley Water Board who have kindly placed a series of borehole records at our disposal and in appropriate parts of the text to many other concerns who have kindly furnished or released information.

K. C. DUNHAM
*Director*

Institute of Geological Sciences
Exhibition Road
London, S.W.7
30th April, 1971.

v

# CONTENTS

(References are listed at the end of each chapter)

# ILLUSTRATIONS

## TEXT-FIGURES

# PLATES

## EXPLANATION OF PLATES

PLATE I   Shining Cliff (seen in profile) shows beds above the Black Bed dipping to the north-east. The ledge marks the position of the base of the Black Bed while the White Bed can be distinguished as a nick in the cliff face. The basal, dark beds of the Eyam Limestones form gently sloping ground above the cliff and are overlain by flat reef forming a feature. The Shale Grit forms a scarp in the background.

PLATE II   A   Well-bedded grey and grey-brown limestones in the Woo Dale Beds form the small cliff in the section and are overlain by a few feet of dark limestones, mainly unexposed, at the top of the subdivision. The uppermost part of the section is in pale massive limestones at the base of the Chee Tor Rock.
           B   Potholed surface, about 110 ft below the base of the Chee Tor Rock, exposed by quarrying. The potholes, up to 1 ft deep, have been mainly cleaned out by rapid weathering of the original clay infilling.

PLATE III  A   Massive pale limestone in the upper part of the Chee Tor Rock shows characteristic vertical jointing. The Lower *Davidsonina septosa* Band lies a little above the higher of the two partings in the middle of the face. The Lower Miller's Dale Lava lies close behind the face.
           B   Massive pale limestones in the Miller's Dale Beds with the Dove Holes Tuff, 6 ft thick, resting on a potholed surface.

PLATE IV   Sections of boreholes in the Michill Bank area showing local variation in the Bee Low Limestones, with the Pindale Tuff, and the lower part of the Monsal Dale Beds.

PLATE V   Sections of boreholes in and near Earle's Quarry showing the development of the Pindale Tuff and lateral variations in the Bee Low Limestones and the lower part of the Monsal Dale Beds. The key map shows the geology of the quarry area and isopachs on the tuff.

PLATE VI   Generalized vertical sections of the Lower Monsal Dale Beds. For explanation see text.

PLATE VII   Generalized vertical sections of the Upper Monsal Dale Beds. For explanation see text.

PLATE VIII   Sections of boreholes in the Bradwell Moor area, showing lateral variation in the upper part of the Monsal Dale Beds.

PLATE IXA   The eastern side of Bradwell Dale, towards the southern end of the dale. Massive flat-reef limestones in the Eyam Limestones rest unconformably on pale cherty limestones in the Upper Monsal Dale Beds. A marked local non-sequence is present near the base of the section.

# LIST OF SIX-INCH MAPS

Geological six-inch maps included wholly or in part in the one-inch map Sheet 99 (Chapel en le Frith) are listed below together with the initials of the surveyors and the dates of survey. The surveyors were: R. A. Eden, W. N. Edwards, G. D. Gaunt, R. H. Price, J. V. Stephens, I. P. Stevenson and B. J. Taylor. Small parts of one-inch Sheet 99 are included on the following six-inch maps which are not yet complete: SJ 97 S.E., SK 07 S.W., SK 07 S.E., SK 17 S.W., SK 17 S.E., SK 27 S.W. Printed maps marked with an asterisk are uncoloured, or, to suit individual requirements, hand-coloured. The remaining maps are available for public reference in manuscript form.

| | | | | |
|---|---|---|---|---|
| SJ 97 N.E.* | Kettleshume .. .. .. | I.P.S., B.J.T. | .. | 1951, 1954–55 |
| SJ 98 S.E.* | Disley and Lyme Handley .. | I.P.S., R.H.P. | .. | 1949, 1952–55 |
| N.E.* | Mellor and New Mills | I.P.S., R.H.P. | .. | 1948–55 |
| SJ 99 S.E.* | Chisworth .. .. .. | J.V.S., I.P.S., R.H.P. | .. | 1926, 1948, 1955 |
| SK 07 N.E. | Dove Holes .. .. .. | I.P.S. | .. .. | 1952–59 |
| N.W.* | Fernilee and Combs .. | I.P.S. | .. .. | 1952–57 |
| SK 08 S.E. | Chapel en le Frith .. | I.P.S. | .. .. | 1952–57 |
| S.W.* | Whaley Bridge and Chinley .. | I.P.S. | .. | 1952–54 |
| N.E. | Kinder Scout .. .. | I.P.S. | .. .. | 1953–62 |
| N.W.* | New Mills and Hayfield .. | I.P.S. | .. .. | 1952–55 |
| SK 09 S.E. | Black Ashop Moor and Cold-harbour Moor | J.V.S., I.P.S., G.D.G. | .. | 1926–61 |
| S.W.* | Charlesworth .. .. .. | J.V.S., I.P.S. | .. | 1927, 1954–61 |
| SK 17 N.E. | Tideswell and Great Hucklow | I.P.S., R.A.E. | .. | 1953–61 |
| N.W. | Peak Forest .. .. .. | I.P.S. | .. .. | 1953–61 |
| SK 18 S.E. | Castleton, Hope and Bradwell | I.P.S., G.D.G. | .. | 1960–61 |
| S.W. | Castleton and Barber Booth | I.P.S. | .. .. | 1956–61 |
| N.E. | Derwent .. .. .. | G.D.G. | .. .. | 1957–60 |
| N.W. | Edale Moor .. .. .. | G.D.G. | .. .. | 1958–61 |
| SK 19 S.E. | Howden .. .. .. | J.V.S., G.D.G. | .. | 1928–58 |
| S.W. | Hope Woodlands .. .. | J.V.S., G.D.G. | .. | 1928–60 |
| SK 27 S.E. | Curbar .. .. .. .. | W.N.E., R.A.E. | | 1939, 1947–54 |
| N.E. | Nether Padley .. .. .. | R.A.E., W.N.E. | .. | 1939–53 |
| N.W. | Eyam and Stoney Middleton | I.P.S., R.A.E., G.D.G., W.N.E. | | 1939–61 |
| SK 28 S.E. | White Path Moss and Millstone Edge .. .. .. .. | W.N.E., I.P.S. | .. | 1939–51 |
| S.W. | Bamford and Hathersage .. | G.D.G., I.P.S., R.A.E., W.N.E. | | 1939–60 |
| N.E. | Hallam Moors .. .. | I.P.S. (see Yorkshire 293 N.E. & S.E.) .. | .. | 1949, 1950–51 |
| N.W. | Bradfield, Derwent and Bamford Moors .. .. .. | I.P.S., G.D.G. | .. | 1949–60 |
| SK 29 S.E. | Agden .. .. .. .. | J.V.S., I.P.S. (see Yorkshire 287 N.E. & S.E.) .. | | 1929–30, 1938–50 |
| S.W. | Bradfield Moors .. .. | W.N.E., J.V.S., I.P.S., G.D.G. | | 1928–58 |

# Chapter I

# INTRODUCTION

## GEOGRAPHICAL SETTING

THIS MEMOIR describes the geology of the district covered by the Chapel en le Frith (99) Sheet of the One-inch Geological Map of England and Wales.[1] This area lies mainly in the Peak District of Derbyshire but also extends into York-shire in the north-east and into Cheshire in the west. The relations of the district to adjacent areas are shown in Fig. 1. Large tracts lie within the Peak District National Park. The region described lies across the axis of the Pennine uplift. The morphology, drainage and pattern of land usage clearly reflect variations in the lithology and structure of the underlying rocks. The principal physical features are illustrated in Fig. 2.

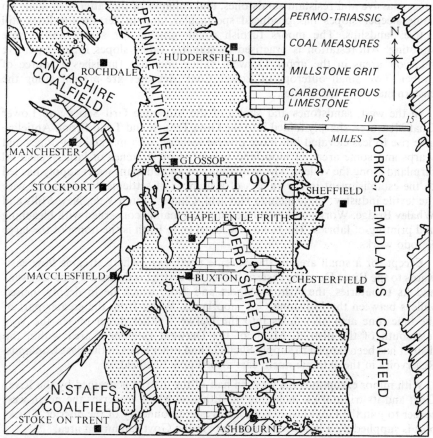

FIG. 1. *Sketch-map showing the general geological relations of the district*

[1]Throughout this memoir the word 'district' refers to this area.

In the south, the Carboniferous Limestone outcrop forms a plateau, ranging in height from 1100 ft to just over 1500 ft on Bradwell Moor. This tract is gently dissected with some more deeply-cut valleys. Much of the drainage of this area is underground and surface streams are limited to valleys near the margin of the outcrop and to areas with perched water tables over impervious igneous rocks. Karst features such as potholes, sink-holes, caves and limestone scars are well-developed, though limestone pavements, typical of certain areas of Yorkshire, are little developed. Numerous spoil heaps and open workings along the lines of veins are the result of centuries of lead mining. In many places the lines of the more important veins stand out by virtue of the trees which have been planted along them. At the present time the working of limestone itself is producing striking modifications to the landscape.

To the north and east of the limestone area lie broad expanses of high, often peat-covered moorland, covering much of the Millstone Grit outcrop and rising to 2088 ft O.D. on the plateau of Kinder Scout, the highest point in the southern Pennines. The alternation of hard sandstones and soft shales produces ridges and hollows along the deeply incised valleys and, where the beds are inclined, a well-developed scarp and dip-slope topography is developed. Tors and craggy escarpments or 'edges' are conspicuous along the outcrops of the more massive and resistant sandstones, and collapse of the latter over shale on some steeper slopes has produced spectacular, and in certain localities still active, landslips. The moors furnish rough grazing for sheep, and grouse shooting. Afforestation is practised to stabilize valley slopes in the extensive reservoir areas in the upper Derwent valley and also furnishes a source of timber. Two small outcrops of Lower Coal Measures are present along the eastern margin of the district.

In the west, sandstones and shales of the Millstone Grit Series and Lower Coal Measures, sharply folded into the Goyt Trough and Todd Brook Anticline give rise to less elevated but more diverse scenery, with marked dip-slopes and scarps and some areas of moorland separated by stretches of largely pastoral farmland along the valleys. The availability of an abundant water supply has led to the establishment in the valleys of a number of settlements associated with the textile industry; these include Hayfield, Chinley, Birch Vale, New Mills and Whaley Bridge. Works in these places are now mainly concerned with bleaching and printing of fabrics. The town of Chapel en le Frith is the home of the large Ferodo works.

Except for a small area in the extreme north-east, where streams flow eastwards to the River Don, the eastern half of the district is drained by the Derwent and its tributaries, the Ashop, Alport and Noe, which lie in deeply incised valleys between the Millstone Grit moorlands. Water from the greater part of the limestone area also flows into the Derwent either direct or via the Wye just south of the district. Much of the Derwent catchment area within the district has been tapped to supply Sheffield and other cities in the East Midlands. Reservoirs in the extreme north-east also supply Sheffield.

With minor exceptions the western half of the district is drained by the River Goyt and its tributaries, including the Etherow which traverses the north-west corner to join the Goyt just beyond the western boundary of the district. Stockport is supplied by a number of reservoirs in the Goyt catchment area.

G.D.G., I.P.S.

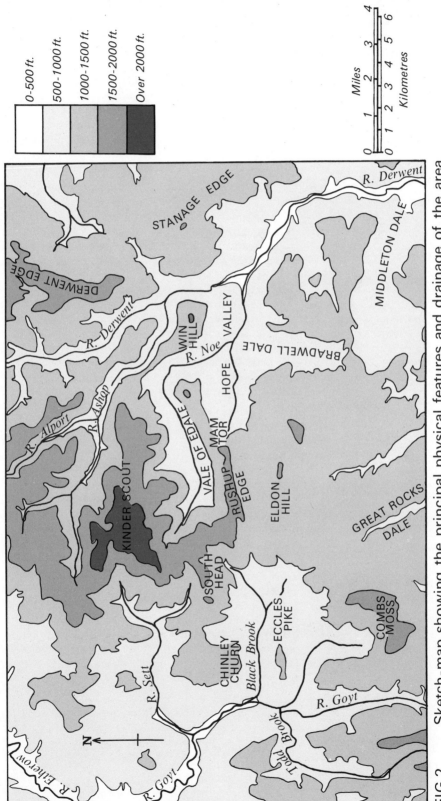

FIG.2    Sketch-map showing the principal physical features and drainage of the area

0-500 ft.
500-1000 ft.
1000-1500 ft.
1500-2000 ft.
Over 2000 ft.

Miles
Kilometres

DERWENT EDGE

R. Derwent

R. Alport

R. Ashop

KINDER SCOUT

R. Sett

SOUTH HEAD

CHINLEY CHURN

Black Brook

ECCLES PIKE

R. Ellerow

N

R. Goyt

Todd Brook

R. Goyt

COMBS MOSS

VALE OF EDALE

WIN HILL

R. Noe

MAM TOR

RUSHUP EDGE

HOPE VALLEY

ELDON HILL

STANAGE EDGE

BRADWELL DALE

GREAT ROCKS DALE

MIDDLETON DALE

R. Derwent

# HISTORICAL

The earliest signs of human occupation in the district are the worked flints, including microliths, of Mesolithic age found beneath the peat on the moorlands (Armstrong 1956, pp. 99–101). Neolithic remains are scarce but include the Bull Ring, a stone circle at Dove Holes and a smaller circle at Green Low, Chapel en le Frith (Armstrong ibid, p. 105; Edwards and others 1962, p. 122). In contrast, the Bronze Age appears to have marked a sharp increase in the settlement of the district. Many hills in or near the limestone area bear tumuli dating from this period and this accounts for the prevalence of the synonym 'low' in place-names in this tract, examples being Bee Low, Hurd Low and Cow Low. Evidence of Iron Age occupation is less widespread but includes the hill forts of Mam Tor and Castle Naze.

The Romans established the town of Aquae Arnemetiae (now Buxton) on account of its thermal waters and linked it with their fort at Anavio (Brough) by a road, Batham Gate. Another Roman road, Long Causeway, ran from Stanage Pole over the Snake Pass to Glossop linking the forts of Templeborough and Melandra Castle (Bartlett and Preston 1956, p. 115). In pre-Saxon times much of the limestone area and the lower Millstone Grit ground was forest, though tree cover was probably sparse in the more exposed areas. Saxon times saw a partial clearance of the forest for agricultural purposes (Edwards and others 1962, pp. 61–2), but after the Norman Conquest the establishment of a royal hunting forest, centred about Peak Forest, stopped further clearance for a time.

The working of minerals in the district is of considerable antiquity. The first mineral to be produced was lead and the Odin Mine at Castleton has been considered by some to have been first worked by the Romans though this is more usually attributed to the Saxons. A Roman pig of lead has been found at Brough (Smythe 1940, pp. 139–45) but may have been produced elsewhere in Derbyshire. Lead working continued in many parts of Derbyshire in the Middle Ages and after the Norman Conquest was sufficiently widespread for the establishment of a special code of laws. These stem from the Inquisition of Ashbourne in 1288 (Fuller 1965, p. 378) which also established special courts, the Barmote Courts, to enforce them; these courts are still in existence today. Smelting of the lead ore was carried out in scattered hearths or 'boles' using local timber and the word 'bole' survives in place names at several localities. More recent working in the seventeenth and eighteenth centuries has been associated with the driving of drainage levels or 'soughs' to de-water the workings, the most important in the present district being Stoke Sough and Moorwood Sough. Raistrick and Jennings (1965, pp. 276–7) state that the maximum production of lead was reached in the mid-eighteenth century.

Coal has been worked for over three hundred years in the area of the Goyt Trough. Bunting (1940, p. 156) states that workings at Fernilee were in operation in 1627. The most extensive coal working in the district was carried out in the nineteenth century at Whaley Bridge and the coal was used partly for lime burning. More recent working has been in conjunction with that of fireclays. The industry is now nearly extinct though considerable reserves remain.

Lime-burning is yet another local industry which has been long established. Bunting (ibid., pp. 297–9) states that fourteen kilns were in operation at Dove Holes in 1650. The older workings took the form of a series of closely spaced shallow excavations situated close to one or more kilns. Many of these areas have been planted with trees.                                    I.P.S.

# GEOLOGICAL SEQUENCE

The formations represented on the one-inch geological map and sections of the district are summarized below:

## SUPERFICIAL FORMATIONS (DRIFT)

RECENT AND PLEISTOCENE

| | |
|---|---|
| Landslips | Head |
| Hill Peat | Glacial Sand and Gravel |
| Alluvium | Boulder Clay |
| River Terrace, undifferentiated | Glacial Lake Deposits |

## SOLID FORMATIONS

CARBONIFEROUS

Thickness in feet
(generalized)

COAL MEASURES (WESTPHALIAN)

Lower Coal Measures
Shales, mudstones and sandstones, including the Milnrow
Sandstone and Woodhead Hill Rock (Crawshaw Sandstone in eastern area) with thin coals and seatearths and
with the *Gastrioceras subcrenatum* Marine Band at base          up to 850

MILLSTONE GRIT SERIES (NAMURIAN)

Rough Rock Group: Yeadonian ($G_1$)
Shales and mudstones with Rough Rock at top and the
*Gastrioceras cancellatum* Marine Band at base          up to 430

Middle Grit Group: Marsdenian ($R_2$)
Shales, mudstones and sandstones including Redmires
Flags, Chatsworth Grit, Roaches and Corbar Grits
(Ashover Grit in south-eastern area) and Heyden Rock
and with the *Reticuloceras gracile* Marine Band at base          up to 1800

Kinderscout Grit Group: Kinderscoutian ($R_1$)
Shales, mudstones and sandstones including Kinderscout Grit, Shale Grit and Mam Tor Beds and including
the upper part of the Edale Shales with the *Homoceras
magistrorum* Band at base          up to ?1700

Beds below Kinderscout Grit Group: Pendleian, Arns-
bergian, Chokierian and Alportian ($E_1$ to $H_2$)
Shales and mudstones with very thin ironstones, limestones, siltstones and 'crowstones' (quartzitic sandstones),
consisting of the main part of the Edale Shales with
*Cravenoceras leion* at base          up to 990

CARBONIFEROUS LIMESTONE SERIES (DINANTIAN)

Viséan

Eyam Group ($P_2$)
Dark, thin-bedded limestones; dark shales in upper part          180

Monsal Dale Group ($D_2$)
Grey and dark grey limestones with Litton Tuff and
Upper Miller's Dale Lava          600

Bee Low Group ($D_1$)
Pale grey massive limestones with Lower Miller's Dale
Lava          550

Woo Dale Group ($S_2$)
Dark grey and grey thin-bedded limestones          230

# GEOLOGICAL HISTORY

The Woo Dale Borehole (Cope 1949) supplies the only information regarding the pre-Carboniferous rocks of the Peak District. The borehole, in the Wye valley just beyond the southern boundary of the Chapel district, proved pyroclastics of probable Pre-Cambrian age overlain unconformably by 892 ft of Lower Carboniferous strata. This unconformity represents a large gap in the geological history for which details are lacking. Following the Dinantian marine transgression, it is possible to follow the development of a stable shelf area of deposition, largely coincident with the present limestone outcrops and a surrounding basinal area in which terrigenous deposits, largely muds, accumulated. Extensive vulcanicity, occurred mainly on the shelf.

Penecontemporaneous movement along the margins of the stable shelf area was associated with the development of the apron-reef at Castleton and subsequently with shallowing and the formation of knoll-reefs and flat reefs in the marginal part of the shelf area. Erosion in the marginal area gave rise to nonsequences and unconformities. In late Viséan times basinal conditions extended to the shelf area though in some areas shoals persisted on which reef-limestones were deposited. At the end of the Viséan carbonate deposition gave way everywhere to that of shales. During the Viséan, vulcanicity resulted in the extensive extrusion of lavas and, more locally, of pyroclastics.

Namurian sedimentation commenced with the deposition of marine muds though, in the case of the lowest strata, non-deposition or strongly condensed deposition occurred in the shelf area. Later the pattern of sedimentation was changed by the advance into the area of a large delta from the north giving the Mam Tor Beds, Shale Grit and Kinderscout Grit. A lesser and shorter-lived delta, from the south, produced the Roaches, Corbar and Ashover grits. Near emergence is shown by the formation of seatearths and coals, though these were better developed in later Namurian times; and less saline conditions are indicated by the presence of mussels in the Heyden Rock ($R_2$) and higher beds. Sedimentation continued with very little change into the Lower Coal Measures, though coal deposition was more regular and frequent.

Igneous activity, probably of late Carboniferous age, resulted in the intrusion of dolerite sills into the Viséan limestones and heralded the Hercynian (Variscan) orogeny which produced the main structural features of the Chapel en le Frith district. There followed a period of mineralization when the veins, flats and pipe-veins were formed. The Hercynian mountain chain suffered peneplanation before the Permian transgression, known from adjacent areas.

Post-Carboniferous solid deposits are absent from the area though the effects of Permo-Triassic conditions are considered to have resulted in deep reddening of certain Upper Carboniferous sandstones in the western part of the area. There is otherwise no vestige of the later Mesozoic or Tertiary history of the district. Tertiary orogenic movements resulted in the uplift and doming of the area and initiated the modern topography.

During the Pleistocene the ice sheets of at least one of the earlier glaciations apparently covered the greater part of the district, but few traces of glacial activity now remain. A later and more clear-cut glaciation affected the lower, western ground, but in most places failed to extend much beyond a height of about 1000 ft on the hill slopes. At this time the remainder of the area suffered

B

periglacial conditions with the formation of valley bulges and the deposition of sheets of solifluxion drift or head. Post-glacial deposits include peat, river-terraces, alluvium and landslip.

## PREVIOUS LITERATURE

In this memoir previous research is summarized under the relevant chapter headings. Nevertheless, certain general works on the geology of the area are worthy of mention here for historical reasons.

Whitehurst (1778) first noted the general succession in the area, with limestone and interbedded toadstones overlain by shale, and the latter by 'millstone grit'.

Mawe (1802) described many features of the geology of the Castleton area and also noted the presence of toadstones. He described the working of Blue John fluorspar and listed the other mineral occurrences in the area. A more detailed description was given by Farey (1811) and this has been widely quoted in the description of the Carboniferous Limestone and Millstone Grit.

The Old Series Geological maps covering much of the area of Sheet 99 were 81 N.E. and S.E. published in 1852 and 81 N.W. and S.W. published in 1864. The first two sheets fell in the area described by Green and others (1887). The two western sheets were described by Hull and Green (1866).

I.P.S.

## REFERENCES

ARMSTRONG, A. L. 1956. Palaeolithic, Neolithic and Bronze ages. *In Sheffield and its region.* Brit. Ass. Sheffield, 90–110.

BARTLETT, J. E. and PRESTON, F. L. 1956. Iron age and Roman period. *In Sheffield and its region.* Brit. Ass. Sheffield, 111–20.

BUNTING, W. B. 1940. *Chapel-en-le-Frith.* Manchester.

COPE, F. W. 1949. Woo Dale Borehole near Buxton, Derbyshire. *Q. Jl geol. Soc. Lond.,* **105**, iv.

EDWARDS, K. C., SWINNERTON, H. H. and HALL, R. H. 1962. *The Peak District.* London.

FAREY, J. 1811. *General view of the agriculture and minerals of Derbyshire,* Vol. 1. London.

FULLER, G. J. 1965. Lead-mining in Derbyshire in the mid-nineteenth century. *E. Midld Geogr.,* 3, 373–93.

GREEN, A. H., FOSTER, C. LE NEVE and DAKYNS, J. R. 1887. The geology of the Carboniferous Limestone, Yoredale Rocks, and Millstone Grit of North Derbyshire. 2nd edit. with additions by A. H. Green and A. Strahan. *Mem. geol. Surv. Gt Br.*

HULL, E. and GREEN, A. H. 1866. The geology of the country around Stockport, Macclesfield, Congleton and Leek. *Mem. geol. Surv. Gt Br.*

MAWE, J. 1802. *The mineralogy of Derbyshire.* London.

RAISTRICK, A. and JENNINGS, B. 1965. *A history of lead mining in the Pennines.* London.

SMYTHE, J. A. 1940. Roman pigs of lead at Brough. *Trans. Newcomen Soc.,* **20**, 139–45.

WHITEHURST, J. 1778. *An enquiry into the original state and formation of the Earth.* London.

# Chapter II

# CARBONIFEROUS LIMESTONE SERIES (DINANTIAN)

## INTRODUCTION

THE CARBONIFEROUS LIMESTONE SERIES crops out between Water Swallows, Castleton and Stoney Middleton. This area is the northern extremity of the Derbyshire Dome (see p. 320). Much of the outcrop is a plateau with relief much less marked than that of the Millstone Grit; dissection of this area has, however, caused the formation of valleys (dales) which provide many of the best sections of the Carboniferous Limestone Series. The presence of marginal apron-reef in the Castleton area (see p. 13) is also associated with strong topographical features.

The term 'Carboniferous Limestone Series' is here used as synonymous with Dinantian or Lower Carboniferous. All the exposed rocks of this age in the present district fall within the upper Dinantian or Viséan. The chief subdivisions of the Carboniferous Limestone used in this account are given in Table 1, together with the major faunal subdivisions on which they are based.

The stratigraphical palaeontology is described separately in Chapter III. The more important elements of the fauna only are given in the details while full fossil lists are quoted in Appendix III.

In the present district the rocks of the Carboniferous Limestone Series consist predominantly of limestones with some interbedded basaltic lavas and tuffs referred to in the older literature as 'toadstones'; the total thickness exposed is about 1500 ft. Just beyond the southern margin the Woo Dale Borehole (Cope 1949, p. iv), starting at about the horizon of the lowest strata exposed in the Chapel en le Frith district, proved a further 892 ft of limestone (much intensely dolomitized) overlying pyroclastics of probable Pre-Cambrian age. In addition to the outcrop, Lower Carboniferous rocks are known at depth from the Edale and Alport boreholes.

## PREVIOUS RESEARCH

Whitehurst (1778, pl. i, fig. 6) first published a generalized section from Grange Mill to Darley Moor showing limestone with three beds of toadstone underlying shale and grit. Though the section lay to the south of the present district it is evident that Whitehurst intended it to apply to the whole of Derbyshire as he refers (ibid., pp. 149–50) to local names for the toadstones; at Castleton ('cat-dirt') and Tideswell ('channel'); the toadstones were considered to be lavas. Whitehurst further (ibid., p. 199) noted the presence of chert and crinoids ('entrochi' or 'screwstones') in the limestone.

7

| Zones | | Shelf Province | | Marginal Province | Basin Province (Alport Borehole) | Zones | |
|---|---|---|---|---|---|---|---|
| Upper *Posidonia* | P₂ pars | Eyam Group | Mudstone — up to 12 ft<br>Eyam Limestones — 0–150 ft<br>UNCONFORMITY IN PLACES | Mudstone 0–15 ft<br>Beach Beds etc. 0–50 ft | Mudstone — 97 ft | Upper *Posidonia* | P₂ |
| Upper *Dibunophyllum* | D₂ | Monsal Dale Group | Upper Monsal Dale Beds with Upper *Girvanella* Band at base — about 250 ft<br>Litton Tuff — 0–40 ft<br>Lower Monsal Dale Beds with Lower *Girvanella* Band at Base — 150–200 ft<br>Upper Miller's Dale Lava — 0–100 ft | Unconformity | Mudstone — 141 ft<br>Limestone — 47 ft<br>Mudstone — 50 ft<br>Limestone and tuff — 56 ft | Lower *Posidonia* | P₁ |
| Lower *Dibunophyllum* | D₁ | Bee Low Group | Miller's Dale Beds with Dove Holes Tuff — 63–142 ft / 0–6 ft<br>Lower Miller's Dale Lava — 0–100 ft<br>Chee Tor Rock — 300–380 ft | Apron-reef of Castleton (B₂–P₁ᵦ) | Bee Low Limestones 600 ft in northern part of outcrop<br>Mudstones with thin limestones — 52 ft<br>Limestone — 485 ft | Upper *Beyrichoceras* | B₂ |
| *Seminula* | S₂ | Woo Dale Group | Woo Dale Beds, including Peak Forest Limestones — 165 ft exposed | Not known | Limestone — 528 ft | Lower *Beyrichoceras* | B₁ |

TABLE 1. *Generalized succession in the Viséan*

(For a discussion of the correlation of coral-brachiopod and goniatite-bivalve zones, see p. 151).

Mawe (1802), in a work devoted largely to the mines of north Derbyshire, followed Whitehurst in stating that the limestone was overlain by shale. He further noted the presence of chert in the limestone at Bradwell and Middleton Dale (pp. 30–31) and of the toadstone outcrops at Wormhill and Castleton (p. 39). A description of a highly fossiliferous stone found at Foolow (pp. 29–30) and used for ornamental purposes probably refers to one of the knoll-reef limestones.

White Watson (1811) added to the knowledge of the Derbyshire Dome, though again his chief published section lay largely to the south of the present district. He also produced a series of inlaid sections across the area (Ford 1960, fig. 4) showing, among other features, the igneous rock underlying the limestones of Middleton Dale and the presence of two lavas between Peak Forest and Wardlow. The work of Farey, which also appeared in 1811, added little to the knowledge of the limestone though it also (pp. 237–9) refers to the presence within it of thick "Basaltic Beds . . . here called Toadstones".

The first Geological Survey map of the district, Old Series 81 N.E., covered most of the limestone outcrop and was published in 1852 (revised 1866), based mainly on the work of J. Phillips. The remainder formed the northern margin of Sheet 81 S.E. by J. Phillips and W. W. Smyth, published in 1852 (revised 1866 and 1867). The explanatory memoir by Green and others (1887) gave the first clear description of the succession and named two of the subdivisions, the "Chee Tor rock" and "Miller's Dale rock" (see p. 22). The presence of a thin 'toad-stone' at Dove Holes was noted and the important sections with a thin coal and seatearth in Coombsdale Quarry described (ibid., pp. 21–2). Further (p. 33) the steep marginal dips in the Castleton area were noted.

Barnes and Holroyd (1897) described the occurrence of rolled-shell beds at Castleton and elsewhere. These were attributed to a shoreline extending from Castleton to Barmoor Clough. In two papers, published in 1894 and 1907, Arnold-Bemrose described first the petrography and later the field relations of the igneous rocks. The two main lavas (the Lower and Upper lavas of Miller's Dale, see p. 23) were considered by Arnold-Bemrose to be widespread over the Derbyshire Dome. Although an over-simplification in other areas (see for example Traill 1939, pp. 859–61, for the Millclose area) Arnold-Bemrose's correlations are, for the most part, valid so far as the present district is concerned. In addition the true pyroclastic nature of the Dove Holes and Litton tuffs was recognized and the lava underlying the latter first described. The vents at Monk's Dale and Castleton (Speedwell Vent) were recorded. Arnold-Bemrose further noted the intrusive nature of the Waterswallows, Peak Forest, Potluck and Tideswell Dale sills.

Hind and Howe (1901, p. 361) first drew attention to the faunal resemblances of the limestones of the Castleton area to those of the Cracoe knolls. They also recorded some details of the section in the railway-cutting (now disused) north of Dove Holes.

Following the publication of Vaughan's classic zonal work in the Avon Gorge (1905), Sibly (1908) was the first to apply the coral-brachiopod zones to the Carboniferous Limestone of the Derbyshire Dome. Sibly recognized three subzones in the Wye valley: the lowest, the *Dibunophyllum* θ ($D_1$) Subzone, included the "Upper Toadstone" and underlying strata and was succeeded by the *Lonsdaleia* ($D_2$) and *Cyathaxonia* ($D_3$) subzones.

Jackson (1908, p. 309) noted the presence of a bed with fish remains in Barmoor Clough Quarry. Later (1925), in a paper concerned with the limestone-shale unconformity, he described sections between Dove Holes and Bradwell and noted the presence of *Goniatites crenistria* at Nun Low. In 1941 (pp. 241–2) he recorded the presence of *Girvanella* at the base of the $D_2$ limestones and of the underlying $D_1$ coral band in Pin Dale; the reef-limestones at the northern end of this section were regarded as the lateral equivalent of the bedded $D_2$ succession.

Morris (1929) gave a detailed description of the succession in Middleton Dale and Coombs Dale. Beneath the exposed beds he recognized the presence of basalt, though he confused this with the Upper Lava of Miller's Dale. Morris further noticed the presence of the Linen Dale knoll and the knoll-affinities of the highest beds in Middleton Dale. The latter were referred to a probable "$D_3$" age and the underlying beds were placed in $D_2$. Morris published extensive faunal lists.

Fearnsides and Templeman (1932) described the succession in a borehole at Hope, including a thick development of pillow lavas beneath basal Namurian.

In a series of papers, published between 1933 and 1939, Cope discussed the succession in the Wye valley just south of the present district. This succession is the type sequence for all but the highest beds of the shelf facies. A particularly important feature of this work was the recognition of the $S_2$ age of the lowest beds.

Shirley and Horsfield (1940) mapped the Castleton-Bradwell area and inter-preted the reef-limestones with a $B_2$ goniatite fauna as lying with marked unconformity against a cliff of standard limestones. The outward dips of the reef-limestones were interpreted as largely depositional. The succession in the standard limestones was also described, the presence of two bands with *Davidsonina septosa* recorded, and the Cave Dale Lava recognized as the Lower Miller's Dale Lava. In the higher beds the presence of a band with *Girvanella* at the base of $D_2$ was noted and near the top of the zone two coral bands, a lower with *Lonsdaleia duplicata* and, some 30 ft higher, an upper band with *Orionastraea* were discovered. The same authors (1945) described the structure and ore deposition of the Eyam area and gave details of the succession in Middleton Dale where the uppermost beds, the Eyam Limestones, of $P_2$ age, were stated to rest unconformably on $D_2$. The sequences in Coombs Dale and Cressbrook Dale were also determined and the presence above the Litton Tuff of an algal band with *Girvanella* nodules, the Upper *Girvanella* Band, noted in the latter.

Shirley and Horsfield's conclusions in the Castleton area were attacked by Parkinson in a series of papers (1943, 1947, 1953). This author argued on faunal grounds and from field relations that the postulated unconformity was absent, or at the most a local non-sequence, and that the $B_2$ and $P_1$ reef-limestones were the lateral equivalents of the $D_1$–$D_2$ standard succession. Ford (1952) described the geology of certain caves in the Castleton area, which supported lateral passage of reef to standard limestone.

Hudson and Cotton (1945a, 1945b) described the succession in the Alport and Edale boreholes. Their work has provided the only description of the beds of the basinal area within the present district and has been quoted in detail below.

As a result of a study of the palaeoecology of the Castleton reef-belt Wolfenden (1958) discovered the presence of wall-like masses of algal limestone at the junction of the fore-reef limestones (see p. 18) with back-reef limestones forming a transition to those of the shelf area. This arrangement clearly showed the lateral passage from the fore-reef limestones to those of the shelf and represented a great advance in the understanding of the morphology of the reef-belt.

The origin of the 'Beach Beds' of the Castleton area has been discussed by Sadler (1964), who concluded that they represented material deposited contemporaneously with the fore-reef limestones at the foot of a submarine channel.

Eden and others (1964) dealt with the detailed stratigraphy of Pin Dale and the area of Earle's Quarry, Hope, and this work has been extensively referred to below. Some of the more important points in this paper are the proof of the presence of a bed of tuff, the Pindale Tuff, beneath both reef and shelf facies (thus giving final proof of the absence of unconformity between these) and the description of detailed stratigraphical variation in the shelf facies adjacent to the reef-belt, including the presence of large knoll-reefs in the upper part of $D_2$.

Parkinson has recently (1965) summarized his own work and that of other authors on the reef-belt and has postulated a break within the $B_2$ apron-reef.

## SEDIMENTATION

There is a general structural division in the present district between the Carboniferous Limestone outcrop, forming the northern part of the broad uplift of the Derbyshire Dome, and the neighbouring down-warped or synclinal region of Kinder Scout and the Goyt–Todd Brook area. The structure of these areas is considered in greater detail on pp. 320–8. It is now generally accepted (see p. 320 and Kent 1966, p. 338) that a stable block of Lower Palaeozoic or Pre-Cambrian rocks, similar to the Alston and Askrigg blocks farther north, underlies the Derbyshire Dome. As in the case of the northern blocks, this is inferred to have exercised a considerable influence on sedimentation during Lower Carboniferous times and this effect continued on into the Namurian (see p. 161).

The above structural entities correspond to two main types of Viséan sedimentation (see Fig. 3) and special local marginal conditions (see below) lead in places to the development of a third.

**Shelf Province.** This, developed in the area of the stable block, is marked by a sequence of predominantly pale massive limestones with a coral-brachiopod fauna comparable with that of the Great Scar Limestone on the Askrigg Block (see for example Rayner 1953, p. 249). Argillaceous sediments are, for the most part, rare or absent. Where the full differences between basin and shelf facies are developed, as in the Widmerpool Formation of the Duffield area, the thickness of these beds is noticeably less than that of corresponding beds of basin facies (see for example *Ann. Rept. Inst. geol. Sci.* for 1967, 1968, p. 82). In the Chapel en le Frith district, however, the thickness variations are not marked (see Fig. 3). The beds of shelf facies (lithofacies and biofacies) are often termed 'standard limestones' and are present over nearly all the outcrop of the Lower Carboniferous of the Chapel en le Frith area.

**Basin Province.** This is marked by a thick development of thin-bedded dark limestones and shales with a goniatite-bivalve fauna. Sediments of this type were formed in negative areas subject to continuing subsidence. Though it has been suggested by Rayner (1953, pp. 248–9) that subsidence rather than depth of water was the prime factor in the formation of basinal deposits there is reason to believe that in the present area the sea-floor of the basin was at least some hundreds of feet deeper than that of the shelf (see p. 18). George (1958, p. 286)

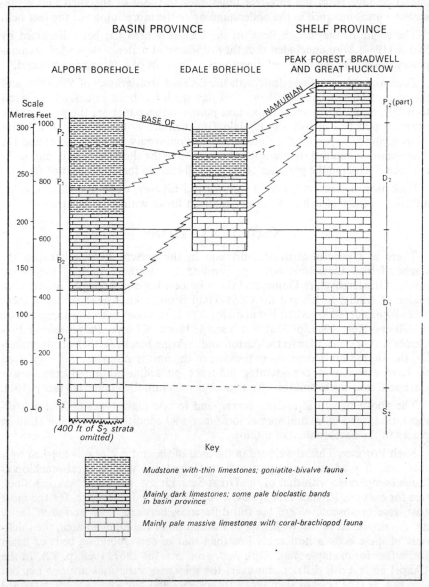

FIG. 3. *Generalized sections of the Viséan strata showing the lateral migration of facies between basin and shelf provinces*

attributes the differences between the two facies to both depth of accumulation and rate of subsidence. Beds of basin facies are known in the present district only at depth in the Edale and Alport boreholes north of the shelf area. To the north-west and west of the shelf area the presence of beds of basin facies may be inferred from indications at the edge of the outcrop at two points and from the proved presence outside the district of beds of this type at outcrop in Stafford-shire (Prentice 1951, pp. 175–90) and in the Gun Hill Borehole (Hudson *in* Hudson and Cotton 1945a, p. 318). The basinal area north of the shelf has been referred to by Kent (1966, p. 337) as the 'Edale Gulf'.

**Marginal Province.** In the present district a marginal reef facies, separating the shelf and basin facies, is developed between Hope and Sparrow Pit, on the northern margin of the shelf area. The fauna, while manifestly shallow-water, shows affinities with that of the basin in the presence of goniatites. The develop-ment of the apron-reef (see pp. 17–8) marks the most extreme effect of the under-lying margin of the block on sedimentation. At higher horizons where the apron-reef is not developed, the margin of the block was reflected in post-apron ($D_2$ and $P_2$) times by the development of knoll-reefs and flat reefs (see p. 18) in or near the marginal area. A fuller discussion of the deposition of reef-limestones is given on pp. 17–9.

The early relations in time of shelf and basin are uncertain, due to lack of knowledge of the lower beds, especially in the basin. The lowest beds ($S_2$) known in detail in both shelf and basin areas, however, both show a continuous or nearly continuous limestone sequence with a coral-brachiopod fauna and only slight difference in lithology. During $B_2$ times the establishment of the apron-reef at Castleton coincided fairly closely with greater variation in sedimentation between shelf and basin, the upper part of these beds in the Alport Borehole bearing shale partings and yielding a goniatite fauna while the lower part shows a lithology and fauna closer to those of the shelf. The greatest facies variation was in $D_2$ times, with the deposition of 'standard limestones' in the shelf area and of thin-bedded dark limestones succeeded by shales in the basin. Near the close of Viséan times, local uplift (see p. 32) was followed by subsidence, which allowed the spread of beds of basin type ($P_2$) on to the shelf area.

**Effects of Contemporaneous Earth Movements.** Earth movements during the deposition of the Viséan rocks of the present district may be inferred from lateral facies variations, from the presence of the reef-belt and from the occur-rence of unconformities. The detailed evidence is presented later under strati-graphical headings but the general sequence of events may be summarized as follows:

(*a*) Deep-seated movement involving the uplift of the margin of the block initiated the formation of the apron-reef. The age of this movement would be either late $S_2$ or early $D_1$ (Cravenian of Hudson and Mitchell 1937).

(*b*) Local intra-$B_2$ ($D_1$) movement is postulated to account for anomalous features in the structure of the apron-reef at Peaks Hill. Parkinson (1965, pp. 172–3) has suggested on faunal grounds that a widespread break in the reef succession is present at this horizon but this cannot be regarded as firmly established. There is, however, some reason to postulate movement of Dirtlow Rake in the region of the reef-belt in late $B_2$ times (see p. 34).

(*c*) Pre-$P_2$ movement; this was acute around the edges of the block resulting in angular unconformity and at one locality (p. 96) in emergence with the local formation of a thin coal and seatearth. Shirley and Horsfield (1945, p. 293) considered that widespread unconformity existed below $P_2$ but the present work suggests that this may have been exaggerated. Movements of this age are responsible for an east–west 'pre-Namurian' fault south of Dove Holes with $P_2$ resting on $D_1$ limestones on its southern side (see Fig. 7).

(*d*) Intra-$P_2$ movement is represented by the unconformity at the base of the $P_2$ shale facies or, where $P_2$ shales are lacking, by the break between the Viséan limestones and the Namurian. There is no evidence of unconformity between the shale facies of the Viséan and Namurian. The stratigraphical evidence shows a sharp break around the edges and over the block and a probable absence of any break in the basinal area.

Where $P_2$ beds are absent it is difficult to distinguish the effect of pre- and intra-$P_2$ movements and it is convenient to refer to the corresponding unconformity as the 'pre-Namurian unconformity'.

## LITHOLOGY AND LITHOFACIES

The lithology of the different rock types of the Carboniferous Limestone Series is summarized below though the account is somewhat restricted by the fact that it has not been found possible to undertake systematic petrographical work on the limestones, although selected thin sections have been studied. In the stratigraphical descriptions the use of petrographical terms has, in general, been avoided as this would imply a full petrographical study. An attempt is made here to provide a correlation between field and petrographical terms.

### THE LIMESTONES

Many of the limestones are bioclastic, crinoid debris and brachiopods being the main constituents. Corals and algae are important in some cases. Usually the main concentrations of macrofossils are on restricted horizons (coral bands and shell bands).

#### SHELF PROVINCE

The **Woo Dale Beds** present a mixed lithofacies showing points in common with both pale and dark facies described below from the higher divisions. The chief rock types include:

(i) Dark grey and dark grey-brown limestones (calcarenites), fine-grained and evenly bedded. In view of the association of rocks of this type with dolomites in the Wye valley (Cope 1933, p. 129), a chemical examination of these rocks in the present district was undertaken: this showed, however, only small MgO percentages (0·22 to 0·46).

(ii) Grey limestones (calcarenites), well-bedded, some bioclastic and including limestones with algal nodules. A large development of pale crinoidal limestones (biocalcarenites) is present at Peak Forest, some of which are partly oolitic.

(iii) Calcilutites (micrites), usually pale-coloured and either in regular beds or in large lenticular masses. The origin of the latter is uncertain though they are best considered as bioherms in which some binding organism (possibly algal) was present but has now disappeared. This would account for the maintenance of the lens form during the deposition of the adjacent well-bedded limestones.

**Bee Low, Monsal Dale and Eyam groups.** A striking feature is the sharp contrast between white, pale grey and grey limestone on the one hand and dark bituminous limestones on the other. Intermediate rock types are relatively rare and the distinction between the two lithofacies has been found most valuable for mapping purposes (see for example Fig. 7).

Among limestones of the pale lithofacies the following types have been recognized:

(i) Pale massive limestones (calcarenites) are developed extensively in both $D_1$ and $D_2$. Some are finely oolitic though, as Wolfenden (1958, p. 881) remarked, ooliths are not very abundant. Bands with algal nodules occur at some horizons. Rubbly-weathering beds in $D_2$ are probably developed by the diagenesis of limestones with some non-calcareous mud content leading to segregation of the latter along cracks.

(ii) Pale cherty limestones show a lithology of the matrix similar to (i). The occurrence of chert is separately dealt with on p. 16.

(iii) Calcilutites (micrites) occur rarely as thin and well-defined beds in $D_1$ and near the top of $D_2$.

(iv) Oolites (other than the imperfectly developed examples noted under (i)) are of rare occurrence but have been found in the back-reef facies near Eldon Hill (see also p. 59) and in the Wardlow Mires No. 1 Borehole (see p. 93).

(v) Coarse-grained limestones (calcirudites) are especially developed in $D_2$ and are due to the presence of coarse organic material, either crinoid debris or shell debris or both. Crinoid debris, particularly where fine, is also common in the other rock types in variable proportions.

(vi) Magnesian limestones are known though uncommon in the pale lithofacies only in one horizon in $D_1$, which may correspond to that of the Lower Miller's Dale Lava, the dolomitization being probably connected with local variations in the composition of the sea-water.

A shallow-water origin for most of the limestones of the pale facies is indicated by the presence of the rare ooliths, by local non-sequences and by rare current-bedding. The last-named is difficult to observe but has been noted in places in $D_1$ limestones. The higher limestones of $D_2$ and the crinoidal facies of $S_2$ (Peak Forest Limestones) often show sorting of coarser and finer constituents. The fauna, especially the presence of algae (cf. Pettijohn 1957, p. 596) and corals (other than zaphrentoids), is also indicative of shallow-water conditions (Ager 1963, p. 41) and the faunal aspect as a whole suggests water which was well oxygenated, allowing the spread of a vigorous fossil community at certain horizons.

The dark lithofacies occurs as both the more continuous development of basin facies and intercalations in the pale facies. Typically the limestones are fine-grained, dark and bituminous. The bedding planes are closely spaced and commonly somewhat undulating. Chert is abundant. Eden and others (1964, p. 93) have remarked that the apparent fine grain of many of these rocks is deceptive; they are in fact fine calcarenites. Some coarser, crinoidal bands occur. Thin shale partings are often present, particularly in $P_2$ where they increase in thickness in the uppermost beds giving a transition to the shale facies.

The dark facies apparently was deposited under reducing conditions with the formation of hydrocarbons and some pyrite and a variable supply of terrigenous mud. These conditions, rather than depth of water would appear to have been the controlling factors so far as deposition in the shelf area was concerned. A similar environment must have been present at greater depths in the basinal areas (the differences in depth being inferred from a consideration of the structure of the reef-belt rather than from facies differences). A marked facies-fauna is associated with these beds and includes in particular zaphrentoids with some inarticulate brachiopods and small gastropods. However, bands with a fauna common to the pale facies also occur.

Stylolites are of common occurrence in both pale and dark lithofacies. The size of the stylolite, however, usually varies with the lithology of the country rock, the largest examples (up to 9 in high) being found in pale fine-grained massive limestones while the dark facies shows only small structures. In some cases, especially in the dark limestones, there is a capping of a solution residue of hydrocarbon or argillaceous material on the stylolites. The pressure-solution origin of stylolites, as put forward by Stockdale (1922), is now generally accepted.

## BASIN PROVINCE

No new work has been done on rocks of this province and the essential features of the succession in the Edale and Alport boreholes are described on pp. 113–9. Certain distinctive features of the lithology are summarized below.

The dark limestones are similar to those already described from the dark lithofacies of the shelf province, though bands of relatively coarse bioclastic material are apparently foreign to the basinal environment and are here considered to be of turbidite origin. This is borne out by Hudson and Cotton's description of 'slumping' at some horizons in the Alport Borehole, for example at 1632–1642 ft (1945a, p. 287). Graded bedding has been noted in thin bands of finely crinoidal limestone in dark limestones in the Castleton Borehole.

## MARGINAL PROVINCE

The apron-reef limestones of the Castleton area show a typical reef lithofacies and vary from calcilutites to bio-calcirudites. These rocks are discussed, together with other reef-limestones, under a general heading below. The overlying Beach Beds are a local deposit of rolled-shell limestones whose relations are discussed on p. 35.

## CHERT

Lenticular, tabular, or nodular chert is, in the shelf area, present in the $D_2$ and $P_2$ limestones of both pale and dark facies at certain moderately well-defined horizons. It also occurs exceptionally in $D_1$ adjacent to the reef-belt (see p. 61). In the basinal area chert also occurs in the $D_1$ and $S_2$ zones and the distribution in these zones resembles that noted by Newell and others (1953, pp. 161–2) in the basin facies of the Permian of the Guadalupe Mountains area, Texas, where chert occurs in the basin and is absent in reef and shelf areas.

Sargent (1921, pp. 267–76) considered that the cherts were consolidated early as bedding was deflected around the nodules. He argued against an organic origin for the silica and attributed its introduction into the sea-water largely to contemporary vulcanism. The present work supports the conclusions of Pettijohn (1957, pp. 439–40) that the cherts are post-depositional and owe their origin to the early replacement of a partly consolidated rock containing diffuse silica, the latter undergoing segregation into the nodules and tabulate bands. The chert is thus in continuity with the surrounding limestones.

In some cases secondary cherts have been produced, probably by a later re-mobilisation of silica; these are most obvious in the vicinity of mineral veins though not completely restricted to this position. The colour of the chert is in most cases governed by that of the surrounding limestone. The later cherts occur as large bulbous masses often showing concentric colour banding and (rarely) as thin veins. The presence of the later cherts is linked with the frequent silicification of fossils at certain horizons. Arnold-Bemrose (1898, p. 178) also noted the presence of quartz crystals in a limestone at the Bull Ring, Dove Holes. Hydrothermal silicification is discussed on p. 310.

## OTHER ROCK TYPES

A thin coal and seatearth are present in the Coombs Dale area (see pp. 96, 98), resting on a potholed surface (see below) of Monsal Dale Beds. Their formation is clearly linked with emergence associated with the unconformity at the base of the Eyam Limestones.

Shale or mudstone is present in the highest $P_2$ strata at outcrop and in addition in the lower $P_1$ beds in the Edale and Alport boreholes and in thin partings in the limestones. The lithology resembles that of the Namurian shales. Other partings, or wayboards, consist of grey-green poorly bedded unfossiliferous and often pyritous clay. Sargent (1912, pp. 409–12) described similar partings from the Crich area and concluded that they were of pyroclastic origin. This view was also taken by Cope (1939, p. 62) and is accepted here. In the present area, clay from a parting in the Chee Tor Rock at Smalldale (see p. 47) is an illite mudstone with coarse pyrite and fine needles of marcasite.

Both clay and mudstone partings in the limestone usually rest on potholed surfaces (see Plate IIB) such as were noted by Sargent (op. cit.) and Cope (1939, pp. 61–2). The maximum depth of the potholes is about 2 ft. Clearly these features represent contemporaneous removal of the uppermost limestone surface either by solution, erosion or both during a halt in sedimentation. Actual emergence at Coombs Dale is proved by the presence of a coal and thin seatearth overlying a potholed surface.

## REEF-LIMESTONES

Following Bond (1950c) the reef-limestones of the Chapel en le Frith district may be subdivided into apron-reef, knoll-reefs and flat reefs. The stratigraphy of these is considered separately on pp. 34–5.

**Apron-reef.** Following Eden and others (1964), the term 'apron-reef' has been extended somewhat from the original definition of Bond (1950 c, pp. 270–1). The apron-reef is a marginal facies developed between the shelf area and the basin and had been divided by Wolfenden (1958) into fore reef, reef (called algal reef in this account) and back reef, the whole being known as the 'reef-complex'.

The fore reef consists of poorly bedded limestones, commonly highly fossiliferous (some hundreds of species being recorded) but in places poorly fossiliferous and fine-grained. In referring a particular outcrop to this facies it is necessary to consider the whole of its characters including lithology, fauna and bedding. The fore reef shows outward dips (towards the basin) of 10°–40° which have been considered by Shirley and Horsfield (1940, pp. 288–9) and by most workers since, to be the original depositional dips only slightly modified by the effects of compaction and minor tectonic movement. Recent work by Broadhurst and Simpson (1967, pp. 443–8) on the inclination of sedimentary infillings of shells and limestone cavities has demonstrated that the dips of the apron are indeed of depositional origin. The difference in height between the top and bottom of the frontal slope is considered to have been about 400 ft. The fauna, though characteristic in itself, shows affinities both with basin, in the local presence of goniatites, and with standard limestones. The limestones at the foot of the reef-slope are often crinoidal. In addition to the fine-grained limestones (calcilutites), breccias occur in places particularly in the eastern outcrop, fragments up to a foot in diameter having been noted by Eden and others (1964, p. 95). Small 'Neptunian dykes' of pale crinoidal limestone have been observed locally at the top (higher) part of the reef. From a study of the palaeo-ecology of the goniatite bed at Cow Low Nick, Ford (1965, pp. 189–90) has concluded that this deposit, situated about two-thirds of the way up the frontal slope of the fore reef, was subject to gentle current action. This and the depositional slope would be incompatible with the accumulation of unconsolidated deposits. Most of the fossils in the fore reef are unworn and must be nearly *in situ*. Following Black (1933), Wolfenden has suggested that the fine matrix of the fore-reef limestones may represent recrystallized algal deposits, the algae having acted as sediment-binders, at least in the upper part within the range of daylight penetration.

The algal reef ('reef' of Wolfenden) occurs as a discontinuous wall-like mass of limestone with abundant stromatolitic algae (see Plate XIII A), sponges and other fossils. Algal reefs have been noted by Wolfenden (1958, pp. 872, 879) at two levels at the point where the strong frontal dip of the reef-complex develops. Wolfenden gives the maximum dimensions as about 30 ft wide and 100 ft high. However, recent exposure of the algal reef on Treak Cliff shows a maximum width of about 250 ft locally. The algal reefs are considered to have originated as barrier-like reefs at two horizons, growth of the lower reef having been inhibited by down-warping, producing a greater depth than the algae could tolerate.

The back reef is intermediate in both lithology and fauna between the fore reef and shelf limestones and usually shows gentle inward dips towards the shelf area. The back-reef limestones are typically well bedded and often show elements of the fore-reef fauna. They are also in general more crinoidal than all but the lowest parts of the fore reef. On the inward side this facies passes gradually into standard limestones and it is the outer facies junction which is the best defined; hence both in Wolfenden's account (ibid., pp. 877, 879–80) and in the present work it has been found necessary to group them with the shelf facies for mapping purposes, though they are separately shown on Plate XI.

**Knoll-reefs.** Since the original description by Tiddeman (1889) from the Craven area there has grown up an abundant literature on knoll-reefs. This has been summarized by Bond (1950a) and Black (1954). Morris (1929, pp. 49–50) first noted the presence of knoll-reefs in the district at Linen Dale. The present work has shown them to be widespread, though restricted to the peripheral area of the limestone (i.e. the margin of the block).

The outstanding features of the knoll-reefs are the presence of original quaquaversal dips and the relatively pale, poorly bedded and fossiliferous character of the rock. A detailed description of knolls in the Pin Dale area has been given by Eden and others (1964, pp. 87–9). Here they were found to be mainly developed at two horizons and to pass laterally into calcirudite bands in the bedded succession. At one locality the formation of a knoll above an earlier one has led to the development of a composite knoll over 50 ft in height.

The knolls pass into the adjacent limestones by lensing or interdigitation. Eden and others (ibid., p. 89) noted current-bedding and some breccia associated with the knolls and concluded that they "stood up as small mounds above the general level of the sea-bed". In some cases, such as the Earle's Quarry knolls and the Dove Holes knoll (see p. 65), where breccia is present in the upper part of the knoll, this is interpreted as evidence that the structure was consolidated and a true knoll, not a bank deposit as has been suggested (Earp and others 1961, p. 44) for the Clitheroe knolls. Recent work by Orme (1970a) on the Earle's Quarry knolls suggests that algae and bryozoans were the chief sediment binders. In the case of the large knoll (see p. 88) in Earle's Quarry a true biohermal origin is also suggested by the superposition of its two component knolls; the higher being more likely to be due to the re-activation of growth of a lower bioherm than to the formation of one bank deposit precisely above another. Like the apron-reef, the knolls have a rich fauna in which brachiopods are abundant.

**Flat reefs.** The term flat reef was suggested by Bond (1950c, pp. 271–2) for large lateral spreads of limestone of reef facies. In many cases they may represent knolls which have grown laterally to a point where they lose their knoll form. Large spreads may be formed by the coalition of several such masses.

In the present district knoll form appears to be lost in reefs over some 50 ft in thickness, upward growth being then largely replaced by lateral growth. This contrasts with the much larger knoll forms in other parts of Derbyshire, for example at High Tor, Matlock (about 100 ft) (Smith and others 1967, p. 23) and the Cracoe area (Bond 1950b, p. 172) where thicknesses of 200 ft or more are common. Probably the knolls of the Castleton–Hucklow area were limited in growth by either gentle uplift or a static sea-floor while the larger Yorkshire knolls were formed during gentle subsidence.

**Formation of reef-limestones.** The conception of facies variation across the reef-complex, and the establishment of an outward-dipping frontal slope requires the postulation of deep-seated movements along the edge of the block with a sharp uplift of the latter to produce the shallow-water conditions allowing the reef fauna to flourish. An alignment of small eruptive centres along the reef-belt is further evidence of deep-seated instability in this zone. The movements produced the initial shallowing which led to the formation of the apron-reef and accord with its linear character.

It is suggested that broad and less acute uplift around the edge of the block was the initial factor which led to the formation of the knoll-reefs and flat reefs. In places (see Eden and others 1964, pp. 79–81) a more restricted local uplift may have controlled knoll formation.

# VOLCANIC ROCKS

Contemporaneous volcanic rocks associated with the outcrop of the Carboniferous Limestone include lavas, tuffs and agglomerates, some of the last two types occurring in vents. In addition interbedded tuffs are known in the basinal area.

**Lavas.** Lavas are known at three horizons within the present district; their stratigraphy (see pp. 23–8) and petrography (pp. 119–23) are discussed separately, The rocks are olivine-basalts typically highly vesicular with calcite and chlorite-filled amygdales. In places the lavas are non-vesicular, harder, darker and more coarsely crystalline. The occurrence of pillow lavas was noted by Fearnsides and Templeman (1932, pp. 102, 107–10) in a borehole at Hope. Pillow structure is, however, not a normal feature and appears in this instance to be connected with the extrusion of lava on the marginal slope between shelf and basin at a greater depth than that at which the other lava flows, on the shelf area, were formed. The conditions under which the lavas of the shelf were extruded are uncertain though the associated sediments suggest a shallow-water origin. The thicker flows may have built up above sea-level though this cannot certainly be established.

A small thickness of clay ('toadstone clay') is in many cases present above or below the lavas. It is invariably highly altered and it is uncertain whether it is derived directly from the overlying flow or whether it is a fine tuff. As noted by Smith and others (1967, p. 12) in the Matlock area, cambering has in places occurred on the clay and is responsible for an apparent decrease in the thickness of the lava at outcrop.

Although vents are present in the area (see below) they nowhere show any evident association with the lavas, the extrusion of which is here considered to have been by eruption along fissures. This is consistent with the uniformity in thickness of the large spreads of lava.

It is usually impossible to determine the number of flows making up a lava-sheet owing to lack of exposures; most are, however, probably composite. The Upper Lava clearly shows this at Hargatewall as it contains a limestone parting in the middle. Similar features are shown by the Lower Lava in Tideswell Dale and in the Litton Dale Borehole.

**Tuffs.** Tuffs, though less widespread than lavas, are known at outcrop at horizons in $D_1$ (Pindale Tuff and Dove Holes Tuff) and $D_2$ (Litton Tuff) and possibly at two lower horizons at depth. Tuffs have also been proved in association with tuffaceous limestones in the Edale and Alport boreholes (see pp. 114, 117).

Most of the tuffs are regularly bedded though depositional form is preserved in the tuff-mound formerly exposed in Earle's Quarry, Hope, which represents a degraded cone (Eden and others 1964, pp. 76–8). Accidental blocks of limestone up to 3 ft in diameter were associated with the tuff-mound. A second mound has also been proved by boreholes at a similar horizon nearby. The Litton Tuff shows features similar to those of the Pindale Tuff, though over a larger area. At Litton it shows its greatest thickness and is locally agglomeratic, indicating proximity to a vent.

The tuffs are in general very altered, the unweathered rocks being much calcitized. They range from coarse lapilli tuffs to fine-grained water-borne rocks of mud grade. Lapilli, where present, are better preserved than the matrix; they are usually cognate but accidental lapilli of limestone are present in the Dove Holes Tuff.

The degree of sorting of the tuffs also varies. For example, drilling on the flanks of the tuff-mound in Earle's Quarry has shown that the unbedded tuff of the mound passes laterally into a well-banded tuff.

Vents are known at three localities. Though structurally intrusive these are clearly associated with explosive episodes when the tuffs were formed. The vents are filled with either fine agglomerate or lapilli tuff. The two vents in Monk's Dale (see Fig. 14) lie on the axis of a sharp monocline and this relation is unlikely to be accidental. It is probable that deep-seated faulting is responsible for both the monoclinal fold and for the location of the two vents on it.

# GENERAL STRATIGRAPHY

## SHELF PROVINCE

### WOO DALE GROUP

(*Seminula* ($S_2$) Zone and basal Lower *Dibunophyllum* ($D_1$) Zone)

#### WOO DALE BEDS

The term Woo Dale Beds is introduced here for the $S_2$ beds of the Wye valley (Cope 1933, pp. 127–30). It is synonymous with the term '*Daviesiella* Beds' though lateral changes of facies and variations in the vertical distribution of *Daviesiella* render this term inappropriate. The type area thus lies to the south of the present district.

In the present district outcrops are confined to the Peak Forest Anticline (Fig. 4) and to exposures farther south along Dam Dale and Hay Dale (Plate IIA). Two distinct lithofacies (see Figs. 4 and 5) are present:

(*a*) The normal facies consists of thinly bedded dark grey and dark brown limestones with some bands of grey limestone and rare oolitic bands. The maximum exposed thickness of these beds is about 165 ft, at the junction of Dam Dale and Hay Dale. The lowest beds contain calcilutites at two horizons, the lower being irregularly bedded and lenticular while the upper is evenly bedded and has served locally as a lithological marker-band. The lowest beds bear *Daviesiella sp.* though the horizon of this brachiopod here does not correspond closely with its vertical range in the Wye valley (Cope 1939, pp. 60–2). The typical $S_2$ brachiopods *Davidsonina carbonaria* and *Linoproductus corrugato-hemisphericus* are also present. The $S_2$–$D_1$ zonal boundary does not correspond exactly to the lithological change at the junction of Woo Dale Beds and Chee Tor Rock. In the Dam Dale–Hay Dale section the uppermost $11\frac{1}{2}$ ft of the Woo Dale Beds bear a $D_1$ coral fauna including *Dibunophyllum bourtonense* and *Palaeosmilia murchisoni*, while at Hernstone Lane Head this fauna occurs in the uppermost $14\frac{1}{2}$ ft.

(*b*) In the northern part of the Dam Dale–Hay Dale section Woo Dale Beds of normal facies pass laterally into a uniform group of pale crinoidal and in places shelly limestones (see Fig. 4) which occupy the greater part of the Peak Forest inlier. These are here called the Peak Forest Limestones though they are not separately shown on the one-inch map. Though good continuous sections of these beds are lacking the total exposed thickness is of the order of 150 ft. They are of shallow-water facies and show marked sorting into bands with coarser and finer bioclastic material, mainly crinoidal; some oolitic bands have also been found. North of the outcrop, a borehole at Eldon Hill (see p. 42) is interpreted as showing 225 ft of Peak Forest Limestones with a 14-ft dolomitic

c

band 45 ft from the base, and underlain by a probable tuff, proved to a thickness of 100 ft. At Castleton, the presence of pale crinoidal limestones of probable $S_2$ age underground beneath $D_1$ limestones has been noted by Ford (1952, pp. 351–2).

The development of beds of shallow-water facies in $S_2$ near the reef-belt suggests that the Peak Forest Limestones developed as the result of a broad uplift along the edge of the Derbyshire Block which heralded the more acute and localized uplift in $D_1/B_2$ times which led to the formation of the apron-reef. $S_2$ beds of comparable facies and in a similar position in relation to the margin of the block have been described by Sadler and Wyatt (1966, pp. 55–64) from the Hartington area.                                                                    I.P.S.

The normal facies of the Woo Dale Beds ranges in the Hay Dale–Dam Dale section (see p. 40) from grey foraminiferal biocalcarenites showing a poor to moderate degree of sorting to fine micritic rocks (calcilutites) showing variations in colour and in the extent of recrystallization. The Peak Forest Limestones are coarser grained, mainly unsorted biocalcarenites and biocalcirudites, largely crinoidal but in places containing lithoclastic debris.                         R.K.H.

## BEE LOW GROUP
### (Lower *Dibunophyllum* ($D_1$) Zone)

The presence of the Lower Miller's Dale Lava to the south of a line between Dove Holes and Ox Low allows the subdivision of this group into Chee Tor Rock below the lava and Miller's Dale Beds above it. Where the lava is absent the limestones are not subdivided and are known as the Bee Low Limestones. All these limestones show a uniform lithology, when compared with both the underlying and overlying beds, being fine-grained, massive and pale. In addition, the absence of chert, with one exception, is striking in view of its abundance in the strata above. Fossils are absent or rare though they are in places abundant at restricted horizons.

#### CHEE TOR ROCK

The Chee Tor Rock consists of pale fine-grained limestones, for the most part very massive. The thickness varies from about 380 ft in the southernmost outcrops in Great Rocks Dale to about 300 ft in the north. Typically the middle and upper parts of the subdivision show the characteristic closely-spaced vertical joints noted by Cope (1933, p. 131) giving a castellated appearance to weathered sections. The lowest beds frequently stand out as strong features. The Chee Tor Rock is separately distinguished along Great Rocks Dale and Doveholes Dale, around the Peak Forest Anticline and south-east of the latter around Peter Dale.

Much of the Chee Tor Rock is relatively unfossiliferous, fossils where present being mostly restricted to well-defined bands with $D_1$ Zone coral-brachiopod faunas or, rarely, with algal nodules. The corals present include *D. bourtonense*, *Koninckophyllum* θ, *Lithostrotion junceum*, *L. pauciradiale* and *P. murchisoni*. The most important faunal bands, here called the Upper *Davidsonina septosa* Band and the Lower *D. septosa* Band, lie near the top of the subdivision. The upper band, lying some 25 ft below the Lower Lava, was originally described as the "*Cyrtina septosa* Band" by Cope (1936, pp. 48–51) in the Wye valley section and in the present district at Buxton Bridge. Cope subsequently (1939, p. 63) noted the presence of a band with a similar fauna, some 10 to 15 ft lower in the sequence; it is uncertain whether this represents a well-developed band with *D. septosa* which occurs in the present district some 35 ft below the upper band and which is here referred to as the Lower *D. septosa* Band. The chief fossil bands in the Chee Tor Rock are indicated in Fig. 6. Despite its monotonous nature certain facts emerge from the detailed study of the stratigraphy:

(*a*) The lowest beds, some 40 ft thick, are usually relatively thin bedded (in 2- to 6-ft posts). A persistent coral band is present at the base on the eastern side of the Peak Forest inlier but has not been located elsewhere. It bears a typical $D_1$ coral fauna in this area but to the north of Peak Forest appears to be represented by a band of limestone with algal nodules. As noted by Cope (1939, p. 62) bryozoa are relatively abundant both in the basal coral band and in the overlying 30 ft or so of beds.

(*b*) The succeeding beds, about 160 ft thick, are frequently very massive near the base. They contain a 6-ft coral band in Peter Dale and a thinner band in the Hay Dale–Dam Dale section (these bands are shown in Fig. 6 in their positions relative to the base of the Chee Tor Rock; if thinning occurs they may represent the same band). A less important coral band occurs at Wheston.

(*c*) The coral band of Laughman Tor, although sporadically developed, is an important horizon, being 10 ft thick at the type locality with a finely bioclastic and somewhat oolitic matrix. A band in this position has also been noted on Middle Hill (south) and near Wheston.

(*d*) Some 80 ft of beds, again usually very massive, separate the last horizon from that of the Lower *Davidsonina septosa* Band. In Duchy Quarry, Smalldale, these beds contain a well-marked algal nodule band about 1 ft thick and 40 ft from the base; this appears to be represented on Middle Hill and at Wheston by a coral band. The uppermost beds locally show crinoid debris including some stems. Some 22 ft higher in Duchy Quarry and Buxton Bridge, a thin band yields *D. septosa*, corals and gastropods. Locally, north of Smalldale these beds are less massive and contain fairly numerous clay wayboards, some with underlying potholed surfaces.

(*e*) The Lower *D. septosa* Band varies from 10 in to 4 ft and although sporadically developed, has been recognized over a wide area.

(*f*) About 35 to 40 ft of limestones with some finely crinoidal bands, separate the Lower from the Upper *D. septosa* Band. At Buxton Bridge these beds contain several thin clay wayboards on potholed surfaces. A thin calcilutite band is present near the top of this group of strata in Bold Venture Quarry.

(*g*) The Upper *D. septosa* Band varies from $1\frac{1}{2}$ to 5 ft. Although widespread it is locally, at Upper End, largely represented by a thick algal band with only rare *D. septosa*.

(*h*) The beds above the Upper *D. septosa* Band are 20 to 25 ft thick and in places very massive. Some finely crinoidal bands are present.

## LOWER MILLER'S DALE LAVA

The term Lower Miller's Dale Lava is here introduced for the Lower Lava of Arnold-Bemrose (1907, pp. 246–7) to distinguish it from the Lower Matlock Lava of Smith and others (1967, pp. 8–12) which is at a higher horizon. The

Lower Miller's Dale Lava[1] occurs within the present district south-east of a line from Dove Holes to Ox Low and as an isolated patch to the north at Cave Dale. The thickness of the lava is 100 ft at Wormhill and 90 ft at Wheston but a marked northerly thinning occurs from these localities until the lava disappears.

The lava is an olivine-basalt (see also p. 120), usually very amygdaloidal and with calcite- and chlorite-filled vesicles. In places non-vesicular basalts are developed and appear to represent the central parts of flows. The lava weathers deeply and natural exposures are infrequent except in swallow-holes which commonly mark its outcrop. The Lower Lava has been penetrated in boreholes at two localities: in Great Rocks Dale a thin uppermost flow and two thicker flows below showed intervening beds of clay and tuff, the whole being 98 ft 2 in thick; the Litton Dale Borehole passed through a thick upper flow, separated from a thinner lower flow by 12 ft of limestone, the whole totalling 90 ft 6 in.

The isolated occurrence of Lower Lava at Cave Dale, Castleton, was left uncorrelated by Arnold-Bemrose (1907, p. 248) and Shirley and Horsfield (1940, p. 275) were the first to show its horizon. It is up to 25 ft thick and its outcrop extends from Hurdlow Barn to Cave Dale. To the east of the latter, drilling has shown that the lava dies out and its place is taken by the Pindale Tuff (Eden and others 1964, pp. 76–8), which stretches from the reef-belt, where it is mainly developed (see p. 61), inwards beneath the back-reef facies before dying out.

### MILLER'S DALE BEDS

The Miller's Dale Beds (see Fig. 6) resemble the Chee Tor Rock in lithology, being pale, fine-grained and massive, and contain similar $D_1$ faunas. Minor differences are the presence locally of limestones almost pure white in colour (a shade not usually found in the Chee Tor Rock) and the generally somewhat less massive character. Fine crinoid debris is more uniformly distributed than in the Chee Tor Rock. A few slightly oolitic bands have been noted. The thickness of the subdivision varies from 142 ft near Dove Holes to 63 ft at Wall Cliff.

The lowest beds are best seen at the north end of Wormhill Moor where they include a few inches of tuffaceous limestone resting on the Lower Lava. The succeeding 30 to 40 ft of strata are richly fossiliferous with an abundant coral fauna and brachiopods including *Davidsonina septosa*. In this vicinity some of these beds show a distinctly reef-like appearance with an irregular banding of possible algal origin. At Wall Cliff the reef-brachiopods *Dielasma hastatum* and *Pugnax pugnus* occur in an 8-ft bed, 32 ft above the Lower Lava. The presence of these two forms may be linked with a shallowing of the sea-floor due to irregularities in the surface of the underlying lava. These more fossiliferous strata include the equivalent of the beds with *Lithostrotion* aff. *maccoyanum* noted by Cope (1933, p. 132) immediately above the Lower Miller's Dale Lava in the Wye valley.

The higher parts of the Miller's Dale Beds call for little comment though they include the Dove Holes Tuff in the western outcrop. The tuff, about 7 ft 6 inches in thickness, lies about 50 ft above the Lower Lava and is a greenish grey clay with common limestone lapilli. Borehole sections show it to be pyritous. It rests on a strongly potholed surface.

---

[1]The name is here in places abbreviated to 'Lower Lava' where no ambiguity exists.

## BEE LOW LIMESTONES

In lithology and other features, such as bedding and the local presence of clay wayboards and potholed surfaces, the Bee Low Limestones, which are about 600 ft around Eldon Hill, are indistinguishable from their equivalents, the Chee Tor Rock and Miller's Dale Beds, and faunas typical of the $D_1$ Zone are present throughout. Detailed correlation is difficult as the discovery of at least four horizons with *D. septosa* means that correlation of a given band is dependent on its position in relation to the top or base of the group.

Bee Low Quarry shows a 300-ft face of massive limestones, lying some 80 ft below the top of the group. Farther east, a coral band 260 ft from the base of $D_1$ probably represents that of Laughman Tor (p. 47). At about the same horizon a $3\frac{1}{4}$-ft porcellanous band is present near Lodesbarn. Near Haddocklow the Upper and Lower *D. septosa* bands have been identified and a higher band with *D. septosa* is tentatively correlated with the *Lithostrotion* aff. *maccoyanum* horizon.

Within some 200 yd of the reef-belt the Bee Low Limestones show an increase in the content of crinoid debris and contain lenses of limestone with reef faunas which are comparable with those from the $B_2$ apron-reef and can be correlated with limestones at about the horizon of the Upper *D. septosa* Band. One of these reef-faunas yielded the goniatite *Bollandoceras sp.* on Gautries Hill. In Eldon Hill Quarry the Upper *D. septosa* Band is reef-like up to 500 yd from the boundary of the apron-reef and the lower beds also contain a lens of reef-limestone. Near the reef margin the back reef (Plate XI) is locally coarsely crinoidal, with debris up to 1 in, and some bands with limestone pebbles and breccia. A $9\frac{1}{2}$-ft oolitic band is also developed north-east of Eldon Hill about 88 ft above the Upper *D. septosa* Band.

At Windy Knoll, back-reef limestones are well developed; they are fossiliferous with reef-brachiopods and also contain some breccia (in addition to the 'Neptunian dykes' noted on p. 110). The back-reef facies is little developed west of Treak Cliff and in The Winnats, but in Cave Dale it is better developed, the lower beds showing interdigitations of reef-limestone. Here the Lower Lava (see p. 60) is succeeded by pale limestones with a coral band, the *L.* aff. *maccoyamum* band of Shirley and Horsfield (1940, p. 275) bearing a sparse fauna of *L.* cf. *aranea* at 39 ft above the lava. A thin tuff band is present just above; this has also been proved in a borehole (Plate IV) in the adjacent Michill Bank area. Shirley and Horsfield considered that the Lower Lava was cut out in the vicinity by an unconformity but its disappearance is here (p. 60) thought to be due to the normal dying out of a flow. Local thinning of the beds between the lava and the base of $D_2$, to about 100 ft, may be due to minor non-sequences or condensed deposition in the back-reef area and shallow-water conditions are indicated by the presence of oolite bands and local cross-bedding in the Cave Dale area.

In Pin Dale some beds of back-reef facies can be recognized but farther east their absence, at least at the present level of erosion, coincides with the general absence of apron-reef of $B_2$–$D_1$ age. The section here shows 82 ft of beds beneath $D_2$ and the Pindale Tuff is calculated from boreholes (see Plate V) to crop just below the exposed beds. An abnormal feature of the section is the presence of nodular chert in some abundance in the lowest 23 ft. The exceptional presence of chert in these limestones is perhaps due to a local silica concentration near the reef; Wolfenden (1959, pp. 566–8) noted the abundance of sponges in the algal reef and the silica may be attributed to this source. The higher beds show wedge-bedding and bands of breccia near the apron-reef (Eden and others 1964, p. 79). The coral band with *Dibunophyllum bourtonense* φ and *Palaeosmilia murchisoni*, noted by Jackson (1941, p. 240) and by Eden and others (1964, p. 78) as lying about 65 ft below the top of $D_1$ is here located at 33 ft from the top, part of the difference being due to the lowering of the base of $D_2$ in the present work (see pp. 72–3).

Pale brown to light grey limestones referred here to $D_1$, have been proved in an underground borehole at Glebe Mine, Eyam (see p. 96).                    I.P.S.

The greater part of the $D_1$ limestones consists of uniformly textured biocalcarenite. The organic content, especially that of crinoid debris, shows some variation. Foraminifera, productoid spines and rare bryozoa occur in addition to coarser bioclasts. The calcite cement shows sporadic recrystallization. The finer limestones (calcilutites) are micritic, but are seldom free from bioclastic material. Authigenic quartz occurs in small quantity in many of the limestones. A clay parting in the Chee Tor Rock (see p. 47) has been found to be a pyritic illite-mudstone with marcasite needles.                    R.K.H.

## MONSAL DALE GROUP
### (Upper *Dibunophyllum* ($D_2$) Zone)

In contrast to the rocks of the $D_1$ Zone these beds are of very variable lithology and include dark beds. They are in addition more uniformly fossiliferous and the faunas are indicative of the $D_2$ Zone. Both pale and dark limestones are cherty at some horizons. The presence of the Litton Tuff or the overlying Upper *Girvanella* Band allows these strata to be subdivided into the Lower and Upper Monsal Dale Beds. Where the Upper Miller's Dale Lava is present this is included in the group following Cope (1937, pp. 192–3). Where the lava is absent there is a sharp break at the base of the Monsal Dale Beds suggestive of a non-sequence, though the extent of this is difficult to assess.

#### MONSAL DALE BEDS BENEATH UPPER MILLER'S DALE LAVA

A small area of these beds, the northern continuation of the Station Quarry Beds of the Wye valley (Cope 1937, p. 180), extends into the Chapel en le Frith district. They are ill exposed at outcrop though $6\frac{1}{2}$ ft of alternating pale and dark limestones were proved below the Upper Lava in the Litton Dale Borehole. Elsewhere these beds are absent, the lava resting directly on beds of $D_1$ age.

In Station Quarry, Miller's Dale, some 800 yd south of the sheet boundary, these beds are 25 ft in thickness and consist of thin-bedded dark grey and grey cherty limestones (Cope ibid., p. 186). The base is strongly erosive and shows a washout at one point.

#### UPPER MILLER'S DALE LAVA

The Upper Miller's Dale Lava[1] is of lesser extent than the Lower in the present district, being developed only between Wormhill Moor and Old Moor

---

[1]The name is here in places abbreviated to 'Upper Lava' where no ambiguity exists.

(north-east of Peak Forest) and Tideswell Dale. It is similar in most respects to the Lower Lava (see p. 23) being an olivine-basalt, usually highly amygdaloidal. The maximum thickness is 100 ft at the south end of the Wormhill Moor outlier and from here the lava thins northwards, eastwards and westwards. In Tideswell Dale a very rapid north-easterly thinning takes place from the junction with Litton Dale, where the thickness is some 60 ft, to the Litton Dale Borehole only 400 yd distant where it has thinned to 8 ft. The rapid easterly disappearance of the lava in the Litton railway-cutting, on the Buxton (111) Sheet has been recorded by Cope (1937, pp. 183–4) whose description suggests a steep snout to the flow; similar conditions probably obtain at Litton Dale.

At Brook Bottom, north of Tideswell, the Upper Lava is replaced by a tuff, the Brook-Bottom Tuff of Arnold-Bemrose (1907, p. 253). At one place (see p. 62), the lava can be seen to overlie the tuff. The tuff locally contains lapilli and is interpreted as a small ash-cone at least partly surrounded and buried by lava.

## LOWER MONSAL DALE BEDS

The original term 'Monsal Dale Beds' of Hudson and Cotton (1945b, pp. 30–1) has been here extended to include the whole of the $D_2$ succession of the Shelf Province. These beds are subdivided for descriptive purposes into a lower group of predominantly dark limestones and an upper group of predominantly pale limestones. Lithological variation of these strata is illustrated in Plate VI. The total thickness varies from about 100 ft on Tideswell Moor to 230 ft in Cressbrook Dale, where it includes the Cressbrook Dale Lava.

The lower dark beds, usually from about 25 to 80 ft in thickness, are developed over most of the outcrop of the Lower Monsal Dale Beds. They are dark grey, usually thin-bedded and in places cherty. Some localities show paler, bioclastic bands, crinoidal or shelly; others show a bed of pale cherty limestone within the dark beds. The Lower *Girvanella* Band, 1 ft 9 in to 9 ft in thickness, is present at many localities near or at the base; 6 ft of dark limestone underlying the Lower *Girvanella* Band in Pindale Quarry is interpreted as occupying a hollow in the underlying $D_1$ limestones. Near the reef margin 10 to 20 ft of pale-coloured coarsely bioclastic limestone ("first calcirudite" of Eden and others 1964, pp. 81–2) occur within the dark beds and extend over the whole area of Earle's Quarry; a similar section is present at Dove Holes. There is, however, in marked contrast to the $B_2$–$D_1$ beds, no uniform development of back-reef facies in the Monsal Dale Beds.

In the absence of the Lower *Girvanella* Band, *Saccamminopsis* sometimes occurs in the lowest few feet of the Lower Monsal Dale Beds, the two fossils appearing to be mutually exclusive. A higher horizon with *Saccamminopsis* is present at Dove Holes and in Earle's Quarry, some 25 ft above the base of $D_2$; this is probably represented in the expanded sequence in Cressbrook Dale by two bands at 39 ft and 50 ft from the base respectively. At Brook Bottom what is probably the same band is well developed near the top of the dark limestones.

In the western outcrops the uppermost part of the dark beds and the lowest part of the succeeding limestones contain knoll-reefs at Dove Holes, Barmoor Clough and Sparrow Pit. The Barmoor Clough knoll has yielded fish teeth in some abundance (see p. 66). The presence of the corals *Zaphrentites enniskilleni* and *Cyathaxonia rushiana* in the dark beds at Barmoor Clough is taken as indicative of the proximity of the basin facies. The latter is further indicated by a small outcrop of $P_1$ shale in the vicinity (see below).

In the outlier of Wormhill Moor and on the main outcrop between White-rake and Old Moor the dark beds are ill developed, consisting only of thin dark bands in predominantly pale massive limestones with some oolitic bands. The thickness of the whole of the Monsal Dale Beds is here only some 100 ft. Near Whiterake, a clay band near the base may represent a thin tuff.

The upper pale beds vary normally from about 90 to 120 ft, an exception being the thin development noted above. These beds are mainly pale massive lime-stones, often crinoidal and with some shell bands. Chert is usually absent but occurs in places in the lowest beds. The most extensive development of these strata is in Cressbrook Dale where a shell bed with *Gigantoproductus sp.* cf. *striatosulcatus semiglobosus* is present 22 ft from the base and the Cressbrook Dale Lava (see below) occurs near the top. *Striatifera striata* occurs close to this horizon both here and elsewhere. In the Pin Dale area, Eden and others (1964, p. 85) have noted the presence of two coarse crinoidal bands (the "second calcirudite" and "third calcirudite") near the middle and at the top of the sub-division. The second calcirudite has been noted to pass laterally into flat reef in the easternmost part of Earle's Quarry. In this vicinity the presence of a local unconformity at the base of the Upper Monsal Dale Beds results in the cutting out of some 30 ft of the uppermost Lower Monsal Dale Beds. The presence of a small knoll-reef near the top of the subdivision on Bradwell Moor is ascribed to shallowing preceding the unconformity.

The presence of knoll-reefs at the junction of this subdivision with the under-lying dark beds in the western outcrops has already been noted. At Barmoor Clough a small thickness of shale at the junction of the main Edale Shales outcrop with limestones of lower $D_2$ age has been found to contain a $P_1$ fauna including the goniatite *Girtyoceras sp.* and is thus the basinal equivalent of part of the upper beds of the subdivision (see also p. 8).

The pale beds are well seen in Victory Quarry, Dove Holes, where they contain a persistent shell band, some 40 ft from their base, with *Giganto-productus dentifer* and *G. edelburgensis* [juv.].

The Cressbrook Dale Lava is an amygdaloidal basalt, up to about 20 ft thick and some 30 ft below the Litton Tuff in the northern part of the dale but dying out rapidly to the south. The lava is much more extensively developed at depth and is known over a wide area from Wardlow Mires to Great Hucklow and Middleton Dale. The greatest thickness proved is in an underground borehole in Glebe Mine (see also p. 95) near Ladywash Shaft: 208 ft of basalt, with the top some 140 ft below the horizon of the base of the Eyam Limestones. The occurrence of igneous rocks at depth north of Cressbrook Dale is illustrated in Fig. 8. It will be noted that the thickest occurrences of lava are overlain by the thinnest developments of high $D_2$ limestone, a fact which may be due in part to the lava building up somewhat near the source, thought in the present state of knowledge, to be near the Ladywash Shaft. As interpreted, the boreholes also show a progressive upward approach of the lava towards the horizon of the Litton Tuff.

The Wardlow Mires No. 1 Borehole proved an unusual shallow-water development of oolitic and crinoidal limestones, 79 ft $5\frac{1}{2}$ in thick, between the

Upper *Girvanella* Band and the Litton Tuff. These are interpreted as having been deposited on the flanks of a cone built up of Litton Tuff.

## LITTON TUFF

The Litton Tuff reaches its maximum thickness of 100 ft at Litton where it is agglomeratic with bombs up to 5 in across. Arnold-Bemrose (1907, p. 254) noted this character together with the presence of accidental bombs of fossiliferous limestone. The tuff thins away from Litton, being up to 42 ft in Cressbrook Dale, where it is poorly exposed, $10\frac{1}{4}$ ft in the Littonfields Borehole and only $2\frac{1}{2}$ ft in the Wardlow Mires No. 1 Borehole, to the north of which it is absent. At outcrop the tuff is more extensively developed, extending for over 2 miles north-west of Litton.

In the Wardlow Mires No. 1 Borehole the Litton Tuff was blue-grey and very fine-grained with gritty bands in contrast to the thick and agglomeratic nature of the tuff at Litton, which suggests a source near the latter locality. It is reasonable to picture the tuff as having formed a cone which suffered progressive burial. The effects of shallow-water sedimentation on the flanks of this cone have already been noted (see p. 28).

## UPPER MONSAL DALE BEDS

The Upper Monsal Dale Beds show a fairly uniform thickness of about 250 ft. The base is marked by the Upper *Girvanella* Band or in its absence by the top of the Litton Tuff. The succession in these beds (see Plate VII) is best shown in Middleton Dale, Cressbrook Dale, Wardlow Mires No. 1 Borehole and in Bradwell Dale. Minor erosion surfaces and clay partings occur locally and are noted in the detailed sections. The more important features are summarized below.

The beds below the Hob's House Coral Band, some 50 to 60 ft thick, are dark limestones with the Upper *Girvanella* Band at or near the base; the band varies from a few inches in thickness to $8\frac{1}{4}$ ft in parts of Earle's Quarry (Eden and others 1964, p. 113). The band has also been found, though thin, in pale limestones in the Hucklow Edge No. 2 Borehole. The overlying beds are of variable lithology. In the main outcrop to the north of Litton and in the Littonfields Borehole they consist of dark cherty limestones while in Cressbrook Dale there is evidence of lateral passage southwards from preponderant dark cherty limestones in the northern outcrops into pale partly bioclastic limestones with a little chert.

The Hob's House Coral Band was originally described by Cope (1933, p. 140), though apparently placed by him too low in the sequence. Shirley and Horsfield (1945, pp. 295, 298) determined the horizon of the band in terms of the Cressbrook Dale section and this work is followed here. The band, up to 5 ft thick, bears an abundant fauna with *Dibunophyllum bipartitum bipartitum*, *Diphyphyllum spp.*, *Lithostrotion pauciradiale* and *L. portlocki*. In the Cressbrook Dale section clisiophylloids are less abundant than at Crossdale Head (Cope, op. cit.). In the northern part of the Cressbrook Dale section the band passes laterally northwards into a shell bed and the latter has been recognized in the Wardlow Mires No. 1 Borehole. The band has also been recognized in the Coombs Dale section (see Plate VII), but the Middleton Dale section fails to show it (see below).

The beds between the Hob's House Coral Band and the Black Bed, some 50 to 60 ft thick, consist of pale massive limestones showing local variation in detail. The Middleton Dale section shows a lower group of well-sorted crinoidal limestones, of evident shallow-water facies, overlain by pale massive limestones with the Lower Shell Bed ("Lower *Giganteus* Bed" of Morris 1929, p. 45). These beds have been described by Orme (1967, p. 27) as crinoid calcirudites and crinoid calcarenites. The Lower Shell Bed is some 6 to 7 ft in thickness and crowded with *Gigantoproductus edelburgensis*, the shells being often silicified. The Wardlow Mires No. 1 Borehole does not show the same development of crinoidal limestones at the base of the group and a bed of pale calcilutite is present in the lower part. The Cressbrook Dale section shows evidence of shallow-water sedimentation in the presence of bands of rolled shells and some oolitic bands at the base of the group.

The Black Bed, which consists of dark bituminous limestone, forms a useful lithological marker band in the middle of the Upper Monsal Dale Beds. It is here named in contrast to the higher White Bed of Morris (1929, pp. 47–8) and includes his *Lonsdaleia* Bed. The present name is preferred as the dark limestone is of wider extent than the faunal band. In Middleton Dale the bed is up to 5 ft thick and it has also been recognized in the Littonfields (see p. 69) and Wardlow Mires No. 1 boreholes. In Cressbrook Dale the Black Bed thickens to 16 ft and this increase is probably related to the basin-ward passage into dark limestones noted by Eden (1954, p. 37) in the eastern end of the Wye valley. *Lonsdaleia floriformis floriformis* is commonly present either just above the Black Bed (*Lonsdaleia* Bed of Morris), at its base or within it (Cressbrook Dale).

The beds between the Black Bed and White Bed are about 55 to 70 ft thick; they consist of pale massive limestones, mostly fine-grained and free from crinoid debris, with abundant chert in the upper one-third and sporadically developed in places below. The Middleton Dale section shows the Upper Shell Bed (*Hemisphericus* Bed of Morris 1929, p. 42) about 1 to 4 ft thick and with abundant *Gigantoproductus crassiventer* and *G. dentifer?* Some 5 ft higher, Orme (1967, p. 28) has noted a thin algal band and a concentration of foraminifera in the limestones (calcarenites) overlying this. In Stoney Middleton, *Orionastraea placenta*, *Diphyphyllum fasciculatum* and other corals occur a little below the White Bed.

The White Bed is not present in Bradwell Dale but its position can be inferred; here the underlying beds are pale massive cherty limestones, crinoidal in the lower part.

The White Bed consists of buff cherty calcilutite up to 7 ft thick. Owing to its pale weathering it forms a conspicuous feature in the Middleton Dale sections where it was first named by Morris (op. cit., p. 42). Work by Orme (1967, pp. 28–9; 1970b) has shown that the White Bed is of somewhat variable texture, part being described as a calcisiltite or fine calcarenite. Insoluble residues contain small silicified gastropods, small *Lingula*, sponge spicules, conodonts and fish debris. The bed is attributed by Orme to very shallow-water conditions.

The beds above the White Bed, some 50 to 75 ft thick, include massive pale cherty limestones and in places some very fine-grained non-cherty beds at the top (including impersistent calcilutites). In Morris's (ibid., p. 48) Shining Tor section a massive fine-grained bed, the "Hackly Bed" is so-called on account of its fine irregular jointing. Immediately above this lies the *Dibunophyllum* aff.

*muirheadi* Bed, in which *D. bipartitum bipartitum* and *D. bipartitum konincki* occur. The band is of only small lateral extent but appears to lie at the horizon of the *Lonsdaleia duplicata* Band, present in Furness Quarry 27 ft below the *Orionastraea* Band. The *Orionastraea* Band, some 30 to 50 ft from the top of $D_2$, is about 2 or 3 ft thick and bears *O. placenta*, *Lithostrotion portlocki* and other corals. A higher band, with a fauna including *Lonsdaleia duplicata duplicata*, is present in some localities a few feet above the *Orionastraea* Band.

The section in Coombs Dale (see Plate VII) shows a general resemblance to that in Middleton Dale. However, much of the middle beds is unexposed. The Hob's House Coral Band has been recognized at the base. In the higher beds it appears likely that the White Bed has passed laterally into dark limestones.

The succession in the Hope-Bradwell Moor area (Plate VIII) shows a broad correspondence to that outlined above. There is a strong development of dark cherty beds in the lower part of the sequence with the Upper *Girvanella* Band recognized at the base. The higher beds are pale and frequently cherty. Knoll-reefs are well developed some 30 ft above the base and these have built up in one place to join a higher reef horizon to give a complex knoll-reef some 50 ft in thickness. The *L. duplicata* Band is well developed in places and can be related to the *Orionastraea* Band, some 28 ft higher, in Intake Dale, and the band has also been noted in the Wardlow Mires No. 1 Borehole and in Cressbrook Dale. Flat-reef limestones are developed on Bradwell Moor between the horizons of these two coral bands.

In Bradwell Dale (Fig. 11) all the exposed high $D_2$ limestones, including beds down to an horizon some 75 ft below the *Orionastraea* Band, can be seen to pass northwards into dark cherty limestones of basin facies. Some evidence of a similar lateral passage has been noted south-east of Little Hucklow.          I.P.S.

A detailed description of the petrography of the Monsal Dale Beds adjacent to the reef-belt has been given by Eden and others (1964, pp. 93–5) and a systematic description of the large range of rock types has not been attempted for the present work. The pale lithofacies consists of biocalcarenites with some lithocalcarenites and biocalcirudites. The White Bed, near the top of the sub-division, has been described (Orme 1967, pp. 28–9) as varying from a calcilutite and calcisiltite to a fine calcarenite (see p. 30). The occurrence of chert in the limestones is noted in the details; scattered authigenic quartz occurs, becoming abundant in the cherty limestones. Foraminifera vary in abundance and a concentration has been noted by Orme (ibid., p. 28) in a bed near the top of the Upper Monsal Dale Beds in Middleton Dale. The dark lithofacies consists mainly of biocalcarenites and some lithocalcarenites, some of which, as noted by Eden and others (ibid., p. 93), appear deceptively fine-grained, though some finer micritic types do occur. The dark limestones are, in general, finer grained than the pale, though some coarser bands, mainly crinoid calcirudites, are present.

R.K.H.

## EYAM GROUP
### (Upper *Posidonia* ($P_2$) Zone)

#### EYAM LIMESTONES

The Eyam Limestones, up to about 150 ft in thickness, consist in their simplest development of dark thin-bedded cherty limestones (Plate X). Locally knoll-reefs and flat reefs are present and the latter may constitute the whole of the subdivision. They contain a rich brachiopod fauna and, at one locality, the $P_2$ goniatite *Sudeticeras sp.* Near Great Hucklow, only grey limestones are present.

The Eyam Limestones rest with a marked break on the Monsal Dale Beds. This unconformity was described by Shirley and Horsfield (1945, p. 293) though its effects have been found to be less than suggested by these authors at some localities. It seems reasonable to regard the junction as an angular unconformity in the marginal areas, where its effects can be observed, and as a non-sequence elsewhere. The junction shows in many places evidence of erosion in the presence of a potholed surface, overlain by a clay band. In Coombsdale Quarry a thin coal and seatearth are present at this horizon, providing clear evidence of emergence. At some localities, for example, near Intake Dale, knoll-reefs developed on the base of the Eyam Limestones and were apparently initiated by shallow-water conditions immediately after the pre-$P_2$ unconformity. Some of these knoll-reefs remain as outliers. In Bradwell Dale some 40 ft of uppermost $D_2$ strata can be seen to be cut out beneath the Eyam Limestones.

In Coombsdale Quarry the thin coal noted above is overlain by a thin shale, the "Bituminous Shale" of Morris (1929, p. 49) which has yielded the coral *Cladochonus sp.* Between Calver and Eyam, the Eyam Limestones show a general arrangement of dark cherty limestones with some knoll-reefs and flat reefs below and a widely developed flat reef above. At Calver Peak, a flat reef at the base of the subdivision is overlain by a smaller well-defined knoll and both show a widely developed patchy silicification probably of epigenetic origin (see p. 310). In Middleton Dale, some 35 ft of dark limestones are present at the base and a 1½-ft coral band, the "*Aulophyllum* Band" of Morris (op. cit., p. 54) occurs at the junction with the overlying flat reef. Morris considered that this bed overlay the flat reef. The latter, up to 100 ft thick and widespread between Stoney Middleton and Eyam, shows the usual faunal and lithological characters of reef-limestones with, in addition, a slight development of chert in places.

In Eyam Dale, some 75 to 80 ft of predominantly dark cherty beds are present below the flat reef, corresponding to a westerly diminution of the latter, with diachronous base. The dark limestones show a bed with *Gigantoproductus giganteus* at 21 ft from the base and a 3-ft band 39 ft higher with *Dibunophyllum bipartitum* subspp. The latter has proved widespread in the Eyam-Foolow area and it is here named the Eyam Dale Coral Band. The overlying flat reef is only some 40 ft in thickness.

To the west of Eyam Dale the flat reef dies out and lateral passage of all but the lowest part of the dark cherty limestones into grey bioclastic limestones takes place. These beds present a more normal shelf facies than the basinal dark beds in the Eyam Limestones and the area in which they are developed is that lying farthest into the shelf area. These beds contain a series of eleven knoll-reefs in their upper part, those in the Grindlow–Foolow area showing a marked elongation parallel to the limestone-shale junction. Lower in the succession a knoll at the junction with beds of the dark lithofacies is well seen in Linen Dale where it was first noted by Morris (1929, p. 52). A bed with *Gigantoproductus edelburgensis* and *G. sp.* cf. *gigantoides* passes laterally into the knoll. The nearby section in the sink-hole near Waterfall Farm shows the Eyam Dale Coral Band.

The Wardlow Mires boreholes proved some 30 ft of dark thin-bedded cherty limestones overlain unconformably by $P_2$ shale (see below). Mudstone partings were a frequent feature of these limestones, especially in their upper part. Abundant *Lingula* and sporadic *Orbiculoidea* and trilobites occurred, especially in the mudstone partings. Inarticulate brachiopods have not been recorded from surface exposures. As elsewhere an erosive base with a mudstone parting is present.

To the south of Great Hucklow the grey lithofacies noted above extends to the base of the Eyam Limestones. To the north-west of the village, however, lateral passage into dark beds again occurs and the grey beds appear to wedge out in the middle of the subdivision. The presence of these dark beds was noted by Shirley and Horsfield (1940), pp. 286, 288) in Jenning's Dale and the fauna compared to that of the dark beds of Cawdor Quarry, subsequently described (Shirley 1959, pp. 414–5) under the name Cawdor Limestones, which contain the $P_2$ fossil *Goniatites granosus*. Shirley and Horsfield also noted the presence of a large knoll-reef in these beds north of Nether Water Farm; this is here considered to have been formed by the junction of two knolls.

In Bradwell Dale a large flat reef forms the whole of the Eyam Limestones. Sections at the south end of the dale show two phases of lateral growth on the margin, an initial stage of restricted spread and a later stage of more vigorous spread. Lateral passage into dark cherty limestones is here well marked. The flat reef reaches a maximum thickness of about 130 ft and the uppermost beds are very coarsely crinoidal in places with debris up to an inch in diameter.

Isolated occurrences of beds referred to the Eyam Limestones in the shelf area are also present in the northern and western outcrops. In Windy Knoll Quarry, they occur as dark fillings to Neptunian dykes in pale massive $D_1$ limestones of back-reef facies. Near Perryfoot, a small outlier of dark limestone is present, resting on $D_1$ back reef. Both these are referred to $P_2$ in view of the presence of the goniatite *Sudeticeras sp.* in similar limestones near Peakshill (see p. 111).

In the westernmost part of the limestone area, the Eyam Limestones are present in a narrow outcrop around Brook House. Fig. 7 shows the relation of this to the Miller's Dale Beds on which they rest with large unconformity, and to the $D_2$ limestones which occupy a similar geographical position farther north. The $P_2$ and $D_2$ outcrops are separated by a fault of general easterly trend and the diagram demonstrates the pre-$P_2$ age of this structure.

The greater part of the Brook House outcrop consists of dark cherty limestones some 30 ft thick. The uppermost beds contain an abundant coral fauna in several bands within 14 ft of strata, some or all of which may represent the Eyam Dale Coral Band. Near Brook House, a flat reef some 30 ft thick occurs at the base and it is on the presence in this of *Sudeticeras sp.*, together with the brachiopod fauna, that the dating of these beds depends. Some 16 ft of grey and dark grey limestones overlie the flat reef.                                               I.P.S.

Little petrographical work has been done on the Eyam Limestones, which typically resemble the dark limestones in the Monsal Dale Beds. In Eyam Dale they are dark biocalcarenites, biomicrites and biocalcirudite-biocalcarenites, and intergranular hydrocarbon occurs sporadically.                          R.K.H.

## SHALES OF $P_2$ AGE

On the shelf, shales of $P_2$ age are known only from the Wardlow Mires outlier. Here some 11 ft of shales have been proved in the Wardlow Mires Nos. 1 and 2 boreholes. No. 1 Borehole yielded *Sudeticeras* cf. *stolbergi* at 80 ft 1 in. The base of the shales is unconformable and in the southern part of the outlier they rest on $D_2$ limestones. At Wardlow Mires itself the shales rest on the lower dark beds of the Eyam Limestones, while west of Castlegate they overlie the higher grey lithofacies. There is no evidence of an unconformable junction with the Namurian.

# MARGINAL PROVINCE

The beds present in this province at surface in the northernmost Viséan outcrop in the district include the reef-complex of the Castleton area and a thin development of overlying strata. The lateral development of the reef-complex is to some extent masked by the unconformable base of the Namurian.

### APRON-REEF

The apron-reef is present nearly continuously from near Sparrow Pit to the vicinity of Bradwell (Plate XI). The lithology and sedimentation have been described on pp. 16–9.

The apron-reef as described here includes the Castleton Limestones ($B_2$) and part of the Nunlow Limestones ($P_1$) of Shirley and Horsfield (1940). The poorly bedded character of the apron-reef limestones makes detailed stratigraphy and correlation difficult to establish, but the general faunal succession is as follows:

*Goniatites falcatus* Zone ($P_{1b}$)

*Goniatites crenistria* Zone ($P_{1a}$)

Upper *Beyrichoceras* Zone ($B_2$)

The faunas, which are rich in brachiopods, gastropods and bivalves and contain occasional goniatites, are further discussed on p. 140. Beds of high $B_2$ age are of widespread occurrence while low $B_2$ and $P_1$ beds are discontinuous and restricted in general to the inner and outer margins respectively of the apron.

The lateral passage of the apron-reef into beds of shelf facies has already been described (p. 18). At the ends of the reef-belt the apron appears to thin and in these areas it can be seen to overlie beds of shelf facies; this is readily interpreted if the lateral passage is considered as taking place partly by interdigitation.

The $B_2$ apron-reef is in general massive and poorly bedded while the $P_1$ reef is more regularly and evenly bedded. Parkinson (1965, p. 164) considers on faunal grounds that a non-sequence of some importance is present between upper and lower $B_2$. This cannot be considered firmly established though a break in about the right position is present near the Blue John Mine. Intra-$B_2$ movements have been invoked during the present work to account for the off-setting of the reef-belt and production of a knoll-like mass of reef-limestone at Middle Hill.

To the east of Pin Dale, apron-reef of $B_2$ age is in general absent at outcrop and it has been suggested (see p. 13) that this is most probably due to penecontemporaneous movement along the line of Dirtlow Rake. Small inliers at the base of Nunlow Quarry and at the extreme eastern end of the reef-belt are, however, of $B_2$ age.

At Jack Bank, reef-limestones of $P_{1b}$ age rest unconformably on $D_1$ shelf limestones with the Pindale Tuff (see p. 108). Strata of $P_{1a}$ age are present locally, probably as a channel-fill (Eden and others 1964, p. 90) in the upper surface of $D_1$. Steep inward dips at the top of the frontal slope of the apron-reef have led the above authors to infer the presence of a $P_{1b}$ knoll, now largely eroded away, at Upper Jack Bank.

### BEACH BEDS

The Beach Beds form a restricted outcrop in a curve in the frontal slope of the apron-reef and at its foot on the west side of Castleton (see Plate XI). They consist for the most part of rounded shell debris with some finer-grained grey beds, grey crinoidal bands and dark bituminous bands of basinal type. A detailed description has been given by Sadler (1964) who concluded that they represented a fan-like deposit at the foot of a channel coinciding approximately with the present-day Winnats gorge. The Beach Beds were considered to be contemporaneous with the adjacent apron-reef and of $B_2$ age.

The above interpretation is not favoured in the present account for the following reasons. The Castleton Borehole (see below) proved the presence of a coral band with a fauna of high $D_2$ age (see p. 111) associated with beds of Beach Beds type. These beds are considered to fall within the $P_2$ of the fully developed basin succession. A widespread unconformity is known to be present in many places, for example at Bradwell, near the margin of the limestone outcrop at the base of the Eyam Limestones and close to the base of $P_2$ (see Fig. 3). If the source of supply lay, as postulated by Sadler, in the back-reef area, it is difficult to account for the abundance of shell debris, as the back-reef is not very fossiliferous behind Treak Cliff. The interdigitation of reef-like limestones within the Beach Beds can be disregarded as proof of their association with the $B_2$ apron in view of the widespread occurrence of flat reef in $P_2$.

The Beach Beds, therefore, are regarded as of probable $P_2$ age though it is possible that they were formed somewhat earlier than the main transgression of the Eyam Limestones.

### DARK LIMESTONES OVERLYING THE APRON-REEF

A small thickness of dark limestones is present above the apron-reef in Castleton village. This has failed to yield diagnostic fossils. However, between Pin Dale and Castleton, Parkinson (1947, p. 112) obtained the $P_{1b}$ fossil *Goniatites* cf. *spirifer* Roemer from a dark limestone at the foot of the frontal slope of the apron and perhaps the basinal equivalent of the highest beds of the apron-reef. Near Peakshill (see p. 111) a few inches of dark limestone, resting on $B_2$ apron-reef, have yielded the $P_2$ goniatite *Sudeticeras sp.*

### $P_2$ SHALE

Shales of $P_2$ age apparently crop out in a small area east and south-east of Treak Cliff, Castleton. The outcrop, obscured by head, has been calculated from the record of the Castleton Borehole (see below). In the Hope Cement Works

Borehole a small thickness of Viséan shales underlies the lowest Namurian; these beds, however, occur at a greater depth than in the Castleton Borehole and have been interpreted as being cut out by unconformable overlap at outcrop.

<h2 style="text-align:center">VULCANICITY</h2>

Vulcanicity in the marginal province resulted in the formation of the Speedwell Vent (see pp. 300–1) from which, however, the surface deposits appear to have been eroded, leaving only the neck. A small thickness of tuff is present in dark post-reef limestones at Castleton, though its source is uncertain. The Pindale Tuff, associated with the tuff-cones in the back-reef area (see pp. 61–2) extends outwards beneath the apron north of Pin Dale and is overlain unconformably by Namurian shales. The Hope Cement Works Borehole proved 149 ft 6 in of pillow lavas with some tuff and a tuffaceous limestone beneath 53 ft of grey to dark grey thin-bedded limestones with *Lonsdaleia duplicata* at the top. The lavas (see also p. 113) are attributed to a local eruptive centre on the line of the reef-belt, perhaps governed by the same deep-seated fracture.

# BASIN PROVINCE

Beds of the basin province are concealed by the Namurian and are known mainly from the Edale (Hudson and Cotton 1945b, pp. 1–35) and Alport (Hudson and Cotton 1945a, pp. 256–95) boreholes (Fig. 3). The succession in these is summarized on pp. 113–9.

*Seminula* ($S_2$) Zone. Beds of this zone are known only from the Alport Borehole where 528 ft were proved. They varied from grey to dark grey limestones with beds of calcilutite and breccia in places. The presence of chert at several levels contrasts with its absence in the shelf area and finds a parallel in the Permian of the Guadalupe Mountains area, Texas (see Newell and others 1953, pp. 161–2). The faunal affinities are with the shelf area with a coral-brachiopod fauna including *Davidsonina carbonaria*.

Lower *Dibunophyllum* ($D_1$) Zone. Beds of this zone, 326 ft thick in the Alport Borehole, consist of grey and grey-brown limestones with some shale partings in their upper part and some chert throughout. These beds bear a $D_1$ coral-brachiopod fauna in the lower part while *Goniatites sp. ?maximus* group was present within the uppermost 59 ft of beds. The uppermost part of the zone is represented by $B_2$ strata (see below). In the Edale Borehole there is no $B_2$ and the $D_1$ Zone is represented by the lowest 77 ft of beds proved: dark grey-brown limestones with a calcilutite band. The fauna of these beds was not diagnostic and their correlation depends on the presence of the Lower *Girvanella* Band at the base of the succeeding strata.

Upper *Beyrichoceras* ($B_2$) Zone. Beds of this zone are present in the Alport Borehole only and consist of 52 ft of mudstone above and 159 ft of limestone, partly thin-bedded or tuffaceous with shale partings, below. Both bear goniatites including *Beyrichoceras* cf. *delicatum*, *Goniatites* aff. *hudsoni* and *G. struppus*.

Upper *Dibunophyllum* ($D_2$) Zone. Beds of this zone are present only in the Edale Borehole as 63 ft of limestones, pale grey with some tuff debris above and dark below, with *Girvanella* in the lowest 10 ft. In the Alport Borehole these strata are represented by beds assigned to $P_1$ (see below).

Lower *Posidonia* ($P_1$. In the Alport Borehole these beds, 294 ft thick, are represented by:

(*a*) Grey and grey-brown limestones, tuffaceous and with tuff bands in their lower part and with a 50-ft mudstone at 56 ft from the base, 153 ft. The fauna includes *G. crenistria* ($P_{1a}$ Zone). The thinner development in the Edale Borehole, 198 ft, also shows mudstones, including some limestones, above and a tuffaceous limestone at the base.

(*b*) Mudstones with some thin limestones representing the *Goniatites falcatus* ($P_{1b}$), *G. sphaericostriatus* ($P_{1c}$) and *G. koboldi* ($P_{1d}$) zones, 141 ft.

Upper *Posidonia* ($P_2$). In the Alport Borehole this consists of 97 ft of beds, mainly mudstone but with 10 ft of pebbly limestone and breccia at the base. Three zones are present: *Mesoglyphioceras granosum* ($P_{2a}$), *Neoglyphioceras subcirculare* ($P_{2b}$) and *Lyrogoniatites georgiensis* ($P_{2c}$). In addition to goniatites, *Caneyella membranacea* is abundant in the mudstones. The Edale Borehole shows 94 ft of $P_2$ strata with 47 ft of grey and dark grey limestones at the base.

**Variations in the basin province.** A comparison of the sections of the two boreholes allows certain conclusions to be drawn regarding variation in the basin area. The lowest beds represent an imperfectly developed basin facies and the faunas recall those of the shelf area, resulting in the use of the coral-brachiopod scheme for zonation. Basin faunas appear first ($B_2$) in the Alport Borehole, farthest away from the shelf area and this also shows a thicker succession than in the Edale Borehole, which is relatively close to the shelf. Another basinal characteristic, the development of thick mudstone, instead of limestones, also occurred earlier ($B_2$) in the Alport than in the Edale Borehole, and where limestones are present at higher levels they tend to be thinner at Alport than at Edale. These facts point to a gradual spread of true basinal conditions outwards towards the shelf from the Alport area, the conclusion of the process being their encroachment on to the shelf area during the formation of the Eyam Limestones.


# DETAILS

## SHELF PROVINCE
### WOO DALE GROUP ($S_2$ and basal $D_1$)

#### WOO DALE BEDS

The main outcrop of the Woo Dale Beds lies in the axial area of the Peak Forest Anticline (Figs. 4–5). Associated small faulted inliers are present in the valley of Dam Dale and Hay Dale[1], south of Peak Forest.

*Dale Head to Hernstone Lane Head.* The Woo Dale Beds in this area show lithological characters closely comparable with those of the $S_2$ beds of the type section in the Wye valley (Cope 1933, p. 127). The section[2] (loc. 1) in the main dale at Dale Head [1234 7662] shows:

---

[1]Dam Dale, Hay Dale, Peter Dale and Monks Dale are parts of the same valley.

[2]Sections from which faunas have been obtained bear a locality number (see also Appendix II) and each individual item is distinguished by a small letter. These letters do not imply any correlation between sections. Where necessary, sections are divided between different stratigraphical headings.

D

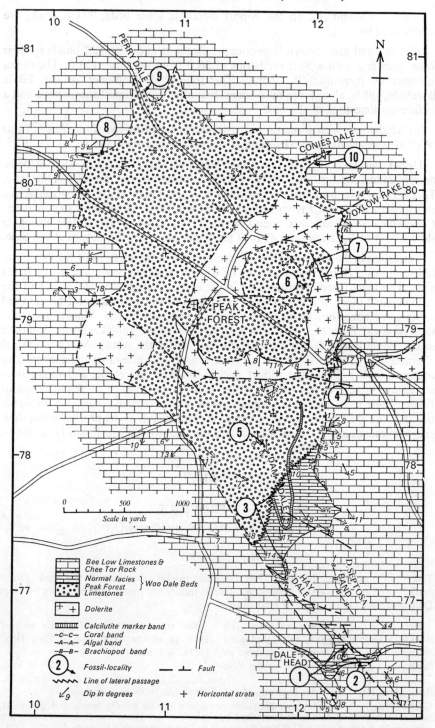

FIG. 4. *Sketch-map of the geology of the Peak Forest Anticline*

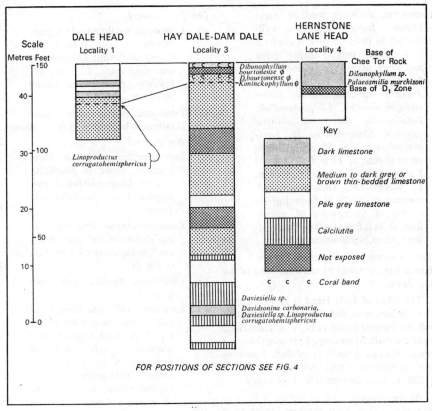

FIG. 5. *Comparative sections of the Woo Dale Beds*

|  | ft |
|---|---|
| Chee Tor Rock (see p. 49) .. .. | — |
| (e) Limestone, medium grey fine-grained; crinoid debris, corals and brachiopods including *Linoproductus sp. hemisphaericus* group .. .. .. | 10 |
| (d) Limestone, dark fine-grained slightly bituminous .. | about 3 |
| (c) Limestone, pale to medium grey | 3 |
| (b) Limestone, dark slightly bituminous fine-grained; scattered crinoid columnals .. .. | 3 |
| (a) Limestone, grey-brown thin-bedded fine-grained; band 20 ft from base with *Lithostrotion martini*, *Dielasma hastatum*, *Linoproductus corrugatohemisphericus*, *Megachonetes sp. papilionaceus* group .. .. | 24 |

The fauna of bed (*a*) is indicative of the $S_2$ Zone. However, the section in the tributary dale at Dale Head (see below), unfortunately separated from this section by faulting, shows a definite $D_1$ fauna in a nearly similar position with regard to the base of the Chee Tor Rock. All the beds (*b*) to (*e*) in this section are accordingly referred to $D_1$.

The section (loc. 2) in the eastern tributary dale at Dale Head [1248 7653] is more complete than that given above:

|  | ft |
|---|---|
| Chee Tor Rock (see p. 49) .. .. | — |
| Beds ill-exposed (*e*); small exposures of dark limestone at 2 ft from base with brachiopods and *L. martini* .. | 18½ |

ft

Limestone, dark to medium grey-brown, finely crystalline, well-bedded in beds up to 2 ft thick; a little crinoid debris, slightly paler and more massive in lowest 12 ft; fauna, at 1¼ to 3¾ ft from top (*d*) *Dibunophyllum bourtonense* φ, *Lithostrotion martini, L. pauciradiale, Palaeosmilia murchisoni, Athyris expansa, Linoproductus sp.* hemisphaericus group, *M. sp.* papilionaceus group; at 17 ft from top (*c*) *Lithostrotion* cf. *aranea, L. martini, L. pauciradiale, Linoproductus* cf. *corrugatus;* at 29 ft from top (*b*) colony of *Syringopora* cf. *geniculata;* at 38 ft from top (*a*) *P. murchisoni, M. sp. papilionaceus* group   ..      **44**

The incoming of *P. murchisoni* in the lowest faunal band is taken to mark the base of the D₁ Zone.

The inlier of Dale Head is faulted off on the northern side, but a further small faulted inlier is present ¼ mile N.N.W. of Dale Head near the main S₂ outcrop. Exposures here are scanty though a few feet of dark limestone are exposed by a small working for calcite [1200 7694] on the west side of the valley.

At the southern end of the main outcrop the section (loc. 3) at the junction of Dam Dale and Hay Dale [1187 7781] (Plate IIA) provides the best sequence of the Woo Dale Beds in the area:

ft

Chee Tor Rock (see p. 50)   ..    ..     —

(*p*) Limestone, dark grey fine-grained; *D. bourtonense* φ, *Lithostrotion martini, L. pauciradiale, L.* cf. *sociale, A.* cf. *expansa*    ..    ..    ..    **3½**

(*o*) *Not exposed*    ..    ..    ..     **3**

(*n*) Limestone, medium brown to grey finely crystalline to fine-grained in beds 2 to 3 ft; top 5 ft with *D. bourtonense* φ, *Koninckophyllum* θ Vaughan (1905, pl. 23, fig. 4), *L. martini, L. pauciradiale* (base of D₁); at 5 ft from base, band with *L.* aff. *martini, M. sp. papilionaceus* group [these beds form a prominent cliff]    ..    ..    ..    **32**

ft

(*m*) *Not exposed*    ..    ..    ..    **15**

(*l*) Limestone, medium to dark grey-brown, in beds ½ to 1½ ft; lowest 10 ft with *Lithostrotion arachnoideum, L. martini,* bryozoa, brachiopods including *Semiplanus sp. nov.*    ..    ..    **24**

(*k*) Limestone, pale grey finely crystalline, in beds up to 1¼ ft    about **7**

(*j*) *Not exposed*    ..    ..    ..    **12**

(*i*) Limestone, grey-brown very finely granular, oolitic in places; *L. martini, Linoproductus* cf. *corrugatus, L. sp. hemisphaericus* group   ..    ..    ..    **15**

(*h*) Calcilutite, brown white-weathering, sharp top and base (this is the lithological marker horizon of Fig. 4)    ..    ..    **3½**

(*g*) Limestone, medium grey fine-grained ..    ..    ..     **2**

(*f*) Limestone, off-white fine-grained with a little fine crinoid debris; top 7 ft more massive with sporadic corals, including *Lithostrotion* cf. *sociale* and indeterminate gastropods; thin-bedded below, with bryozoa, *A. expansa, Linoproductus sp. hemisphaericus* group, *M. sp. papilionaceus* group, *Semiplanus sp. nov* ..    ..    ..    **10½**

(*e*) Calcilutite, grey to pale grey poorly bedded; band at 9 ft from base with *L. sp. hemisphaericus* group    ..    ..    **13½**

(*d*) Limestone, dark grey fine-grained in beds up to 2 ft; band at 3 ft from base with *Daviesiella sp., Linoproductus sp.*    ..    ..    **5½**

(*c*) Calcilutite; upper 3 ft dark, in 3 beds; lower part brown and irregularly bedded; *Carcinophyllum vaughani, Davidsonina carbonaria, Daviesiella sp., L. corrugatohemisphericus, L. sp. hemisphaericus* group ..    ..    **5½**

(*b*) Limestone, pale brown fine-grained, somewhat oolitic, shelly in top 10 in and at base; brachiopods including *M. sp. papilionaceus* group    ..    ..    **10**

(L 164)

A. Cliff at junction of Hay Dale and Dam Dale:
Woo Dale Beds overlain by Chee Tor Rock                PLATE II

B. Potholed surface in Chee Tor Rock: Smalldale

(L 170)

ft

(a) Calcilutite, brown; brachiopods
including *L.* aff. *corrugatohemisphericus, L.* cf. *corrugatus, L. sp. hemisphaericus* group,
*M. sp. papilionaceus* group    3

Traced laterally, the calcilutites of the lower
part of the section (though not the 'marker
horizon') are found to be very irregular and
lenticular in character, swelling out and
thinning rapidly. It should also be noted
that the horizon of the beds with *Daviesiella*
in this section differs substantially from that
found by Cope (1939, pp. 61–2) in the Wye
valley. The present section yielded *Daviesiella*
only between a level 130 ft below the base of
the Chee Tor Rock and the lowest beds
exposed some 23 ft below.

The most northerly section (loc. 4) attributable to the normal facies of the Woo Dale
Beds is at Hernstone Lane Head [1220 7876]:

ft

Chee Tor Rock (see p. 50).    ..    —

(d) Limestone, dark; corals in upper
5¼ ft, thin-bedded and shelly
below; *Dibunophyllum sp.,
Lithostrotion junceum, L. pauciradiale, L.* cf. *sociale, P. murchisoni,* bryozoa, brachiopods
including productoids, *M. sp.
papilionaceus* group    ..    11¼

(c) Limestone, dark grey fine-
grained; *L. martini, P. murchisoni,* bryozoa, *Dielasma hastatum, M. sp. papilionaceus* group,
*Streptorhynchus senilis* (base of
D₁ Zone)    ..    ..    3¼

(b) *Not exposed*    ..    ..    ..    4

(a) Limestone, pale grey, massive in
lowest 6 ft; *C. vaughani, L.
martini, M. sp. papilionaceus*
group ..    ..    ..    ..    15½

*Peak Forest.* The Peak Forest Limestones
present a facies distinct from that already
described, being pale, more massive and
crinoidal. They show less contrast in lithology
to the overlying Chee Tor Rock than the
normal Woo Dale Beds but are still quite
distinct. The lateral passage between the two
facies takes place quite sharply along a
north-north-easterly line running from near
Hernstone Lane Head through a point a little
north of the large section in Dam Dale

described above. The relationships of the two
facies are made clear when the calcilutite
marker band noted in the Dam Dale section
is traced northwards into the outcrop of the
Peak Forest Limestones. Fig. 4 shows the
positions of the outcrops of the two facies
and of the calcilutite marker. The lateral
passage appears to occur at a higher horizon
at Hernstone Lane Head than in Dam Dale.

The Peak Forest Limestones are well
exposed on the west side of Dam Dale
[1173 7845]: rather dark grey fine-grained
limestone 5½ ft, on 24 ft of grey-brown
fine-grained limestone in posts up to 3 ft
thick, with some very crinoidal bands and
some productoids. On the opposite side of
the valley the most northerly exposure
showing the calcilutite marker band is seen
[1194 7846] 210 yd E. of the last exposure:
pale grey limestone 4 in, calcilutite 1 ft 6 in,
on fine-grained pale grey limestone 1 ft 2 in.
An exposure (loc. 5) [1170 7809] east of
Loosehill Farm showed 2 ft 6 in of pale
brown finely granular slightly crinoidal
limestone with brachiopods including *M. sp.
papilionaceus* group, smooth spiriferoids and
orthotetoids.

Around Peak Forest village the Peak
Forest Limestones occur as an outlier forming
part of the roof of a large dolerite sill (see p.
295). An exposure (loc. 6) [1198 7928] showed
6 ft of buff crinoidal and oolitic limestone
with *Koninckopora inflata, Linoproductus* cf.
*corrugatohemisphericus, L.* cf. *corrugatus* and
*M. sp. papilionaceus* group. A small exposure
(loc. 7) [1202 7946] yielded *Lithostrotion
martini* [close to the S₂ form of the species].
Partly altered beds overlying the dolerite are
seen in another section [1163 7877] near
Damside Farm:

ft

Dolomite, brown saccharoidal, with
calcite veins    ..    ..    about 4
*Not exposed* (except for 6 in of dark
brown fine-grained limestone in
middle)    ..    ..    ..    6
Limestone, pale grey-brown finely
crinoidal, with a 9 in by 5 ft irregular
dark silicified patch    ..    ..    7
Limestone, dark brown finely crystalline, with some fine crinoid debris    6

ft

Limestone, grey-white fine-grained slightly marmorized, with some crinoid columnals and fine shell debris; chert in irregular masses up to 1 ft, bedding rather irregular ..   4

Limestone, grey to grey-white marmorized with dark chert in lenses up to 2 in by 10 in and irregular masses up to 3 ft 3 in; some silicified crinoid columnals; calcite in a 2-in pocket at top, bedding irregular, expanding around chert masses ..   11½

*Not exposed* ..   ..   ..   about 5

Dolerite, hard dark ..   ..   ..   5½

In view of the general absence of dolomites in the Peak Forest Limestones, the dolomite in the above section is considered to be secondary. This and the other features of this section due to alteration by the dolerite are discussed on p. 295.

To the north-west of Peak Forest a quarry [1090 7952] near Chamber Farm shows a 21-ft section of the Peak Forest Limestones, finely granular pale grey-brown, rather massive and with very crinoidal bands. Some ½ mile to the north-west these beds have been dug in the past in a series of small workings for lime-burning in and around Hartle Plantation. A typical exposure [1053 7992] shows 4 ft of off-white shelly bioclastic and oolitic limestone, and another [1055 7999] nearby shows 12 ft of off-white granular crinoidal shelly limestone. A quarry (loc. 8) to the north of Hartle Plantation [1040 8020] shows beds near the top of the Peak Forest Limestones:

ft

(*d*) Limestone, pale grey crinoidal, with some coarser bands ..   11

(*c*) Limestone, pale grey oolitic and finely crinoidal ..   ..   2½

(*b*) Limestone, pale grey; *C. vaughani, Lithostrotion martini, L.* aff. *portlocki*, bryozoa, *Linoproductus corrugatohemisphericus, L. sp. hemisphaericus* group, *M. sp. papilionaceus* group ..   1

(*a*) Limestone, pale grey-brown, with very crinoidal bands ..   7

The fauna is indicative of the $S_2$ Zone, the top of the Peak Forest Limestones corresponding to the $S_2/D_1$ boundary.

Exposures in Perry Dale show beds near the top of the Peak Forest Limestones, closely overlain by the Bee Low Limestones, though the actual junction is not exposed. A section (loc. 9) [1071 8069] of the uppermost exposed Peak Forest Limestones is as follows: off-white thin-bedded limestone with *L. sp. hemisphaericus* group and *M. sp. papilionaceus* group 1 ft, on pale grey to brown well-bedded finely crinoidal limestone with more coarsely crinoidal bands 12 ft. The fauna, though not diagnostic, is probably $S_2$.

A section [1078 8025] in an old quarry south of Nether Barn shows beds lower in the Peak Forest Limestones than those described in preceding sections: pale brown well-bedded crinoidal limestone 21 ft.

On the north-eastern side of the outcrop a section in Conies Dale (loc. 10) [1201 8019] lies at or near the junction of the Peak Forest Limestones and the Bee Low Limestones: finely crinoidal limestone a few inches, on pale brown fine-grained limestone with *Lithostrotion martini*, bryozoa and *Linoproductus sp. ?hemisphaericus* group 1 ft 9 in. Between this locality and Conies Farm the underlying limestones, pale brown and crinoidal, are exposed.

*Eldon Hill and Castleton.* A borehole at Eldon Hill Quarry[1] [1128 8156], after passing through about 280 ft of beds referred to the Bee Low Limestones (see p. 58), entered a series of predominantly grey and pale grey crinoidal limestones. The only major lithological variations in these beds were the presence of grey non-crinoidal limestones between about 405 ft and 427 ft and of grey fine-grained limestones with a little dolomite between about 446 ft and 460 ft. From 505 ft to the bottom of the borehole at 605 ft feldspar, clay minerals and abundant iron oxides occurred in the samples indicating the presence of igneous rock; this is tentatively interpreted as a tuff. Tuffs in a similar geographical relation to the reef-belt (see p. 61), though at a higher horizon, have been described by Eden and others (1964, pp. 76–8).

---

[1] Information from driller's log and from examination of cutting-samples by Dr. D. C. Knill. The samples were loaned by Messrs. Eldon Hill Quarries Limited.

At Castleton, the presence underground of limestone "indistinguishable from $S_2$ limestones of Peak Forest" was recognized by Ford (1952, pp. 351–2) in the Speedwell Mine. These limestones were met in the Speedwell Level (the main crosscut) between its intersection with Faucet Rake at the 'Bottomless Pit' [1400 8235] and its furthermost point [1395 8215]. Ford records a 2-in bed of clay "at the end of the artificial level". The height of the level is given as 742 ft O.D. and the horizon of the surface here, at about 1260 ft O.D., is a few feet above that of the Lower Lava. Assuming an average thickness of about 300 ft for the Chee Tor Rock, the clay parting must lie some 200 ft below the top of $S_2$ and therefore at or close to the horizon of the igneous rock met with in the Eldon Hill Quarry Borehole.

In Peak Cavern, Castleton, Ford noted "rather dark crystalline limestone" in the entrance, overlain by pale calcite-mudstone of $D_1$ age. The lowest beds here were also referred by Ford to the $S_2$ Zone and they become pale and crinoidal when traced into the inner part of the cave beneath the upper end of Cave Dale, only ¼ mile S.E. of the end of the Speedwell Level.

## BEE LOW GROUP ($D_1$): SOUTHERN AREA

### CHEE TOR ROCK

*Great Rocks Dale.* Extensive quarrying of the Chee Tor Rock has provided excellent sections, particularly in the upper third of the subdivision; the more important of these are quoted below.

On the southern margin of the district all but the uppermost few feet of the Chee Tor Rock is exposed in the inaccessible face of Tunstead Quarry [096 743]. The total thickness is about 380 ft. A generalized section[1] of the uppermost beds, those falling within the present district, is as follows:

|  | ft | in |
|---|---|---|
| Limestone (large *Davidsonina septosa*[2] at top) .. .. .. | 40 | 0 |
| Clay, on potholed surface .. .. | | 10 |
| Limestone .. .. .. .. | 35 | 0 |
| Potholed surface .. .. .. | — | — |
| Limestone .. .. .. .. | 6 | 0 |
| Clay .. .. .. .. .. | | 6 |
| Limestone .. .. .. .. | 34 | 0 |
| Clay .. .. .. .. .. | | 3 |
| Limestone .. .. .. .. | 6 | 0 |
| Potholed surface .. .. .. | — | — |
| Limestone, very thick-bedded .. | 70 | 0 |

The lowest beds are about 180 ft above the base of the Chee Tor Rock and the highest near the horizon of the Upper *D. septosa* Band. The very massive character of the lowest beds in the section is a particularly striking feature.

At the "north end of Tunstead Quarry", about 35 ft and 50 ft respectively below the top of the Chee Tor Rock, Cope (1939, pp. 61–2) recorded potholed surfaces beneath clay partings. The potholes were from 3 to 4 ft in diameter and 2 to 3 ft in depth. The possible origin of these features, which have been found during the present work to be more extensive both laterally and in horizon, is discussed on p. 17.

To the north of Tunstead Quarry most of the sections in the Chee Tor Rock have now been obscured by tipping. Faulting has resulted in the preservation of several patches of Lower Lava in this area (see p. 52). A section [1004 7493], near the railway and starting below the lava, is as follows:

|  | ft |
|---|---|
| Clay, weathered Lower Lava .. | — |
| Limestone, top with potholes up to 1 ft deep; top 10 ft strongly marmorized; a 1½-in clay parting at 9¼ ft from top .. .. .. .. | 26 |

The exceptional extent of the marmorization at this locality is difficult to explain, particularly as the overlying rock is a lava. About ¼ mile to the north of this locality the Upper *D. septosa* Band, 2 ft thick, is exposed [0990 7532] in a railway-cutting. The same band is better developed (loc. 11) [1011 7512]

---

[1]Supplied by Mr. P. F. Dagger with permission of Imperial Chemical Industries Limited.

[2]Fossil names in un-numbered surface sections refer to field identifications.

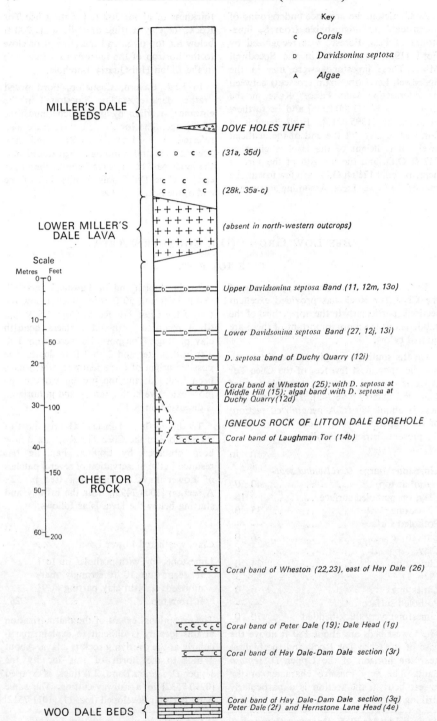

Key

c    Corals

D    Davidsonina septosa

A    Algae

MILLER'S DALE BEDS

DOVE HOLES TUFF

(31a, 35d)

(28k, 35a-c)

LOWER MILLER'S DALE LAVA

(absent in north-western outcrops)

Scale

Metres  Feet

Upper Davidsonina septosa Band (11, 12m, 13o)

Lower Davidsonina septosa Band (27, 12j, 13i)

D. septosa band of Duchy Quarry (12i)

Coral band at Wheston (25); with D. septosa at Middle Hill (15); algal band with D. septosa at Duchy Quarry (12d)

IGNEOUS ROCK OF LITTON DALE BOREHOLE

Coral band of Laughman Tor (14b)

CHEE TOR ROCK

Coral band of Wheston (22,23), east of Hay Dale (26)

Coral band of Peter Dale (19); Dale Head (1g)

Coral band of Hay Dale-Dam Dale section (3r)

Coral band of Hay Dale-Dam Dale section (3q) Peter Dale (2f) and Hernstone Lane Head (4e)

WOO DALE BEDS

FIG. 6. *Generalized section of $D_1$ strata, showing horizons of the more important fossil-localities*

near Great Rocks Lees as a 4-ft band with
*Koninckopora sp.* and gastropods.

The old quarry [0980 7576] at Buxton
Bridge shows a good section of the Chee Tor
Rock, commencing about 12 ft below the
top:

| | ft | in |
|---|---|---|
| Limestone, pale fine-grained some-<br>what crinoidal; scattered fossils | 12 | 0 |
| Upper *D. septosa* Band: lithology<br>as below; *D. septosa* and produc-<br>toids .. .. .. .. | 2 | 6 |
| Limestone, near white fine-grained;<br>scattered fossils .. .. .. | 9 | 0 |
| Clay parting, orange and white up<br>to 3 ft thick on north side of<br>quarry, 6 in on south, on erosion<br>surface with potholes up to 21 in<br>deep .. .. .. average | 1 | 9 |
| Limestone, pale fine-grained; scat-<br>tered corals .. .. .. | 5 | 3 |
| Clay parting on slightly eroded sur-<br>face .. .. .. 0 in to | | 2 |
| Limestone, as above; scattered fos-<br>sils include *?D. septosa* fragments<br>and a gastropod .. .. .. | 4 | 10 |
| Clay parting, on slightly potholed<br>surface .. .. .. 0 in to | | 3 |
| Limestone, massive fine-grained,<br>with some crinoid debris; produc-<br>toids at 5 ft from base; gastropods<br>in a 6-in band at 13 ft from base | 15 | 0 |
| Clay parting, on slightly potholed<br>surface .. .. .. 0 in to | | 1 |
| Limestone, pale massive fine-<br>grained; scattered productoids<br>near top .. .. .. .. | 4 | 0 |
| Lower *D. septosa* Band: lithology as<br>above; algae and *D. septosa* .. | 1 | 6 |
| Limestone, as above .. .. | 4 | 0 |

At Peak Dale, Duchy Quarry (loc. 12)
[0938 7681] (Plate IIIA) 280 yd E. of Peak
Forest Station the following section is shown
immediately underlying the Lower Lava:

| | | ft | in |
|---|---|---|---|
| (*p*) | Limestone, pale fine-grained | 8 | 0 |
| (*o*) | Limestone, as above, locally<br>crinoidal .. .. .. | 4 | 0 |
| (*n*) | Limestone, as above, with<br>corals and scattered brachio-<br>pods .. .. .. .. | 11 | 6 |

| | | ft | in |
|---|---|---|---|
| (*m*) | Upper *D. septosa* Band: lime-<br>stone as above; *D. septosa*,<br>*Gigantoproductus sp.*, *Lino-<br>productus sp.* .. .. | 1 | 6 |
| (*l*) | Limestone, pale fine-grained,<br>sparsely crinoidal in top 14<br>ft, finely crinoidal below with<br>rare brachiopods .. .. | 28 | 0 |
| (*k*) | Limestone, some crinoid debris<br>in upper part (part coarse);<br>algae, *D. septosa* and a few<br>productoids .. .. .. | 5 | 2 |
| (*j*) | Lower *D. septosa* Band; pale<br>fine-grained limestone with<br>*Chaetetes depressus, Dibuno-<br>phyllum bourtonense, Litho-<br>strotion martini, P. murchi-<br>soni, Syringopora spp., David-<br>sonina septosa, Delepinea<br>comoides, Gigantoproductus<br>sp. edelburgensis* group, *M.<br>sp. papilionaceus* group .. | | 10 |
| (*i*) | Limestone, pale rather cri-<br>noidal, with coarsely crinoi-<br>dal lenses; a few *Davidsonina<br>septosa* and productoids .. | 8 | 0 |
| (*h*) | Parting; brown clay on pot-<br>holed surface (with potholes<br>up to 9 in deep) .. 1 in to | | 10 |
| (*g*) | Limestone, pale grey to off-<br>white with stylolite band up<br>to 9 in at base; rare corals<br>and productoids; marked<br>irregular bedding plane at<br>base .. .. .. .. | 10 | 9 |
| (*f*) | Limestone, pale grey fine-<br>grained; *L. martini, P. mur-<br>chisoni* and *Delepinea sp.* .. | 2 | 3 |
| (*e*) | Limestone, pale grey to off-<br>white fine-grained; rare<br>corals and productoids .. | 20 | 0 |
| (*d*) | Limestone, pale fine-grained<br>algal; 'Girvanella' nodules<br>encrusting shell fragments,<br>*C. depressus, Davidsonina<br>septosa* [with algal coating,<br>rare] .. .. .. .. | 1 | 0 |
| (*c*) | Limestone, pale fine-grained | 6 | 6 |
| (*b*) | *Not exposed* .. .. 1 ft to | 2 | 0 |
| (*a*) | Potholed surface on quarry<br>floor, with potholes up to 7<br>in beneath quarry floor .. | — | — |

On the west side of Great Rocks Dale the section in Bold Venture Quarry [0890 7655] is closely similar to that in Duchy Quarry but does not extend down so far, showing the uppermost 58 ft of the Chee Tor Rock, the top of the section being close to the horizon of the base of the Lower Lava. The section differs from that in Duchy Quarry mainly in the local development of a band of pale calcilutite 1 to 1¼ ft thick and 6½ ft below the Upper *D. septosa* Band.

Over half a mile west of the last section, Upperend Quarry (loc. 13) [0833 7707] also shows the uppermost part of the Chee Tor Rock, the highest beds lying 2 to 3 ft below the Lower Lava:

|  |  | ft | in |
|---|---|---|---|
| (*t*) | Limestone, off-white fine-grained .. .. .. | 5 | 0 |
| (*s*) | Limestone; algal nodules encrusting shell fragments, *K. inflata, P. murchisoni?, D. septosa, D. septosa transversa, G. edelburgensis, Linoproductus sp. hemisphaericus* group, *Megachonetes sp.*, smooth spiriferoids .. | 1 | 0 |
| (*r*) | Limestone, white fine-grained; sporadic brachiopods including *D. septosa*, a few large crinoid columnals .. .. | 5 | 6 |
| (*q*) | Parting of yellow clay .. | | 3 |
| (*p*) | Limestone, pale fine-grained, slightly crinoidal .. .. | 7 | 0 |
| (*o*) | Upper *D. septosa* Band: limestone as above; algae [enclosing shell fragments], *Dibunophyllum bourtonense* φ, *Lithostrotion martini, S.* cf. *geniculata,* bryozoa, *Davidsonina septosa, D. septosa transversa, Linoproductus sp. hemisphaericus* group, *M. sp. papilionaceus* group, smooth spiriferoids, *Bellerophon costatus, Straparollus sp.* (fauna mainly in top 1½ ft and bottom 2 ft) .. .. .. | 5 | 0 |
| (*n*) | Limestone, pale fine-grained, slightly crinoidal; scattered corals and brachiopods; clay parting 0 to 3 in at 8 ft from base; stylolites at base .. | 16 | 0 |

|  |  | ft | in |
|---|---|---|---|
| (*m*) | Limestone, somewhat algal with sporadic crinoid stems [some encrusted by algae]; algae very abundant in a 14 in band at 2 ft 10 in from base, brachiopods and corals common just above this band; *Dibunophyllum bourtonense* φ, *P. murchisoni, Davidsonina septosa, M. sp. papilionaceus* group, *Bellerophon sp.* .. .. .. | 5 | 0 |
| (*l*) | Limestone, pale fine-grained slightly crinoidal; sporadic productoids and rare *D. septosa* .. .. .. | 5 | 4 |
| (*k*) | Clay parting, on potholed surface .. .. .. 12 in to | | 0½ |
| (*j*) | Limestone, pale fine-grained; scattered productoids and gastropods .. .. .. | 4 | 0 |
| (*i*) | Lower *D. septosa* Band: limestone as above; rare *P. murchisoni*, common *D. septosa* | 4 | 0 |
| (*h*) | Limestone, pale massive fine-grained; stylolite band 5 ft from top; 2-ft band at 7 ft from top with scattered productoids, *D. septosa* and a few large crinoid columnals | 13 | 0 |
| (*g*) | Clay parting .. .. .. | | 0½ |
| (*f*) | Limestone, pale fine-grained | 5 | 0 |
| (*e*) | Limestone, pale fine-grained shelly; common productoids | 3 | 0 |
| (*d*) | Clay parting, on potholed surface .. .. .. 0 in to | | 0½ |
| (*c*) | Limestone, massive pale fine-grained, slightly crinoidal in upper part; sporadic productoids and rare *D. septosa* .. | 7 | 3 |
| (*b*) | *Not exposed* .. .. about 1 | | 0 |
| (*a*) | Limestones with potholed surface (in quarry floor) .. | — | — |

The long section in Newline Quarry starts in the lowest level of the quarry [0917 7732] north of Peak Forest Station, and finishes ¼ mile to the north-west [0888 7778]; the uppermost beds lie a little below the Upper *D. septosa* Band:

|  | ft | in |
|---|---|---|
| Limestone, pale fine-grained; a few crinoid stems in lowest 5 ft .. | 18 | 0 |

A.  DUCHY QUARRY, SMALLDALE: CHEE TOR ROCK

(L168)

B.  HOLDERNESS QUARRY, DOVE HOLES: MILLER'S DALE
BEDS WITH DOVE HOLES TUFF

PLATE III

(L 172)

ft in

Lower *D. septosa* Band: lithology as
above; productoids, *D. septosa*,
gastropods and crinoid debris  ..  2  10

Limestone, pale fine-grained  ..  4  9

*Not exposed* ..  ..  ..  .. 12  0

Limestone, pale fine-grained; productoids and *D. septosa* in top
4 ft  ..  ..  ..  ..  7  0

Clay parting, on potholed surface
with potholes up to 1½ ft deep about  1

Limestone, pale massive fine-
grained, with a 2½-ft band with
corals, *D. septosa*, productoids
and gastropods 14 ft from top;
stylolite band up to 9 in, 25 ft
from top ..  ..  ..  .. 37  0

*Not exposed* ..  ..  ..  ..  4  0

Clay parting, on strongly potholed
surface with dolomite crust 2 in to  9

Limestone, grey to dark grey variegated, fine-grained with irregular
clay partings in places  ..  1 ft to 3  0

Clay parting[1], on potholed surface
(floor of top lift of quarry)  0 in to  6

Limestone, pale massive fine-
grained; dark-mottled in top 2 ft;
1-ft algal band 15 ft above base  21  0

Clay parting, on potholed surface up to  6

Limestone, pale grey-brown (slightly
darker than overlying limestones)  3  6

Clay parting  ..  ..  0 in to  4

Limestone, off-white fine-grained;
some darker mottling in top 2 ft;
stylolite bands at top and base  ..  5  3

Clay parting  ..  ..  0 in to  4

Limestone, off-white fine-grained,
with scattered coarse crinoid
debris; gastropods, corals and
productoids in lowest 3 ft  ..  7  6

Limestone, off-white fine-grained ..  2  0

Clay parting, grey-green where unweathered, on slightly potholed
surface  ..  ..  .. 2 in to  5

Limestone, pale grey-brown about 1  8

Clay parting, on potholed surface 1 in to  2

ft in

Limestone, off-white fine-grained;
with thin clay wayboards at
4 ft 3 in and 6 ft 6 in from base;
thin clay parting in places at base  16  6

Limestone, grey-brown massive fine-
grained; rare *P. murchisoni* and
productoids  ..  ..  .. 15  6

Limestone, pale grey-brown massive
fine-grained  ..  .. about  7  6

*Smalldale, Peter Dale and Monk's Dale.*
Between Laughman Tor and Peter Dale an
expanse of flat-lying or gently undulating
Chee Tor Rock forms a broad outcrop. On
the plateau surface exposures are numerous,
but large sections are lacking. The tendency
for fossils in the Chee Tor Rock to be
restricted to more or less well-defined horizons enables small exposures of these in
many cases to be correlated by calculating
their position in relation to the top or base
of the Chee Tor Rock.

At Laughman Tor the section (loc. 14)
[1015 7799] is as follows:

ft

(*b*) Limestone, pale grey well-bedded
fine-grained; abundant corals;
*Dibunophyllum bourtonense* φ,
*Lithostrotion* aff. *martini*, *L.
pauciradiale*, *P. murchisoni*, *S.*
cf. *geniculata*, *A. expansa*,
*Gigantoproductus* sp., *Linoproductus* sp., *M.* sp. *papilionaceus* group, *Pugilis?*, smooth
spiriferoids  ..  ..  .. 10

(*a*) Limestone, grey-white massive
fine-grained, part finely bioclastic and oolitic; top 20 ft
sparsely fossiliferous; foraminifera, *Dibunophyllum* sp.,
*Lithostrotion* aff. *martini*, *L.
pauciradiale*, *M.* sp. *papilionaceus* group  ..  ..  .. 30

The coral bed forms the craggy top of the
tor. It lies about 140 ft below the top of the
Chee Tor Rock (Fig. 6).

[1] A thin section (E36118/P) shows a microcrystalline aggregate of pale brown clay mineral showing orientation along several directions, patchy weak anisotropism and abundant inclusions. The last-named comprise sub-microscopic dust, discrete heavy minerals (zircon, apatite, tourmaline) and an abundance of opaque sulphide needles, up to 80 microns long, generally tapering (up to 4 microns wide), geniculate, individual or stellate, and slightly curved in places. A polished thin section showed the needles to be anisotropic and they are identified as marcasite. They may have originated in microscopic tension cracks in the clay.

To the south-east of Laughman Tor the Chee Tor Rock forms a gentle dip-slope, and at the Marvel Stones [1045 7780] weathering of this has given rise to a limestone pavement, an infrequent feature in the area. A quarry [1065 7810] in Smalldale Plantation shows a typical section: brown-white fine-grained limestone, rather crinoidal 10 ft, on pale grey or grey-white massive fine-grained limestone 13 ft.

Near Gorsey Nook a band bearing a diffuse brachiopod-coral fauna is exposed [1042 7685] and has been traced laterally for a distance of about a quarter of a mile. The band lies about 100 ft below the top of the Chee Tor Rock at or near the horizon of the algal band near the base of the Duchy Quarry section. An outlier of the same bed is present at the western end of Middle Hill, an exposure (loc. 15) [1079 7698] showing pale fine-grained limestone, sparsely fossiliferous near base 2 ft 2 in, on pale fine-grained limestone with *Caninia sp.* [ *?C.* cf. *densa* Lewis of Hudson and Cotton 1945a, p. 306], *Clisiophyllum sp.* [juv.], *Lithostrotion martini, Davidsonina septosa, Delepinea* aff. *comoides, Linoproductus* aff. *corrugatus* and *M. sp. papilionaceus* group. On the east side of the hill lower beds are exposed, the section (loc. 16) [1129 7706], with the uppermost beds [1124 7696] lying about 10 ft below the band with *D. septosa* of the previous locality, being:

|     |                                                                                                                                                                                                                                                                                                | ft   |
| --- | --- | --- |
| (e) | Limestone, off-white massive fine-grained in beds about 3-ft thick                                                                                                                                                                                                                            | 36   |
| (d) | Limestone, grey-white fine-grained; coral band with *Lithostrotion pauciradiale, L. sp. aranea* group, *Megachonetes sp.*                                                                                                                                                                      | 3    |
| (c) | Limestone, off-white, rather well-bedded fine-grained; scattered brachiopods and corals     ..                                                                                                                                                                                                | 9½   |
| (b) | Limestone, pale brown to off-white fine-grained; abundant corals in upper 3¼ ft and corals and brachiopods in lower 2 ft; fauna includes *Caninia* cf. *densa* [of Hudson and Cotton], *Clisiophyllum keyserlingi, Dibunophyllum bourtonense* φ, *Koninckophyllum* θ, *L. junceum, L. martini, L. pauciradiale, L.* cf. *sociale, P. murchisoni,* brachiopods including *Gigantoproductus sp. maximus* group | 8    |
| (a) | Limestone, off-white, fine-grained                                                                                                                                                                                                                                                            | 15   |

The thickness, rich coral fauna and estimated stratigraphical position of the lower band suggest equivalence to the coral band of Laughman Tor.

The upper coral band of the last section is more strongly developed (loc. 17) [1059 7721] on the north-west side of Middle Hill where it is 6 ft thick and has yielded *Lithostrotion* cf. *aranea, L.* aff. *maccoyanum, L.* aff. *martini* and *P. murchisoni.*

Nearly 1½ miles S.E. of the last locality two fossil-bands are present locally at closely similar horizons to those of Middle Hill. The upper band is the better developed and is exposed [1231 7561] north-east of Hargate Hall in the following section (loc. 18):

|     |                                                                                                                                                             | ft | in |
| --- | --- | --- | --- |
| (d) | Limestone, near white very finely crystalline, with a little crinoid debris; a 9-in band at 2 ft from base with corals including *L. martini, G. sp. edelburgensis* group about | 5  | 0  |
| (c) | Clay, yellow on irregular pot-holed surface     ..     about                                                                                               | 1  |    |
| (b) | Limestone, pale fine-grained, with irregular fracture, lower half fossiliferous; *L. martini, A. expansa, G. sp. edelburgensis* group, *Linoproductus sp. hemisphaericus* group     .. | 4  | 9  |
| (a) | Limestone, pale massive fine-grained; rare corals including *Lithostrotion martini* 3½ ft below top     ..     ..     ..                                   | 6  | 6  |

A band some 40 ft lower in the succession and 4¾ ft thick is exposed nearby [1241 7553] and yields brachiopods and corals; this appears to be about 150 ft below the top of the Chee Tor Rock. The two bands crop out on the south-western limb of a small sharp anticline traceable for a distance of half a mile and trending N.W.–S.E. The Monk's Dale Vent, first noted by Arnold-Bemrose (1907, p. 251), more fully described on p. 300, lies at the south-eastern end of the anticline near the axis and a further vent has been found ¼ mile to the north-west and lying close to the anticline. The association of these two vents with a sharp anticline, foreign to the tectonic pattern of the limestone massif, suggests strongly that the structure was of penecontemporaneous origin.

The relationship of the vents to the anticline is illustrated in Fig. 14.

In Peter Dale the lower part of the Chee Tor Rock is exposed nearly continuously. Beds low in the subdivision stand out as a marked cliff of massive limestones showing the characteristic vertical jointing at Dale Head. The massive beds forming the cliff are underlain by some 40 ft of thinner-bedded limestones. A coral band up to 4 ft thick is present at the base of the Chee Tor Rock in the eastern tributary dale at Dale Head above the section (loc. 2) [1248 7653] of Woo Dale Beds described on pp. 39–40:

|  |  | ft |
|---|---|---|
| (g) | Limestone, medium to pale grey-brown massive fine-grained; *Koninckophyllum sp.* [compound] 4 ft below top; *L. martini* and *M. sp. papilionaceus* group 10 ft below top .. | 25 |
| (f) | Coral band, lithology as above except for 4-in darker limestone at base; *D. bourtonense* ♀, *Koninckophyllum* θ, *L. junceum?*, *L. martini*, *L. pauciradiale*, *P. murchisoni*, *Antiquatonia* aff. *insculpta*, *Athyris expansa*, *Dielasma hastatum*, *Gigantoproductus? sp. nov.* [wrinkled concentric ornament], *Linoproductus sp. hemisphaericus* group, *M. sp. papilionaceus* group .. .. | 3½ |
| Woo Dale Beds | .. .. .. | — |

At Dale Head the coral band (f) disappears laterally but it reappears farther north in the long section at the junction of Hay Dale and Dam Dale (p. 50). The section (loc. 1) in the main dale at Dale Head [top of section: [1232 7630] shows higher beds in the Chee Tor Rock:

|  |  | ft |
|---|---|---|
| (i) | Limestone, pale fine-grained massive (forming crags); band 5 ft from base with *Carcinophyllum vaughani*, *Lithostrotion martini*, *A. expansa*, *M. sp. papilionaceus* group .. .. .. .. | 24 |
| (h) | Beds poorly exposed. Some small exposures of grey limestone; one at 15 ft from top with *Dibunophyllum bourtonense* ♀, *Dielasma hastatum*, *G. sp. maximus* group, *G.? sp. nov.* [wrinkled concentric ornament], *Linoproductus sp. hemisphaericus* group, *M. sp. papilionaceus* group .. .. | 26 |
| (g) | Limestone, pale massive (forming crags); 6-ft coral band 15 ft from base (fauna see loc. 19 below); at 8 ft from base, *Lithostrotion martini* .. .. | 26 |
| (f) | Limestone, pale rather thin-bedded fine-grained; at 20 ft above base, *M. sp. papilionaceus* group .. .. .. .. | 40 |
| Woo Dale Beds (see p. 39) .. | | — |

Some 55 ft above the base of the Chee Tor Rock is a well-developed coral band, bed (g) of the preceding section. This is best seen (loc. 19) in a small cliff on the west side of Peter Dale [1222 7626], where it is 6 ft thick and yielded *C. vaughani*, *D. bourtonense* ♀, *Lithostrotion martini*, *L. pauciradiale*, *L.* cf. *sociale*, *P. murchisoni*, *A. expansa*, *Gigantoproductus? sp. nov.* [wrinkled concentric ornament], *Linoproductus sp. hemisphaericus* group and *M. sp. papilionaceus* group. The overlying limestones, exposed in Peter Dale between the coral band and the vents previously noted, show their usual lithological characters but no fossil bands were found. Near the Monk's Dale vents, however, a 2-ft band overlying 80 ft of unfossiliferous beds on the eastern side of the dale (loc. 20) [1304 7539] yielded *Lithostrotion martini*, *Linoproductus sp. hemisphaericus* group and *M. sp. papilionaceus* group. The band appears to lie about 140 ft below the top of the Chee Tor Rock and to be the equivalent of the lower of the two fossil bands exposed north-east of Hargate Hall. The higher horizon is probably present in an exposure (loc. 21) [1347 7521] near Monksdale House as 3 ft of grey limestone with *Caninia benburbensis* and *D. bourtonense* ♀.

Between Monk's Dale and Tideswell Dale the Chee Tor Rock crops out over a broad plateau which, however, shows only minor exposures.

To the east of the outcrop the only information regarding the underground extension of the Chee Tor Rock, is provided by the Derwent Valley Water Board's trial borehole at Litton Dale [1599 7498]. The borehole was not cored and the record is from the borers, supplemented by examination of cutting-samples by Professor F. W. Shotton. The base of the Lower Lava was found at 243 ft and from thence to 340 ft the hole passed through pale fine-grained limestones with some pyrite in the uppermost 5 ft. These beds represent the topmost part of the Chee Tor Rock. Below 340 ft the succession is abnormal, the beds to the bottom of the borehole at 385 ft being either lava with calcitization in places or interbedded lava and tuff. Extrusive igneous rock at this horizon is unknown elsewhere in the district, though this is not the earliest evidence of vulcanicity in the area.

*Wheston to Ox Low.* Between Wheston and Dale Head faulting brings up beds low in the Chee Tor Rock to crop out on the plateau surface. An exposure (loc. 22) [1270 7637] west of Wheston Hall (see Fig. 6) showed a coral band 2 ft thick and about 100 ft above the base with *Lithostrotion portlocki, Syringopora* cf. *distans, Gigantoproductus dentifer, Linoproductus sp. hemisphaericus* group and *M. sp. papilionaceus* group. A nearby exposure (loc. 23) [1273 7630] yielded *Lithostrotion martini* and *P. murchisoni* from the same band.

Higher up the slope, towards Wheston, an exposure (loc. 24) [1286 7623] shows a 21-in band with *Lithostrotion martini, L. pauciradiale, P. murchisoni?, Davidsonina septosa* and *Linoproductus sp. hemisphaericus* group. This exposure lies some 150 ft below the top of the Chee Tor Rock and is at or near the horizon of the coral band of Laughman Tor (see p. 47).

A higher coral band is present over a distance of 300 yd farther up the slope and lies some 100 ft below the top of the Chee Tor Rock. An exposure (loc. 25) [1295 7618] shows the band some 8 ft thick with *Dibunophyllum sp., Lithostrotion decipiens?, L. martini, P. murchisoni* and *Linoproductus sp. hemisphaericus* group. These beds rest on a potholed surface with yellow clay which, if equivalent to that developed north-east of Hargate Hall, suggests that the lower part of the band has failed.

At the junction of Hay Dale and Dam Dale the following section (loc. 3) in the lowest part of the Chee Tor Rock, continues that of the Woo Dale Beds given on p. 40 [top of section 1190 7727]:

|  | ft |
|---|---|
| Limestone, grey-white to brown-white massive fine-grained; 44 ft from base (*s*) *Lithostrotion junceum;* 34 ft from base (*r*) *Dibunophyllum bourtonense, Lithostrotion martini, L. portlocki, P. murchisoni,* bryozoa, *A.* cf. *expansa, G. dentifer, Linoproductus sp. hemisphaericus* group, *M. sp. papilionaceus* group, gastropods; 2 ft from base a 1-ft band (*q*) yields *Caninia* cf. *densa* [of Hudson and Cotton], *D. bourtonense* φ, *Koninckophyllum* θ, *Lithostrotion junceum, L. martini, L. pauciradiale, L.* cf. *sociale, P. murchisoni,* bryozoa, *Dielasma hastatum, Gigantoproductus spp.,* including *G.? sp. nov.* [wrinkled concentric ornament], *Linoproductus sp. hemisphaericus* group, *M. sp. papilionaceus* group, gastropods .. .. .. .. | 45 |
| Woo Dale Beds .. .. .. — | |

The coral band at the base is the same as that developed in this position at Dale Head (loc. 2, p. 49). Higher beds in the Chee Tor Rock, uniformly pale, massive and fine-grained, are exposed on the east side of Hay Dale for a distance of a quarter of a mile above this section; beyond here, however, they are faulted off.

About 100 ft from the base of the Chee Tor Rock, on the slopes east of Hay Dale (loc. 26) [1225 7721] a band (traceable for some quarter of a mile) 5 ft thick yields *Lithostrotion martini, P. murchisoni,* bryozoa, *Delepinea* aff. *comoides* and *Linoproductus sp. hemisphaericus* group. The band appears to be that developed as a coral band west of Wheston (loc. 22, see above).

Between Hay Dale and Hernstone Lane Head the coral band at the base of the Chee Tor Rock is developed in places. At Hernstone Lane Head the section (loc. 4) of the lowest part of the subdivision is continuous with that of the Woo Dale Beds given on p. 41 [top of section 1226 7876]:

ft

(*i*) Limestone, pale grey-brown massive fine-grained in beds up to
  3½ ft  ..    ..    ..    10
(*h*) *Not exposed*    ..    .. about 14
(*g*) Limestone, pale brown finely
  granular; top 1 ft with *Lithostrotion* aff. *martini, M. sp.
  papilionaceus* group ..    .. 6
(*f*) Limestone, pale grey-brown very
  fine-grained, white-weathering
                  3 in to ½
(*e*) Limestone, pale grey finely crinoidal, corals and brachiopods
  in pockets near base; *Dibunophyllum sp., L. junceum, L.
  martini, P. murchisoni*, bryozoa,
  *G. sp. maximus* group ..    .. 19
Woo Dale Beds        ..    .. —

Farther east the Chee Tor Rock forms a
broad outcrop with sporadic exposures only.
The Lower *D. septosa* Band, some 60 ft below
the top, is exposed in a small limestone scar
and can be traced for some hundreds of
yards at the crest of a small hill west of
Potluck House. An exposure (loc. 27)
[1294 7805] shows the band, 2 ft thick, with
corals including *Caninia sp. subibicina* group,
*P. murchisoni, Davidsonina septosa,
Delepinea* aff. *comoides, Linoproductus sp.
hemisphaericus* group.

At Wall Cliff the uppermost beds of the
Chee Tor Rock are exposed in the following
section (loc. 28) [1424 7738]:

                  ft  in
(*j*) Lower Lava (see p. 53)    .. — —
(*i*) Limestone, pale to medium grey
  well-bedded    fine-grained;
  *Dibunophyllum bourtonense* φ  6  0
(*h*) Calcilutite, pale to medium grey
  finely banded    ..    ..  6

ft  in

(*g*) Limestone, pale and dark grey
  variegated, slightly tuffaceous    ..    ..    ..  3  3
(*f*) Calcilutite, grey-white, part
  finely banded    ..    ..  2  1
(*e*) Limestone, grey-white    ..  1  4
(*d*) Limestone, pale fine-grained
  with 'hackly' weathering in
  beds up to 1 ft; *Lithostrotion
  junceum* at 1 ft 4 in from base  4  6
(*c*) Calcilutite, white weathering,
  ?algal    ..    .. 0 in to  6
(*b*) Limestone, pale fine-grained;
  *Lithostrotion martini, Lino-
  productus sp. hemisphaericus*
  group    ..    ..    ..  1  3
(*a*) Limestone, massive pale fine-
  grained; 'hackly' weathering,
  with a little darker variegation in lowest 13 ft ..    .. 33  0

The presence of calcilutite bands high in the
Chee Tor Rock is a striking feature of the
section and recalls that at Bold Venture
Quarry (see p. 46).

At the northern end of the outcrop the
Chee Tor Rock is well seen on Ox Low. A
shell band is present locally on the south-
west side of the hill and is seen on the
north side of old open workings on Oxlow
Rake (loc. 29) [1260 8002] in the top 6 ft of a
40-ft section. This bed yielded *P. murchisoni,
Gigantoproductus sp.* and *Linoproductus sp.*
Some 90 ft higher in the succession, the
Upper *D. septosa* Band is exposed [1265
8014] as a 2-ft band with brachiopods
(including *D. septosa*) and gastropods. This
band lies about 30 ft below the base of the
Lower Lava and the total thickness of the
Chee Tor Rock hereabouts is about 300 ft.

To the north of Ox Low, with the disappearance of the Lower Lava the Chee Tor
Rock and Miller's Dale Beds are considered
together as the Bee Low Limestones.

## LOWER MILLER'S DALE LAVA

*Water Swallows, Great Rocks Dale and
Dove Holes.* The outcrop of the Lower Lava
enters the district half a mile south of Waterswallows but is almost immediately cut out by
the transgressive Waterswallows Sill. The

latter has, however, been cut through in
Waterswallows Quarry [0860 7504] to show
up to 6 ft of pale calcitized amygdaloidal lava
resting on limestone and with an irregular
junction with the overlying dolerite. Near this

junction the lava usually contains pyrite veins and pyrite amygdales. The altered lava has been recently compared by Dr. F. Moseley (*in* discussion Dunham and Kaye 1965, p. 275) to the White Whin of the Durham area.

To the east of Water Swallows, faulted outliers of the lava are present north of Tunstead Quarry though they are much obscured by tipping. A section [0979 7538] near the railway shows 13 ft of amygdaloidal lava in a down-faulted wedge.

On the eastern side of Great Rocks Dale the Lower Lava has been passed through in a number of boreholes[1]. The greatest thickness was proved in the southernmost hole T/B/21 [1034 7547], which gave the following abridged succession:

|  | Thickness ft in | Depth ft in |
|---|---|---|
| Limestone (Miller's Dale Beds) .. .. .. | 60 3 | 60 3 |
| Clay, yellow-brown soft ('toadstone clay') .. | 5 10 | 66 1 |
| Lava, soft and with pyrite to 67 ft 9 in; hard and amygdaloidal below .. .. .. | 6 11 | 73 0 |
| Clay .. .. .. | 2 2 | 75 2 |
| Lava, dark, some amygdales, to 78 ft 6 in, mainly non-vesicular to 97 ft 7 in; amygdaloidal to 104 ft 2 in; non-vesicular, with hematite specks to 107 ft .. | 31 10 | 107 0 |
| Clay, reddish brown .. | 2 | 107 2 |
| Lava, grey-green with hematitic joints to 110 ft 6 in; part vesicular to 129 ft 3 in; dark and non-vesicular to base | 37 6 | 144 8 |
| Tuff, pale grey-green pyritous with chlorite specks .. .. .. | 13 9 | 158 5 |
| Limestone .. .. | 27 7 | 186 0 |

The lava shows a clear division into three flows separated by the clay partings at 75 ft 2 in and 107 ft 2 in. A similar but thinner section was proved farther north in borehole T/B/23A [0985 7614] which started just below the top of the lava and showed 74 ft 8 in of it.

To the west and south-west of Upper End the Lower Lava forms a broad outcrop over the crest of a shallow anticline; in places it is capped by small outliers of Miller's Dale Beds. In this area the outcrop bears frequent swallow-holes, many of which provide sections of the lava. For example, north-east of Water Swallows [0836 7557] where 14 ft of amygdaloidal lava are exposed. Near Batham Gate the lava is 50 to 60 ft in thickness. An exposure [0840 7644], again in a sink-hole, shows 12 ft of amygdaloidal lava. The most northerly exposure on this outcrop of lava in a more or less unaltered state is [0801 7753] ¼ mile S.E. of Dove Holes station. Between Dove Holes and Lodes Marsh the lava thins steadily, the outcrop showing as a depression with bright yellow clay visible in places. The lava finally disappears [0865 7836] at Lodes Marsh.

*Wormhill Moor, Tunstead and Hargatewall.* To the east of Great Rocks Dale the Lower Lava crops out around the high ground forming Wormhill Moor. Near Rock Houses the thickness of the lava is about 75 ft but it thins northwards to 30 ft at the northern end of the outcrop near Smalldale. Sections are infrequent though the lowest beds, reddish weathering and amygdaloidal, are well seen close above the abandoned Great Rocks Quarries [0956 7633].

Around the hamlet of Tunstead and at Hargatewall the Lower Lava reaches its maximum thickness of about 100 ft. It is well exposed, for instance, in the sides of the lane [1106 7471] 350 yd S. of Tunstead. Locally, hard dark compact non-amygdaloidal basalt of rather coarse grain is present in small masses within the Lower Lava, as at [1132 7467] south-east of Tunstead and is considered to be the middle part of a flow.

*Tideswell Dale to Bradwell Moor.* The outcrop of the Lower Lava enters the district at Tideswell Dale but is here complicated by the presence of the Tideswell Dale Sill at the base. On the east side of the dale [154 744] features suggest the splitting of the Lower

---

[1]Published by permission of Imperial Chemical Industries Limited.

Lava with the development of a thin lime-stone parting, too small to be shown on the one-inch map. The most northerly outcrops in Tideswell Dale are faulted and may not represent the whole of the Lower Lava; the sequence is best illustrated by the Litton Dale Borehole [1599 7498] (see also p. 55). This is as follows:

|  | Thick-ness | Depth |
|---|---|---|
|  | ft | ft |
| Base of Miller's Dale Beds .. | — | 152½ |
| Tuff .. .. .. .. | 2½ | 155 |
| Lava .. .. .. .. | 64 | 219 |
| Limestone, light grey and dark grey; silicified near base .. | 12 | 231 |
| Lava .. .. .. .. | 12 | 243 |
| Chee Tor Rock (see p. 50) .. | — | — |

From Tideswell Dale the outcrop of the Lower Lava is faulted westwards to Heathy-dale Ward. Around Wheston the thickness of the lava is about 90 ft, but to the north it thins, only some 40 ft being present around Potluck. Near here an exposure [1303 7712] shows

non-vesicular lava 6 ft on amygdaloidal lava 7 ft.

At Wall Cliff the section (loc. 28, see also p. 51) shows the Lower Lava [1433 7731] to be 35 ft thick. It is poorly exposed but where seen shows its normal amygdaloidal charac-ter. Farther north, at the intersection of the outcrop with White Rake, strike-faulting brings the Upper Lava and the Lower Lava into juxtaposition. An exposure of the latter, deeply weathered, is visible to a depth of 7 ft in an openwork on the vein [1452 7815].

The most northerly exposures worthy of note are at The Cop; these show amygda-loidal lava [1287 7968] and, 60 yd farther north, non-vesicular lava. The total thick-ness here is some 30 ft. To the north of The Cop the Lower Lava thins rapidly and disappears [1355 8108] near Hazard Mine.

The isolated occurrences of Lower Lava at Cave Dale and of tuff at a comparable horizon at Pin Dale are, for geographical reasons, best considered with the Bee Low Limestones.

## MILLER'S DALE BEDS

*Water Swallows and Dove Holes.* On the southern margin of the district the Miller's Dale Beds form a broad outcrop west of Water Swallows. A section in an old quarry [0707 7457] on Fairfield Common, shows 18 ft of pale massive limestone, shelly in the lowest 1 ft and with sporadic productoids and corals in a 3½-ft band at 5½ ft from the base. To the east of the main outcrop, faulted outliers of Miller's Dale Beds overlie the Lower Lava in places on a shallow ridge between Hardy-barn and Batham Gate. Between Batham Gate and Dove Holes the Dove Holes Tuff is present within the lower half of the sub-division.

Between Batham Gate and Bibbington the Miller's Dale Beds have been proved in a number of boreholes for quarry develop-ment. Borehole No. 11 of this series [0749 7674] proved the full thickness of the Miller's Dale Beds[1]:

|  | Thick-ness | Depth |
|---|---|---|
|  | ft | ft |
| Base of Monsal Dale Beds .. | — | 140 |

|  | Thickness | Depth |
|---|---|---|
|  | ft | ft |
| Limestone, grey and light grey, thick-bedded; sporadic large productoids .. .. | 88 | 228 |
| Dove Holes Tuff: grey pyri-tous clay with limestone lapilli at base .. .. | 5 | 233 |
| Limestone, as above but nearly porcellanous in places .. | 49 | 282 |
| Lower Lava .. .. .. | — | — |

The outcrop of the Dove Holes Tuff is traceable for a stretch of 300 yd to the east of Peak House. To the south of a line between Peak House and the site of borehole No. 11 this tuff cannot be located at outcrop and it has also been proved absent by several bore-holes. No. 13 Borehole [0780 7628] near Field Farm, however, showed 7 ft 6 in of 'greenish ash with pyrites' with base at 89 ft on 11 ft of 'blue limestone' overlying the Lower Lava. This is the only record of a tuff at this horizon in the area.

The upper part of the Miller's Dale Beds is well exposed in Victory Quarry (loc. 30) [0774 7693] ¼ mile N.W. of Peak House, the

---

[1]Information from examination of cores by Professor F. W. Cope.

E

lowest beds lying within a foot or two of the Dove Holes Tuff:

|  | ft | in |
|---|---|---|
| Monsal Dale Beds (see pp. 63, 65) .. | — | — |
| (j) Limestone, pale fine-grained; 3-in band with productoids 8 ft from base .. .. | 22 | 0 |
| (i) Limestone, pale fine-grained; *Lithostrotion junceum, L. martini, P. murchisoni, Linoproductus sp. hemisphaericus* group .. .. about | 1 | 4 |
| (h) Limestone, pale massive fine-grained, with scattered crinoid debris .. .. .. | 18 | 0 |
| (g) Limestone, pale very fine-grained .. .. .. | 4 | 0 |
| (f) Clay parting, on potholed surface, both dying out laterally up to | 1 | |
| (e) Limestone, pale massive, with a little fine crinoid debris; shelly in top 2 ft .. .. | 7 | 0 |
| (d) Shale, on irregular surface of limestone .. .. about | 2 | |
| (c) Limestone, pale fine-grained; gastropods in top 2 ft and productoid fragments 3 ft from top .. .. .. | 11 | 0 |
| (b) Limestone, grey-brown finely crinoidal .. .. .. | 4 | 0 |
| (a) Limestone, pale brown very fine-grained .. .. .. | 16 | 0 |

A section complementary to the above is that in Dove Holes Dale Quarry (loc. 31) [0782 7748]. This shows the Dove Holes Tuff and some of the underlying beds:

|  | ft | in |
|---|---|---|
| (f) Limestone, pale grey well-bedded finely crinoidal .. | 7 | 0 |
| (e) Limestone, pale grey banded fine-grained, finely crinoidal and slightly oolitic; productoids including *L. sp. hemisphaericus* group .. .. | 11 | |
| (d) Limestone, pale grey massive fine-grained .. .. .. | 5 | 4 |
| (c) Dove Holes Tuff: much grassed over, grey clay with rounded limestone lapilli visible at 2½ ft up, yellow clay at base on strongly potholed surface | 5 | 0 |

|  | ft | in |
|---|---|---|
| (b) Limestone, grey-brown massive fine-grained .. .. .. | 8 | 0 |
| (a) Limestone, pale massive fine-grained, fossiliferous in top 8 ft; *Dibunophyllum bourtonense* φ, *Lithostrotion* cf. *martini, P. murchisoni, Davidsonina septosa* and *Linoproductus sp. hemisphaericus* group .. .. .. | 24 | 0 |

The Dove Holes Tuff was originally described by Arnold-Bemrose (1907, p. 253) from a quarry near the Bull Ring, but this is now covered by tip. Farther east it is well seen in the section in Holderness Quarry (loc. 32) (Plate IIIB) [0842 7817]:

|  | ft |
|---|---|
| (e) Limestone, medium grey-brown well-bedded fine-grained, with sporadic shells; *G. edelburgensis* | 13 |
| (d) Dove Holes Tuff: grey clay with limestone lapilli up to ½ in, thinning to 3½ ft at south end of section; surface with potholes up to 15 in deep at base .. | 6 |
| (c) Limestone, very fine-grained massive off-white; corals throughout include *Dibunophyllum bourtonense* φ, *Lithostrotion* cf. *martini;* brachiopods in upper part include *Davidsonina septosa* and *Linoproductus sp. hemisphaericus* group .. .. | 40 |
| (b) *Not exposed* (position of quarry floor) .. .. .. about | 2 |
| (a) Limestone, near-white, massive fine-grained; top 1½ ft with productoids and a few corals .. | 13½ |

The lowest beds of this section lie within a foot or two of the top of the Lower Lava.

The uppermost part of the Miller's Dale Beds is visible in the section (loc. 33, bed a) [0799 7828] near the Bull Ring where 9½ ft of pale grey fine-grained thick-bedded limestone with *Gigantoproductus crassiventer* and other brachiopods are overlain by Monsal Dale Beds (see p. 65). To the north of Near Ridgeclose the Miller's Dale Beds and Chee Tor Rock are grouped together as the Bee Low Limestones.

*Wormhill Moor and Hargatewall.* To the south of Hargatewall the Miller's Dale Beds form a broad outcrop, approximating to a dip-slope. The best section [1207 7482] near Hargate Hall, shows 11 ft of off-white fine-grained and finely crinoidal limestone with gastropods near the top. Another section [1267 7475] near Wormhill Hall shows 5 ft of pale crinoidal and slightly oolitic limestone with shell debris.

To the north-west of Hargatewall a tongue-like area of Miller's Dale Beds is present beneath and around the outlier of Wormhill Moor. On the south-east side of this area the subdivision is thinly developed, only some 75 ft being present, while at the northern end, near Smalldale, the thickness is about 100 ft. The lowest beds, a few feet above the Lower Lava, are exposed in a roadside quarry north-west of Hargatewall (loc. 34) [1120 7598]: 12 ft of off-white massive fine-grained limestone, slightly crinoidal with, in the top 6 ft, a fauna including *Caninia benburbensis, Lithostrotion martini, P. murchisoni, Linoproductus sp., M. sp. papilionaceus* group and gastropods. On the western side of Wormhill Moor the Miller's Dale Beds are little seen, except for poor exposures of the lowest beds. At the northern end of the outcrop, however, near Smalldale, they are both well exposed and very fossiliferous. A section in a small quarry (loc. 35) [0977 7700] is as follows:

|  | ft | in |
|---|---|---|
| (d) Limestone, white massive fine-grained with columnar jointing, scattered fossils throughout but more fossiliferous in top 4 ft; *Caninia* cf. *densa* [of Hudson and Cotton], *Lithostrotion portlocki, A.* cf. *expansa, Davidsonina septosa, D. septosa* towards *transversa, Delepinea* aff. *comoides, Linoproductus sp. hemisphaericus* group | 23 | 0 |
| (c) Limestone, white massive fine-grained; *Dibunophyllum bourtonense, Lithostrotion martini, L. pauciradiale, P. murchisoni, Linoproductus sp. hemisphaericus* group | 11 | 2 |
| (b) Limestone, very pale grey; *D. bourtonense* φ, *Lithostrotion martini, L. pauciradiale, P. murchisoni* .. about | 1 | 10 |

|  | ft | in |
|---|---|---|
| (a) Limestone, soft tuffaceous Clay, yellow; top of Lower Lava .. .. .. .. | 0 in to — | 4 — |

*Tideswell to Old Moor.* To the south of Tideswell the Miller's Dale Beds form a broad spread between Heathydale Ward and Tideswell Dale. They are well exposed in the latter south of its junction with Litton Dale. The Litton Dale Borehole [1599 7498] (see also pp. 50–53) proved the Miller's Dale Beds as pale limestones between the base of the Station Quarry Beds at 55 ft and the top of the Lower Lava at 152 ft 6 in.

To the west of Tideswell, exposures of pale massive fine-grained limestone of the Miller's Dale Beds are fairly frequent. An exposure (loc. 36) [1402 7574] near Crossgate, starting some 30 ft above the Lower Lava, showed 18 ft of pale massive fine-grained limestone with *Saccamminopsis?* 3 ft from the top.

Between Wheston and Wall Cliff, the Miller's Dale Beds form a broad outcrop showing small exposures only of uniformly pale massive fine-grained limestone. These beds give rise to a marked feature at Wall Cliff where the section (loc. 28) (overlying the Lower Lava at about 1440 7726) continues that given on pp. 51–53:

|  | ft |
|---|---|
| Tuff (horizon of Upper Lava) .. | — |
| (p) *Not exposed* .. .. about | 5 |
| (o) Limestone, massive, fine-grained with a little irregular silicification; *L. martini, L. pauciradiale, G. sp. maximus* group .. .. | 12 |
| (n) *Not exposed* .. .. .. | 6 |
| (m) Limestone, buff massive; *Aulina furcata, L. junceum, L. pauciradiale, Dielasma hastatum, Pugnax pugnus* .. .. | 8 |
| (l) *Not exposed* .. .. .. | 10 |
| (k) Limestone, grey-white massive fine-grained; coral band 2¾ to 4⅓ ft thick at 4 ft above base and corals at 11 ft 6 in above base; *A. furcata, L.* cf. *maccoyanum, L. martini, L. pauciradiale, G. sp. maximus* group | 22 |
| (j) Lower Lava .. .. .. | — |

The total thickness, 63 ft, of the Miller's Dale Beds is the minimum found in the area. The coral band at the base of the section is correlated with the *Lithostrotion* aff. *maccoyanum* band found above the Lower Lava at Castleton (Shirley and Horsfield 1940, p. 276) and in the Wye valley (Cope 1933, p. 132).

Between Wall Cliff and White Rake the Miller's Dale Beds are cut out by a strike-fault. They are, however, present between White Rake and The Cop, though displaced somewhat by Shuttle Rake and Moss Rake. A section [1380 7875] near the base of the subdivision on Tideswell Moor shows 9 ft of pale fine-grained limestone.

To the north of Batham Gate the lowest part of the Miller's Dale Beds shows its typically fossiliferous character. A section [1351 7939], over ¼ mile W.N.W. of The

Holmes, shows 2 ft of near-white fine-grained limestone with corals, *Davidsonina septosa* and productoids. Farther to the north-west, "at Cop Round" Shirley and Horsfield (1940, p. 276) record a coral band with *Lithostrotion* cf. *arachnoideum* and *L.* aff. *maccoyanum* about 20 ft above the base of the Miller's Dale Beds. The feature of Cop Round [1302 8002] is formed by the lowest part of the Monsal Dale Beds and it would appear that Shirley and Horsfield were referring to a locality a little west of this. The thickness of the Miller's Dale Beds hereabouts is about 80 ft.

To the north and north-west of Old Moor the Miller's Dale Beds are not separately recognized.

# BEE LOW GROUP (D$_1$): NORTHERN AREA

## BEE LOW LIMESTONES

*Bee Low and Gautries Hill.* The Bee Low Limestones are well displayed on Bee Low and, three-quarters of a mile to the east, on Lower Bee Low. To the east again of Lower Bee Low, the lowest beds are ill exposed. On Lower Bee Low a section [1019 7920], starting about 200 ft above the base of the subdivision, showed 83 ft of pale massive fine-grained limestone. Farther west and probably slightly higher in the succession the following section (loc. 37) [0970 7909] is exposed:

|   |   | ft |
|---|---|---:|
| (e) | Limestone, massive pale fine-grained, with scattered corals; *Dibunophyllum?* and *L. martini* .. .. .. .. | 14 |
| (d) | Coral band with *L.* cf. *martini* and *L. pauciradiale* .. .. | 2 |
| (c) | Limestone, as above; at 25 ft above base, *Dibunophyllum bourtonense* φ, *L. arachnoideum*, *L. junceum*, *L. martini*, *Palaeosmilia murchisoni*, *M. sp. papilionaceus* group .. .. | 29 |
| (b) | Coral band, with *D. bourtonense* φ, *Lithostrotion junceum*, *L. martini*, *L. pauciradiale*, *L.* cf. *sociale*, *P. murchisoni*, *Athyris expansa*, *Linoproductus sp. hemisphaericus* group, *M. sp. papilionaceus* group .. .. | 3 |

|   |   | ft |
|---|---|---:|
| (a) | Limestone, as above; at 4 ft from top, *Lithostrotion martini*, *L. pauciradiale*, *G. dentifer*, *M. sp. papilionaceus* group; at 13 ft from top *D. bourtonense* φ, *L. pauciradiale*, *Gigantoproductus? sp. nov.* [wrinkled concentric ornament], *Linoproductus sp. hemisphaericus* group; at base, *Lithostrotion* cf. *martini* | 25 |

The lower of the two coral bands in this section is thus about 260 ft above the base of the Bee Low Limestones and the band is correlated with at least the lower part of the coral-bearing beds of Laughman Tor (see p. 47).

Separated from the above section by a fault and perhaps 50 ft higher stratigraphically, is a band [0938 7903], 1 ft 3 in thick, with *Davidsonina septosa* and productoids. This band appears to be in that part of the succession where the Upper and Lower *D. septosa* bands occur, though it cannot be precisely correlated owing to faulting. The same band has also been located at the eastern end of the Bee Low Quarry section though again in the fault-zone. The section in the quarry (loc. 38) [0922 7915] is as follows:

ft

(*l*) Limestone, pale massive; white and very fine-grained in top 4 ft with rare gastropods and corals .. .. .. .. 12

(*k*) Clay parting, on potholed surface up to 1½

(*j*) Limestone, pale massive about 65

(*i*) Limestone, medium grey (conspicuous in face) .. .. 3½

(*h*) Limestone, pale massive; clay parting locally up to 1 ft thick at 6 ft above base .. .. 15

(*g*) Clay parting, on potholed surface —

(*f*) Limestone; *P. murchisoni* .. 4½

(*e*) Limestone, rather crinoidal, with scattered coarser columnals; some large *D. septosa* and productoids .. .. .. 8

(*d*) Limestone, pale fine-grained, with scattered fine crinoid debris .. 8

(*c*) Limestone, fine-grained bioclastic, oolitic in places; occasional large crinoid stems and small algal colonies; *P. murchisoni* .. .. .. .. 12

(*b*) Limestone, pale fine-grained, with coarse crinoid debris in top 6 ft .. .. .. .. 45

(*a*) Limestone, pale fine-grained (not all accessible); at about 50 ft above base, *L. martini* and *P. murchisoni* .. .. .. 120

The field evidence suggests that the uppermost beds lie some 80 ft below the base of $D_2$. The medium grey limestone (*i*) is considered to lie at the horizon of the Lower Lava. The underlying occurrence of *D. septosa* would then be at the horizon of the Upper *D. septosa* Band.

To the south-east of Bee Low, an exposure [0953 7868] near Lodesbarn shows fine-grained finely crinoidal and slightly oolitic limestone 2¼ ft on brown calcilutite 3¼ ft. The stratigraphical position of this exposure is difficult to estimate though it must be near the middle of the Bee Low Limestones. A similar calcilutite band has already been noted (see p. 46) near the top of the Chee Tor Rock at Upper End.

Exposures on both sides of the main road between Sparrow Pit and Peak Forest provide a good cross-section of the Bee Low Limestones. The lowest beds, perhaps 370 ft thick,

are well exposed south-east of Harratt Grange and show in general a lithology typical of this part of the $D_1$ succession, though the lowest beds are somewhat oolitic [e.g. 1026 7966]. Higher beds are exposed in Haddocklow Plantation, the section [0984 8024] being:

ft

Limestone, pale massive slightly crinoidal .. .. .. .. 16½

Coral band, with *P. murchisoni* (horizon of Upper *D. septosa* Band) .. 2½

Limestone, as above .. .. .. 6

*Not exposed* .. .. .. about 22

Limestone, pale massive fine-grained, with scattered productoids, *D. septosa* and corals (horizon of Lower *D. septosa* Band) .. .. .. 10½

The upper band yielded *D. septosa* in the south-eastern corner of Haddocklow Plantation [0992 8001]. The correlation of the two bands is on the basis of their position within the $D_1$ sequence.

Some 250 ft of limestones, intermittently exposed, overlie the Upper *D. septosa* Band. These contain, in Gautries Plantation [0937 8058], a 16-in crinoidal band with sporadic *D. septosa*. This is tentatively correlated with the higher of the two *D. septosa* bands on Eldon Hill (see below).

The lowest part of the Bee Low Limestones is well exposed in Perry Dale where some 300 ft of these beds are present between the base of the subdivision and the point where lateral passage into apron-reef takes place. The lowest beds are finely crinoidal, though less so than the underlying Peak Forest Limestones. Within 200 yd of the reef margin the Bee Low Limestones develop thick lenses of shelly, poorly bedded limestone of reef type; such beds are referred to the transitional back-reef facies (see Plate XI) and an exposure of these is seen (loc. 39) [1005 8087] on Gautries Hill:

ft

(*c*) Limestone, pale fine-grained; *L. martini*, orthocone nautiloid fragments and *Bollandoceras sp. nov.* .. .. .. .. 9

ft

(b) Shell band; *L. martini, Antiquatonia* cf. *insculpta, Dielasma hastatum, Linoproductus* sp. *?hemisphaericus* group, *M.* sp. *papilionaceus* group, *Productus productus* .. .. .. 2½

(a) Limestone, pale fine-grained .. 9

Both in their geographical relation to the reef-belt and their well-bedded character, the limestones of this exposure should clearly be referred to the shelf succession. However, the rich brachiopod fauna and particularly the presence of a goniatite shows faunal affinities with the reef-belt.

On the south side of the summit of Gautries Hill the inward dips towards the shelf area, frequent at the margin of the reef-belt, are present. Such dips are, however, not developed in Perry Dale.

*Eldon Hill to Windy Knoll.* On Eldon Hill much of the Bee Low Limestones is present. The lower beds, 280 ft thick, were proved in the water borehole at Eldon Hill Quarry [1128 8156] as pale fine-grained limestones. Higher beds are exposed in the quarry (loc. 40) [1128 8134]:

ft

(f) Limestone, pale massive well-bedded .. .. .. .. .. 40

(e) Limestone, medium to rather dark grey .. .. about 5

(d) Clay parting .. .. .. —

(c) Limestone, pale massive well-bedded; 1 ft shelly band 9 ft from top; scattered brachiopods and gastropods in lowest 20 ft .. .. .. .. 60

(b) *D. septosa* band; *Dibunophyllum bourtonense, Lithostrotion martini, L. portlocki, Antiquatonia* cf. *insculpta, Athyris expansa, Davidsonina septosa,* gastropods, nautiloid, trilobite, ostracods .. .. .. .. 12

(a) Limestone, pale grey; scattered fossils including *L. martini*; a pocket of very fossiliferous grey-brown limestone at about 50 ft above base in the northeastern corner of the quarry [1135 8148], with *L. junceum, L. martini, Antiquatonia* cf.

ft

*insculpta, Athyris expansa, Dielasma hastatum, Gigantoproductus edelburgensis, G. sp.* cf. *moderatus, Overtonia fimbriata,* gastropods and bivalves .. 110

The *D. septosa* band of this section lies some 300 ft above the base of $D_1$. The fossiliferous lenses are indicative of the nearness of the reef-belt.

Some 150 ft above the *D. septosa* band of the quarry, another band is present on the eastern side of Eldon Hill. This is well seen in an old quarry [1218 8102]:

ft

Limestone, pale well-bedded fine-grained .. .. .. .. 4¼

Minor erosion surface .. .. —

Limestone, as above; with 1-ft band with *Davidsonina septosa* and corals at top .. .. .. .. 6

Over much of Eldon Hill, however, the higher band is ill defined. It was considered by Shirley and Horsfield (1940, p. 275) to be at or about the horizon of the Upper *D. septosa* Band of the Miller's Dale succession. The band or its horizon cannot, owing to probable displacement by Watt's Grove Vein, be traced into the proved succession at Ox Low, but the general geometry of the outcrop suggests rather that the *D. septosa* band of the quarry is the Upper *D. septosa* Band of the standard succession, and that the higher band, on Eldon Hill, represents the *Lithostrotion* aff. *maccoyanum* horizon.

On the south-west side of Eldon Hill the following section (loc. 41) [1124 8077] shows beds at and below the Upper *D. septosa* horizon:

ft in

(j) Limestone, medium to dark grey; slightly oolitic in top 5½ ft, very fine-grained below (horizon of Upper *D. septosa* Band); *Lithostrotion martini, L. pauciradiale, P. murchisoni, A. expansa, M.* sp. *papilionaceus* group .. .. 6 6

(i) *Not exposed* .. .. about 27 0

(h) Limestone, pale algal; *Linoproductus* sp. *hemisphaericus* group .. .. .. 1 6

|  | ft | in |
|---|---|---|
| (g) Limestone, pale finely bioclastic | 2 | 0 |
| (f) Limestone, pale algal; L. martini .. .. .. .. | 1 | 7 |
| (e) Limestone, pale very fine-grained .. .. .. | 1 | 0 |
| (d) Limestone, pale algal, rare crinoid debris; A. expansa, M. sp. papilionaceus group .. |  | 8 |
| (c) Limestone, pale fine-grained, with some crinoid debris .. | 2 | 9 |
| (b) Limestone, pale crinoidal; L. martini .. .. .. | 2 | 2 |
| (a) Limestone, pale fine-grained | 1 | 9 |

The concentration of algae below the Upper *D. septosa* Band recalls the same association as that in other localities, for example Upperend Quarry (see p. 46).

Beds above and including the Upper *D. septosa* Band are exposed on the flank of a hill east of Snels Low [1157 8164]:

|  | ft |
|---|---|
| Limestone, pale fine-grained [1168 8171] .. .. .. .. .. | 10 |
| *Not exposed* .. .. .. .. | 6 |
| Limestone, pale oolitic, with some fine bioclastic material and a little fine crinoid and shell debris; *D. septosa* and some corals .. .. .. | 9½ |
| *Beds poorly exposed;* loose blocks indicate coral band, up to about 3-ft thick, 4 ft from top .. .. | 26 |
| Limestone, pale fine-grained .. .. | 2 |
| *Not exposed* .. .. .. .. | 24 |
| Limestone, pale fossiliferous, somewhat oolitic in top 1 ft; scattered crinoid debris, algae and shells, fragmentary *D. septosa* .. .. | 7½ |
| Limestone, pale massive very fine-grained .. .. .. .. | 24 |
| Upper *D. septosa* Band: limestone, pale, with some crinoidal lenses, *P. murchisoni* and a few large brachiopods; fragmentary *D. septosa* at base .. .. .. .. .. | 4 |
| Limestone, pale massive fine-grained | 8¾ |

The coral band near the top of this section was correlated by Shirley and Horsfield (1940, p. 276) with the *Lithostrotion* aff. *maccoyanum* band but is here considered to lie at a somewhat higher horizon. Wolfenden (1958, p. 873) noted the presence of oolitic limestones "near the *Cyrtina* band" in this vicinity.

To the south and south-east of Middle Hill the Bee Low Limestones, mostly above the horizon of the Lower Lava, form a broad plateau. Some 100 ft above the horizon of the Lower Lava a quarry (loc. 42) [1299 8194] 265 yd S.W. of Rowter Farm shows 18 ft of pale finely crinoidal limestone with a 1-ft band 12 ft above the base, with *L. junceum* colonies in position of growth and *Plicochonetes sp. buchianus* group. The band lies at or near the horizon of the higher *D. septosa* band of Eldon Hill.

Between Perryfoot and Eldon Hill the back-reef facies, often showing inward dips, is usually well developed. The Bull Pit [1065 8142], east of Perryfoot, shows 55 ft of off-white to grey massive fine-grained limestone with coarsely crinoidal lenses and some corals and brachiopods. The same facies, here showing variable gentle dips towards the reef, is exposed in a quarry (loc. 43) [1125 8163] opposite Eldon Hill:

|  | ft |
|---|---|
| (b) Limestone, pale weathering fine-grained, somewhat algal .. | 2 |
| (a) Limestone, pale grey-brown with very crinoidal bands with columnals up to 1 in diameter; some bands with limestone pebbles; evenly bedded in posts from 6 in to 1 ft 6 in except at base; large colony of *L. junceum* at 6 to 8 ft from base; other fauna includes *P. murchisoni*, abundant reef brachiopods and a specimen of *Weberides sp.* .. | 24 |

Farther to the north-east a quarry (loc. 44) [1208 8236] on the south side of Middle Hill also shows the back-reef facies:

|  | ft |
|---|---|
| (b) Limestone, pale grey very poorly bedded, fine-grained, with very fossiliferous pockets; *L. pauciradiale*, abundant reef brachiopods including *Acanthoplecta mesoloba, Antiquatonia antiquata, 'Camarotoechia' sp., Dielasma hastatum, G. edelburgensis, Productus productus* cf. *hispidus, Pugnax sp.* .. .. | 16 |
| (a) Limestone, pale grey finely crinoidal; *L. martini, Athyris* cf. *expansa, Plicatifera* cf. *plicatilis* .. .. .. .. | 6 |

From the vicinity of Oxlow House the back-reef facies extends northwards along the limestone margin to Windy Knoll. Here a quarry [1261 8300] shows 32 ft of limestone, much of it crinoidal and with some breccia; the lowest 9 ft are very fossiliferous, with reef brachiopods and *Davidsonina septosa*. The top 6 ft of the limestone are impregnated by elaterite (see p. 308). The presence of Neptunian dykes at this locality has been noted on pp. 110–1. The beds with *D. septosa* were considered by Shirley and Horsfield (1940, p. 275) to be the lower of the two *D. septosa* bands of Eldon Hill, which is here correlated with the Upper *D. septosa* Band.

*Castleton.* To the west of Treak Cliff and in the Winnats gorge the back-reef facies is little developed or absent. The Bee Low Limestones in this area show their normal lithology, being pale, massive and fine-grained. In the Winnats a 1-ft band with sporadic *D. septosa* is developed [1344 8264] at a horizon some 170 ft below that of the Lower Lava. The same fossil occurs sporadically in the succeeding 60 ft of strata. The band has been correlated by Wolfenden (1958, pp. 873–4) with the 'Cyrtina band' (i.e. the Upper *D. septosa* Band), but all the occurrences noted here appear to lie below this horizon.

To the south of Castleton the section in Cave Dale shows the most complete development of the Bee Low Limestones. The lowest beds are in the back-reef facies [149 825] just south-east of Peveril Castle; these show relatively strong inward dips near the reef margin but are usually less crinoidal than the back-reef facies near Eldon Hill. A fauna of reef brachiopods is sparsely developed in these beds in places. The extent of the back-reef facies is shown on Plate XI.

Some 100 ft of Bee Low Limestones of normal facies are present beneath the Lower Lava in Cave Dale.

The Lower Lava, absent over much of the northern outcrops, reappears locally in Cave Dale and continues westwards from thence to near Hurdlow Barn. In Cave Dale it is up to 25 ft thick and shows its normal amygdaloidal character. The Lower Lava thins rapidly northwards and disappears on the east side of Cave Dale but has been proved by boreholes to extend eastwards for a short

distance (see Plate IV). As the lava disappears its place is taken by the Pindale Tuff described by Eden and others (1964, pp. 76–8; see p. 61 of this account). The disappearance of the Lower Lava was attributed by Shirley and Horsfield (1940, pp. 275–6 and 287) to its being cut out by an unconformity at the base of the "Miller's Dale Beds". This explanation is not accepted here as, although minor non-sequences are frequent near the reef margin, there seems no need to postulate anything other than the normal dying out of a flow, frequently observed in the district. The *L. aff. maccoyanum* band at the base of the "Miller's Dale Beds", on which the above authors' conclusion was based, appears to be patchily developed and is difficult to trace laterally.

In Cave Dale two sections show the beds above the Lower Lava. The first (loc. 45) [1475 8215] ½ mile S.S.W. of Castleton church is as follows:

|   |   | ft |
|---|---|---|
| (h) | Limestone, pale massive fine-grained .. .. .. .. | 16 |
| (g) | Clay parting, thin, on marked bedding-plane .. .. .. | — |
| (f) | Limestone, pale, with finely crinoidal patches and bands; some irregular shell-debris in coarser patches .. .. | 3¼ |
| (e) | Limestone, pale fine-grained; irregular fracture .. .. | 5¼ |
| (d) | Limestone, granular, finely crinoidal and oolitic .. .. | 3½ |
| (c) | Tuff, pale brown gritty-textured, soft weathered .. 0 to | 1½ |
| (b) | Limestone, pale; *L.* cf. *aranea* at 4½ ft from top .. .. about | 28 |
| (a) | *Not exposed* .. .. about | 11 |
| Top of Lower Lava | | — |

The coral band does not appear to persist laterally.

The second section [1489 8219] shows a full development of the beds between the lava and the Monsal Dale Beds:

|   | ft |
|---|---|
| Monsal Dale Beds .. .. .. | — |
| Limestone, pale massive .. .. | 45 |
| Limestone, granular crinoidal .. | 4 |
| *Not exposed* .. .. .. .. | 0¾ |
| Limestone, cross-bedded, granular crinoidal .. .. .. .. | 1 |

|  | ft |
|---|---|
| Limestone, pale well-bedded fine-grained .. .. .. .. | 28 |
| *Not exposed* .. .. .. about | 5 |
| Lava, part amygdaloidal .. .. | 11 |
| *Not exposed* .. .. .. .. | 5 |
| Lava .. .. .. .. .. .. | — |

In the higher part of Cave Dale much of the beds above the Lower Lava are exposed. Near the top and some 60 ft below the base of the Monsal Dale Beds the following section (loc. 46) [1416 8178] is exposed:

|  | ft |
|---|---|
| (d) Limestone, off-white finely bio-clastic .. .. .. .. | 0½ |
| (c) Limestone, pale grey .. .. | 7½ |
| (b) Limestone, as above; *Saccam-minopsis sp.*, *L. junceum* and brachiopods .. .. .. | 0½ |
| (a) Limestone, pale grey .. .. | 6 |

These beds lie some 60 ft below the base of $D_2$ and the fossil band would thus appear to be at or near the horizon of the coral band of Pin Dale (see below).

On Michill Bank the upper part of the Bee Low Limestones has been proved extensively. Borehole AF 4[1] [1476 8188], showed the following succession:

|  | Thickness | | Depth | |
|---|---|---|---|---|
|  | ft | in | ft | in |
| Monsal Dale Beds, (see p. 72) .. .. .. | — | — | 154 | 6 |
| Limestone, mostly pale grey fine-grained, oolitic in places .. .. | 78 | 9 | 233 | 3 |
| Limestone, grey crinoidal | 2 | 11 | 236 | 2 |
| Tuff, pale greyish green | | 4 | 236 | 6 |
| Limestone, pale grey fine-grained with some pyrite and calcite veining below 241 ft .. | 15 | 10 | 252 | 4 |
| Tuff, pale grey-green with calcite veining .. | | 9 | 253 | 1 |
| Limestone, pale grey fine-grained; with some nodules of pale chert; partly oolitic towards base .. .. .. | 16 | 11 | 270 | 0 |

A fuller development of the Pindale Tuff has been proved by other boreholes in the area (see Plate IV). The thin tuff at a slightly higher horizon is at about the position of that occurring above the Lower Lava in Cave Dale.

*Pin Dale and Earle's Quarry*. In this area the most important section is that in Pindale Quarry (loc. 47) [1594 8231]:

|  | ft |
|---|---|
| Monsal Dale Beds (see p. 72) .. .. .. | — |
| (f) Limestone, pale grey massive, bioclastic and part crinoidal; a 1-ft coral band 4½ ft from base with *Caninia* cf. *densa* [of Hudson and Cotton], *C.* cf. *buxtonensis*, *D. bourtonense* φ, *L.* cf. *aranea*, *L. portlocki*, *P. murchisoni* .. .. .. | 37 |
| (e) Limestone, pale grey thin-bedded (beds from 9 in to 2 ft 6 in) fined-grained, with some clay partings .. .. .. | 17 |
| (d) *Not exposed* (quarry floor) about | 6 |
| (c) Limestone, pale crinoidal, with a little pale nodular chert .. | 5 |
| (b) Clay parting 1 in .. .. | — |
| (a) Limestone, pale fine-grained, with some fine crinoid debris, nodular pale chert .. .. | 18 |

Certain marginal features of the succession and the lateral passage to apron-reef are discussed on p. 108. The presence of chert in the $D_1$ limestones in this section is remarkable in view of its total absence in beds of this age elsewhere in north Derbyshire. The coral band is nearer to the top of $D_1$ than stated by Eden and others (1964, p. 78) but this is due in part to the placing of dark beds below the Lower *Girvanella* Band in $D_2$ in this account.

In Earle's Quarry the uppermost $D_1$ limestones have been extensively worked on the Lower Bench. Near the crusher house [1612 8196] some 20 ft of massive grey limestones are exposed. Eden and others (ibid., p. 79) described these beds as having a maximum thickness of 53 ft in the quarry area. These beds overlie a group of darker grey limestones and overstep these in the vicinity of the tuff mound (see below) only some 10 ft of $D_1$ beds crossing the axis of the mound.

The Pindale Tuff was formerly seen in Earle's Quarry where it occurred as a cone

---

[1]See also Plate IV. Cores examined by Dr. D. V. Frost; record published by permission of Messrs. Earle Limited.

with dips up to 30° on the flanks and reached to a height of 70 ft (Shirley and Horsfield 1940, p. 282). This cone is now obscured by tip but extensive drilling has shown the widespread nature of the tuff beneath the quarry area and also the presence of a second cone, lying to the north-west of the first. The borehole records are summarized in Plate V, which also shows generalized isopachs on the tuff. To the east of the quarry area the Pindale Tuff is represented in attenuated form in a cutting (see p. 109) within the reef-belt.

## MONSAL DALE GROUP (D₂)

### MONSAL DALE BEDS BELOW UPPER MILLER'S DALE LAVA

Near the southern margin of the district a thin development of dark limestone appears to be present locally beneath the Upper Lava, though exposures are lacking. These beds are in the position of the Station Quarry Beds of Miller's Dale (Cope 1937, p. 180); their presence is inferred from the topography, from debris on the outcrop, and from the relationships of the Station Quarry Beds to the Upper Lava on Hammerton Hill (One-inch Sheet 111) south of the present district. The Station Quarry Beds are of very limited extent within the Chapel en le Frith district, being faulted off near the junction of Tideswell Dale and Litton Dale and not developed again to the north of this point. They are included with the Monsal Dale Beds on the one-inch map. The examination of cutting samples from the Litton Dale Borehole [1599 7498] (see also p. 55) is interpreted as showing the presence of pale grey and black limestones beneath the Upper Lava, between the depths of 8 and 14½ ft. In view of this thin development the Station Quarry Beds may be expected to disappear at depth only a short distance north of the borehole.

### UPPER MILLER'S DALE LAVA

*Wormhill Moor.* In this area the Upper Lava crops out around the base of an outlier of Monsal Dale Beds forming high ground between Bole Hill and Withered Low [100 765]. The lava shows its maximum thickness of 100 ft at the southern edge of the outlier, but thins northwards to 10 to 15 ft on Withered Low. The southern outcrops are unusual in that two beds of limestone are present within the lava; the lower some 30 ft from the base near Hargatewall and the higher, near the top, at Bole Hill. The limestone bands are interesting in view of the difficulty in distinguishing different flows within the outcrops of both Upper Lava and Lower Lava.

Exposures showing the normal amygdaloidal character of the lava are lacking in this area though yellow clay or fragments of weathered lava have been observed in places. At Bole Hill [1059 7543] a roadside exposure shows dark crystalline non-vesicular basalt interpreted, as in the case of similar features in the Lower Lava (see p. 52), as the central part of a flow.

*Tideswell Dale to Old Moor.* The Upper Lava crops out on the high ground east of Tideswell Dale though a short distance beyond the southern margin of the district it fails. The rapid dying out of the lava in Litton railway-cutting (One-inch Sheet 111) has been described by Cope (1937, p. 179). Near the junction of Tideswell Dale and Litton Dale [1563 7484] the lava shows its normal amygdaloidal character. In this vicinity it appears to be at least 60 ft thick but it must thin rapidly eastwards for the Litton Dale Borehole [1599 7498], starting near the top, showed only 8 ft of weathered lava (purple to grey clay).

The Upper Lava is faulted off at the junction of Tideswell Dale and Litton Dale but it reappears west of Tideswell though it is there rather thin and not exposed. Between Crossgate and Wall Cliff the only exposure [1367 7676], near Wheston, shows dark crystalline non-vesicular lava. On the south-west side of Wall Cliff the lava thins and disappears, though it reappears farther to the east, where a small thickness of Upper Lava is present locally passing southwards into tuff. This was described by Arnold-Bemrose (1907, p. 253), as the Brook-Bottom Tuff. A section [1438 7709] north-west of Highfield House shows amygdaloidal lava 1½ ft on banded dark calcareous tuff. The tuff thickens southwards and another section (1444 7702] shows 10 ft of dark banded calcareous tuff with dark cognate lapilli up to ¾ in diameter. The tuff disappears rapidly eastwards across the valley and the lava is interpreted as having been cut out by strike-faulting between Wall Cliff and White Rake. On the south side of White Rake the lava is again present and 11 ft of it, deeply weathered, were exposed in an open working [1480 7810] on the vein.

To the north of White Rake exposures of the Upper Lava are rare, though an excavation made in a sink-hole near the base of the bed [1388 7909] near The Holmes showed yellow clay with blocks of tuffaceous limestone up to 1 ft across at from 7 to 15 ft below ground level. The weathered state of the exposure makes it uncertain whether these beds represent a bedded tuff at the base of the flow or whether the tuff and limestone were dragged up by the flow and incorporated in it. To the north of this locality the Upper Lava is little seen. At Cop Round the outcrop is conspicuous as a 'slack' beneath the scarp of the lowest part of the Monsal Dale Beds. The thickness here is about 20 ft. The lava is last seen on the north wall of an open working on the vein-plexus of Oxlow Rake [1350 8087]. This shows the lowest part of the Upper Lava, much weathered, as 4½ ft of yellow clay, resting on limestone. Traced northwards from here, the lava thins rapidly and disappears.

## LOWER MONSAL DALE BEDS

*Dove Holes.* At Dove Holes (Fig. 7) the Lower Monsal Dale Beds are present between Peak House and Near Ridgeclose. The best section, in Victory Quarry (loc. 30) [top at 0748 7702], continues that described on p. 54:

|  | ft | in |
|---|---|---|
| (y) Limestone, grey-white massive fine-grained to finely granular, some fine crinoid debris; colonies of *Syringopora sp.* at base | 10 | 0 |
| (x) Limestone, very fine-grained, calcilutite in places, shelly with *Gigantoproductus dentifer, G. edelburgensis* [juv.], *Linoproductus sp.* hemisphaericus group | 4 | 7 |
| (w) Limestone, pale grey massive fine-grained | 4 | 0 |
| (v) Limestone, grey-white fine-grained, with some fine crinoid and shell debris; more shelly in upper half | 4 | 5 |
| (u) *Not exposed* .. about | 3 | 0 |
| (t) Limestone, off-white massive fine-grained; somewhat shelly in top 10 ft, scattered shells below; coral band 1½ ft at 10½ ft down and band with sporadic corals 3½ ft thick 2 ft from base; *Caninia benburbensis, Dibunophyllum bipartitum, Diphyphyllum lateseptatum, Lithostrotion* | | |

|  | ft | in |
|---|---|---|
| *junceum, L. martini, G. dentifer, G. edelburgensis* [juv.], *Linoproductus sp.* hemisphaericus group | 25 | 0 |
| (s) Limestone, medium grey massive fine-grained, somewhat crinoidal; *Antiquatonia hindi wettonensis, 'Brachythyris' planicostata, G. dentifer, G. edelburgensis, L. sp.* hemisphaericus group (bed absent on south side of quarry) 0 to | 3 | 0 |
| (r) Limestone, dark thin-bedded; irregular thin clay wayboard at base | 5 | 0 |
| (q) Limestone, grey massive cherty; coral in position of growth at 4 ft up | 7 | 0 |
| (p) Limestone, dark massive, with scattered chert | 5 | 0 |
| (o) Limestone, dark grey to grey-brown massive fine-grained; *Dibunophyllum sp.* [juv.], *Lithostrotion?, A. insculpta, G. crassiventer, Productus sp.* | 6 | 0 |
| (n) Clay parting | | 2½ |
| (m) Limestone, grey to dark grey, bituminous in places, sometimes finely crinoidal, thin-bedded; a 6-in band with *Saccamminopsis sp.* 3 ft above base | 6 | 2 |
| (l) Limestone, grey finely crinoidal and slightly oolitic; some shells | 1 | 2 |

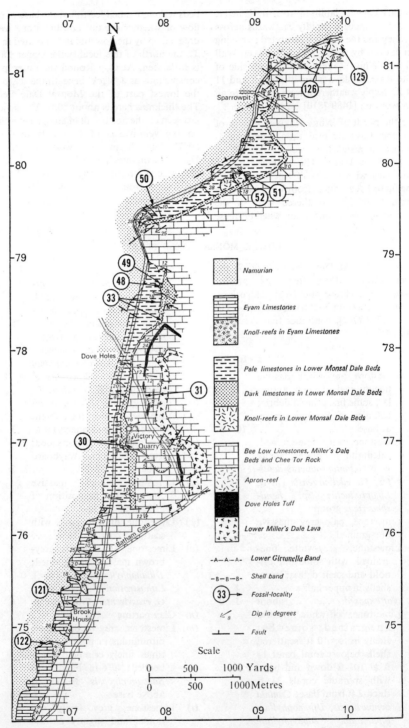

FIG. 7. *Sketch-map of the western margin of the Carboniferous Limestone outcrop*

ft in

(*k*) Limestone, grey finely granular;
darker in upper 4 ft 4 in;
slight erosion surface at base      6   6
Miller's Dale Beds   ..     ..     ..  — —

The shell bed at the top of the section is traceable for some 400 yd northwards from this locality, cropping out again [0748 7734] 110 yd N. of Lower Bibbington. The dark limestones at the base of the Monsal Dale Beds are a feature of the whole of the outcrop at Dove Holes.

The section [0799 7828] (loc. 33) along the old Peak Forest railway at Dove Holes shows the lowest part of the Monsal Dale Beds:

ft

(*f*) Limestone, grey-brown finely
granular, crinoidal, weathering
dark; slight erosion surface at
base   ..     ..     ..     ..    3
(*e*) Limestone, pale brown massive
fine-grained; fragmentary productoids, *L. portlocki* ..    ..   25
(*d*) Limestone, dark very fine-grained,
'hackly' fracture; corals in
top 1 ft, *L. junceum*, *L. martini*,
*P. murchisoni*; abundant *Saccamminopsis     fusulinaformis*
from 4½ ft to top of bed; also
'*Brachythyris*' *planicostata* and
*Semiplanus sp. latissimus* group   12
(*c*) Limestone, dark bituminous,
thin-bedded with shaly partings   ..     ..    12½ ft to   15
(*b*) Limestone, dark grey massive
abundant '*Girvanella*' nodules
in lowest 1¾ ft; sharp base   ..    4
Miller's Dale Beds (see p. 54)   ..  —

When traced northwards the uppermost dark limestones of this section and the lowest 8 ft or so of the overlying beds are replaced laterally by a knoll-reef (loc. 48) [0797 7866] west of Near Ridgeclose; this shows up to 24 ft of pale grey often coarsely crinoidal limestone with a rather poor fauna, including *L. martini*, *A. hindi wettonensis*, *G. crassiventer* and *G. dentifer*. To the north of the knoll a coral band (loc. 49), probably in the position of that at the top of bed (*d*) of locality 33, yields *Diphyphyllum lateseptatum* and *L. martini*. The overlying limestones yield

*Gigantoproductus giganteus* cf. *crassus* and *G. crassiventer*.

*Barmoor Clough and Sparrow Pit*. This outcrop (Fig. 7) is nearly continuous with that of Dove Holes and shows a similar development of dark limestones, up to about 20 ft in thickness, at the base. The higher beds are pale, well-bedded limestones with two knoll-reefs. The dark beds are well seen in small exposures near the road junction at Barmoor Clough. A short distance to the east a section (loc. 50) [0789 7949] at the limestone-shale boundary, is as follows:

ft   in

(*e*) Shale, fissile, black; decalcified
mudstone at base (referred to
Edale Shales)   ..     ..    2   0
(*d*) Limestone, dark grey ..     ..      10
(*c*) *Not exposed*  ..     ..     ..    2   0
(*b*) Mudstone, dark grey passing
down into platy shale    ..    2   0
(*a*) Limestone, medium to dark
grey; *Martinia* cf. *glabra*,
*Rugosochonetes sp.*, *Cane-
yelia membranacea*, *Posidonia corrugata*, *Rineceras
sp.*, *Girtyoceras sp.* ..     ..    2   6

The occurrence of the $P_1$ goniatite *Girtyoceras* shows the presence of a small area of beds of basin facies at this locality.

The old Barmoor Clough Quarry, east of Bennetston Hall, shows the top of the dark limestones and the succeeding beds; a knoll-reef, developed in the latter on the eastern side of the quarry, complicates the section. At the northern end of the quarry[1] (loc. 51) [0881 7988] the sequence is as follows:

ft   in

(*i*) Limestone, very pale grey
granular with crinoidal bands
near top; at 6 ft from top,
productoid fragments inclu-
ding *Productus sp.* ..     ..   16   0
(*h*) Limestone, pale grey massive;
top 3 ft near-white with *Rotiphyllum costatum*, *Zaphrentites enniskilleni*, *Gigantoproductus crassiventer*, *G. dentifer*, *G. edelburgensis*, *Productus productus hispidus*  ..   10   0

---

[1]Now filled in.

                          ft    in

(g) Shale .. .. .. 0 to   0½

(f) Limestone, grey, crinoidal in lower half; *D. lateseptatum, Lithostrotion pauciradiale, Lonsdaleia floriformis laticlavia* .. .. .. 4   3

(e) Shale, hard calcareous, with shell debris .. .. ½in to 1

(d) Limestone, grey crystalline, somewhat crinoidal; *Diphyphyllum furcatum, Lithostrotion portlocki, Gigantoproductus giganteus* .. 1   9

(c) Limestone, pale finely granular; shell fragments .. .. 3   4

(b) *Not exposed* .. .. .. 6   0

(a) Limestone, dark (top only seen) — —

The highest beds form part of the western margin of the knoll-reef. In the middle part of the quarry a non-sequence develops within bed (h) and results in the cutting out of 7 ft of the underlying beds. Another section of the knoll nearby [0878 7977] shows 10 ft of pale grey strongly banded reef-limestones, the top 4 ft conglomeratic, resting on grey normal-bedded cherty limestones, 3¼ ft. A small quarry (loc. 52) [0890 7981] at the eastern end of the knoll, shows the following section:

                          ft    in

(c) Limestone, grey-brown coarsely granular; *Cyathaxonia rushiana, Cladodus sp.* .. .. 10

(b) Limestone-breccia, pale brown; *Caninia sp. subibicina* group 1   0

(a) Limestone, grey-brown granular .. .. .. .. 2   6

The presence of the fish *Cladodus* suggests that this is the locality from which Jackson (1908, p. 309) recorded "*Psammodus rugosus*" and "*Psephodus magnus*".

The greatest thickness of Lower Monsal Dale Beds in the vicinity is some ½ mile south of Sparrow Pit where it is about 200 ft. The highest beds are exposed in a small quarry [0900 8025] some ¼ mile S. of Sparrow Pit: grey-brown thin-bedded cherty limestone, somewhat crinoidal and shelly, algae investing shell debris in top 3½ ft, on pale grey-brown well-bedded granular oolitic limestone, slightly crinoidal, 3½ ft. The position of the algal horizon compares with the

average position of the Upper *Girvanella* Band (125 ft above base of Monsal Dale Beds at Pin Dale and 300 ft up in Cressbrook Dale) with which it is tentatively correlated.

To the north of the last locality the thickness of the exposed Lower Monsal Dale Beds decreases steadily. The lower, dark beds persist but the higher strata pass laterally into crinoidal knoll-reef limestones forming the crest of a small hill at Sparrow Pit. An exposure [0915 8082] 175 yd N.E. of the crossroads at Sparrow Pit shows grey-brown poorly bedded shelly reef limestone 5 ft, on dark bituminous crinoidal limestone. A little to the north [0918 8098] the Monsal Dale Beds are cut out by the unconformity at the base of the Namurian.

*Wormhill Moor.* In this area the lowest 50 ft of the Monsal Dale Beds forms an outlier. Here pale limestones usually overlie the Upper Lava though in one place [1040 7624] a foot of dark grey limestone is exposed in this position.

A quarry (loc. 53) [1059 7552] on Bole Hill provides the only good section:

                          ft

(f) Limestone, pale crinoidal .. 3

(e) Limestone, off-white fine-grained, shelly; *Striatifera striata* .. 1¼

(d) Limestone, off-white massive slightly crinoidal; *Lithostrotion decipiens, G. sp. edelburgensis* group, *Linoproductus sp. hemisphaericus* group .. .. 14½

(c) Limestone, off-white fine-grained, sporadic shells     10 in to 1

(b) Limestone, off-white massive, slightly crinoidal and oolitic .. 6¾

(a) Limestone, pale grey-brown finely granular, many small shells including *Athyris expansa* .. 2¼

*Cressbrook Dale.* This dale provides the best, though not continuous, section of the beds below the Litton Tuff in the southern part of the area. The lowest beds consist of dark, thin-bedded cherty limestones. These strata are generally rather poorly exposed. Shirley and Horsfield (1945, p. 297) give the thickness as 80 ft, which is accepted here. The main outcrop lies south of where White Rake crosses Cressbrook Dale and the dark beds also occur as a faulted inlier in the lower part of Tansley Dale and for some

300 yd from its confluence with the larger dale. Shirley and Horsfield also noted the presence of *Saccamminopsis sp.* at two horizons in the dark limestones, the lower at the base being the better exposed, while the upper band "about 50 feet higher" was known only at the northern end of Litton Frith. The lower band was noted in an exposure by a footpath [1734 7435] on the eastern side of the dale. The section (loc. 54) [1716 7437] at Litton Frith shows the lowest part of the Monsal Dale Beds:

|   |   | ft |
|---|---|---|
| (k) | Limestone, dark fine-grained; *Saccamminopsis* in uppermost 6 in .. .. .. .. | 2 |
| (j) | Limestone, medium to pale grey fine-grained, with some chert .. .. .. .. about | 2 |
| (i) | Limestone, dark fine-grained .. | 2 |
| (h) | *Not exposed* .. .. .. | 8 |
| (g) | Limestone, dark grey, with some fine crinoid debris; *Saccamminopsis* in uppermost 2 in .. | 3½ |
| (f) | *Not exposed* .. .. about | 1½ |
| (e) | Limestone, dark to medium grey, thin-bedded, with some chert; *L. junceum* 1½ ft from base .. | 6 |
| (d) | Limestone, dark fine-grained with a little fine crinoid debris; a silicified colony of *L. junceum* 3 ft from base .. .. .. | 8 |
| (c) | *Not exposed* .. .. .. | 3 |
| (b) | Limestone, dark grey finely crinoidal .. .. .. .. | 2½ |
| (a) | Limestone, dark grey with scattered shell debris and fine crinoid debris; *Saccamminopsis* throughout .. .. .. | 6¾ |
| Miller's Dale Beds .. .. .. | — |

This section shows both the lower and higher *Saccamminopsis* horizons noted by Shirley and Horsfield, the latter 45 ft from the base of the Monsal Dale Beds. There is also a similar band 39 ft from the base. The succeeding beds, some 120 ft thick, consist of pale chert-free limestones. Some 22 ft above the base of the pale beds the shell bed (loc. 55c), noted by Shirley and Horsfield (1945, p. 297) as containing *Striatifera striata*, can be traced from near the southern margin of the district along both sides of the dale. This fossil has not been confirmed in the shell bed though it has been found just above (loc. 55d).

Near Peter's Stone a thin lava, whose presence was noted by Arnold-Bemrose (1907, p. 255), is present for a distance of some 400 yd along the dale sides close beneath the Litton Tuff. The most complete section (loc. 55) of the Lower Monsal Dale Beds hereabouts [1725 7490] about ½ mile S.S.W. of Littonfields, is as follows:

|   |   | ft |
|---|---|---|
| (q) | Litton Tuff (see p. 75) .. .. | 42 |
| (p) | *Not exposed* (features suggest limestone) .. .. .. | 28 |
| (o) | Limestone, pale grey; shell debris | 1 |
| (n) | *Not exposed* (limestone) .. | 2 |
| (m) | *Not exposed* (outcrop of lava) .. | 6 |
| (l) | Limestone, pale grey massive; a few *Gigantoproductus sp.* .. | 8 |
| (k) | Limestone, pale grey crinoidal; a few rolled shells .. .. | 1½ |
| (j) | Limestone, pale grey massive; some crinoid debris .. .. | 21 |
| (i) | *Not exposed* .. .. .. | 5 |
| (h) | Limestone, pale crinoidal .. | 2 |
| (g) | *Not exposed* .. .. .. | 6 |
| (f) | Limestone, pale grey, crinoidal in part .. .. .. .. | 5 |
| (e) | *Not exposed* .. .. .. | 10 |
| (d) | Limestone, pale grey very crinoidal in places, part shelly; *Striatifera striata* .. .. | 16 |
| (c) | Limestone, pale grey crinoidal, shelly in top 1 ft, less so below (shell bed of text); *Gigantoproductus sp.* cf. *semiglobosus* | 4 |
| (b) | Limestone, grey to pale grey massive crinoidal; a little chert in lowest 4 ft .. .. .. | 22 |
| (a) | Limestone, dark grey below, grey above, thin-bedded, chert lenses (upper part of lower, dark limestones of text) .. .. .. | 30 |

The following section [1729 7506] supplements that given above so far as the relationship of the lava to the Litton Tuff is concerned:

|   | ft |
|---|---|
| Litton Tuff .. .. .. .. | — |
| Limestone, dark to medium grey poorly bedded though with some thin argillaceous limestone partings; some shell debris .. .. .. | 7 |
| *Not exposed* .. .. .. .. | 10 |
| Lava, dark amygdaloidal .. seen | 10 |

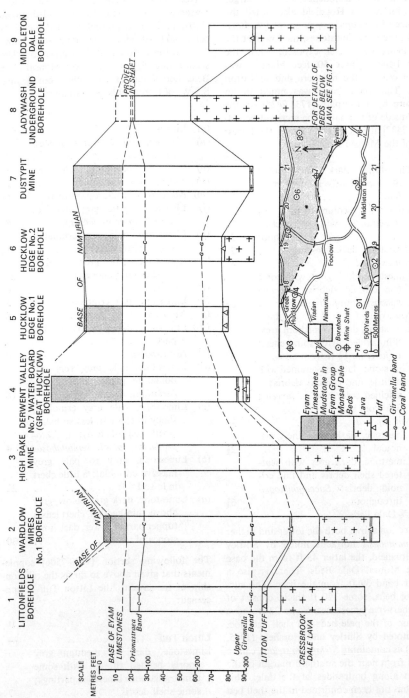

FIG. 8. *Shaft and borehole sections showing the occurrence of the Cressbrook Dale Lava*

Shirley and Horsfield (1945, p. 295) show the lava as reaching a thickness of 20 ft. In that part of Cressbrook Dale south of Tansley Dale the lava is absent. A section here [1754 7465], starting some 75 ft above the horizon of the shell bed, is as follows:

|  | ft |
|---|---|
| Litton Tuff (*not exposed*) .. .. | — |
| Limestone, grey coarsely crinoidal; some shells .. .. .. .. | 5 |
| Limestone, grey shelly, rather rubbly weathering .. .. .. .. | 3 |
| Limestone, grey; occasional shells .. | 15 |
| *Not exposed* .. .. .. .. | 75 |
| Shell bed (bed *c* of loc. 55, p. 67) .. | — |

*Litton, Tideswell and Brook Bottom.* In this area the beds below the Litton Tuff are in general poorly exposed. The dark beds at the base are present on the southern edge of the district south-west of Litton, but die out rapidly northwards. Debris of dark limestone with corals was seen [1582 7443] in this area. In Litton Dale the dark limestones are apparently absent but they reappear in the northern part of Tideswell village (see below). As in Cressbrook Dale the overlying beds consist of pale and apparently everywhere non-cherty limestones though these are seen only in scattered small exposures. Some 50 ft above the Upper Lava a coral band is exposed (loc. 56) [1584 7468] in Litton Dale; this consists of a 9-in coral-bearing bed below and, 3 ft higher, a similar 12-in bed. Both beds yield *L. junceum.* The strata above are seen in small exposures of predominantly pale crinoidal limestones south-east of Litton Dale [e.g. 1658 7497]. Between Litton and Tideswell some darker beds make their appearance in the middle of the Lower Monsal Dale Beds; these also are only seen in small exposures [e.g. 1585 7522].

The upper part of the Lower Monsal Dale Beds was passed through in the Littonfields Borehole[1] [1751 7595]:

|  | Thick-ness ft | Depth ft |
|---|---|---|
| Upper Monsal Dale Beds (see pp. 84–5) .. .. .. | — | 249 |
| Mudstone .. .. .. | 0½ | 249½ |
| Limestone .. .. .. | 20 | 269½ |

|  | Thick-ness ft | Depth ft |
|---|---|---|
| Litton Tuff (see also p. 75) .. | 10½ | 280 |
| Limestone, tuffaceous, with pyrite .. .. .. | 3 | 283 |
| Limestone, pale grey .. | 4 | 287 |
| Limestone, black banded crinoidal .. .. .. | 3 | 290 |
| Limestone, pale grey above, white below .. .. | 19 | 309 |
| Tuff, pale green .. .. | 2 | 311 |
| Cressbrook Dale Lava; basalt, amygdaloidal with a probable tuff parting at 389 ft .. | 109 | 420 |
| Limestone, pale grey .. | 45½ | 465½ |
| Mudstone, dark grey .. | 2½ | 468 |
| Limestone, pale grey .. | 12 | 480 |

To the west of the Tideswell–Brook Bottom valley beds low in the $D_2$ Zone crop out in an elongated tract. The nature of the lowest beds is uncertain south-west of Tideswell though 3 ft of pale limestone were seen 30 ft above the base [1461 7556] near Summer Cross. On the western edge of the outcrop, dark beds make their appearance near Crossgate [1435 7597] where 1½ ft of dark limestone are present 5 ft above the Upper Lava. A section (loc. 57) [1425 7610] near Crossgate shows the junction of the dark limestones and overlying beds:

|  |  | ft |
|---|---|---|
| (*d*) | Limestone, fine-grained grey-brown; 5-in band at 2 ft from base bearing *L. junceum?* and *Gigantoproductus sp.* .. .. | 4 |
| (*c*) | Limestone, grey-brown bioclastic; abundant bryozoa and shell fragments .. .. | 0¼ |
| (*b*) | Limestone, dark grey-brown thinbedded fine-grained .. .. | 4 |
| (*a*) | *Not exposed* .. .. .. | 12 |
|  | Upper Lava .. .. .. | — |

To the north of the last locality the dark limestones may be traced as far as Water Lane [1384 7680], north of which they disappear, probably by lateral passage. The beds overlying the dark limestones are exposed in a small quarry (loc. 58) [1383 7669]:

---

[1]Record from examination of cutting samples by Professor F. W. Shotton.

F

ft

*b*) Limestone, pale thinly bedded; *Caninia juddi*, *L. junceum*, *L. martini*, abundant brachiopods including *A. insculpta*, *G. edelburgensis*, *G. sp.* cf. *moderatus*, *M. sp. papilionaceus* group, *P. productus hispidus* .. .. 8

(*a*) Limestone, medium to dark grey 2½

Higher beds, grey to pale grey limestones [e.g. 1473 7598] and pale crinoidal limestones [e.g. 1448 7631], both chert-free, form a marked dip-slope between the last described exposures and the lip of the Brook Bottom valley.

In Brook Bottom the dark limestones at the base of the Monsal Dale Beds are traceable from the outskirts of Tideswell [1501 7604] to a point [1449 7717] a little north of Highfield House where they pass laterally into pale limestones. The best section (loc. 59) [1473 7638] is as follows:

ft

(*k*) Limestone, buff, with scattered crinoid and shell debris in places; shell band at top with *G. edelburgensis*, *Pugilis pugilis* .. 16

(*j*) Limestone, dark to medium grey fine-grained shelly; *G. edelburgensis*, *G. sp.* cf. *bisati*, *Productus productus hispidus* .. 2¼

(*i*) Limestone, dark grey fine-grained, with a little chert; gastropods at 7 ft above base; *G. sp.* cf. *bisati* at top .. .. 13¼

(*h*) *Not exposed* .. .. .. 2

(*g*) Limestone, dark fine-grained cherty, in beds up to 1 ft: *G. sp.* cf. *moderatus* .. .. .. 8

(*f*) Limestone, dark, fine-grained, poorly exposed .. .. 12

(*e*) *Not exposed* .. .. .. 3

(*d*) Limestone, dark grey bituminous, in beds up to 9 in; thin partings of bituminous shale .. 2½

(*c*) *Not exposed* .. .. .. 1¼

(*b*) Calcilutite, pale brown finely banded; some wavy banding .. 1½

(*a*) Limestone, grey-brown fine-grained .. .. .. 2

The position of the lowest 3½ ft of beds in this section is in some doubt, but these have been referred on lithological grounds to the pale facies of the Lower Monsal Dale Beds; they are, perhaps, the first indication of the southward failure of the dark beds at Tideswell which has already been noted (p. 69). Farther north, the dark beds thin; a shaft [1454 7663] to the sough at the eastern end of Edge Rake, which started near the top of these, proved only 20 ft of limestones overlying 'toadstone'. To the south-west of Highfield House the dark beds bear a *Saccamminopsis* band at their top [between 1448 7673 and 1445 7683]. This band has also been found [1450 7703] north of Highfield House in the following section (loc. 60):

ft

(*b*) Limestone, grey .. .. .. 2½

(*a*) Limestone, dark grey thinbedded; *Saccamminopsis sp.* abundant in lowest 3 ft and sporadic above .. .. 6

To the south of Wall Cliff, a section [1377 7723] shows pale beds close above the Upper Lava:

ft

Limestone, pale finely crinoidal, with fine shell debris, some invested with ?algae .. .. .. .. 1

Limestone, pale fine-grained, with a 7-in silicified lens .. .. .. 4¼

*Wall Cliff to The Holmes.* Between Wall Cliff and White Rake the lowest part of the Lower Monsal Dale Beds is apparently cut out by a strike fault. Some exposures of higher beds close beneath the Litton Tuff are present south of Tides Low. These show grey limestones [e.g. 1503 7758] and grey to pale grey crinoidal limestones [e.g. 1488 7754] both non-cherty. Open workings in White Rake show the following section (loc. 61) [1474 7814] in the north wall:

ft

(*d*) Limestone, pale fine-grained; small shells .. .. .. 2½

(*c*) Limestone, pale to medium grey, fine-grained; sparse *Saccamminopsis sp.* .. .. .. 7¼

(*b*) Limestone, pale to medium grey thin-bedded fine-grained 3½

(*a*) Limestone, off-white to pale grey massive fine-grained to finely granular .. .. .. 23

The base of the section lies some 10 ft above the Upper Lava. Immediately to the north-west of this locality a parting of grey clay, possibly weathered tuff, perhaps some 3 to 5 ft in thickness, has been proved in auger holes. The base lies some 15 ft above the Upper Lava and the clay bed extends laterally for some 250 yd.

Between Maiden Rake and Shuttle Rake the lowest Monsal Dale Beds form a strong scarp above the Upper Lava. A section of these beds (loc. 62) [1436 7860] west-north-west of Bushyheath House is as follows:

|    |    | ft |
|----|----|----|
| (d) | Limestone, medium to pale grey thin-bedded slightly crinoidal; *L. junceum, L. maccoyanum, G. dentifer, M. sp. papilionaceus* group, *Striatifera striata* and gastropods; slight erosion surface at base .. .. .. | 15 |
| (c) | Limestone, pale massive finely oolitic .. .. .. .. | 13 |
| (b) | Limestone, pale grey very crinoidal with some shells .. | 3 |
| (a) | Limestone, grey fine-grained, with a small included mass of darker limestone (7 in × 1 in) .. | 2½ |

Near Shuttle Rake a small exposure [1416 7903] shows 1 ft of dark limestone some 20 ft above the Upper Lava but there is sufficient exposure in the vicinity to show that this is exceptional. Higher beds, seen in a number of exposures north of Bushyheath House, consist of pale grey limestones, non-cherty and often massive.

Near The Holmes the Lower Monsal Dale Beds appear to be only some 100 ft in thickness. A section (loc. 63) near the base [1403 7918] is as follows:

|    |    | ft |
|----|----|----|
| (d) | Limestone, pale crinoidal; *C. benburbensis, P. murchisoni, Linoproductus sp. hemisphaericus* group .. .. .. | 9 |
| (c) | *Not exposed* .. .. .. | 2 |
| (b) | Limestone, massive pale crinoidal | 7 |
| (a) | Limestone, medium to dark grey fine-grained .. .. .. | 2½ |

A section (loc. 64) starting some 50 ft above the Upper Lava exposed nearby [1411 7920] shows 6 ft of pale grey massive limestone with a 1½ ft band 3 ft from base, bearing *Dielasma*

*hastatum, L. sp. hemisphaericus* group and *Productus productus.*

*Bradwell Moor.* At Cop Round the basal Monsal Dale Beds form a conspicuous scarp and are well exposed. Here [1302 8002] 11 ft of near-white fine-grained limestones crop out close above the Upper Lava. Another section of the lowest beds (loc. 65) [1336 8060] is: grey-brown finely bioclastic and oolitic limestone with a colony of *L. maccoyanum* 1¼ ft, grey-brown fine-grained limestone 1½ ft, on grey-brown finely bioclastic and oolitic limestone 2 ft.

To the north of Dirtlow Rake the recognition of the Upper *Girvanella* Band in boreholes and surface exposures has made the stratigraphical relations of the Lower Monsal Dale Beds clearer than in the part of Bradwell Moor to the south. The Lower *Girvanella* Band is probably represented in a small exposure [1417 8149] showing grey-brown finely bioclastic limestone 5 ft thick with algae at 2 ft from base. The section on the south-east side of Cave Dale [top of section 1493 8218], continuing that given on p. 60, shows the lowest beds:

|    | ft |
|----|----|
| Limestone, pale grey-brown finely crinoidal cherty .. .. .. | 4 |
| *Not exposed* .. .. .. .. | 6 |
| Limestone, grey, crinoidal at top .. | 8½ |
| Limestone, variegated dark and pale, thin-bedded, finely crinoidal at top and coarsely so at base .. .. | 6 |
| Bee Low Limestones .. .. .. | — |

Farther east, beds near the base of the Monsal Dale Beds are also seen just north of Dirtlow Rake [1526 8211]: pale crinoidal limestone, coarse in lower half, 3 ft 10 in, on pale massive limestone with a little crinoid debris, finely banded and sorted towards top, 10 ft 6 in. The only other exposures worthy of note in this area lie a little below the Upper *Girvanella* Band. These include a small knoll-reef (loc. 66) [1440 8149] showing 10 ft of pale massive fine-grained limestone with some calcite veining; the fauna includes *Lithostrotion decipiens*, a large colony of *L. junceum* and brachiopods including *Eomarginifera tissingtonensis cambriensis*. The knoll is apparently on the horizon of the "third pale coloured calcirudite" noted by Eden and others (1964, p. 85) close below the Upper

*Girvanella* Band in Earle's Quarry. A nearby exposure [1436 8147] of beds just underlying the knoll, shows dark bituminous limestone 2 ft, on pale to medium grey limestone with shells and corals, some silicified, 1¼ ft. Dark beds in this position are not frequent in the Lower Monsal Dale Beds but have been noted (ibid., p. 80) in Earle's Quarry No. 4 Borehole.

The most complete section of the Lower Monsal Dale Beds in the area north of Dirtlow Rake is provided by Messrs. Earle's Borehole AF 4[1] (loc. 67):

| | Thickness | | Depth | |
|---|---|---|---|---|
| | ft | in | ft | in |
| Soil, etc. .. .. .. | 1 | 8 | 1 | 8 |
| Upper *Girvanella* Band: limestone, dark, thin-bedded, with (*b*) '*Girvanella*' at 4 ft 6 in to base .. .. .. | 10 | 6 | 12 | 2 |
| Limestone, grey-brown finely oolitic .. | 2 | 10 | 15 | 0 |
| Limestone, pale coarsely crinoidal .. .. | 7 | 3 | 22 | 3 |
| Limestone, grey-brown in top 4 ft, grey to dark grey below .. .. | 10 | 3 | 32 | 6 |
| Limestone, pale coarsely crinoidal; somewhat oolitic; a little chert | 26 | 6 | 59 | 0 |
| Limestone, grey-brown somewhat oolitic .. | 3 | 4 | 62 | 4 |
| Limestone, grey to dark grey cherty; (*a*) '*Girvanella*' ? .. .. .. | 5 | 2 | 67 | 6 |
| Limestone, pale grey-brown, oolitic in places | 28 | 6 | 96 | 0 |
| Limestone, grey to grey-brown partly oolitic, darker towards base; a little chert .. .. | 32 | 0 | 128 | 0 |
| Limestone, pale grey cherty .. .. .. | 15 | 8 | 143 | 8 |
| Limestone, grey to dark grey fine-grained, with dark partings (position of Lower *Girvanella* Band) .. .. .. | 10 | 10 | 154 | 6 |
| Bee Low Limestones (see p. 61) .. .. .. | — | — | — | — |

*Pin Dale and Earle's Quarry, Hope.* The Lower Monsal Dale Beds have been proved in a large number of boreholes by Messrs. Earle and are also well seen in Pindale Quarry and Earle's Quarry (see Plate V). These beds show non-sequences at several horizons, more acute as the margin of the reef-belt is approached, and there is also much lateral variation within individual beds. The main features of the stratigraphy have been described by Eden and others (1964).

The section in Pindale Quarry (loc. 47) (Shirley and Horsfield 1940, pp. 284–6) shows much of the Lower Monsal Dale Beds (see also p. 61). The section refers, for the greater part, to the southern end of the quarry [top at 1581 8216] where the sequence is most complete. Most of the section, except for the lowest and some of the middle beds, is inaccessible and this part has been compiled from observation of the quarry face, from borehole records[2] close behind the face and from the account by Eden and others (1964):

| | | ft |
|---|---|---|
| (*s*) | Limestone, pale very coarsely crinoidal shelly ("third calcirudite"); slight non-sequence at base .. .. .. .. | 10 |
| (*r*) | Limestone, grey, slightly irregularly bedded, somewhat crinoidal .. .. .. .. | 20 |
| (*q*) | Limestone, dark, thin-bedded cherty .. .. .. 3 to | 7 |
| (*p*) | Limestone, grey and dark grey; large shells near base .. .. | 9 |
| (*o*) | Limestone, pale; top 3 ft with abundant crinoid debris, some coarse, and fairly common large shells; less crinoidal and with scattered shells below ("second calcirudite"); thickens in depression in bed below; non-sequence at base .. 9½ to | 20 |
| (*n*) | Limestone, dark grey; mainly crinoidal in lower half, passing into grey crinoidal limestone at south end of quarry .. | 13 |
| (*m*) | Limestone, grey, shelly in upper half .. .. .. .. | 4 |

---

[1] See also Plate IV. Cores examined by Dr. D. V. Frost.
[2] Earle's Quarry No. 7, see Eden and others (1964, pp. 114–6); Earle's Quarry No. 8.

|  |  | ft |
|---|---|---|
| (*l*) | Limestone, dark crinoidal | 2 |
| (*k*) | Limestone, pale crinoidal; some shells | 11 |
|  | Unconformity (absent in south end of quarry) | — |
| (*j* | Limestone, dark crinoidal, uppermost 3 ft passing laterally into pale limestone at south end of quarry; some shells | 6 |
| (*i*) | Limestone, grey to dark grey, somewhat crinoidal and shelly | 7 |
| (*h*) | Lower *Girvanella* Band: limestone, dark thin-bedded cherty; '*Girvanella*' (especially common at top) | 9 |
| (*g*) | Limestone, dark thin-bedded crinoidal | 6 |

The lowest item (*g*) appears to be a local development, probably filling a slight hollow in the underlying Bee Low Limestones, as it is not elsewhere developed. Beds (*j–m*) constitute the "first calcirudite" of Eden and others. The unconformity within the latter (noted by Shirley and Horsfield 1940, p. 284, fig. 1) cuts out all the lower $D_2$ beds at the north end of the quarry. The overlying beds of the "first calcirudite" (*k–m*) pass northwards into pale shelly flat-reef limestones as the reef-belt is approached. Compared with the unconformity the two non-sequences noted are slight, the lower cutting out up to 3 ft of the underlying bed and the upper up to 1 ft.

To the east of the entrance to Pin Dale the lateral passage of beds low in the $D_2$ sequence into apron-reef limestones (see p. 108) takes place at the crest of the slope above Upper Jack Bank Quarry. A little to the south the large excavation of Earle's Quarry again shows the base of the Monsal Dale Beds. A section (loc. 68) at the south-east end of Upper Jack Bank Quarry [1620 8208] (Eden and others 1964, pp. 99, 105) shows the lowest beds:

|  |  | ft |
|---|---|---|
| (*b*) | Limestone, grey well-bedded; some shelly bands | 6 |
| (*a*) | Lower *Girvanella* Band: limestone, dark thin-bedded with partings of dark shale; some shelly bands; 9-in band 3 ft from base, with abundant *Girvanella staminea* | 6 |
| Bee Low Limestones | | — |

The irregular nature of the Lower *Girvanella* Band is shown by its absence in a section (loc. 69) [1620 8202] only 60 yd to the south:

|  |  | ft |
|---|---|---|
| (*d*) | Limestone, medium to pale grey massive | 4 |
| (*c*) | Limestone, dark; *Saccamminopsis sp.* | 2 |
| (*b*) | Limestone, dark grey coarsely crinoidal | 2¼ |
| (*a*) | Limestone, dark thinly but irregularly bedded | 2¼ |
|  | Erosion surface | — |
| Bee Low Limestones | | — |

In the main quarry (loc. 70) the succession in the Lower Monsal Dale Beds has been described in detail by Eden and others (1964, pp. 81, 85) and may be summarized as follows:

|  |  | ft |
|---|---|---|
| Base of dark limestones with Upper *Girvanella* Band | | — |
| (*g*) | Limestone, grey ("calcarenite") with some crinoidal bands; occasional shells; Middle Bench [1592 8206] contains a 1-ft coral band, with base 5 ft from top, bearing *Dibunophyllum bipartitum konincki* and *Diphyphyllum* cf. *lateseptatum* 10 to | 15 |
| (*f*) | Limestone, pale coarsely crinoidal and with rolled shell fragments ("third calcirudite") .. about | 10 |
| (*e*) | Limestone, grey below, rather dark ("dark calcarenite") in upper part near reef-belt but becoming grey away from it, fine-grained; in Middle Bench a non-sequence is present at the base of the dark limestones and cuts out up to 10 ft of the underlying beds; a band of large productoids is present in middle of subdivision 25 to | 50 |
| (*d*) | Limestone, pale crinoidal ("second calcirudite"); in Middle Bench [1608 8210] the top 3 ft is a flat reef with *P. productus hispidus;* away from reef-belt the lower part passes laterally into finer-grained beds ("pale calcarenite") and the upper into a rolled-shell bed .. | 12 |

ft

(c) Limestone, usually rather dark well-bedded ("calcarenite") with coarsely crinoidal bands and lenses; 8 ft below the top [1609 8209] *P. productus* and *P. productus hispidus* .. 20 to 35

(b) Limestone, pale coarsely crinoidal ("first calcirudite"); shelly; in eastern part of quarry the top few feet pass laterally into flat-reef limestones (back-reef); with breccia locally near reef-belt [1613 8196] associated with a non-sequence in area of Middle Bench (cf. Pindale Quarry).. .. .. 10 to 20

(a) Limestone, dark, with Lower *Girvanella* Band; some crinoidal bands .. .. about 10

The above general succession is supplemented by the selected borehole records shown graphically in Plate V. In addition two sections in the lowest beds are worthy of note. The first (loc. 71) was measured by R. A. Eden in a small pillar [1622 8191] (ibid., pp. 99, 105), left by quarrying adjacent to the tuff mound and now tipped over:

ft

(l) Limestone, pale very crinoidal; 1½ ft shell bed at base .. .. 8

(k) Limestone, grey, mottled in part, fine grained; crinoidal bands and shelly bands .. .. 3

(j) Limestone, rather dark grey in lower part and paler above, fine-grained; some crinoidal debris .. .. .. .. 7

(i) Limestone, fine-grained dark; *C. benburbensis, L. pauciradiale;* 3 ft from base, *Draffania sp., Girvanella sp.* .. .. .. 4½

(h) Clay parting, thin, on slight erosion surface .. .. —

(g) Limestone, dark coarse-grained crinoidal .. .. .. 1

(f) Clay parting, thin, on erosion surface .. .. .. —

(e) Limestone, grey, somewhat crinoidal .. .. .. .. 4

(d) Limestone, grey to dark grey, with crinoid debris in fine-grained matrix; some shells .. 6

ft

(c) Limestone, dark, fine-grained, with clay partings .. .. 0½

(b) Limestone, grey massive, somewhat crinoidal; some shells 6

(a) Clay, brown, with lenses of fine-grained dark shelly limestone 0¾

Bee Low Limestones .. .. —

Beds (j–l) represent the "first calcirudite". The underlying beds are somewhat thicker than normal; it may be that, as has been suggested for the Pindale Quarry section, the two lowest beds fill a slight hollow in the underlying strata.

To the south-east of Wilson's Monument and the site of the tuff mound (see pp. 61–2) important unconformity is present at the horizon of the Upper *Girvanella* Band. This cuts out up to 30 ft of the underlying beds and is most acute in the vicinity of the tuff mound. The unconformity is attributed by Eden and others (ibid., p. 78) to later folding on the axis of the mound. This interpretation agrees with the relations in Monk's Dale where two vents are present on the line of a local sharp fold (see p. 323).

The easternmost section (loc. 72) in Earle's Quarry [1643 8189] shows the lowest part of the Monsal Dale Beds adjacent to the reef-belt:

ft

(j) Limestone, pale coarsely crinoidal (flat reef) [1650 8193] .. .. 8

(i) Clay parting .. .. 0 to 0¼

(h) Limestone, grey to dark grey massive fine-grained; some crinoid debris .. .. .. 3½ to 6

(g) Limestone, dark thin-bedded fine-grained, with shale partings; abundant large shells .. 3 to 3¾

(f) Limestone, grey, crinoidal 0 to 3

(e) Limestone, dark massive; *Saccamminopsis* in middle, some large shells in lowest 2½ ft .. 7

(d) Clay parting, thin .. .. —

(c) Limestone, pale grey crinoidal 4¼ to 10

(b) Limestone, dark well-bedded, finely crinoidal and bioclastic; *L. junceum* 1 ft from base .. 6

(a) Limestone, dark finely crinoidal with partings of black shale and small limestone pebbles; *Saccamminopsis sp.* throughout 5

Bee Low Limestones .. .. —

The highest bed (*j*), representing the "second calcirudite" appears to join the overlying "third calcirudite" to form a composite mass of flat-reef limestones which can be followed for some 200 yd eastwards from this section. Northwards the flat reef passes laterally into apron-reef. The "first calcirudite" is represented in the section by bed (*c*).

To the south of the quarry area an outcrop of Lower Monsal Dale Beds is present though the limits of this are uncertain. With the exception of Nether Cotes boreholes NC6 to NC8 (see Plate V) there is little information on the detailed stratigraphy of this tract and the relationships are complicated by the lateral passage of the standard limestones into basin facies (see also p. 91) with the development of dark cherty limestones.

## LITTON TUFF

*Cressbrook Dale.* The outcrop of the Litton Tuff is well marked though exposures are poor and infrequent. Due east of Litton the tuff (loc. 55*q*) overlies the beds listed on p. 67, and is some 42 ft thick; the northernmost part of the outcrop [1732 7536] shows only some 25 ft of tuff. The landslip feature of Peter's Stone is due to the slipping of the overlying beds on the Litton Tuff.

*Litton.* The tuff reaches its maximum development of 100 ft on the slopes north and north-east of the village, the overlying limestones giving rise to a strong feature. Exposures are present in the roadside between Peep o'Day and Litton village, one [1682 7518] close below the top showing 12 ft of weathered brown tuff. A trench [1676 7517] in the roadside ¼ mile E.S.E. of Sterndale House showed greenish brown calcareous tuff with vesicular cognate lapilli and bombs up to 5 in diameter. The thickness of the tuff at Litton and the presence of the coarse ejectamenta strongly suggest proximity to the source of the deposit.

To the north-east of Litton, the Littonfields Borehole [1752 7596] proved the tuff between 269½ and 280 ft. The uppermost 8½ ft were not recovered, but the lowest 2 ft were a pale grey fine-grained tuff. Wardlow Mires No. 1 Borehole [1850 7553] (see pp. 91, 93) proved grey-blue fine-grained tuff 2 ft 6½ in thick at 452 ft; records of boreholes and shafts north of here show the tuff to be absent.

*Tideswell to Tideswell Moor.* From the type locality the outcrop of the Litton Tuff extends north-westwards and then northwards as a conspicuous slack which is traceable as far as Bushy Heath where the tuff dies out. In general exposures are lacking, though there are shows of yellow clay in places. In this area the Litton Tuff has been seen in trenches. In the main A 623 road [1525 7663 to 1529 7661] weathered brown igneous rock was encountered. Another section on a side-road [1529 7641 to 1521 7639] showed weathered tuff, very fine-grained and limonitic in places.

## UPPER MONSAL DALE BEDS

*Middleton Dale.* Here much of the upper part of the Upper Monsal Dale Beds is continuously exposed (see Fig. 9). Detailed sections at three places are given below to illustrate the succession, and these and supplementary sections are shown on Fig. 10. The sequence was first described by Morris (1929, pp. 37–67), some of whose stratigraphical terms are retained. Two shell bands, here termed the Lower and Upper Shell beds, are traceable over much of the dale and form convenient datum-horizons. The bed of white calcilutite termed the "White Bed" by Morris (ibid., p. 47–8) is also a useful datum, and near the top of these beds the *Orionastraea*

Band has been followed over part of the section (see Fig. 10). Towards the western end of the dale the section (loc. 73) in Furness Quarry [2085 7596] is as follows:

|  | ft |
|---|---|
| Eyam Limestones (see p. 99), clay parting at base [2100 7608] .. .. | — |
| (*x*) Limestone, grey-white massive fine-grained, with scattered fine crinoid debris and some shells | 8¾ |
| (*w*) Limestone, as above but cherty | 5 |
| (*v*) Clay parting, on slight erosion surface .. .. .. 0 to | 0¼ |

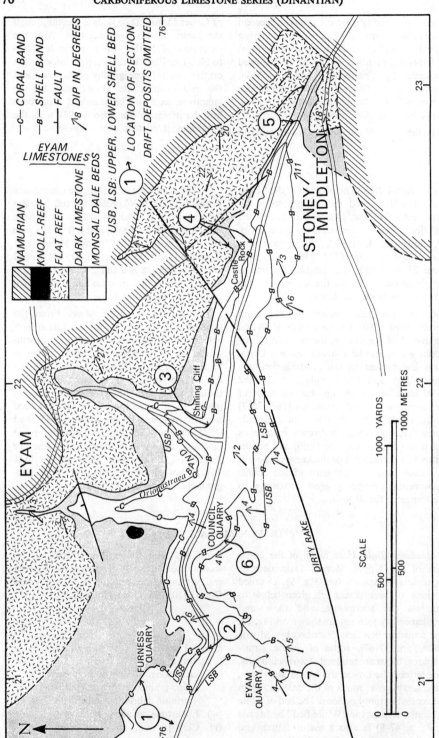

Fig. 9. Sketch-map of the geology of Middleton Dale.

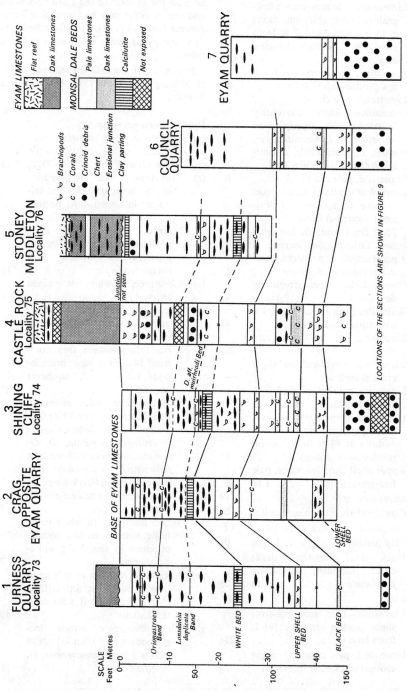

FIG. 10. *Sections of the Upper Monsal Dale Beds and Eyam Limestones in Middleton Dale*

ft

(*u*) Limestone, pale grey-brown fine-grained; fairly abundant chert; a 15-in coral band 2 ft from base with *Lonsdaleia duplicata duplicata* .. .. .. 3½

(*t*) Limestone, pale very cherty; a few productoids near base .. 4

(*s*) Limestone, pale cherty .. .. 1½

(*r*) *Orionastraea* Band: pale cherty limestone with *Diphyphyllum lateseptatum, Lithostrotion junceum, L. pauciradiale, L. portlocki, Orionastraea placenta*[1] .. 2¼

(*q*) Limestone, pale cherty .. .. 6

(*p*) *Lonsdaleia duplicata* Band: pale, massive limestone with scattered chert; 1-ft bed with corals 30 ft from base, *D. lateseptatum, Lithostrotion junceum, L. pauciradiale, L. portlocki, Lonsdaleia duplicata duplicata* .. 51

(*o*) Chert, dark, rather irregularly developed .. .. 4 in to 0¼

(*n*) White Bed: calcilutite, pale grey, with some chert .. .. 4½

(*m*) Limestone, pale grey, with rare chert .. .. .. .. 20½

(*l*) Clay, thin wayboard on slight erosion surface .. .. —

(*k*) Limestone, pale fine-grained, with a little dark chert .. .. 7½

(*j*) Limestone, grey-brown massive; an impersistent line of chert nodules at 4¾ ft from top (in north-east of section) .. 8½

(*i*) Upper Shell Bed: limestone, pale fine-grained shelly .. 1 to 3

(*h*) Limestone, grey .. .. .. 11½

(*g*) Clay, on slight erosion surface 0 to 0¼

(*f*) Limestone, as above .. .. 12

(*e*) Clay parting .. .. 1 in to 0¼

(*d*) Black Bed: limestone, dark; corals including *Lonsdaleia floriformis* .. .. 5 in to 0¼

(*c*) Clay .. .. .. .. thin

(*b*) Limestone, pale grey massive; slight erosion surface at 9½ ft from base .. .. .. 28½

(*a*) Limestone, pale crinoidal, slight erosion surface 2 ft from base 6½

The two erosion surfaces in items (*a*) and (*b*) lie near the horizon of the Lower Shell Bed and one or other may be responsible for its absence (see also p. 81). The Black Bed (*d*) is thin compared with the sections quoted below.

Farther down the dale, the section (loc. 74) at Shining Cliff [2184 7582] (Plate I) is as follows:

ft

Eyam Limestones (see p. 98) .. .. —

(*u*) Limestone, pale massive fine-grained oolitic, with a little fine crinoid debris [2193 7592] 6

(*t*) Limestone, pale well-bedded cherty; finely granular and crinoidal in lowest 4 ft, passing up into buff calcilutite; fossiliferous near top, with *Rotiphyllum costatum, Gigantoproductus edelburgensis, Productus productus* .. .. 16

(*s*) Limestone, off-white fine-grained cherty .. .. .. .. 1¾

(*r*) Limestone, buff massive fine-grained cherty; 3-in band of pseudobreccia 15½ ft from base; impersistent thin coral band 14 ft from base; brachiopods including *P. productus hispidus* .. .. .. 25¾

(*q*) *Dibunophyllum* aff. *muirheadi* Bed: limestone, grey-white fine-grained shelly; *Dibunophyllum bipartitum bipartitum, D. bipartitum konincki, Lithostrotion junceum, L. pauciradiale, Palaeosmilia regia, G. edelburgensis* 1

(*p*) Limestone, off-white fine-grained cherty .. .. .. .. 3

(*o*) White Bed: buff to white calcilutite, with bands and nodules of chert in top 3 ft; white-weathering .. .. .. 7

(*n*) Limestone, medium to pale grey-brown, darker at top, a little chert; lowest 12 ft with scattered shells .. .. .. 18

(*m*) Limestone, pale brown fine-grained; many silicified shells in top 3 ft; *G. crassiventer, G. dentifer* .. .. .. .. 11

---

[1]The collections from this and the overlying coral band were made from the access-road on the north-east side of the quarry.

ft

(*l*) Upper Shell Bed: massive pale limestone with abundant *G. crassiventer?* and *G. dentifer?*  1½

(*k*) Limestone, pale massive, brown at base; 6-in band at 3 ft from base with *L. junceum, L. pauciradiale, Lonsdaleia floriformis floriformis* (*Lonsdaleia* Bed of Morris); *Lonsdaleia floriformis floriformis* at 7 ft above base: *Diphyphyllum lateseptatum, Lithostrotion junceum, L. pauciradiale, L. portlocki* and brachiopods at 8 to 11 ft from base; gigantoproductoids at 14½ ft from base  ..  ..  23½

(*j*) Black Bed: limestone, dark bituminous; scattered shells; *Lonsdaleia floriformis floriformis*  ..  4¾

(*i*) Limestone, pale grey, variegated darker in top 3 in; abundant corals in top 1 ft 2 in; *D. lateseptatum* and *Syringopora sp.*  5

(*h*) Not exposed  ..  ..  ..  2

(*g*) Limestone, pale grey massive, with scattered crinoid debris  3¾

(*f*) Lower Shell Bed: off-white to buff fine- to very fine-grained limestone with abundant shells, some silicified; a little chert in lowest 2½ ft; *G. sp. edelburgensis* group, *G. sp.* cf. *moderatus*  7

(*e*) Limestone, grey-white fine-grained; 5-in bed of fine shell debris 1 ft from base; prominent slightly irregular bedding plane 4 ft from base  ..  5½

(*d*) Limestone, grey-brown, finely crinoidal at base, less crinoidal above  ..  ..  ..  ..  7½

(*c*) Limestone, buff to grey-white finely crinoidal; scattered productoids in upper half include *G. edelburgensis*  ..  ..  17½

(*b*) Not exposed  ..  .. about  12

(*a*) Limestone, buff crinoidal, with some sorting into coarse and fine bands  ..  ..  ..  7

This appears to be the type section for Middleton Dale described in detail by Morris (1929, pp. 45–8); Shirley and Horsfield (1945, p. 294) suggest that the succession is a condensed one. Fig. 10 shows that the local thinning of the upper part of the succession is restricted to the beds just above the White Bed and is best explained by the presence of a local non-sequence at this horizon, though this is difficult to observe in any one section (see also p. 81).

The third detailed section in Middleton Dale is that at Castle Rock (loc. 75) [224⁷ 7569]:

Eyam Limestones( see p. 99)  ..  ..  —

(*z*) Limestone, buff; shelly, including *Gigantoproductus sp.*  12

(*y*) Limestone, pale fine-grained  1

(*x*) Limestone, pale crinoidal shelly, 'rubbly-weather'; *Antiquatonia* cf. *hindi*, *tovia spinulosa*, *Overtoni imbriata, Semiplanus sp. issimus* group [juv.] near b  ..  8

(*w*) Limestone, pale mass fine-grained cherty; 3 soft-weathering band at h e  ..  6

(*v*) Limestone, buff well-ded fine-grained with rare c rt  ..  9

(*u*) Not exposed  ..  .. about  6

(*t*) Limestone, pale assive fine-grained ..  ..  ..  3

(*s*) Limestone, buff ely crinoidal, rubbly in plac at base  ..  6½

(*r*) Limestone, pa grey-brown fine-grained; ma ed bedding plane (?erosiona with traces of clay 3 ft from h e ..  ..  ..  4¾

(*q*) Limestone le cherty (especially at base)  ..  ..  ..  1

(*p*) Limestor pale massive finely biocla c, some chert about 5 ft f n base; scattered corals 4 t ½ ft from base; *L. juncei L. martini* and *L.* cf. *portlo*  ..  ..  ..  ..  8¼

(*o*) Cl arting  ..  ..  .. thin

(*n*) Li stone, pale brown fine-ained irregular-weathering forms gash in cliff face]  ..  6½

ight erosion surface  ..  ..  —

(*m*) Limestone, pale grey massive fine-grained, scattered shells and sporadic corals near base  12¼

Upper Shell Bed: buff fine-grained to finely granular shelly limestone; *G. crassiventer*  ..  4¼

ft

(k) Limestone, pale grey-brown massive fine-grained, with a few shells; prominent bedding plane at base .. .. .. 6¼

(j) Limestone, buff massive fine-grained; silicified coral 1½ ft from top .. .. .. 7¾

(i) Limestone, pale grey-brown massive finely crinoidal; scattered shells .. .. .. .. 4¾

(h) Limestone, as above, with a little chert at base; *Lonsdaleia floriformis floriformis* 6 in from base 4½

(g) Black Bed: limestone, dark fine-grained, top 9 in rubbly weathering; colony of *Lithostrotion portlocki* 3 ft from base .. .. .. .. 5¼

(f) Clay parting, thin, on prominent bedding plane .. .. .. —

(e) Limestone, dark fine-grained; corals including *D. lateseptatum* in top 1 ft .. .. .. 2

(d) Limestone, pale grey-brown massive; some fine crinoid debris; 1 ft with scattered shells 5 ft from base and a few shells above this .. .. .. 8¼

(c) Lower Shell Bed; pale limestone with abundant *G. edelburgensis* 6

(b) Limestone, pale fine-grained; scattered corals at base .. 1¾

(a) Limestone, pale grey to buff massive finely granular; 4-in shelly band with coral 6 ft from base 17

Shirley and Horsfield (ibid., p. 294) record *Orionastraea* cf. *indivisa* from "90 feet above the roadway"; this is about 45 ft above the Upper Shell Bed.

The cliff section (loc. 76) in Stoney Middleton [2279 7560] ½ mile E.N.E. of Highfields Farm shows the highest beds in a more accessible position than at Castle Rock:

ft in

Eyam Limestones (see p. 98), potholed surface at base .. .. — —

(n) Limestone, pale granular .. 2 0

(m) Clay parting, on slightly eroded surface .. .. 0 to 1

(l) Calcilutite, buff to white .. 2 6

ft in

(k) Limestone, grey-white and buff granular bioclastic somewhat 'reef-like' in lowest 4½ ft; coarsely crinoidal lens at 4½ ft from base; uppermost 2½ ft mainly fine-grained .. .. .. 7 0

(j) Limestone, pale massive cherty 12 6

(i) Limestone, pale cherty bioclastic, part finely crinoidal, rubbly-textured, bedding irregular .. .. .. 14 0

(h) Limestone, pale cherty; some corals in lowest 1 ft; *L. junceum* .. .. .. .. 11 0

(g) *Orionastraea* Band: pale limestone with some chert, *L. junceum*, *L. portlocki*, *O.* aff. *indivisa* .. .. .. 1 9

(f) Limestone, pale shelly, 'reef-like', thinning out in places maximum 5 8

*?Slight erosion surface* .. — —

(e) Limestone, pale massive cherty (especially in lower two-thirds) .. .. .. 13 10

(d) Limestone, buff fine-grained cherty; finely bioclastic in upper part with scattered corals; shell and crinoid debris .. .. .. 2 8

(c) White Bed: pale grey to pale buff cherty calcilutite .. 4 4

(b) Limestone, pale cherty, irregularly bedded and 'reef-like' with many small shells; a 9-in coral band at 3½ ft from base with *Dibunophyllum bipartitum konincki*, *Diphyphyllum fasciculatum*, *L. junceum* .. .. .. 4 6

(a) Limestone, pale buff massive; irregular chert band at 9¾ ft from base; corals 10 to 11 ft from base; *L. junceum*, *L. portlocki*, *O. placenta* .. 16 0

The section starts about 2 ft above the Upper Shell Bed.

The most important features of the succession in Middleton Dale are summarized below and should be read in conjunction with the detailed sections already given (see also Fig. 10):

(1) Beds below the Lower Shell Bed. These are present in the floor of the dale from the vicinity of Hanging Flat to Castle Rock. The greatest thicknesses are seen at Shining Cliff (see p. 78) and in Eyam Quarry [211 756]. The latter shows some 46 ft of limestone, mainly pale massive and crinoidal, below the Lower Shell Bed. The relations of these beds to the underlying but unexposed basalt of the Middleton Dale Borehole are discussed on p. 96.

(2) The Lower Shell Bed "Lower *Giganteus* Bed" of Morris (1929, p. 45) is recognizable throughout the dale east of the longitude of Farnsley Farm. To the west it cannot be recognized with certainty, but it is probably thinly developed in a section west of Hanging Flat.

(3) The beds between Lower and Upper Shell beds are about 35 to 40 ft thick, and consist of pale massive limestones with a bed of dark bituminous limestone up to 5 ft at, or just below, the middle; this may be conveniently termed the 'Black Bed'. The "*Lonsdaleia* Bed" of Morris (ibid., pp. 42, 46) lies some 3 ft above the dark limestone but appears to be only locally developed.

(4) The Upper Shell Bed ("*Hemisphaericus* Bed" of Morris ibid., p. 46) is traceable around most of the dale but may be seen to fail in Eyam Quarry. In Furness Quarry also the bed was absent, but farther west it probably reappears.

(5) The beds between the Upper Shell Bed and the White Bed, 20 to 25 ft thick, are pale massive limestones showing the incoming of chert in the upper half, most of the beds above this horizon being cherty. The appearance of chert occurs at about the same horizon throughout the considerable distance covered by the dale.

(6) The White Bed, where fully developed, is up to 7 ft thick, and is so called from its white weathering, a feature which causes it to be very evident on the cliff sections. In many places this bed weathers more rapidly than those above and below and it forms a horizontal gash in the natural sections. The White Bed is absent at Castle Rock where Shirley and Horsfield (1945, p. 294) considered that it had been cut out by the overlying "Hackly Bed". The uppermost part of the bed may be seen to fail before the lower on the west side of Castle Rock, which

indeed suggests the cutting out of the bed. The section at Castle Rock shows no evidence of the "conglomerate" noted by Shirley and Horsfield (ibid., p. 294) and the record may be due to the deceptive appearance of fragments of the overlying bed resting on a slightly eroded surface or on a stylolitic bedding plane. Farther east, the White Bed reappears in the cliff at Stoney Middleton.

(7) The beds above the White Bed are 75 ft thick where fully developed, though only 53 ft 7 in at Shining Cliff. They are massive cherty limestones with an irregularly weathering bed (the irregularities being due to a large number of random small joints), the "Hackly Bed" of Morris (op. cit., p. 48), at the base. The *Orionastraea* Band, normally some 50 ft above the base, has been traced from Furness Quarry to Eyam Dale. It has not been found at Shining Cliff though Shirley and Horsfield (1945, p. 294) record *O. placenta* from this horizon "at a point intermediate between this (Castle Rock) and Shining Cliff and about 20 ft above the 'White Bed' of Morris". The *Lonsdaleia duplicata* Band is present in Furness Quarry 27 ft below the *Orionastraea* Band and may be represented at Shining Cliff by the *Dibunophyllum* aff. *muirheadi* Bed of Morris, though this does not appear to continue far along the section. Near the top a higher band with *Lonsdaleia duplicata duplicata* is present locally a few feet above the *Orionastraea* Band and would appear to be the lateral equivalent of the higher of the two *Orionastraea* bands of the Coombs Dale section (see p. 82). The thickness variation noted above appears to affect the lowest part of these beds only and is attributed to the presence of a non-sequence rather than to the condensation of the sequence suggested by Shirley and Horsfield. The non-sequence cannot, however, be directly observed in the sections. At the eastern end of Middleton Dale the bedding of this group of strata becomes rather irregular, individual beds thinning and thickening, though this does not affect the overall pattern of the group as a whole.

Cutting out of beds beneath the Eyam Limestones cannot be observed in Middleton Dale though a strongly potholed surface is present at the junction in Stoney Middleton.

*Coombs Dale and Middleton Moor.* In the

part of Coombs Dale falling within the present district, exposure of the Upper Monsal Dale Beds is much less complete than in Middleton Dale. The best sections are at the junction of a southern tributary dale with the main dale. The highest beds are seen at a cliff (loc. 77) on the eastern side of the tributary dale [2239 7429]:

| | | ft |
|---|---|---|
| Eyam Limestones (see p. 98) | .. | — |
| (f) Limestone, grey-white, fine-grained cherty .. .. .. | | 3½ |
| (e) Limestone, buff fine-grained; *Aulophyllum fungites cumbriense, Diphyphyllum lateseptatum, D.* cf. *parricida, Lithostrotion junceum, O. placenta, Antiquatonia insculpta, Productus productus hispidus* .. .. | | 3 |
| (d) Limestone, buff fine-grained cherty; *Antiquatonia hindi, A. insculpta, Avonia thomasi, Eomarginifera lobata* aff. *laqueata* .. .. .. .. | | 22 |
| (c) *Orionastraea* Band: limestone, pale grey-brown cherty; *Dibunophyllum bipartitum bipartitum, Diphyphyllum lateseptatum, L. junceum, L. portlocki, O. placenta, Palaeosmilia murchisoni, Zaphrentites derbiensis, Alitaria panderi, Antiquatonia antiquata, A. hindi wettonensis, Echinoconchus punctatus, Spirifer (Fusella) triangularis* .. | | 10 |
| (b) Not exposed .. .. about | | 17 |
| (a) Limestone, dark poorly exposed about | | 12 |

The lower coral band (c) appears to represent the main *Orionastraea* horizon of Middleton Dale and Bradwell Dale. Some 75 ft of little-exposed beds separate the dark limestone (a) from the lowest beds exposed in the floor of the dale (loc. 78) [2222 7436]:

| | | ft | in |
|---|---|---|---|
| (m) Limestone, dark fine-grained | | 3 | 0 |
| (l) Limestone, medium to pale grey-brown, massive at top, fine-grained; *G. crassiventer, G. edelburgensis* .. .. | | 8 | 0 |
| (k) Limestone, dark grey fine-grained slightly bituminous and very cherty; shelly in places .. .. .. | | 5 | 4 |

| | | ft | in |
|---|---|---|---|
| (j) Limestone, pale grey-brown; some shells and broken shell debris .. .. .. | | | 6 |
| (i) Limestone, pale brown fine-grained; shelly in top 4 in; *G. crassiventer, G. edelburgensis* .. .. .. | | 1 | 9 |
| (h) Not exposed .. .. .. | | 5 | 0 |
| (g) Limestone, pale brown shelly; *G. crassiventer* .. .. | | 3 | 4 |
| (f) Limestone, pale brown fine-grained cherty, with shelly bands .. .. .. | | 5 | 8 |
| (e) Calcilutite, pale brown, with abundant chert .. .. | | 1 | 3 |
| (d) Limestone, pale grey-brown cherty, with thin bands of silicified shells; *G. edelburgensis* .. .. .. | | 2 | 8 |
| (c) ?Hob's House Coral Band: limestone, medium to pale grey and grey-brown fine-grained cherty; with abundant corals, sporadic shells also in top 2 ft; *Caninia juddi cambrensis, Dibunophyllum bipartitum bipartitum, D. bipartitum konincki, Diphyphyllum lateseptatum, L. junceum, L. martini, L. sp.* cf. *pauciradiale* [with cerioid tendency] .. .. .. | | 5 | 10 |
| (b) Limestone, thin-bedded fine-grained cherty; some shells near top; *G. edelburgensis* .. *Prominent irregular bedding plane* .. .. .. .. | | 4 — | 6 — |
| (a) Limestone, pale brown finely crinoidal, massive except for top 2½ ft; 9-in shell band 5 ft 8 in from base, with *G. crassiventer, G. sp.* cf. *semiglobosus, Linoproductus sp. hemisphaericus* group .. | | 15 | 4 |

The dark limestone beneath the main *Orionastraea* Band is developed on both sides of Coombs Dale for ¼ mile north and to the west of the junction with the tributary dale.

Between Coombs Dale and Middleton Dale high Monsal Dale Beds (apparently all in the pale cherty limestones of the Middleton Dale sequence) occupy an undulating plateau.

This shows few sections of importance, though recent excavations at Messrs. Glebe Mines' Cavendish Mill showed the following section (loc. 79) [2054 7534]: Eyam Limestones (see p. 101) on 16 ft of pale limestones, part cherty with a 1½-ft band at 8 ft from base with *Diphyphyllum lateseptatum, L. junceum* and *L. pauciradiale*. The same band, or one close to this horizon, is exposed (loc. 80) near the head of Middleton Dale [1980 7616]: pale cherty limestone 5 ft, with abundant corals, including *Dibunophyllum bipartitum bipartitum* and *D. bipartitum craigianum*, in top 1¾ ft. The horizon is about 25 ft below the Eyam Limestones and the band is referred to the main *Orionastraea* horizon.

*Cressbrook Dale*. In this dale the Upper Monsal Dale Beds can be divided naturally into two groups at the horizon of the Hob's House Coral Band. The lower group consists of some 60 ft (see Littonfields Borehole, p. 84) of limestones, dark at the base and in general poorly exposed. These beds correspond to a more extensive development of dark limestones north of Litton (see p. 85). The Upper *Girvanella* Band at the base was recorded by Shirley and Horsfield (1945, p. 297) "on the west side of the wall at the foot of the slope" (loc. 81) [1731 7537]. The thickness of the exposed part of the band is here about 1 ft, the lower beds immediately above the Litton Tuff being unexposed. During the present study the Upper *Girvanella* Band, 5½ ft thick and consisting of dark limestones with *Girvanella*, was seen (loc. 82) close above the Litton Tuff [1715 7426] ⅓ mile S.W. of Littonfields.

On the east side of Cressbrook Dale the following section (loc. 55) continues that given on p. 67 [top of section at 1736 7512]:

ft
(v) Hob's House Coral Band: limestone, pale grey massive; corals including *Dibunophyllum bipartitum bipartitum, Diphyphyllum lateseptatum, L. pauciradiale, L. portlocki* .. .. 5
(u) Limestone, grey to pale grey, darker towards base, with irregular bedding and some shaly partings; sporadic shells and rare corals; *Lithostrotion junceum, Lonsdaleia floriformis laticlavia* .. .. .. 20

ft
(t) *Not exposed* (except some small exposures of rubbly grey-brown limestone in top 6 ft) .. .. 15
(s) Limestone, grey rather crinoidal in parts, rare chert lenses .. 12
(r) Limestone, dark to medium grey, much lenticular dark chert; sporadic shells and corals; *L. floriformis floriformis* .. 6
(q) Litton Tuff (see p. 75) .. .. —

The thickness of dark limestones in this group of strata is less than in the Littonfields Borehole and they decrease farther southwards and are difficult to recognize on the southern limit of the district west of Wardlow Hay.

The beds above the horizon of the Hob's House Coral Band are exposed in the north-eastern part of the dale (loc. 83) [1736 7538]:

ft
Eyam Limestones (see p. 102) [top of section 1788 7550] .. .. .. —
(z) Limestone, pale, part fine-grained, part slightly crinoidal; irregular fracture in lowest 4 ft .. .. 15
(y) Limestone, pale with irregular fracture; 2-in soft-weathering bands at top and bottom; band of broken compound corals 0 to 6 in at top; *D. lateseptatum, Lithostrotion portlocki* .. 0¾
(x) Limestone, pale .. .. 3 ft to 5
(w) Limestone, pale rather fine-grained, with some dark chert lenses; a 3-in chert band at 1¾ ft above base .. .. .. 5½
(v) Limestone, pale; irregular fracture in top 6 in and lowest 2½ ft; a few corals in lowest 2½ ft; *L. junceum, L. pauciradiale* .. 11
(u) Limestone, pale, with much fine crinoid and shell debris, some chert in lowest 3 ft; scattered corals and thin band of corals 3 ft from base; *D. lateseptatum, L. pauciradiale, L. portlocki, Palaeosmilia regia* [horizon of *Orionastraea* Band] .. .. 9
(t) Limestone, pale fine-grained cherty, with nodules up to 1½ ft thick; top 2 ft with much fine shell and crinoid debris .. 16

|  |  | ft |
|---|---|---|
| (s) | Limestone and clay, mixed, forming a rubbly soft-weathering band | 0⅓ |
| (r) | Limestone, pale; lower 3 ft weather differentially giving a hollow; *Lonsdaleia duplicata duplicata* | 6 |
| (q) | Limestone, pale cherty, with scattered shells; bands of shells in lowest 8 ft; an impersistent band up to 9 in thick with *Lithostrotion portlocki, Nemistium edmondsi, P. regia* and other corals at 20 ft from base [horizon of *Lonsdaleia duplicata* Band] | 27 |
| (p) | Limestone, pale fine-grained | 1¾ |
| (o) | Limestone, pale, with some chert and a few shells; common silicified corals; *Lithostrotion portlocki, Lonsdaleia floriformis floriformis* | 1½ |
| (n) | Limestone, pale; scattered corals | 10 |
| (m) | Limestone, pale medium- to coarse-grained, partially silicified and dark weathering; abundant small shells, some rolled; *L. floriformis floriformis* 2 ft to | 2½ |
| (l) | Black Bed; limestone, medium to dark grey; a few shells, some rolled; corals frequent in top 5 ft; *Lithostrotion pauciradiale, L. portlocki, Lonsdaleia floriformis floriformis* | 16 |
| (k) | *Beds mainly unexposed* (a few small exposures of pale limestone) | 10 |
| (j) | Limestone, grey, finely granular, crinoidal | 3 |
| (i) | Limestone, pale grey, finely granular, with silicified shells | 1½ |
| (h) | Limestone, pale | 1 |
| (g) | Limestone, brown, soft-weathering | 0¾ |
| (f) | Limestone, pale, very finely bioclastic; *Lithostrotion portlocki* and a few shells including '*Brachythyris*' *planicostata* | 8 |
| (e) | Limestone, pale, with some shell and crinoid debris; *L. martini, Nemistium edmondsi*; some bands of larger rolled shells in | |

|  |  | ft |
|---|---|---|
| | lowest 15 ft; top 5 ft oolitic in places | 26 |
| (d) | Limestone, grey-brown, soft weathering; some shells and crinoid debris | 2 |
| (c) | *Not exposed* | 1 |
| (b) | Limestone, grey; broken shell debris and coral fragments | 1 |
| (a) | Hob's House Coral Band: limestone, grey to dark grey, with a little chert, some argillaceous lenses; *Caninia benburbensis, Dibunophyllum bipartitum bipartitum, D. bipartitum konincki, Diphyphyllum lateseptatum, L. martini, L. pauciradiale, P. murchisoni, 'B.' planicostata* | 9½ |

The Hob's House Coral Band is underlain by 11 ft of pale massive limestone and then by 40 ft of dark limestones, cherty in part and containing the Upper *Girvanella* Band of locality 81 at their base.

Shirley and Horsfield (1945, p. 297) considered that a "bed 4 feet thick, with weathering along irregular small joints" at "about 14 feet" below the Eyam Limestones represented the "Hackly Bed" and *D.* aff. *muirheadi* Bed of Middleton Dale; this correlation is rejected here (see Plate VII).

On the eastern side of Cressbrook Dale a crag (loc. 84) [1737 7494] yielded a fish-tooth, *Psephodus?*, from about 5 ft below the Hob's House Coral Band.

*Litton and Hucklow Moor.* In this area the Upper Monsal Dale Beds occupy a broad tract with few large exposures. The subdivision into a lower group of dark limestones and a higher group of pale limestones made in Cressbrook Dale (see p. 83) still holds, and the outcrops of these subdivisions are considered separately below. Nearly all of the Upper Monsal Dale Beds was passed through in the Littonfields Borehole and all of the group in two percussive boreholes[1] also made by the Derwent Valley Water Board, near Great Hucklow. The record of Littonfields Borehole (loc. 85) [1751 7595] may be summarized:

---

[1] Information from examination of cutting samples by Professor F. W. Shotton.

|  | Thickness | Depth |
|---|---|---|
|  | ft in | ft in |
| No core .. .. .. | 11 4 | 11 4 |
| Limestone, pale grey cherty, with corals; (c) *L. junceum* .. | 1 3 | 12 7 |
| Limestone, pale grey granular; some crinoid debris to 19 ft; (b) *L. portlocki* at 18 ft 3 in .. | 16 3 | 28 10 |
| Limestone, pale cherty; (a) *A. insculpta?* and and '*B.*' *planicostata* at 31 ft .. .. .. | 12 2 | 41 0 |
| Limestone, pale fine-grained, some patchy replacement by chert .. | 5 0 | 46 0 |
| Limestone, pale cherty .. | 57 0 | 103 0 |
| Limestone, pale .. .. | 8 9 | 111 9 |
| Calcilutite, dark (Black Bed) .. .. .. | 1 10 | 113 7 |
| Limestone, pale (little core recovered) .. | 76 5 | 190 0 |
| Limestone, dark (little core recovered); horizon of Upper *Girvanella* Band at base .. | 59 0 | 249 0 |
| Lower Monsal Dale Beds (see p. 69) .. .. | — — | — — |

The section compares closely with that of the Wardlow Mires No. 1 Borehole. The lowest item is the lower, dark subdivision. As the dark bed at 113 ft 7 in is correlated with the Black Bed of Middleton Dale, the contention of Shirley and Horsfield (1945, p. 297) that the uppermost 40 ft of the Shining Cliff succession are missing in Cressbrook Dale cannot be accepted.

At outcrop the lower group of dark limestones, about 60 ft thick, is traceable over the whole of the tract. Exposures are mainly of the lowest beds in the feature of Litton Edge above the Litton Tuff. The lithology varies from grey well-bedded limestones [1659 7538] to dark crinoidal and shelly limestones [1605 7583] and dark cherty limestones [1590 7644]. Exposures of these beds can be seen in open workings along Tideslow Rake [1520 7798 to 1564 7792] showing dark thin-bedded cherty limestones with some crinoidal and shelly bands. Farther north a section (loc. 86) [1512 7940] on the north side of Shuttle Rake shows beds near the top of the dark limestone group:

G

|  |  | ft |
|---|---|---|
| (c) | Limestone, dark thin-bedded cherty .. .. .. .. | 12 |
| (b) | Limestone, grey shelly, with some chert; some crinoid debris; shelly fauna includes *E. lobata* aff. *laqueata*, *G. crassiventer*, *G. edelburgensis*, *P. productus hispidus* .. .. .. | 3½ |
| (a) | Limestone, medium to dark grey poorly bedded fine-grained cherty .. .. .. .. | 9½ |

The upper pale beds are cherty at the top and mainly non-cherty below. For the most part, good sections are rare in this area, the best being near the top of the Monsal Dale Beds in the small dale extending north-westwards from the site of the Littonfields Borehole. A section here [1717 7639] is as follows:

|  | ft |
|---|---|
| Limestone, grey well-bedded cherty rubbly-weathering .. .. .. | 3 |
| Limestone, grey massive, fine-grained, abundant chert; some wedge-bedding .. .. .. .. | 18 |
| Limestone, grey massive .. .. | 4 |

The lower part of the pale beds is less well exposed. The lithology varies from grey limestone with rolled shell debris [1625 7750] to grey well-bedded limestone [1677 7632]. Farther north, old open workings on High Rake show a number of sections, mainly on the northern (downthrow) side exposing grey cherty limestones in the upper part of the pale beds [1673 7785; 1632 7778]. To the north of Tideslow Rake [1573 7825] a section shows some of the lower part of the pale beds: grey granular limestone with bands of shells, some rolled, 11 ft.

Beds close below the Eyam Limestones are seen in a section (loc. 87) in a dry valley south-east of Little Hucklow [1650 7813]:

|  |  | ft |
|---|---|---|
| (l) | Limestone, dark, some shells; a little chert in lowest 1 ft .. | 8 |
| (k) | Limestone, grey, weathering nodular and soft .. .. | 1½ |
| (j) | Clay, grey, on prominent bedding plane .. .. .. .. | thin |

ft

(i) Limestone, grey, with a little chert; nodular weathering at base; poor clisiophylloids and *L. junceum* in top 2 ft; *Caninia juddi, D. bipartitum* [juv.], *L. junceum* at base .. .. 10

(h) Clay parting, impersistent, thin .. —

(g) Limestone, grey; a 6-in bed in middle, with thin clay partings above and below; weathers soft and nodular .. .. .. 12½

(f) Not exposed .. .. .. 1

(e) Limestone, grey .. .. .. 3

(d) Not exposed .. .. .. 1

(c) Chert, pale .. .. .. 1

(b) Limestone, grey, fine-grained .. 6

(a) Limestone, grey fine-grained cherty, with some crinoid debris 9

The presence of dark limestones, high in the Monsal Dale Beds, should be noted here; their incoming recalls a similar development at the north end of Bradwell Dale (see p. 91). The coral band is the *Orionastraea* Band (Shirley and Horsfield 1945, plate xviii).

To the north of Shuttle Rake the upper pale limestones bear some chert in their lower part, for example near Berrystall Lodge [1508 7989]. Higher beds are well seen in the upper part of Intake Dale [1614 7949] as pale cherty limestones. A section (loc. 88) in a small dale joining Intake Dale [1626 7986] shows the *Orionastraea* Band:

ft

(i) *Orionastraea* Band: limestone, pale, in three thin beds; *L. portlocki, O. placenta* and *Pugilis pugilis* .. .. .. .. 2

(h) Limestone, pale massive .. 2¾

(g) Not exposed .. .. about 7

(f) Calcilutite, brown-white, with chert up to 11 in at top .. 6

(e) Not exposed .. .. .. 3

(d) Limestone, pale irregularly jointed ('hackly') .. .. 4¾

(c) Limestone, pale massive fine-grained .. .. .. .. 3

(b) *Lonsdaleia duplicata* Band: limestone, with *D. bipartitum bipartitum, L. duplicata duplicata, L. floriformis floriformis* .. .. 1

(a) Limestone, pale massive fine-grained .. .. .. .. 2

The section is important as being the locality where the relationships of the *Orionastraea* and *L. duplicata* bands can be best established (see also Shirley and Horsfield 1940, p. 277). The occurrence here of *L. floriformis floriformis* in the *L. duplicata* Band is exceptional (see p. 138).

On the southern side of Intake Dale the relation of the *Orionastraea* Band to the Eyam Limestones is shown in the following section [1635 7966]:

ft

Eyam Limestones (knoll-reef, see p. 103) .. .. .. .. .. —

Limestone, pale grey massive fine-grained .. .. .. .. 8

Limestone, near-white fine-grained, with some chert in lower part .. 10

Limestone, pale brown crinoidal .. 2

Limestone, pale massive, finely crinoidal in places .. .. .. 16

*Orionastraea* Band: limestone, sorted, fine and coarser shelly and crinoidal, with scattered corals .. .. 6

Limestone, pale cherty finely crinoidal, with some shells .. .. .. 4½

Limestone, fine-grained cherty .. 11

The *Orionastraea* Band has evidently been subjected to considerable winnowing by current action here.

To the north of Intake Dale the *Orionastraea* Band has been traced to Earl Rake where an exposure (loc. 89) [1619 8007] yielded *Lithostrotion portlocki* (a common associate of *Orionastraea*). Farther north a series of exposures of high Monsal Dale Beds is provided by Hartle Drake and Jenning's Dale; these show pale limestones, cherty in places and with some crinoidal bands. An exposure (loc. 90) [1616 8046] at the western end of Hartle Dale is as follows:

ft

(e) Limestone, pale massive, some chert .. .. .. .. 1

(d) Limestone, as above, with corals, disappearing laterally; *L. junceum, L. pauciradiale* .. .. 1

(c) Limestone, pale massive, with a little chert; some shells at base and at 10½ to 14 ft above base 15

(b) Limestone, fine-grained, pale grey, with silicified productoids 1½

(a) Limestone, pale grey massive fine-grained, with a little chert .. 4¾

The coral band was considered by Shirley and Horsfield (1940 plate xviii), to be the *Orionastraea* Band but it is certainly at a lower horizon and may represent the *Lonsdaleia duplicata* Band.

*Bradwell Moor.* Over much of this area exposure is poor. On the western side the presence of some dark limestones at the base of the subdivision is indicated by scattered exposures. These beds, with the Upper *Girvanella* Band, are well seen on the northern side of Dirtlow Rake (loc. 91) where a section [1449 8155] in old open workings south-east of Hurdlow Barn shows the following:

|  |  | ft | in |
|---|---|---|---|
| (*f*) | Limestone, dark to grey slightly bituminous .. .. .. | 3 | 0 |
| (*e*) | Limestone, grey crystalline; *Antiquatonia hindi, A. sulcata, Eomarginifera lobata* aff. *laqueata, E. tissingtonensis cambriensis, G. edelburgensis, G. sp.* aff. *moderatus* | 6 | 6 |
| (*d*) | Clay parting .. .. up to | 2 |  |
| (*c*) | Limestone, grey finely crinoidal, abundant '*Girvanella*' nodules in lowest 1 ft .. | 5 | 0 |
| (*b*) | Limestone, grey fine-grained; abundant '*Girvanella*' nodules .. .. .. .. | 1 | 2 |
| (*a*) | Limestone, pale grey .. .. | 2 | 9 |

The Upper *Girvanella* Band, 2 ft thick, is also exposed (loc. 92) [1484 8186] at a point a little north of Dirtlow Rake. In the area north of Dirtlow Rake, the band is succeeded by some 20 ft of dark limestones and these by a small knoll-reef [145 817].

The beds between the lowest dark limestones and the *L. duplicata* Band are little exposed, and knowledge of them is largely from boreholes. A detailed log of a typical hole (No. BM 27) and a partial log of another (No. BM 28) have already been published (Eden and others 1964, pp. 107–10) and these and other typical records are shown graphically in Plate VIII. An exposure (loc. 93) [1424 8045] about 100 ft above the base in the central part of Bradwell Moor showed 19 ft of pale limestone with shelly and crinoidal bands and large chert masses in the lowest 4 ft; the fauna included *G. crassiventer, G. dentifer* and *G. giganteus* cf. *crassus.*

Beds from the *L. duplicata* Band upwards form an oval outcrop in the central area of Bradwell Moor. The *L. duplicata* Band has been proved in Borehole BM 32B [1475 8121] (see Appendix I, p. 362) and lies some 190 ft above the Upper *Girvanella* Band. It is exposed (loc. 94) [1471 8114] in an old quarry:

|  |  | ft |
|---|---|---|
| (*b*) | Limestone, pale fine-grained .. | $2\frac{3}{4}$ |
| (*a*) | Limestone, pale; *Diphyphyllum furcatum, Lithostrotion portlocki, Lonsdaleia duplicata duplicata, Nemistium edmondsi,* brachiopods and rare trilobites | $2\frac{1}{2}$ |

On the eastern side of the outcrop the *L. duplicata* Band is again exposed (loc. 95) [1525 8123] and Eden and others (1964, pp. 99, 106) list *L. duplicata duplicata* and brachiopods from this locality. Near this exposure a knoll-reef is developed just above the *L. duplicata* Band.

Farther east Shirley and Horsfield (1940, p. 281) noted the presence of the *L. duplicata* Band near Bird Mine. The following section (loc. 96) [1565 8118] was measured during the present work:

|  |  | ft |
|---|---|---|
| (*c*) | Limestone, pale coarsely crinoidal | $1\frac{1}{4}$ |
| (*b*) | *L. duplicata* Band: limestone, crinoidal and cherty; *D. lateseptatum, L. duplicata* [rare], *P. regia, A. insculpta, Productus productus hispidus, Pugilis pugilis* .. .. .. .. | $2\frac{3}{4}$ |
| (*a*) | Limestone, pale fine-grained cherty .. .. .. .. | 5 |

To the north of Bradwellmoor Barn an area nearly $\frac{1}{4}$ mile wide and extending for $\frac{1}{2}$ mile parallel to Dirtlow Rake on its south side is strewn with large blocks of silicified limestone. The largest of these [1543 8199] reaches a thickness of 15 ft and shows traces of purple fluorite. The disposition of the blocks makes it clear that they result from the erosion of a horizon which has suffered preferential silicification adjacent to Dirtlow Rake. The horizon was probably not far from that of the *L. duplicata* Band (see also p. 313).

*Earle's Quarry.* In the quarry the lowermost 100 ft or so of the Upper Monsal Dale Beds with the Upper *Girvanella* Band are present. These beds are predominantly dark limestones (Eden and others 1964, pp. 85–6) with six well-defined but impersistent paler calcirudite bands up to 15 ft in thickness. The calcirudite bands in places pass laterally into knoll-reefs in the vicinity of which they are at their coarsest; towards the reef-belt these beds show an increase in the rounding of their constituents and in extreme cases constitute rolled-shell limestones. The dark limestones also show some lateral variation, becoming more crinoidal towards the reef-belt.

The section [1592 8206] on the Upper Bench (1962) of the quarry (loc. 97) on the northern side of the large knoll-reef is described below (see also Eden and others 1964, pp. 99–103, 106):

|     |     | ft |
|-----|-----|----|
| (j) | Limestone, dark cherty (thinning and passing over top of knoll-reef) .. .. .. .. | 25 |
| (i) | Limestone, dark cherty (passing into top part of knoll-reef) .. | 10 |
| (h) | Limestone, fine-grained pale cherty, with coarsely crinoidal bands; *Palaeosmilia murchisoni*, abundant brachiopods including *G.* cf. *dentifer*, *Productus productus hispidus* (passing first into grey to pale grey non-cherty limestone before merging with the knoll-reef) .. | 8 |
| (g) | Limestone, pale rather crinoidal; large chert nodules at base; passing southwards first into thin-bedded dark limestones and then into knoll .. .. | 10 |
| (f) | Limestone, medium to dark grey crinoidal, with partings of dark shale (position of base of large knoll) .. .. .. .. | 1 |
| (e) | Limestone, medium to dark grey, lowest 6 ft cherty in northern part of section; some shells in lowest 6 ft .. .. .. | 11 |
| (d) | Limestone, dark to medium grey thin-bedded crinoidal; bands of dark chert nodules in top 5 ft; occasional clay partings .. | 11 |
| (c) | Clay parting .. .. about | 0¼ |
| (b) | Limestone, grey .. .. .. | 4½ |

|     |     | ft |
|-----|-----|----|
| (a) | Upper *Girvanella* Band: limestone, dark to medium grey, with some dark chert, sporadic shells; lowest 2½ ft rather crinoidal with dark shale partings and some breccia fragments near the reef margin; abundant *Girvanella staminea* .. .. | 10 |

The large knoll-reef [1576 8180] referred to (Eden and others 1964, pp. 87–9) as the "main knoll" is the equivalent of beds (f) to (i) inclusive, reaches a thickness of 55 ft and appears to result from the combination of two horizons of minor knolls, the lower some 30 ft above the Upper *Girvanella* Band, the higher some 50 ft above it. It is composed of unbedded vughy calcilutite with intercalations of dark calcarenite at the edges. Some breccia and a little crinoid debris are present. The fauna of the knoll (loc. 98) includes the goniatite *Bollandites sp.* in addition to a rich brachiopod fauna; *Palaeosmilia murchisoni* with *P. regia* is abundant in a 6-in band near the top. Another knoll [1582 8171] of large dimensions is present 100 yd S.E. of the "main knoll" and lies at the same horizon.

At the south-eastern end of the quarry [163 818] a marked unconformity develops at the base of the Upper *Girvanella* Band as the tuff mound (see p. 74) is approached.

*Smalldale and Bradwell.* To the south-east of Earle's Quarry the relations of the lower part of the Upper Monsal Dale Beds are complicated by the lateral passage of pale limestones into dark thin-bedded limestones near the margin of the basinal area (see also p. 91). This phenomenon has already been noted in the Lower Monsal Dale Beds. The two subdivisions are indistinct as a result.

The greater part of the beds described in the quarry passes rapidly southwards into coarsely crinoidal limestones. A borehole, NC1 [1627 8152] proved these limestones to a depth of 87 ft, underlain by dark crinoidal limestones to 99 ft and grey and dark grey bioclastic limestones with 'Girvanella' (the Upper *Girvanella* Band) to 106 ft. The pale crinoidal limestones, although not completely reef-like, show some affinities with a large flat reef in their development. A typical exposure [1648 8132] at Smalldale Head shows 12 ft of pale coarsely crinoidal limestone on 5 ft of pale cherty limestone. The crinoidal limestones extend southwards to, and a little beyond, Moss Rake, though displaced by the latter. At Outlands Head a quarry [1661 8080] shows 30 ft of pale shelly and often crinoidal limestone. South of Outlands Head the top of the crinoidal beds must lie not far below the *Orionastraea* Band.

Towards the margin of the limestone outcrop, pale cherty limestones down to about 100 ft below the *Orionastraea* Band persist locally on the south side of Bradwell. To the south-east, these limestones can be observed to pass into dark limestones and the junction with dark limestones in Bradwell village may also be diachronous.

*Bradwell Dale.* The main features of the stratigraphy of this dale are summarized in Fig. 11. Two detailed sections are quoted below and other sections are shown graphically in the figure. Variations in the succession are discussed below.

The most complete section measured (loc. 99) is near the southern end of the dale[1] [1714 8039]:

| | ft |
|---|---|
| Eyam Limestones (see p. 104) .. | — |
| (*f*) Limestone, pale fine-grained, very finely crinoidal; some wedge-bedding near top; colony of *Lithostrotion decipiens* in position of growth at 9 ft from base .. .. about | 12 |
| (*e*) Limestone, pale irregularly bedded cherty, with a little crinoid debris; 9-in band 1¼ ft from top with *Lithostrotion junceum*, *Lonsdaleia duplicata* and productoids .. .. | 22 |
| (*d*) Limestone, pale shelly crinoidal, in three 8- to 10-in beds; *Antiquatonia* cf. *antiquata*, *A. insculpta*, *Avonia thomasi*, *A. youngiana*, *E. lobata laqueata*, *Semiplanus sp. latissimus* group | 2¼ |
| (*c*) *Orionastraea* Band: limestone, pale cherty; *Lithostrotion decipiens*, *L. junceum*, *O. placenta*, *Antiquatonia hindi*, *A. insculpta*, '*B.*' *planicostata*, *Krotovia spinulosa* .. .. .. | 3¾ |
| (*b*) Limestone, pale massive fine-grained cherty, in 3- to 4-ft beds; scattered crinoid debris, finely crinoidal in lowest 2 ft .. | 24 |
| (*a*) Limestone, pale grey-brown crinoidal (some columnals up to 1 in diameter), with bands of pale chert, well-bedded; scattered productoids .. .. | 12 |

At the northern end of Bradwell Dale the following section (loc. 100) [1741 8077] was measured:[2]

| | ft |
|---|---|
| Eyam Limestones (see p. 104) .. | — |
| (*k*) Limestone, pale thin-bedded; *A. insculpta* .. .. .. | 1 |
| (*j*) *Orionastraea* Band: limestone, pale thin-bedded, with a little chert; *Clisiophyllum keyserlingi*, *Dibunophyllum bipartitum bipartitum*, *Diphyphyllum lateseptatum*, *L. decipiens*, *L. junceum*, *L. portlocki*, *Palaeosmilia sp.* .. .. .. | 3¾ |
| (*i*) Limestone, pale cherty, with scattered crinoid stems and shells | 4¾ |
| (*h*) Limestone, grey-brown finely crinoidal, with chert .. .. | 4¼ |
| (*g*) Limestone, pale irregularly-bedded cherty, with some coarse crinoid debris; productoids .. .. .. .. | 5¼ |
| (*f*) Limestone, pale massive fine-grained, with some chert; limestone pebbles up to ¾ in diameter; *G. edelburgensis*, *G. sp.* cf. *bisati* .. .. .. | 5¾ |
| (*e*) Limestone, pale grey evenly bedded cherty (especially in lowest 6 ft), finely crinoidal in places | 13¾ |
| (*d*) Limestone, dark to medium grey thin-bedded cherty, with fine wavy banding .. .. .. | 8¾ |

[1]"Morton's quarry" of Shirley and Horsfield.

[2]The lettering of beds on this implies no correlation with the preceding section.

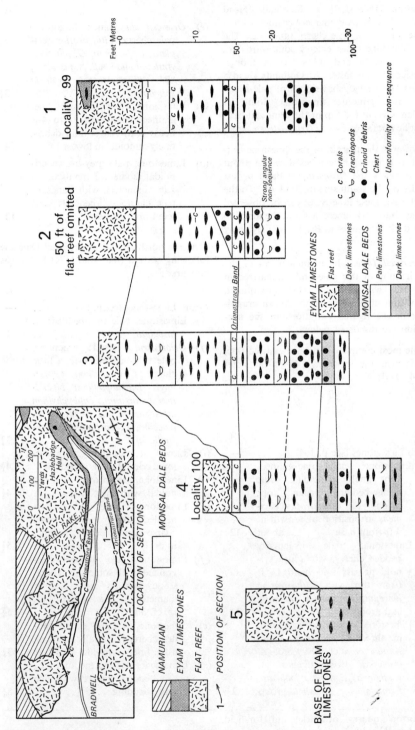

Fig. 11. Sections illustrating the relationship of Eyam Limestones and Upper Monsal Dale Beds in Bradwell Dale.

(L 179)

PLATE IX

A. Bradwell Dale: Upper Monsal Dale Beds and Eyam Limestones

B. Furness Quarry, Middleton Dale: Eyam Limestones resting on
Upper Monsal Dale Beds

(L 181)

ft

(c) Limestone, pale massive, with coarse crinoid debris, cherty in lowest 2½ ft .. .. .. 6

(b) Limestone, medium to pale grey massive cherty, crinoidal; chert in lenses up to 2 ft thick; giganto-productoids at top and base and *G. crassiventer* at 4 ft from top .. .. .. .. 17½

(a) Limestone, medium to dark grey thin-bedded and irregularly banded; much dark chert; very dark at base .. .. .. 5

The general features of the succession in Bradwell Dale may be summarized as follows:

1. Beds below the *Orionastraea* Band. These show two non-sequences. The lower is present locally in the northern half of the dale in the position of bed (*f*) of loc. 100 as a pebbly horizon some 20 ft below the *Orionastraea* Band; it can be seen as an angular break some 100 yd S. of loc. 100. A higher non-sequence, about 7 ft below the *Orionastraea* Band, is well seen (see Plate IXA) at the southern end of the dale [1722 8046]; beneath lie some 10 ft of massive crinoidal and shelly limestone, bed (*a*) of loc. 99. The lower beds are folded into an east–

west anticline, the upper 8 ft or so being cut out by the non-sequence.[1] At this locality the uppermost 3 ft of beds below the *Orionastraea* Band also bear corals.

To the north of loc. 100 the uppermost part of these beds, down to a horizon a few feet below the *Orionastraea* Band, is cut out progressively by the unconformity at the base of the Eyam Limestones. The underlying beds pass northwards into thin-bedded dark cherty limestones, which persist at some horizons, e.g. beds (*a*) and (*d*) of loc. 100, southwards for some distance into beds of normal shelf facies (see Fig. 11).

By comparison with the section in Intake Dale a sufficient thickness of strata is exposed for the *Lonsdaleia duplicata* Band to be present in Bradwell Dale. However, no corals have been seen at this horizon.

2. *Orionastraea* Band and overlying strata. At the southern end of Bradwell Dale exposures do not allow the joining of the outcrops of the marker-band on the two sides of the dale; otherwise it can be traced around the sides of the dale until it is cut out by the unconformity at the base of the Eyam Limestones. The beds above the *Orionastraea* Band consist of pale limestones, some cherty, with rather 'reef-like' shelly and crinoidal beds in places.

## CONCEALED MONSAL DALE BEDS

*Wardlow Mires, Great Hucklow and Eyam.* The Wardlow Mires No. 1 Borehole (loc. 101) [1850 7553] proved Monsal Dale Beds, overlain by Eyam Limestones, between 113½ ft and the base of the borehole at 632 ft 2 in. The correlation is illustrated in Fig. 12; the sequence is summarized below:

| | Thickness ft in | Depth ft in |
|---|---|---|
| Eyam Limestones (see p. 102) .. .. .. | 113 6 | 113 6 |
| Limestone, pale, cherty in places; *Orionastraea* Band 1 ft 7 in at 150 ft, with *Dibunophyllum sp.*, *Diphyphyllum sp.* and *Lithostrotion portlocki* | 57 2 | 170 8 |
| Calcilutite, cherty .. | 2 7 | 173 3 |

| | Thickness ft in | Depth ft in |
|---|---|---|
| Limestone, pale, with rare chert; *L. duplicata* Band at 193 ft .. .. | 68 5 | 241 8 |
| Limestone, dark fine-grained (Black Bed of Middleton Dale section) .. .. .. | 4 4 | 246 0 |
| Limestone, pale; *Lonsdaleia floriformis* at 249 ft 4 in; *Lithostrotion portlocki* at 252 ft 2 in | 47 9½ | 293 9½ |
| Calcilutite, buff .. .. | 2 0½ | 295 10 |
| Limestone, pale, with some crinoidal bands; partly oolitic from 356 ft 7 in to 361 ft 4 in; | | |

[1]The direction of folding and lithological similarity with the lowest beds of loc. 100 are taken to indicate that the structure is anticlinal rather than of knoll form.

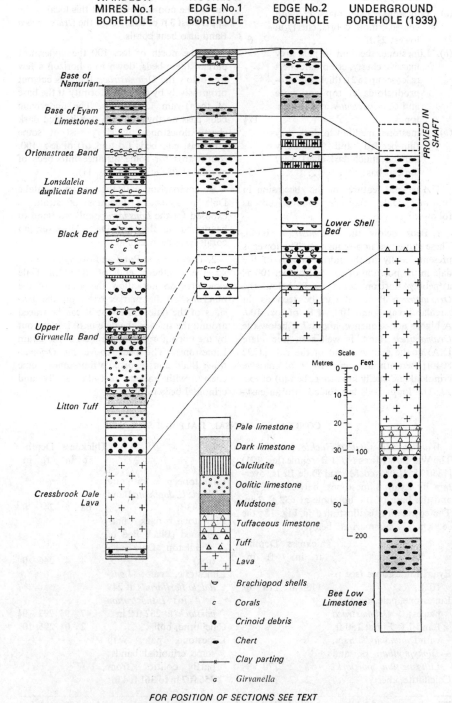

FIG. 12. *Sections of boreholes in the Monsal Dale Beds and associated strata*

| | Thickness | Depth |
| --- | --- | --- |
| | ft in | ft in |
| bioclastic and shelly at base .. .. .. | 68 10 | 364 8 |
| Limestone, dark, with 'Girvanella' (Upper Girvanella Band) .. .. | 5 4 | 370 0 |
| Limestone, pale crinoidal, shelly in places, partly oolitic near top .. | 12 10 | 382 10 |
| Limestone, pale predominantly oolitic, some chert in places, crinoidal at base .. | 44 9 | 427 7 |
| Limestone, partly crinoidal, with some chert; shelly in lower part .. | 7 1 | 434 8 |
| Limestone, dark cherty; Lingula in places; crinoidal below 446 ft 6 in; lowest ½ in tuffaceous .. .. .. | 14 9½ | 449 5½ |
| Tuff, grey fine-grained (Litton Tuff) .. .. | 2 6½ | 452 0 |
| Limestone, dark .. | 6 1½ | 458 1½ |
| Limestone, pale bioclastic and finely oolitic .. | 9 5½ | 467 7 |
| Limestone, dark .. | 5 5 | 473 0 |
| Mudstone, dark calcareous .. .. .. | 2 5 | 475 5 |
| Limestone, pale buff; crinoidal below 486 ft | 32 5 | 507 10 |
| Lava, mostly amygdaloidal but in places compact and non-vesicular (Cressbrook Dale Lava) .. .. .. | 105 5½ | 613 3½ |
| Tuff .. .. .. | 1 2½ | 614 6 |
| Limestone, grey to dark grey .. .. .. | 7 6 | 622 0 |
| Mudstone, dark calcareous .. .. .. | 3 | 622 3 |
| Limestone, grey to dark grey .. .. .. | 9 | 623 0 |
| Limestone, grey crinoidal in part .. .. .. | 3 6 | 626 6 |
| Limestone, dark bituminous .. .. .. | 2 6½ | 629 0½ |
| Mudstone, blue-grey .. | 7 | 629 7½ |
| Limestone, grey-brown fine-grained .. .. | 2 6½ | 632 2 |

The presence of oolites in abundance between the Upper Girvanella Band and the Litton Tuff is abnormal and has been discussed on p. 28. The absence of the greater part of the dark beds above the Upper Girvanella Band is remarkable in view of their persistence at outcrop. The recognition of the Black Bed (see p. 81) of the Middleton Dale section provides checks on the correlation of the higher coral bands. The correlation of coral occurrences indicates a section closely comparable to that in Furness Quarry (see p. 75) and disproves the contention of Shirley and Horsfield (1945, p. 297) that some 40 ft of the top beds of D₂ have been cut out in the nearby Cressbrook Dale section by the unconformity at the base of the Eyam Limestones (see Plate VII). Wardlow Mires No. 2 Borehole (loc. 102) [1825 7586] (see Appendix I, pp. 378–82) proved the base of the Eyam Limestones at 107 ft 9 in and pale D₂ limestones from thence to the bottom of the hole at 133 ft 6 in.

The colony of *L. portlocki* at 129 ft 3 in represents the coral band lying just above the *Orionastraea* Band in the Wardlow Mires No. 1 Borehole.

Hucklow Edge No. 2 Borehole (loc. 103) [2025 7760] north-east of Shepherd's Flat proved the following abridged section of the Monsal Dale Beds beneath the Eyam Limestones:

| | Thickness | Depth |
| --- | --- | --- |
| | ft in | ft in |
| Eyam Limestones (see p. 101); base at .. .. | | 344 6 |
| Limestone, pale fine-grained; with a little chert below 362 ft 11 in | 26 11 | 371 5 |
| Calcilutite, buff cherty .. | 7 10 | 379 3 |
| Limestone, pale, with a little chert; *D. bipartitum bipartitum, Diphyphyllum furcatum* and *L. portlocki* from 386 ft 3 in to 389 ft 1 in (*Orionastraea* Band) .. | 17 10 | 397 1 |
| Calcilutite, pale grey-white cherty .. .. | 9 4 | 406 5 |
| Limestone, pale, with finely crinoidal bands | 64 9 | 471 2 |
| Limestone, buff shelly, with silicified productoids (Lower Shell Bed of Middleton Dale) .. | 8 3 | 479 5 |
| Limestone, pale, with some finely crinoidal bands .. .. .. | 23 7 | 503 0 |

| | Thickness ft in | Depth ft in |
|---|---|---|
| Limestone, pale cri-noidal; Upper *Girvanella* Band in lowest 11 in .. .. .. | 1 11 | 504 11 |
| Limestone, buff, some-what crinoidal and shelly .. .. .. | 6 6 | 511 5 |
| Limestone, buff crinoidal bioclastic, shelly in lowest 1 ft 6 in .. | 21 7 | 533 0 |
| Limestone, grey-brown to buff .. .. .. | 8 10 | 541 10 |
| *No core* (recorded by driller as 'limestone') | 3 8 | 545 6 |
| Tuff, pale grey fine-grained, pyritous in part .. .. .. | 4 8 | 550 2 |
| Lava, mostly dark green amygdaloidal, non-vesicular in places, some purple mottling (Cressbrook Dale Lava) | 60 6 | 610 8 |

The Cressbrook Dale Lava lies considerably nearer the base of the Eyam Limestones than in the Wardlow Mires No. 1 Borehole (see Fig. 12) and the Upper *Girvanella* Band is less strongly developed. This latter is probably due to an adverse environment as the band is not normally developed in pale limestones. Many of the higher beds are readily corre-lated with the Middleton Dale section as indicated above. The Black Bed has, however, disappeared and the Upper Shell Bed has not been recognized. As in the case of the Middleton Dale sections chert is restricted in occurrence to the uppermost D₂ strata.

The Derwent Valley Water Board No. 7 (Great Hucklow) Borehole [1777 7762] at Great Hucklow, proved the following abridged succession[1]:

| | Thick-ness ft | Depth ft |
|---|---|---|
| Eyam Limestones (see p. 102) | — | 146 |
| Limestone, mainly pale grey cherty; darker coloured 262 to 269 ft .. .. .. | 123 | 269 |
| Limestone, pale to medium grey; some chert in lowest 30 ft .. .. .. | 61 | 330 |
| Limestone, dark cherty .. | 10 | 340 |
| Limestone, white to grey, cherty .. .. .. | 22 | 362 |
| Limestone, pale to dark grey cherty .. .. .. | 18 | 380 |
| 'Shale', ashy .. .. .. | 1 | 381 |
| Limestone, pale to dark grey | 12 | 393 |
| Ash, grey pyritous .. .. | 6 | 399 |
| Basalt, hematite-stained amyg-daloidal (Cressbrook Dale Lava) .. .. .. | 6½ | 405½ |

The Broad Low Borehole [1753 7821] north of Great Hucklow proved 257 ft of limestones below the Eyam Limestones without reaching igneous rock: pale cherty limestones 133 to 295 ft, pale to dark grey non-cherty limestones to 310 ft, medium to pale grey limestones with a little chert to 335 ft, grey to dark grey cherty limestones to bottom of hole at 390 ft. The beds from 295 to 335 ft represent the middle non-cherty subdivision and the beds below 335 ft most of the lower, dark part of the Upper Monsal Dale Beds. The Little Hucklow Borehole [1684 7863] proved Monsal Dale Beds from the base of the Eyam Limestones at 34 ft to the bottom of the hole at 270 ft.

In addition to the borehole records given above, igneous rocks, all probably the Cress-brook Dale Lava, have been proved in mining at three places in this area:

1. Milldam Mine [1766 7796], Great Huck-low (Pumping Engine Shaft); shale to 30 ft, limestone to 420 ft on "toadstone" (Green and others 1887, p. 136). The toadstone was stated to be amygdaloidal and so was certainly a lava. Of the overlying limestones, a comparison with the Great Hucklow Bore-hole section suggests that the lower 300 ft should be referred to the Monsal Dale Beds and the remainder to the Eyam Limestones.

2. Dusty Pit Mine [2075 7707], Eyam, proved shale to 12 ft, limestone to 372 ft and amygdaloidal "toadstone" to 375 ft (ibid., p. 138).

3. Watergrove Sough entered 'toadstone' below the top end of Middleton Dale [197 762], some 220 ft below the base of the

---

[1]Percussive, information from examination of cuttings by Professor F. W. Shotton.

Eyam Limestones. The latter figure indicates a rapid thinning of the $D_2$ beds above the lava when compared with the Wardlow Mires No. 1 borehole section.

*Glebe Mine.* The Glebe Shaft [2192 7640], Eyam, commencing at the limestone-shale junction, continued in limestone to a depth of 283 ft. The estimated thickness of the Eyam Limestones in the vicinity is about 125 ft and from this it can be inferred that the bottom of the shaft is at a horizon a little below that of the Lower Shell Bed. In the main (280-ft level) crosscut linking Glebe and Ladywash shafts, the Lower Shell Bed ("Lower *Giganteus* Bed") was considered to be present by Shirley (*in* Dunham 1952, p. 86) as a band 8-ft thick in that part of the main crosscut which follows Ashton's Pipe. The main concentration of shells is in a 2-ft band within the thicker and more sparsely shelly bed. Between the junction with Ashton's Pipe and Glebe Shaft the crosscut follows the bedding. A more detailed section by the junction with Ashton's Pipe [2181 7667] is as follows:

|  | ft | in |
|---|---|---|
| Limestone, dark rubbly, with a prominent bedding-plane .. .. | 10 | |
| Limestone, as below, scattered shells | 3 | 9 |
| Limestone, grey to buff, finely granular; shelly, especially in top 1 ft; cerioid *Lithostrotion* at top .. | 2 | 0 |

The dark limestones at the top of this section are correlated with the Black Bed.

The following succession was proved below the Lower Shell Bed in a vertical borehole [2181 7663] "820 ft north of the shaft" (Dunham 1952, p. 86)[1]:

|  | ft |
|---|---|
| Limestone .. .. .. .. | 102 |
| Toadstone .. .. .. .. | 250 |
| Limestone .. .. .. .. | 33 |
| Hard toadstone .. .. .. | 5 |
| Limestone .. .. .. .. | $31\frac{1}{2}$ |

The top of the toadstone appears to be some 50 ft higher stratigraphically than in the Hucklow Edge No. 2 Borehole.

Successively higher $D_2$ strata are present in the crosscut as it is followed north-eastwards from Ashton's Pipe. These beds

are described by Dunham (ibid., p. 86) as "calcite mudstones and medium-grained limestones, cherty at the top". A recent examination shows that calcilutites are restricted to an 8-ft cherty bed [2194 7696] some 530 ft south of Old Edge Vein. The calcilutite is overlain by the Eyam Limestones, at the base of which a 2-in to 6-in shale parting is present. A thin clay way-board is also present beneath the calcilutite. The bed is here equated with the White Bed of Middleton Dale and the correlation is strengthened by the disposition of chert which is present above the White Bed but absent after a few feet below the bed (cf. Middleton Dale section p. 77).

From the last locality Eyam Limestones are present in the crosscut until a point 120 yd north of Old Edge Vein [2203 7721] is reached (see also p. 99). To the north of the latter tract $D_2$ beds reappear and the crosscut continues in these to the vicinity of Lady-wash Shaft. The *Orionastraea* Band was found by Shirley (*in* Dunham 1952, p. 86) at the point (loc. 104) [2182 7743] where the branch to Ladywash Shaft leaves the main crosscut. The band, 2 ft 6 in thick and in pale cherty limestone, yielded the following fauna: *Diphyphyllum gracile*, *L. junceum*, *L. portlocki* and *O. placenta*.

A borehole [2184 7756] near Ladywash Shaft (Schnellman and Willson 1947, p. 5) put down in 1939 at 455 ft O.D. from the 280-ft crosscut level proved:

|  | Thick-ness ft | Depth ft |
|---|---|---|
| Medium-coloured limestone, cherty in upper part, fossiliferous in lower .. .. | 108 | 108 |
| Brecciated limestone .. | 4 | 112 |
| Lava .. .. .. .. | 208 | 320 |
| Irregular beds of pyritic tuff and wayboards, alternating with beds of crinoidal limestone .. .. .. | 35 | 355 |
| Medium-coloured bedded crinoidal limestone .. .. | 99 | 454 |
| Black limestone, very fine-grained, some chert .. | 40 | 494 |
| Light brown to light grey crystalline limestone .. .. | 72 | 566 |

[1] On the crosscut and N.23°W. of the shaft.

The dark limestones from 454 to 494 ft are equated here with the dark beds at the base of $D_2$ and the underlying pale limestones are referred to $D_1$.

*Middleton Dale.* The presence of an unexposed vesicular basalt was noted by Morris (1929, pp. 42, 44) at Hanging Flat at a horizon some 100 ft below the "Lower *Giganteus* Bed" (Lower Shell Bed). More recently the basalt has been proved in a borehole by the Derwent Valley Water Board at this locality [2052 7600][1]:

| | Thick-ness ft | Depth ft |
|---|---|---|
| 'Clay with stones' .. .. | 9 | 9 |
| Limestones, light grey; cherty in lowest 5 ft .. .. | 39½ | 48½ |
| Limestone, grey and brown crinoidal .. .. .. | 15 | 63½ |
| Limestone, dark bituminous some shale in lower part .. | 25 | 88½ |
| Tuff, grey calcareous and pyritous .. .. .. | 5 | 93½ |
| Basalt, amygdaloidal except between 223 and 234 ft and below 249 ft .. .. | 164 | 257½ |

# EYAM GROUP ($P_2$)

## EYAM LIMESTONES

*Calver and Coombs Dale.* Between Calver and Calver Peak the Eyam Limestones occupy a broad sloping tract approximating to a dip-slope. They are dark limestones with extensive but apparently disconnected masses of flat-reef limestones at Calver Peak and in Calver Peak Quarries (see below). At Calver Peak, the flat reef is capped by a smaller mass of reef-limestone showing knoll form; this is frequently shelly and also shows an irregular patchy silicification, developed over large areas of the rock.

The Eyam Limestones were formerly much quarried at Calver Peak Limekilns. A section (loc. 105) of the higher beds [2358 7466] is:

ft

(*d*) Limestone, pale grey thin-bedded fine-grained .. .. about   8

(*c*) Limestone, pale grey, with several thick shell bands with *E. lobata* aff. *laqueata, G. edelburgensis, Pugilis pugilis*; a band about 2 ft from base with *Lithostrotion junceum, Lonsdaleia floriformis floriformis, Palaeosmilia regia* ..   25

ft

(*b*) Limestone, pale grey to grey-white, very crinoidal, with scattered pockets of shells; *Antiquatonia hindi wettonensis, Avonia youngiana, E. lobata* aff. *laqueata* .. .. ..   23

(*a*) Limestone, dark grey crinoidal, paler coloured at top; scattered chert nodules, a few shells   4½

The above beds form a transition between the normal dark facies of the Eyam Limestones and the easternmost of the two large flat reefs noted above. A complete lateral passage into reef-limestone takes place within some 50 yd to the south-east. In the middle of the quarry (loc. 106) [2365 7459] the flat reef consists of about 75 ft of pale massive crinoidal and shelly limestone with some finer grained bands. The fauna includes *Alitaria panderi, Antiquatonia insculpta, Avonia youngiana, Echinoconchus punctatus, Eomarginifera lobata* aff. *laqueata* and *K. spinulosa.*

The lowest part of the Eyam Limestones is present in Coombsdale Quarry. A section [2330 7478] in the middle part of the quarry (loc. 107) is as follows:

---

[1]Information from examination of cutting samples by Professor F. W. Shotton.

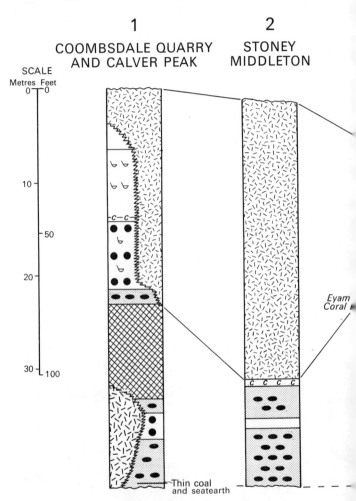

1
COOMBSDALE QUARRY
AND CALVER PEAK

2
STONEY
MIDDLETON

SCALE
Metres Feet

Eyam
Coral

Thin coal
and seatearth

*Positions of sections shown on Fig. 13*    Unconforr

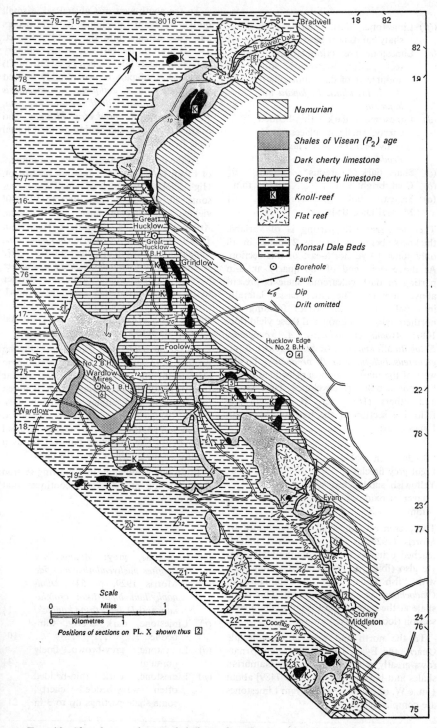

Namurian

Shales of Visean (P₂) age

Dark cherty limestone

Grey cherty limestone

K  Knoll-reef

Flat reef

Monsal Dale Beds

⊙  Borehole

Fault

↙₅  Dip

Drift omitted

Scale
Miles
0      1
0   Kilometres   1
Positions of sections on PL. X shown thus ☐2

FIG. 13. *Sketch-map showing lithological variation within the Eyam Limestones*

98 CARBONIFEROUS LIMESTONE SERIES (DINANTIAN)

ft

(f) Limestone, dark cherty, with
shaly bands near base .. .. 4½
(e) Limestone, grey crinoidal (knoll-
reef passing westwards into
about 2 ft of dark limestone);
A. cf. insculpta, E. lobata aff.
laqueata .. .. .. 8
(d) Limestone, dark thin-bedded
cherty, shelly in places, especially near top; A. cf. insculpta,
Productus productus .. .. 17
(c) Shale, dark calcareous .. about 0¼
(b) Coal, bright ½ in .. .. .. thin
(a) Seatearth, buff .. .. .. 0¼
Monsal Dale Beds .. .. —

The shale/seatearth parting is of variable
thickness, being as little as 2 inches in all
only some 10 yd north-east of this section.
At the eastern end of the quarry an 8-in
parting of dark calcareous shale is present
above item (e). The latter passes laterally
into pale rubbly limestones forming the
northern edge of a knoll and these yield (loc.
108): *Avonia* cf. *thomasi*, *A. youngiana*,
*E. lobata* aff. *laqueata*, *G. edelburgensis*, *G.* cf.
*giganteus inflatus* and *P. productus*. To the
west of the main section dark-bedded cherty
limestones only are developed. Here Green
and others (1887, pp. 21–2) noted the
following section at the base of the Eyam
Limestones:

ft in

"Bright clean coal .. .. .. 2
Light-grey fireclay .. .. .. 1 6
Yellowish sandy clay, with a thin
layer of oxide of iron at the bottom .. .. .. .. 3"

The presence of the coal was confirmed by
Morris (1929, p. 55) who stated that it
reached 6 inches in the Coombs Dale area.
He also (ibid., p. 49) recorded a thin shale
with "fish spines and an abundance of
*Cladochonus sp.*" above the coal. This is the
shale at the base of the Coombsdale Quarry
section (loc. 107) given above.

To the north and west of Coombsdale
Quarry the Eyam Limestones outcrop narrows greatly owing to overlap by Namurian
shales and beyond a point [2283 7489] about
¼ mile W. of the quarry the Eyam Limestones
are completely cut out.

To the south-west of Calver Peak a small
outlier of Eyam Limestones flat reef occurs
and part extends into the present district.
The section[1] on the east side of the tributary
to Coombs Dale (loc. 77, bed *g*; see also p. 82)
shows 5½ ft of pale brown roughly bedded
limestones with *A. antiquata*, *Echinoconchus
punctatus*, *Eomarginifera lobata laqueata*,
*K. spinulosa*, *P. productus hispidus* and
*S. sp. latissimus* group resting on the Monsal
Dale Beds.

To the north of Coombs Dale, two outliers
of Eyam Limestones flat reef are present on
High Fields. An area of small exposures
some 80 yd across (loc. 109) [2159 7490]
yielded *Antiquatonia hindi*, *A. insculpta*,
*Avonia davidsoni*, *A. thomasi*, *A. youngiana*,
*Dielasma hastatum*, *Echinoconchus punctatus*,
*Eomarginifera lobata* aff. *laqueata*, *Girtyella?
sacculus*, *Pleuropugnoides pleurodon*, *Pugnax
pugnus* and *Schizophoria resupinata*. Farther
to the west, a third outlier, on Moisty Knowl,
shows a small thickness of dark cherty limestones. An excavation [2090 7511] showed
3 ft of beds of this type.

*Middleton Dale and The Cliff*. On the
western outskirts of Stoney Middleton
[2269 7536] the Eyam Limestones appear
from beneath the Namurian cover. Between
Shining Cliff and Stoney Middleton dark
beds at the base form steep grassy or wooded
slopes above the cliffs of Upper Monsal
Dale Beds. The overlying flat reef forms a
feature on the northern lip of the dale.

In Stoney Middleton the following section
(loc. 76) [top at 2300 7550] continues that
given on p. 80:

ft

(s) Limestone, grey: *Aulophyllum
fungites pachyendothecum* (*fide*
Morris 1929, p. 54), *Dibunophyllum bipartitum craigianum* and *D. bipartitum konincki* 1¼
(r) Limestone, dark cherty, thin-
bedded .. .. .. .. 10
(q) Limestone, grey-brown finely
granular .. .. .. 3¼
(p) Limestone, dark thin-bedded
often wavy-bedded cherty,
some shale partings up to 3 in 21

---

[1]Top of section [2246 7422] just south of southern edge of district.

ft

(*o*) Clay parting on potholed surface
(potholes up to 1 ft deep, filled
with clay)  ..    ..  0 to   0¼
Monsal Dale Beds  ..   ..   —

The coral band lies at the base of the flat reef
and appears to be a local development only.

Farther west the Castle Rock section (loc.
75, see also p. 79) [top at 2245 7583] is continued upwards in the Eyam Limestones as
follows:

ft

(*dd*) Limestone, pale massive fossiliferous, with some coarse crinoid debris (flat reef); *A. hindi
wettonensis*, '*B.*' *planicostata*,
*P. productus hispidus* ..    ..  6
(*cc*) Limestone, pale, much pale chert;
few fossils  ..    ..    ..  6
(*bb*) *Not exposed* (flat reef)  ..    5
(*aa*) Limestone, dark (poorly exposed)
about  40
Monsal Dale Beds  ..   ..   —

To the north-east of this section flat-reef
limestones form an extensive outcrop with a
steep dip-slope, near the junction with the
Namurian; the dip may have a slight depositional element. The maximum thickness of
the flat reef is of the order of 100 ft. Farther
west, dark limestones of the Eyam Group
overlie the Monsal Dale Beds of the Shining
Cliff section.

In Furness Quarry the section (loc. 73)
[top at 2100 7608], continues that of p. 75:

ft

(*z*) Limestone, dark thin-bedded,
with shaly partings and thin
rather regular bands of chert;
gastropods and *Gigantoproductus spp.* ..    ..    ..    ..  14
(*y*) Clay, grey rusty-weathering (thinning to south) ..    .. ½ to   1¼
Monsal Dale Beds  ..   ..   —

*Eyam and Foolow.* The upper part of Eyam
Dale provides an excellent section (loc. 110)
of the Eyam Limestones [top at 2200 7634]:

ft

(*j*) Limestone, light grey thin-
bedded; *A.* cf. *insculpta*    ..  4

ft

(*i*) Limestone, dark thin-bedded
cherty; 3-ft coral band (Eyam
Dale Coral Band) at 6 ft from
top; *D. bipartitum bipartitum,
D. bipartitum craigianum, D.
bipartitum konincki*  ..    ..  10
(*h*) Limestone, grey to dark grey thin-
bedded cherty ..    ..    ..  1¼
(*g*) Limestone, dark grey to grey
massive slightly crinoidal, with
a little chert  ..    ..    ..  9½
(*f*) Limestone, brownish grey thin-
bedded ..    ..    ..    ..  4
(*e*) Limestone, dark fine-grained thin-
bedded; 9-in shelly band at 3¾
ft from base  ..    ..    ..  8
(*d*) Limestone, grey finely crinoidal    1¼
(*c*) Limestone, dark fine-grained
thinly and irregularly bedded,
cherty; *G. giganteus, G. sp.
edelburgensis* group  ..    ..  10
(*b*) Limestone, grey to pale grey
cherty; scattered shells and a
1-ft shell band at 1¼ ft from
base; *G. giganteus* cf. *crassus,
P. productus hispidus* ..    ..  6½
(*a*) Limestone, dark fine-grained
poorly bedded, with irregular
chert  ..    ..    ..    ..  9½

The lowest beds are some 10 ft above the base
of the Eyam Limestones while the highest
are a few feet below the base of the flat reef
(see p. 98). A good section of the latter is in
an old quarry [2208 7623]: pale grey even-
and thin-bedded fine-grained to finely
crinoidal limestone, more crinoidal and shelly
at base and with some chert, 10 ft.

To the north of Eyam, the Eyam Limestones have been found underground in two
places in a crosscut joining Glebe and
Ladywash shafts (Dunham 1952, p. 86). Pale
massive shelly and crinoidal reef (probably
flat reef) limestones at the base of the group
were passed through in the vicinity of the
intersection with the Old Edge Vein [2194
7696 to 2203 7721]; these are overlain by
dark cherty limestones south-east along Old
Edge Vein [2210 7704]. North of here a gentle
southerly dip takes the base of the Eyam
Limestones above the crosscut but they
reappear at the Ladywash end where, 38 yd
from the shaft, the roof of the crosscut shows
dark fine-grained cherty limestone with
limestone pebbles in lowest 2 to 3 in, resting

on a thin clay parting over pale fine-grained limestones of the Monsal Dale Beds (see p. 95). By Ladywash Shaft (loc. 111) [2189 7754] the lowest beds of the Eyam Limestones yielded *Avonia* aff. *davidsoni* and *P. productus hispidus*. In Eyam village the Eyam Limestones are about 125 ft thick and include about 40 ft of flat reef at the top. Underground, progressive erosion at the base of the Namurian occurs and at Ladywash Shaft some 30 ft only of dark thin-bedded limestones are present (Dunham ibid., p. 86).

To the west and south-west of Eyam the lower, dark limestones show a progressive increase in thickness. Correspondingly the base of the flat reef appears at higher and higher horizons and it disappears altogether about half a mile west of the village. All but the lowest beds of the dark limestone pass laterally into grey limestones, the passage taking place along a sinuous line running in a general south-westerly direction from Eyam View. Small knoll-reefs are present at two horizons, the lower at the base of the Eyam Limestones near Hanging Flat [203 761] and on the west side of Cucklet Dale [215 761]. The higher knolls, four in number, lie in the grey limestones between Linen Dale and Eyam View. Their disposition shows a close geographical relationship to the flat reef.

Linen Dale shows a long section of part of the lower dark limestones overlain by grey limestones. The upper end of the dale provides a cross-section of a knoll whose base lies in the dark beds about 6 ft below the gigantoproductoid band noted below, but which extends upwards into the grey limestones. This knoll was first described by Morris (1929, pp. 49–50 and 52). A section (loc. 112) [1984 7688] near Waterfall Farm showed 25 ft of pale massive shelly and crinoidal reef-limestone with *Alitaria panderi, Antiquatonia hindi, A. hindi wettonensis, A. insculpta, A.* cf. *sulcata, Avonia thomasi, A. youngiana, Echinoconchus punctatus, Eomarginifera lobata* aff. *laqueata, Girtyella? sacculus* and *K. spinulosa*. The total height of the knoll is about 35 ft. On the northern flank of the knoll (loc. 113) [1989 7689] the following beds are exposed:

                                    ft

(*e*)   Limestone, grey to dark grey finely granular    ..    ..    4

                                    ft

(*d*)   *Not exposed*    ..    ..    ..    4

(*c*)   Limestone, pale grey-brown, thinly but irregularly bedded; *Alitaria panderi, Antiquatonia* cf. *insculpta, Avonia thomasi, A. youngiana, Pugnax pugnus* [small form]    ..    ..    3

(*b*)   Limestone, grey-brown fine-grained    ..    ..    6½

(*a*)   Limestone, grey; abundant large shells; *A.* cf. *hindi, Gigantoproductus edelburgensis, G. sp.* cf. *gigantoides*    ..    ..    5

The lowest item passes laterally into the knoll; the higher beds appear to thin and to pass upwards over it, though some lateral passage into the knoll in the lowest part of these beds cannot be excluded.

A more extensive section (loc. 114) of the beds above the gigantoproductoid bed (*a*) was measured [1988 7693] on the western side of Linen Dale:

                                    ft

(*g*)   Limestone, dark grey thin-bedded fine-grained, with bands of chert up to 11 in thick    ..    10½

(*f*)   Limestone, grey irregularly bedded fine-grained; some chert in middle and at top    ..    6

(*e*)   Limestone, pale brown crinoidal (some columnals up to 1-in diameter); lowest 6 in thin-bedded; *L. floriformis floriformis* and brachiopods including *A.* cf. *insculpta*    ..    ..    4

(*d*)   Limestone, grey to dark grey fine-grained, with a little chert    ..    5¾

(*c*)   *Not exposed*    ..    ..    ..    2

(*b*)   Limestone, grey fine-grained    ..    3

(*a*)   Limestone (gigantoproductoid bed of previous section) dark grey, with *G. sp.* cf. *gigantoides*    2

Near the junction with the Namurian, a large sink-hole (loc. 115) [1988 7704] north-west of Waterfall Farm showed the following section:

ft

(*f*) Limestone, grey to dark grey thin-bedded cherty; a 2-ft coral band 4 ft from base, with *A. fungites pachyendothecum, D. bipartitum craigianum, D. bipartitum konincki* .. .. .. 9½

(*e*) Limestone, pale grey-brown crinoidal; *Avonia thomasi, E. lobata* aff. *laqueata* .. .. 6

(*d*) Limestone, dark grey thin-bedded cherty, paler at top .. .. 4½

(*c*) Limestone, dark grey massive cherty; very cherty in lowest 2 ft; shell band 3½ ft, 5 ft from base with *G. giganteus* .. 15

(*b*) Limestone, dark grey thin-bedded very cherty .. .. .. 5

(*a*) Limestone, dark grey thick-bedded, with a little chert .. 14½

To the north of the last section, the Eyam Limestones have been proved, beneath Namurian cover, in the Hucklow Edge No. 2 Borehole (loc. 103) [2025 7760]:

| | Thickness ft in | | Depth ft in | |
|---|---|---|---|---|
| Base of Namurian .. | — | — | 265 | 6 |
| Limestone, pale brown crinoidal .. .. | 1 | 0 | 266 | 6 |
| *No core* .. .. .. | 8 | 5 | 274 | 11 |
| Limestone, pale grey to pale brown, very cherty in top 2 ft 5 in and with irregular silicification below .. .. .. | 3 | 10 | 278 | 9 |
| Limestone, pale grey to grey with some darker coloured bands and bioclastic bands .. | 5 | 3 | 284 | 0 |
| Limestone, buff to pale grey with some darker bands, mostly fine-grained but with some finely bioclastic bands, some chert; *Lithostrotion junceum* at 294 ft 8 in .. .. .. | 20 | 8 | 304 | 8 |
| Limestone, blue fluoritized .. .. .. | | 8 | 305 | 4 |
| *No core* .. .. | 12 | 1 | 317 | 5 |

H

| | Thickness ft in | | Depth ft in | |
|---|---|---|---|---|
| Limestone, mainly dark grey, with some paler bands; silicified and with bands of chert in places above 329 ft 9 in Monsal Dale Beds (see p. 93) .. .. .. | 27 | 1 | 344 | 6 |
| | — | — | — | — |

*Water Grove and Burnt Heath.* To the south of Middleton Dale the Eyam Limestones, some 120 ft in thickness, occupy a broad tract extending from Burnt Heath to Water Grove. The dark limestones at the base are present as far as the south-eastern extremity of this outcrop, but disappear by lateral passage into the overlying grey limestones south and south-east of Castlegate Farm. The section at Cavendish Mill [2054 7530] (see also p. 83) showed 4 ft of dark cherty limestone resting on the Monsal Dale Beds. To the south of Castlegate Farm two knoll-reefs are present at the base of the Eyam Limestones. One of these, some 14 ft in height, (loc. 166) [1953 7503] showed pale fine-grained poorly bedded reef-limestone, crinoidal in places, and with *Antiquatonia hindi, A. hindi wettonensis, A. insculpta, A. sulcata, Avonia thomasi, A. youngiana, Echinoconchus punctatus, Eomarginifera lobata* aff. *laqueata* and *Girtyella? sacculus.* Farther south, three knoll-reefs (one largely beyond the southern edge of the district) occur as outliers resting on the surface of the Monsal Dale Beds.

To the west of Castlegate Farm the outcrop of the Eyam Limestones is concealed by the shale outlier of Wardlow Mires. The lowest, dark beds are, however, seen in the axis of a sharp N.N.W.–S.S.E. anticline which crosses the main road between Water Grove and Housley. The following section (loc. 117) [1911 7583] was measured on the north side of the road:

ft

(*h*) Chert, pale brown [1902 7581] .. 1

(*g*) *Not exposed* .. .. .. 2

(*f*) Limestone, buff fine-grained much silicified; some shells; corals at base .. .. .. 4

(*e*) Limestone, dark .. .. .. 12

(*d*) Limestone, grey massive .. 2

ft

(c) Limestone, pale grey-brown crinoidal: *D. bipartitum konincki*, *L. junceum*, *Palaeosmilia* cf. *murchisoni*, *P. sp.* [cerioid form?] .. .. 3½

(b) Limestone, pale grey-brown crinoidal; shelly in top 3 ft and with *Lithostrotion junceum* and *Lonsdaleia floriformis floriformis;* 2 ft very coarsely crinoidal 3½ ft from base; finely crinoidal in lowest 3 ft .. .. .. 11

(a) Limestone, dark fine-grained irregularly bedded ('hackly'); some paler variegation towards top; some large crinoid stems in upper half .. .. .. 1¾

Item (a) marks the top of the lowest dark limestones; these form three inliers (the northern two being small in extent) along the axis of the anticline. Information from a section of Watergrove Sough shows the coral band (c) to lie some 80 ft above the base of the Eyam Limestones. It is uncertain whether this represents the Eyam Dale Coral Band, the underlying beds having expanded, or whether it is at a higher horizon.

*Wardlow Mires and Great Hucklow.* Just to the west of Wardlow Mires poorly exposed dark limestones at the base of the Eyam Limestones overlie the Monsal Dale Beds of the Cressbrook Dale section [top at 1788 7550] (see also p. 83). Wardlow Mires No. 1 Borehole (loc. 101) [1850 7553] proved the overlying Viséan shales to 82 ft 10 in (see p. 91). The shales rested, with slightly irregular base, on dark thin-bedded limestones of the Eyam Limestones. These beds had frequent mudstone partings in their upper part and were cherty below: *Lingula*, *Orbiculoidea* and small trilobites were common, *Lingula* especially so in the shale partings. The Eyam Limestones rested on an irregular surface of Monsal Dale Beds at 113½ ft with a 3-in mudstone with fish debris at the junction. Wardlow Mires No. 2 Borehole (loc. 102) [1825 7586] (see p. 378) showed a closely comparable section, though here a ½-in carbonaceous band was present in addition, in the mudstone parting at the base.

When traced north from Wardlow Mires the lower, dark beds persist along Silly Dale at least to a point ¼ mile south of Grindlow

where higher beds come in. However, on the western edge of the outcrop near Stanley Lodge the grey lithofacies extends rapidly downwards to the base of the Eyam Limestones. From here to Great Hucklow all the exposed part of the Eyam Limestones is grey.

The section in Silly Dale shows the uppermost part of the lower, dark limestones overlain by the grey. A small knoll-reef is present at the base of the grey limestones on the west side of the dale, at the south end [181 766]. At the northern end of the valley a section (loc. 118) [1818 7697] at about the same horizon, is as follows:

ft

(e) Limestone, pale shelly, with a little chert .. .. .. 2¾

(d) Limestone, grey, cherty and with some dark bands in upper part; abundant *L. junceum*, some silicified, in lowest 3 ft.. .. 7

(c) Limestone, dark bituminous .. 11

(b) Limestone, grey to pale grey well-bedded finely bioclastic; some shells .. .. .. .. 2½

(a) Limestone, dark bituminous .. 2

By comparison with the record of the Great Hucklow Borehole (see below) the coral band lies 61 ft above the base of the Eyam Limestones and is correlated with the Eyam Dale Coral Band.

Between Foolow and Grindlow a series of 6 knoll-reefs, up to some 20 ft in height, is present at or near the top of the Eyam Limestones. The larger of these (up to 450 yd in length) showed a N.W.–S.E. orientation. At Great Hucklow [175 778] another, apparently unorientated, knoll-reef is present at a similar horizon.

The Great Hucklow (No. 7) Borehole [1777 7762] (see also p. 94) showed the following generalized section:

ft

'Clay and stones' .. .. .. 8

Limestone, cherty, mainly pale but with 4 ft of dark limestone at 16 ft and 5 ft of dark limestone at 35 ft; 3 ft dark and non-cherty at 49 ft .. 85

Limestone, dark cherty; thin shale band at base .. .. .. 129

Limestone, pale, with a little chert .. 143

Limestone, dark cherty, with shale partings .. .. .. .. 146

Monsal Dale Beds .. .. .. —

The general division into upper pale and lower dark limestones holds here, but there are also signs of the incoming (129 to 143 ft) of pale limestones near the base.

*Little Hucklow and Bradwell.* To the north of High Rake the grey lithofacies of the Eyam Limstones (see p. 32) extends for a short distance only, and then the subdivision consists of dark cherty limestones, with small knolls in places. One of the latter, elongated N.W.–S.E., and lying at about the same horizon as the Grindlow knolls, is developed on Broad Low.

Exposures of the dark beds are in general poor. A section [1702 7819] north-east of Poyntoncross House shows dark fine-grained limestone, cherty and with lenses of grey limestone up to 1½-ft thick with many corals. These beds appear to pass rapidly eastwards into limestones of the grey lithofacies. To the north of Great Hucklow the Derwent Valley Water Board Broad Low (No. 6) Borehole [1753 7821] (see also p. 94) passed through the full thickness of the Eyam Limestones[1]:

|  | Thickness ft | Depth ft |
|---|---|---|
| Landslip and Edale Shales .. | 25 | 25 |
| Limestone, pale to dark grey, with 10 in of chert and some black shale .. .. .. | 5 | 30 |
| Limestone, white .. .. | 17 | 47 |
| Limestone, grey to black shaly | 3 | 50 |
| Limestone, pale grey very cherty .. .. .. | 29 | 79 |
| Limestone, dark grey cherty | 26 | 105 |
| Limestone, pale grey, with rare chert .. .. .. | 10 | 115 |
| Limestone, pale grey to black, some chert .. .. .. | 17 | 132 |
| Shale, black .. .. about | 1 | 133 |
| Monsal Dale Beds .. .. | — | — |

The record appears to show beds transitional between the grey and dark lithofacies.

The Little Hucklow Borehole [1684 7863] showed the following section:

|  | Thickness ft | Depth ft |
|---|---|---|
| Soil .. .. .. .. | 1 | 1 |
| Limestone, black cherty .. | 7 | 8 |
| Limestone, grey, with dark chert; some dark shale partings near base .. .. | 14 | 22 |
| Limestone, black cherty, dark shale partings near base .. | 11 | 33 |
| Clay, brown (*fide* driller) .. | 1 | 34 |
| Monsal Dale Beds (see p. 94) | — | — |

A little to the north of Nether Water a large knoll-reef is present in the upper part of the Eyam Limestones. The knoll has been cut through in Stanlow Dale, in all but the southernmost sections, to show the underlying dark limestones. Dips suggest that the knoll is composite and that two knolls, on the east and west sides of the dale respectively, have coalesced. Shirley and Horsfield (1940, p. 288) recognized the presence of this knoll and recorded a fauna as follows: "*Productus (Gigantella)* sp., *P. (G.) auritus* Phillips, *P. (G.). latissimus* Sowerby, *P. hindi* Muir-Wood, *Spirifer* aff. *bisulcatus* Sowerby and occasional specimens of *Caninia*".

To the west of the main outcrop of the Eyam Limestones an outlier of knoll-reef is present near Intake Farm. A section (loc. 119) [1641 7956] 150 yd N. of the farm shows nearly the whole thickness: pale very crinoidal limestone with columnals up to 1 inch in places, shelly with *Acanthoplecta mesoloba, Alitaria panderi, Antiquatonia hindi, A. insculpta, Avonia davidsoni, A. thomasi, A. youngiana, Echinoconchus punctatus, Eomarginifera lobata* aff. *laqueata, G.? sacculus, Plicochonetes buchianus* and *Spirifer (Fusella) triangularis.* An almost identical fauna was obtained from an exposure (loc. 120) [1646 7965] at the north-eastern end of the knoll.

To the north of the main Tideswell–Bradwell road an extensive flat reef is developed within the Eyam Limestones, and to the north of an east–west line through Hartlemoor Farm this occupies the whole of the subdivision. Dark limestones below the flat reef are exposed near the junction of

---

[1]Information from the examination of cutting samples by Professor F. W. Shotton.

Jenning's Dale and Green Dale [1698 7992] where "230 yd west of the farm" (Hazlebadge Hall) Shirley and Horsfield (1940, pp. 286, 288) record a fauna including "*Cyathaxonia cornu, Zaphrentis enniskilleni derbiense*, brachiopods and *Phillipsia sp.*"

The relations of dark limestones to the flat reef are most easily observed in Bradwell Dale. A section on the east side of this valley at the south end shows a small development of flat reef overlying the Monsal Dale Beds. The flat reef first grew strongly upwards and later spread laterally at a horizon some 30 ft above the base of the Eyam Limestones. The total thickness of the flat reef here is about 130 ft, of which the lowest 75 ft are exposed in the dale side; the uppermost beds are well seen farther east in the side of a fluorite working [1731 8035] where they are often very coarsely crinoidal. At the southern end of the dale, the continuation of the section (loc. 99) [1714 8039] of p. 89 is as follows:

|  |  | ft |
|---|---|---|
| (*h*) | Limestone, dark bituminous, passing northwards into flat reef; brachiopods, gastropods and the fish *Petalodus acuminatus* .. .. .. .. | 6½ |
| (*g*) | Limestone, pale fine-grained shelly; *Alitaria panderi, Antiquatonia* cf. *antiquata, A. insculpta, E. lobata laqueata* and *K. spinulosa* .. .. .. | 22 |
|  | Monsal Dale Beds .. .. | — |

The section shows both flat reef and normal dark facies and closely matches that on the east side of the dale.

At the north end of the dale the section (loc. 100) [1741 8077] given on p. 89 is continued by flat reef as follows:

|  |  | ft |
|---|---|---|
| (*l*) | Limestone, pale crinoidal, massive in lowest 7 ft, more thin-bedded above; *Acanthoplecta mesoloba, Antiquatonia hindi, A. hindi wettonensis, A. insculpta, E. punctatus, K. spinulosa, Pleuropugnoides pleurodon, Pugnax pugnus* .. .. .. | 11 |

*Brook House*. A narrow outcrop of beds referred to the Eyam Limestones on faunal grounds is present along the Viséan-Namurian junction west of Water Swallows. This outcrop is bounded on the east by the Miller's Dale Beds and on the west by the Namurian as is that of the Monsal Dale Beds at Dove Holes. The two outcrops, however, are separated by an east–west fault of pre-$P_2$ age and the relations of the two are shown in Fig. 7.

The greater part of the Eyam Limestones of this area, some 30 ft thick, are dark, thin-bedded and cherty. A quarry [0716 7602] near Blackedge Farm showed 15 ft of grey to dark grey fine-grained thin-bedded cherty limestone with small brachiopods and occasional corals. Farther south, the section (loc. 121) [0690 7534] is as follows:

|  |  | ft |
|---|---|---|
| (*h*) | Limestone, dark well-bedded fine-grained cherty; rare corals, *Cyathaxonia sp. cornu/rushiana* group, *Rotiphyllum* aff. *rushianum* .. .. .. .. | 2 |
| (*g*) | Limestone, as above; abundant corals; *Aulophyllum fungites cumbriense, Caninia sp. nov.* aff. *buxtonensis, D. bipartitum bipartitum, D. bipartitum konincki, Koninckophyllum interruptum, K. magnificum* .. | 1 |
| (*f*) | Limestone, grey to dark grey thin-bedded (beds up to 5-in thick); corals above 5 in from base; *Aulophyllum fungites cumbriense, Dibunophyllum bipartitum bipartitum, D. bipartitum craigianum, D. bipartitum konincki, Diphyphyllum fasciculatum, Alitaria panderi* .. | 1¼ |
| (*e*) | Limestone, grey to dark grey shelly and crinoidal; *Alitaria panderi, Avonia thomasi, A. youngiana, E. lobata laqueata, K. spinulosa, P. productus hispidus, S. (F.) triangularis* ½ to | 1¾ |
| (*d*) | Limestone, grey well-bedded .. | 1 |
| (*c*) | Clay parting .. .. up to | 0¼ |
| (*b*) | Limestone, grey well-bedded .. | 2 |
| (*a*) | Limestone, dark grey-brown finely crystalline cherty; *Aulophyllum fungites cumbriense, D. bipartitum bipartitum, G.* |  |

ft

*giganteus crassus* throughout;
*A. fungites cumbriense, D. bi-
partitum konincki, Alitaria panderi, Avonia youngiana, Pugilis
pugilis* in pocket at 2 ft from
base .. .. .. .. 6

The section shows the highest exposed post-
$D_1$ strata in the vicinity.

Near Brook House a lenticular mass of
knoll-reef limestone makes its appearance at
the base of the Eyam Limestones and
thickens southwards. The section (loc. 122)
[0673 7484] is as follows:

ft

(*d*) Limestone, grey-brown irregularly bedded shelly; *Alitaria
panderi, Antiquatonia antiquata,
A. hindi, A. insculpta, Avonia
thomasi, A. youngiana, K. spinulosa* .. .. .. .. 5½

ft

(*c*) Limestone, grey to dark grey
massive fine-grained .. .. 2½

(*b*) Limestone, grey, darker towards
base, shelly throughout and
with corals in top 6 in; *D. bipartitum bipartitum, Productus
productus hispidus* .. .. 8

(*a*) Limestone, grey-brown massive
(knoll-reef); *Acanthoplecta mesoloba, Antiquatonia antiquata,
A. hindi, A. insculpta, Avonia
davidsoni, A. thomasi, A.
youngiana, Echinoconchus punctatus, Eomarginifera lobata* aff.
*laqueata, G. ? sacculus, K. spinu-
losa, S. (F.) triangularis, Sudeticeras sp.* .. .. .. 30

The last item represents about the full
thickness of the knoll-reef.

## SHALES OF $P_2$ AGE

These beds are known in the shelf area
only from the Wardlow Mires outlier.
Material from a trench (loc. 124) [1892
7540] collected by Dr. J. Shirley, showed a
probable $P_2$ fauna with *Chonetes sp.* [s.l.],
*Crurithyris sp., Posidonia corrugata* and
*Weberides* cf. *mailleuxi.* Similar shelly mudstones are known from the tips of Watergrove Mine (loc. 123*b*) [1882 7577] and from
a 13-ft excavation [1810 7558] at Wardlow
Mires. In the Wardlow Mires No. 1 Borehole (loc. 101) [1850 7553] (see also p. 102) the

base of the Namurian is considered to be at a
depth of 67 ft 9 in. The Namurian is underlain by $P_2$ shales extending to the slightly
irregular junction with the Eyam Limestones
at 82 ft 10 in. *Sudeticeras* cf. *stolbergi* Patteisky occurred at 80 ft 1 in. Wardlow Mires
No. 2 Borehole (loc. 102) [1825 7586] proved
a closely comparable section of $P_2$ shales.
Here the recognition of *Cravenoceras leion*
Bisat provided a clear base to the Namurian
at 65 ft 4 in and the shales extended down to
a depth of 77 ft 7½ in.

## MARGINAL PROVINCE

### THE APRON-REEF

*Perryfoot and Windy Knoll.* The westernmost exposures of apron-reef limestones lie
¼ mile N. of Sparrow Pit. Here the apron-reef
is narrow but, traced north-eastwards, it
expands laterally to an average width of
some 300 yd. The steep frontal slope is
normally well developed and shows dips of
from 15°–30°. A line of sink-holes is present
at the junction of the apron-reef and the
Namurian and most of these show good
sections of the upper part of the reef.

Perry Dale shows a cross-section of the
apron-reef. On the inward side, the algal
reef is well developed to the south-west of the
dale [1021 8102 to 1029 8110] where differential erosion has caused it to stand out. The
algal reef has a maximum width of about
90 ft and a minimum height of about 60 ft,
the base not being seen. A lesser development of the algal reef is present on the
north-east side of the dale. Fossiliferous
fore-reef limestones (see also Table 5, p. 142)

are exposed between the algal reef and the large sink-hole by Tor Top. The latter (loc. 126) [0989 8126] shows the following generalized section:

|  | | ft |
|---|---|---|
| (b) | Limestone, pale massive reef-type, with coarsely crinoidal pockets and shelly patches, matrix fine-grained; large reef fauna including *Goniatites sp.* *crenistria* group  .. .. | 20 |
| (a) | Limestone, darker and finer-grained than above; large reef fauna including *Productus productus* cf. *hispidus*, abundant *Streblopteria laevigata* and *Beyrichoceras sp.* (*rectangularum* or *delicatum*)  .. .. .. | 18 |

The upper item (b) is of $P_{1a}$ age from the goniatite evidence, and is indicated separately on Plate XI.

Over the half-mile stretch of apron-reef north-east of Perryfoot typical exposures [1014 8144 and 1078 8177] occur in the sink-holes at the limestone edge. An exposure [1059 8146], near the Bull Pit, shows a poorly developed algal reef at the inner margin of the apron; this was noted by Wolfenden (1958, p. 877) who considered that it belonged to the lower of the two levels of algal reef development recorded by him. The present mapping, however, would suggest that this algal reef is at a stratigraphically higher level than that of Perry Dale. The algal reef is also exposed at the top of Snels Low [1134 8178] and in the small valley [1159 8205] between the same feature and Middle Hill. The width of the algal reef in all these occurrences does not appear to exceed 20 ft.

The valley between Snels Low and Middle Hill shows a good cross-section of the apron-reef between the algal reef (see above) and the outer edge [1154 8224]. The section shows the normal flattening of the dip on the inward side and a frontal dip of 20 degrees at the north (outward) end. A rich fauna has been obtained (loc. 127) here though no goniatites were found.

On Middle Hill and Peaks Hill[1] [118 827] the structure of the apron-reef is more complex than is usually the case. The normal apron relationships are shown on Middle

Hill, with small occurrences of algal reef on the inward side and strong frontal dips (see Plate XI). At the point where the latter develop there is a well-defined valley in which Giant's Hole [1194 8268] lies and north-west of this lies Peaks Hill which appears to 'adhere' to the outer margin of the reef. The slopes on the southern side of the Giant's Hole valley yield a reef-fauna which is not diagnostic of age (locs. 132–3, see Plate XI and Table 5, p. 142) though Parkinson (1943, p. 126) records a high $B_2$ fauna with *Goniatites* aff. *crenistria* and *G. maximus* from "Middle Hill" and the same author (1947, p. 107) lists the following $B_2$ goniatites from "the vicinity of the gorge" by Giant's Hole: *Beyrichoceras* aff. *delicatum*, *Beyrichoceratoides sp.*, *Bollandoceras* aff. *micronotum*, *Goniatites hudsoni* group, *G.* aff. *antiquatus* Bisat and *G.* aff. *maximus* Bisat.

Peaks Hill itself is formed of a knoll-like mass of reef-limestones elongated parallel to the reef margin and showing quaquaversal dips. These beds have yielded (loc. 130) [1176 8267] the goniatites *Goniatites crenistria* [early form] and *G.* cf. *hudsoni* indicative of the uppermost part of $B_2$. A nearby exposure (loc. 128a) [1170 8267] yielded *Goniatites wedberensis* Bisat in addition (see also p. 149). The exposure (loc. 131) [1179 8266] on the south side of Peaks Hill with *Palaeosmilia regia* was first noted by Shirley and Horsfield (1940, p. 279). The relations of Peaks Hill to the main mass of the apron-reef are here attributed to uplift in late $B_2$ times along and parallel to the reef margin with the consequent offsetting of the later part of the apron-reef in relation to the earlier. The unusually high inward dips at Peaks Hill would thus be interpreted as a reflection of the uplift. As noted elsewhere (p. 162), late Viséan movements in the neighbouring limestone salient near Mam Tor appear to have been particularly acute.

On the eastern side of Middle Hill the reef margin swings northwards and at Windy Knoll the apron-reef is entirely concealed by the Namurian. A small outcrop [1253 8292] of algal reef is present a little south of the latter locality.

*Treak Cliff and Winnats.* In this area the apron-reef occurs as a belt trending N.N.W.–

---

[1]Not named on six-inch or one-inch maps though the name is used in literature.

S.S.E. At the northern end it makes a westward turn, just before disappearing under the Namurian, which suggests a junction at depth of the two outcrops of apron-reef on the sides of the $D_1$ salient near Mam Tor.

On Treak Cliff the inner margin of the reef-complex is well marked by the presence of a number of well-developed algal reefs (Wolfenden 1958, p. 879). The most northerly of these lies close to the shale boundary and is noteworthy as the locality [1328 8334] where Shirley and Horsfield (1940, p. 280) recorded signs of unconformity at the junction of reef and standard limestone. Wolfenden's confirmation of Parkinson's (1953, p. 254) contention that the apron-reef passed laterally into limestones of the standard sequence makes it likely that any break at this locality is more of the order of a non-sequence; small Neptunian dykes filled with crinoidal limestone are evidence of this in the algal reef.

The best developed of the algal reefs forms the crest of Treak Cliff (Plate XIIIA) [1343 8322 to 1345 8283]. This has been found by Ford (1964, pp. 31–2) to be considerably wider than stated by Wolfenden (ibid., pp. 876–7) being at its maximum some 250 ft across; the height is given as 50 ft. The general characters of the algal reefs are described on p. 18 and the Treak Cliff occurrence is typical. Apart from the characteristic stromatolitic algae, sponges and pockets of small reef brachiopods occur (loc. 136c–k), and at a point (loc. 136h) [1344 8298] nearly ¼ mile S.E. of Blue John Cavern entrance colonies of Lithostrotion martini and L. pauciradiale up to 6 ft across and in the position of growth are to be seen. Some 5 to 10 yd N. of this locality Dibunophyllum bipartitum and a restricted fauna of reef brachiopods were collected. Original cavities between algal colonies have been filled with calcite in a characteristic cuspate form.

The highly fossiliferous fore-reef limestones of Treak Cliff show outward dips of from 25° to 35°. The essentially depositional nature of these was stressed by Shirley and Horsfield (1940, pp. 288–9) and this has been recently confirmed by Broadhurst and Simpson (1967, pp. 443–8) by demonstrating that shells and limestone cavities showed horizontal or near horizontal sedimentary infillings. One example is an orthocone nautiloid obtained [1329 8344] near the Odin

Mine. The most marked change across this belt is the increase in the amount of crinoid debris; on the lower slopes some of this is quite coarse [e.g. 1365 8300]. The results of detailed collecting across the outcrop are summarized on p. 150. This area has formed the basis of most of the previous references to the Castleton area. Bisat (1934, pp. 285–6) records a goniatite fauna with Bollandites castletonensis (Bisat) from Treak Cliff, and Shirley and Horsfield (1940, p. 279) noted Goniatites cf. hudsoni. These records indicate an upper $B_2$ age for the fore reef here. A fauna indicative of the top of the zone was obtained by Parkinson (1947, p. 107; 1953, p. 258) from coarsely crinoidal limestones above the 'Odin Fissure' [1342 8345]; this included Goniatites crenistria and Beyrichoceratoides truncatus Phillips sp. (see also loc. 135, p. 149). Ford (1952, p. 353) recorded Goniatites maximus from a point [1353 8303] underground in the Treak Cliff Cavern. Parkinson (1952, p. 201) has described a thin lenticular bed with abundant Dielasma hastatum from near the summit of Treak Cliff apparently within the algal reef [1343 8317].

In the Winnats gorge the transition from shelf facies to apron-reef is well seen. Here Wolfenden (1958, pp. 875–6) recognized the presence of a lower algal reef somewhat inside (geographically) the position of the upper algal reef of Treak Cliff. The present work confirms the presence of the lower algal reef [1350 8270], though it appears more nearly in line with the upper than shown by Wolfenden.

*Long Cliff and Goosehill.* On the slopes between Long Cliff and Goosehill the apron-reef dips northwards at from 15° to 32°. On the inner side an algal reef has been recognized in this tract only at Longcliff Plantation [1410 8226], though it is less well developed than at Treak Cliff. On the outer side the apron-reef is overlain by Beach Beds as far east as Goosehill but beyond here dark limestones of basin facies rest on the apron-reef (see p. 110).

The best cross-section of the fore reef is provided by the gully below Cow Low (Cow Low Nick). This contains, at a point (loc. 138) [1420 8241] about half way down, a pocket (about 1½ ft × 4 ft) of fine-grained limestone crowded with goniatites. The occurrence was first noted by Bisat (1934, p. 291). Ford (1965, pp. 186–91) lists the fauna of which *Beyrichoceras rectangularum*, *Bollandoceras submicronotum* (Bisat) and *Goniatites maximus* are the commonest forms with less common *B. vesiculiferum* (de Koninck) and *Prolecanites* cf. *P. discoides* (Foord and Crick); the age is upper B₂. Ford has suggested that the goniatite lens accumulated in a surge channel or submarine cave in the fore reef.

On Cow Low, an exposure [1427 8235] near the inner edge of the apron-reef shows tuffaceous limestone. This is interpreted as indicating a former extension of the nearby outcrop of the Lower Lava into the reef-belt.

At Goosehill, massive fore-reef limestones are well seen, though inaccessible, in the gorge [1487 8263] below Peveril Castle. This shows a vertical cliff of some 300 ft in these beds. Shirley and Horsfield (1940, p. 279) recorded *Goniatites maximus* "from the slopes above Goosehill".

*Cave Dale and Siggate.* The section in Cave Dale shows the transition from apron-reef to standard limestones. At the junction there is no algal reef and the transition appears to take place by interdigitation of the two facies in the lowest part of the dale. At a higher horizon, however, the apron-reef overlies limestones of shelf facies indicating a slight inward movement of the apron as it built up.

The apron-reef is well seen in the quarry [1507 8267] at the entrance to Cave Dale which shows some 40 ft of poorly-bedded fossiliferous limestone. No diagnostic fossils have been obtained from this exposure.

Between Cave Dale and Siggate the frontal slope of the apron-reef is well developed with dips of 15° to 23° (Plate XIIA). As at Treak Cliff, the beds at the foot of the apron are commonly crinoidal [e.g. 1559 8260]. From a roadside exposure [1587 8248] on Siggate, Parkinson (1947, p. 110) recorded *Goniatites crenistria* and *Beyrichoceratoides truncatus* (or a near form) indicative of a low P₁ₐ horizon. The goniatites listed as *G. crenistria* by Parkinson have been re-identified as *G. maximus*, indicating an upper B₂ age (Smith and Yü 1943, p. 49). Shirley and Horsfield (1940, p. 279) collected *G.* cf.

*maximus* from an unspecified locality at Siggate.

*Pin Dale and Mich Low.* The relations in this area differ from those in the remainder of the reef-belt in that fore-reef limestones of P₁ₐ to P₁ᵦ age overlie D₁ limestones of shelf facies in most places. B₂ reef-limestones are present at Nunlow Quarry.

The section in Pin Dale shows D₁ and D₂ standard limestones passing laterally into apron-reef. Both shelf and reef facies are underlain by a tuff (see also p. 61). At Pindale Mine [1629 8256] the shaft passed through Edale Shales to 90 ft and continued in tuff to 300 ft (Green and others 1887, pp. 129–30). Here the apron-reef limestones or their equivalents have been cut out by the pre-Namurian unconformity. Some original thinning of these beds may also be assumed, corresponding to the thick development of tuff beneath. The section in Pindale Quarry (loc. 47) [1594 8231] shows shelf limestones of D₁ and early D₂ age passing north-eastwards at the inner edge of the apron-reef into limestone-breccia with "angular fragments up to a foot or more in length in a matrix which is commonly seen to be markedly oolitic" (Eden and others 1964, pp. 79, 84, and fig. 2).

In Upper Jack Bank Quarry (loc. 142) [1616 8220] 30 ft of pale massive shelly reef-limestone rest on 4 ft of grey massive limestone. The lower item is perhaps an interdigitation of standard facies in the reef. Here the reef-limestones have yielded *Goniatites striatus* some 2 ft from the base and also near the top of the section (Eden and others 1964, pp. 90, 102, 105); *Productus productus hispidus* also occurs. There is an unusually steep inward dip of up to 20° here and this is explained (ibid.) as being due to the presence of a knoll, now largely eroded away, at the edge of the reef-belt. It is here suggested that these relations could also be explained by the effects of uplift along the reef margin.

The section in Lower Jack Bank Quarry (loc. 141) [1616 8234], near the foot of the frontal slope of the apron-reef, has been described in detail by Eden and others (ibid., pp. 79, 90–1) and includes their localities 9 and 11; it may be summarized as follows:

(L 190)

A. Frontal slope of apron-reef east of Castleton

PLATE XII

B. The Winnats, Castleton: a gorge cut through
REEF-COMPLEX LIMESTONES

(L 195)

A. ALGAL-REEF LIMESTONE: TREAK CLIFF, CASTLETON

PLATE XIII

B. BEACH BEDS: NEAR SPEEDWELL MINE, CASTLETON

ft

Limestone (fore reef) massive, usually crinoidal and with bands of breccia; fauna of reef brachiopods including *Productus productus hispidus* and *Orbiculoidea nitida* near top (*d*); *Goniatites sp. crenistria* group in lower part (*c*) .. .. up to 60

*Marked unconformity* .. .. —

(*b*) Limestone, grey-brown well-bedded fine-grained (shelf $D_1$ lithofacies), with *Gigantoproductus sp. maximus* group .. 30

(*a*) Tuff, formerly seen in quarry floor —

The uppermost apron-reef limestones were considered to lie on the same horizon as the $P_{1b}$ limestones of Upper Jack Bank Quarry, the presence of *P. productus hispidus* providing a faunal link. The lower ($P_{1a}$) beds, with cf. *G. crenistria* were considered probably to occupy an erosional hollow in the underlying $D_1$ limestones; elsewhere in the quarry these beds are absent, $P_{1b}$ limestones resting directly on $D_1$.

In Nunlow Quarry (loc. 143) [1651 8226] ¼ mile S.E. of Pindale Farm, Eden and others (1964, p. 91) describe a sequence which is in many respects similar to that in Lower Jack Bank Quarry:

ft

(*b-c*) Pale unbedded reef-limestones; these have yielded (*c*) *Productus productus* towards *hispidus*, near the top (?$P_{1b}$); (*b*) *Goniatites crenistria* ($P_{1a}$) has been recorded from fallen material probably from the lower part (Jackson 1925, p. 272) about 120[1]

(*a*) Pale massive fine-grained limestones and calcilutites, with a $B_2$ fauna of reef brachiopods about 40

To the south-east of the last exposure, the frontal slope of the apron-reef, clear of quarrying operations, is again visible. There appears to be a progressive south-easterly thinning of the apron. On Mich Low the apron-reef merges into a knoll-like mass forming the crest of the hill. Dips on the outer side of this cannot be observed but exposures on the inner, southern side, show dips of 37° and 25°. The underlying limestones are apparently of $D_1$ shelf facies.

On the north-west side of Mich Low a small railway-cutting [1693 8195] shows the following section of beds referred to the Bee Low Limestones:

ft

Limestone, grey massive irregular-bedded, locally oolitic; brachiopods, productoid spines and crinoid debris near base .. .. .. .. 32

Limestone, tuffaceous, and grey tuff, interbanded .. .. .. 0 to 3¾

Tuff, grey, soft and brown-weathering 4

The limestones of this section are referred to the back-reef facies.

The overlying reef-limestones are exposed on the south side of Mich Low (loc. 144) [1699 8180] where they yield *Antiquatonia* cf. *antiquata*, *Avonia davidsoni*, *A. sp.* [juv.] *youngiana* group, *Gigantoproductus edelburgensis*, *G. sp.* cf. *moderatus*, *Krotovia spinulosa*, *Linoproductus sp. hemisphaericus* group, *Pleuropugnoides pleurodon*, *Productus productus hispidus*, *Pugnax pugnus* [large form]. Parkinson (1947, p. 111) has recorded *Goniatites striatus* from Mich Low, proving the $P_{1b}$ age of these beds.

To the north of Bradwell, a group of small inliers of reef-limestone is present. This appears to be an easterly continuation of the outcrops of Mich Low, the more southerly exposures dipping south towards the shelf area. An exposure (loc. 145) [1740 8187] of 10 ft of poorly-bedded reef-limestone yielded a shelly fauna which is probably of low $P_1$ age. However, on the east side of the road here (loc. 147) [1749 8185] 1 ft of dark limestone contains an abundant shelly fauna of upper $B_2$ age.

---

[1]Thicknesses added to published description.

POST-REEF DEPOSITS

The small outcrops of these beds, although at least in part earlier than the main mass of the Eyam Limestones ($P_2$), have been included with them on the map.

*Castleton.* Beds here considered to rest on the apron-reef are present between the lower slopes of Treak Cliff and Castleton village. The Beach Beds, consisting for the greater part of rolled-shell limestones, occupy much of the outcrop. A detailed description of the lithology and fauna of the outcrops has been published by Sadler (1964) though these are differently interpreted here (see p. 35).

A section [1393 8284] in a quarry by the main road shows $6\frac{1}{2}$ ft of pale, roughly bedded limestone with crinoidal bands and some broken shells. Sadler (1964, p. 368) recorded beds of basin facies beneath this: dark thinly bedded coarse-grained limestone with some brachiopods 2 ft, on dark massive coarsely crinoidal limestone with "*Zaphrentis* cf. *enniskilleni*", 4 ft. From a nearby quarry [1384 8290] 200 yd N.N.W. of the entrance to the Speedwell Mine, Jackson (1941, p. 243) records *Archaeocidaris urii* and the fishes *Petalodus acuminatus*, *Strepsodus sp.*, *Ctenopetalus lobatus* and *Petalorhynchus sp.* (teeth).

The Beach Beds are well exposed near the Speedwell Mine. A section [1385 8277] on the north side of the road shows 20 ft of rolled-shell limestone with worn shell fragments (up to 2 in diameter) and some crinoid debris; the rock is well bedded and is coarse-grained in the lowest $2\frac{1}{2}$ ft. On the south side (Plate XIIIB) of the Speedwell Mine [1400 8270], 6 ft of rubbly rolled-shell limestone contains limestone pebbles up to 2 in. An exposure [1394 8267] 50 yd S. of the Speedwell Mine shows beds near the base of the Beach Beds:

|  | ft |
|---|---|
| (*d*) Limestone, pale massive irregularly bedded, finely crinoidal .. | $4\frac{1}{4}$ |
| (*c*) Limestone, grey fine-grained   $1\frac{1}{2}$ to | 2 |
| (*b*) Limestone, dark grey thin-bedded; finely crystalline and somewhat bituminous. Sadler (1964, p. 366) recorded *Semenewia* aff. *concentrica* (de Koninck), *Dictyoclostus sp.*, *Buxtonia sp.* and *Athyris sp.* .. | 1 |
| (*a*) Limestone, pale massive finely crinoidal   ..   ..   .. | 2 |

The exposure is important in showing the presence of intercalations of beds of basin type in the Beach Beds.

The Beach Beds are exposed in two old quarries between the Speedwell Mine and Goosehill. The more westerly of these [1423 8264] shows 20 ft of rolled-shell limestones, part crinoidal and showing rough sorting into coarser and finer bands. The more easterly [1440 8261] shows rolled-shell limestone $1\frac{1}{4}$ ft on well-sorted crinoidal limestone with some shell debris 13 ft.

At a point 40 yd N.E. of the last exposure a small thickness of dark bituminous limestone, in part crinoidal, crops out. This lies at the western end of the small outcrop which extends eastwards to Goosehill. By Goosehill Hall, exposures in a lane [1474 8272] show dark bituminous limestones interbedded with grey crinoidal or shelly limestone at this horizon.

On the east side of Peakshole Water another area of dark beds overlying the apron-reef is present. The beds at the base are exposed in an old quarry [1488 8277], where dark shelly limestone with small crinoids $2\frac{1}{2}$ ft, overlies soft buff-coloured tuff $2\frac{1}{2}$ ft, on apron-reef (loc. 139).

A roadside exposure (loc. 148) [1492 8282] in Castleton shows beds at a slightly higher horizon; dark bituminous limestone with *Rotiphyllum costatum*, *Zaphrentites* cf. *derbiensis* and *Antiquatonia insculpta*. Parkinson (1947, p. 112) noted the presence of a small exposure [1586 8265] of dark limestones of probable $P_{1b}$ age, between Castleton and Pin Dale, yielding *Goniatites* cf. *spirifer* Roemer and resting on the upper surface of the reef. The extent of the outcrop of these beds (ibid., pl. viii) would, however, appear to have been exaggerated.

*Windy Knoll and Perryfoot.* This area shows remnants of dark limestone later than the apron-reef. These deposits are considered to be of $P_2$ age on the basis of fossil evidence from near Peakshill Farm (see below).

At Windy Knoll Quarry [1261 8300] deposits of the above type occur as 'Neptunian dykes' in the Bee Low Limestones.

The fillings of these are dark bituminous limestone, grey crinoidal limestone and a breccia of blocks of the country rock in a dark bituminous matrix. Parkinson (1965, p. 176) considered that these infillings were contemporaneous or nearly so with the country rock but the sharp contrast in lithology makes this unlikely.

Farther west, at the reef margin an exposure (loc. 128b) [1170 8267] 130 yd E.15°

N. of Peakshill showed a few inches of dark limestone with the $P_2$ goniatite *Sudeticeras sp.* resting on the upper surface of the reef-limestone and overlain by shale.

Near Perryfoot, a small outcrop of dark limestone is present, resting on beds low in the Bee Low Limestones. An exposure [1062 8133] by the main road shows 4 ft of these beds.

## BOREHOLES IN THE MARGINAL PROVINCE

The **Castleton Borehole** (loc. 149) [1410 8293], after passing through Namurian and Viséan shales, reached limestone at 114 ft 6 in. The section is as follows:

| | Thickness ft in | Depth ft in |
|---|---|---|
| Base of Namurian | — — | 99 7 |
| Mudstone, dark, with a 7-in silty calcareous band at 101 ft 1 in; slump structures 101 ft 2 in; *Caneyella membranacea* (McCoy), *Dunbarella* aff. *persimilis* (Jackson), *Posidonia* cf. *corrugata* (Etheridge jun.), *Lyrogoniatites georgiensis* Miller and Furnish, *Sudeticeras sp.*, conodonts ($P_{2c}$) | 6 11 | 106 6 |
| Mudstone with thin calcareous bands and two coarser-grained thin limestones (both showing graded bedding) ½ in at 110 ft 1½ in and 1 in at 110 ft 6 in; *C.* aff. *membranacea*, cf. *Mesoglyphioceras granosum* (Portlock), *Lyrogoniatites sp.*, *Sudeticeras?* ($P_2$) | 4 1 | 110 7 |
| Mudstone, dark, with calcareous bands; 6-in band at 113 ft 8 in shows calcareous debris with graded bedding; *C.* aff. *membranacea*, *Posidonia sp.*, *Lyrogoniatites sp.* | | |
| at 110 ft 11 in to 111 ft 9 in ($P_2$) | 3 11 | 114 6 |
| Limestone, buff-grey with dark patches, very coarsely crinoidal and shelly at top, becoming finer grained and less crinoidal below; *Brachythyris* cf. *ovalis*, *Eomarginifera?*, orthotetoids, *Spirifer sp. bisulcatus* group, *Reticularia?* | 2 3 | 116 9 |
| Mudstone, dark, in irregular parting about | ¼ | 116 9¼ |
| Limestone-breccia of fine-grained fragments up to 2 in set in a finely crinoidal matrix; *Avonia* aff. *davidsoni*, *Buxtonia sp.*, cf. *Crurithyris sp.*, *Fluctuaria sp.* [juv.], *Linoproductus sp.*, cf. *Martinothyris lineata*, *Productus productus*, *Reticularia?*, *Spirifer sp.* | 2 5¾ | 119 3 |
| Limestone, grey compact, shelly in places; thin shale parting at 121 ft 9 in: *Antiquatonia sp.*, *Buxtonia sp.*, *Linoproductus sp.*, *P. productus*, *Reticularia sp.*, *Spirifer sp. bisulcatus* group | 2 7½ | 121 10½ |
| Limestone, rubbly | 10½ | 122 9 |
| Mudstone, black | 1 | 122 10 |

| | Thickness ft in | Depth ft in |
|---|---|---|
| Limestone, grey-brown granular, shelly, with darker bands above 127 ft; grey-brown finely crinoidal with some shells below: *Buxtonia?*, *Linoproductus sp.*, *Pleuropugnoides pleurodon*, *Productus productus*, *Reticularia sp.*, *S. sp. bisulcatus* group .. | 5 4 | 128 2 |
| Limestone, grey-brown coarsely crinoidal, with a 1-in mudstone parting at 134 ft 1 in; marked stylolite with mudstone coating at base: *Antiquatonia hindi?*, *Dictyoclostus?*, *Gigantoproductus sp.*, *Spirifer sp.* .. | 11 1 | 139 3 |
| **Beach Beds** | | |
| Limestone, grey-brown, crinoidal and shelly (many worn shells): some intraclasts of grey limestone: *A. insculpta?*, *Gigantoproductus sp.*, *Spirifer sp.* .. | 4 0 | 143 3 |
| Limestone, grey-brown, shelly (some shells worn), coarse crinoid debris in places; *Dibunophyllum sp.*, *Echinoconchus sp.*, *Gigantoproductus spp.*, including *G. striatosulcatus* (Schwetzow) subspp., *Spirifer sp.* .. | 14 3 | 157 6 |
| Limestone, grey-brown coarsely crinoidal, with shells, some worn: *Gigantoproductus sp.*, *S. (F.) trigonalis* .. | 4 0 | 161 6 |
| Limestone, grey coarsely crinoidal and shelly; *Gigantoproductus sp.*, *S. sp. bisulcatus* group | 5 6 | 167 0 |

| | Thickness ft in | Depth ft in |
|---|---|---|
| Limestone, grey to grey-brown crinoidal, with some shells: some darker grey intraclasts; *Echinoconchus cf. elegans*, *Linoproductus?*, *Spirifer sp.*; two bands with corals, one from 167 ft 3 in to 167 ft 9 in with *Lithostrotion portlocki* and *Nemistium edmondsi;* the other from 169 ft 5 in to 169 ft 7 in with *Aulophyllum fungites*, *Dibunophyllum?* and *Diphyphyllum lateseptatum* | 5 0 | 172 0 |
| Limestone, dark grey, with some worn shell debris and zaphrentoids; *Zaphrentites sp.*, trilobite fragment; passing into .. | 6 | 172 6 |
| Limestone, pale grey with scattered shells, some crinoid debris; bands of worn shell debris; *Antiquatonia?*, *Eomarginifera sp.*, *S. (F.) trigonalis*, fish fragment | 31 0 | 203 6 |

The borehole record confirms the association of the Beach Beds with beds of basin facies. The presence of *L. portlocki* with *N. edmondsi* at 167 ft 9 in suggests a position high in the $D_2$ shelf succession perhaps near the $P_1$–$P_2$ boundary of the basin. *N. edmondsi* is restricted in the present area to the Upper Monsal Dale Beds and two of the three records are from or near the horizon of the *L. duplicata* Band. It is uncertain whether the limestones above 139 ft 3 in have been included at outcrop with the Beach Beds or whether they thin out and disappear in that direction.

The **Hope Cement Works Borehole** (loc. 150) at Salter Barn [approximate position 1678 8228][1] was described by Fearnsides and Templeman (1932, pp. 105–10). Templeman's unpublished record may be summarized:

---

[1]Now covered by Earle's Cement Works.

| | Thickness ft | Depth ft |
|---|---|---|
| Base of Namurian (see p. 200) | | 280 |
| Mudstone, dark, with calcareous bands; *Posidonia corrugata* at 285 ft, indeterminate goniatites throughout; fish and plant debris 282 ft 6 in to 287 ft .. .. | 12 | 292 |
| Mudstone, dark platy with phosphate and pyrite nodules; some fish scales .. | 5 | 297 |
| Limestone, dark; indeterminate goniatites .. .. | 1 | 298 |
| Limestone, of very varied lithology, including grey to dark grey thin-bedded limestones with shale partings, limestone breccias and tuffaceous beds; coarsely bioclastic bands are shelly or crinoidal; *Lonsdaleia duplicata duplicata* 299 to 309 ft; *Productus productus* at 306 ft; 8 in of dark limestone with *Amplexizaphrentis ?*, *Claviphyllum eruca* (McCoy) and *Rhopalolasma sp.* at 310 ft 8 in; *Weberides sp.* at 342 ft .. .. .. | 52 | 350 |
| Tuff .. .. .. .. | 1 | 351 |
| Lava, predominantly with pillow structure; some thin | | |

| | Thicknesss ft | Depth ft |
|---|---|---|
| tuff bands and sporadic thin partings of limestone .. | 101½ | 452½ |
| Tuff .. .. .. .. | 3 | 455½ |
| Limestone, tuffaceous .. | 15 | 470½ |
| Tuff .. .. .. .. | 19½ | 490 |
| Lava, poor pillow structure .. | 9½ | 499½ |

The lavas are all recorded as vesicular. The figure by Fearnsides and Templeman (ibid., p. 108) of a length of core from 378 to 383 ft shows an arrangement of vesicles becoming finer towards the outside of each pillow. As these authors remark, the succession in the borehole cannot be correlated in detail with the known sequence. Eden and others (1964, p. 76) suggested that the igneous rocks lay on the horizon of the Pindale Tuff. This correlation does not hold as (*a*) the volcanic rocks of the borehole are overlain by limestone bearing *L. duplicata duplicata* and their tuffaceous partings show them to be closely associated, (*b*) beds of $P_1$ age are present on the apron-reef front near this locality and there is no reason to suppose a sudden flattening of the outward dip, (*c*) there is no evidence of lava at this horizon under the quarry area. The igneous rocks are here interpreted as of later $P_1$ (upper $D_2$) or $P_2$ age and lying near an unexposed eruptive centre.

# BASIN PROVINCE

The succession in the **Edale Borehole** [1078 8493] at Barber Booth, was described by Hudson and Cotton (1945b, pp. 1–36). Viséan beds were passed through between 325 ft and the bottom of the hole at 757 ft. The following description is based on that of Hudson and Cotton with minor revision of fossil names:

| | Thickness ft | Depth ft |
|---|---|---|
| Base of Namurian (see p. 186) | — | 325 |
| Beds of $P_2$ age | | |
| Shales with thin argillaceous limestones; black pyritous limestone, part crinoidal, | | |

| | Thickness ft | Depth ft |
|---|---|---|
| more abundant crinoid debris 348 to 357 ft; *Neoglyphioceras subcirculare* (Miller) and *Sudeticeras* cf. *stolbergi* at 330 to 336 ft; *N. subcirculare* at 344 to 348 ft. Other fossils include *Zaphrentites* cf. *oystermouthensis* (Vaughan), *Chonetes laguessianus* de Koninck, *Martinia* aff. *glabra*, *Caneyella membranacea*, cf. *Posidonia corrugata* .. | 47 | 372 |

| | Thick-ness | | Depth | |
|---|---|---|---|---|
| | ft | in | ft | in |

Limestone, cherty, grey crystalline shelly to 377 ft; bioclastic with limestone and tuff fragments, with *Hyalostelia smithii* Young and Young, to 382 ft; alternating pale bioclastic and dark limestone with foraminifera and algae including '*Girvanella*' to 410 ft; dark fine-grained limestone to 419 ft. Other fossils include *Zaphrentites* cf. *disjunctus* (Carruthers), *Eomarginifera setosa*, *Productus concinnus* J. Sowerby and *Krotovia spinulosa* .. .. .. **47    419**

Beds of upper P₁ age

"Lower Shales-with-limestones"; black crinoidal limestone with paler limestone pebbles and some chert pebbles to 420 ft; tuff, re-worked with "granular carbonates" to 422 ft; shales with bullions and a few argillaceous limestones with *Posidonia becheri* Bronn [abundant], *Eomarginifera tissingtonensis*, *Goniatites elegans* Bisat (at 441 to 445 ft) to 445 ft; limestone, black crinoidal and shelly to 449 ft; mudstone, pyritous with black crinoidal limestones to 458 ft; shale with some grey crinoidal limestone and fragments of black mudstone to 464 ft; calcareous mudstone with bullions, argillaceous limestones and some shelly limestones with *P. becheri* [common], *G. falcatus* Roemer (490–493 ft); other fossils include *Plicochonetes waldschmidti* Paeckelmann and *G. striatus* group to 493 ft .. **74    493**

"Pebbly and crinoidal shell limestones"; coarsely bioclastic shelly and crinoidal

limestones, dark except from 500 to 505 ft; 5 ft shale and black limestone with *G. striatus* group at 500 ft; 2 ft shale with some black limestone at 507 ft; fauna includes *Linoproductus hemisphaericus*, *Antiquatonia insculpta*, *A. sulcata* and *Echinoconchus elegans* .. **19    512**

Beds of lower P₁ age

Calcareous mudstone with crinoid debris .. .. **4    516**

Limestone, grey coarsely crinoidal and shelly limestone with some calcareous mudstone .. .. .. **10    526**

Fragmental shelly limestone, slightly tuffaceous, with some blue and black chert **4    530**

"Upper tuff and tuffaceous limestones", interbedded, the tuffs pyritous in places and grey or green in colour, the limestones shelly, crinoidal and tuffaceous .. **30    560**

Limestone, dark tuffaceous, cherty and crinoidal to 570 ft; shelly with mudstone partings with *Nomismoceras* and *Prolecanites* to 578 ft; black mudstone with *G. crenistria* (s.l.) to 580 ft; dark argillaceous cherty and tuffaceous to 585 ft; pale grey shelly and tuffaceous to base .. .. .. **28    588**

Tuff and tuffaceous limestone **6    594**

Limestones, dark grey, some crinoid and shell debris, chert, thin shale partings; 2 ft pebbly at 599 ft and 3 ft pebbly at 609 ft; 2 ft with "nodules enclosing shell fragments" (probably algal) at 611 ft; argillaceous limestone and shale with occasional volcanic fragments 611–614 ft; shale with a little bedded tuff to 617 ft .. **23    617**

Thick- Depth
ness
ft    ft

[In addition to the fossils listed above, the beds from 512 to 617 ft yielded a fauna including *Nomismoceras* cf. *spirorbis* (Phillips), *Antiquatonia antiquata*, *A. sulcata*, *Echinoconchus punctatus*, *Linoproductus hemisphaericas*, *Pustula pustulosa* (Phillips), *Striatifera striata*.]

Beds of $D_1$ and $D_2$ age[1]

Limestone, pale grey foraminiferal, slightly crinoidal in places with bands of limestone breccia; tuff pebbles between 632 and 637 ft .. 43    660

Limestone, dark grey often shelly or crinoidal, with shale partings in upper part; some pebbly bands; 3 ft dolomitic at 665 ft; 3 ft with tuff fragments at 670 ft; fossiliferous throughout, especially 660 to 662 ft; fossils include '*Girvanella*' (especially 670 to 680 ft), *Palaeosmilia murchisoni*, *Echinoconchus punctatus*, *Gigantoproductus maximus*, *Plicatifera plicatilis*, *Sinuatella sinuata* .. .. 29    689

"Calcite-mudstones", some brecciated .. .. .. 23    712

Limestone, dark grey to dark brown; finely crystalline and dolomitic to 730 ft; finely fragmented and ?dolomitic below .. .. .. 45    757

[The beds between 689 ft and 757 ft yield a poor fauna composed mainly of productoids].

Fossils from this borehole are not included in the appendices. The stratigraphy of the borehole is discussed on p. 36.

The **Alport Borehole** [1360 9105] near the junction of Swint Clough and the River Alport was also described by Hudson and Cotton (1945a, pp. 254–311). The following description is abridged from their account, though the fossil names have been brought up to date and certain of the goniatite determinations and zonal boundaries revised.

Thick- Depth
ness
ft    ft

Base of Namurian (see pp. 195–7) .. .. .. — 1099

*Lyrogoniatites georgiensis* ($P_{2c}$) Zone

Mudstone, calcareous with nodular and calcareous bands; *Caneyella membranacea*, *Dimorphoceras sp.*, *Girtyoceras costatum* (Ruprecht), *G. shorrocksi* Moore, *G.* cf. *waitei* Moore, *Lyrogoniatites georgiensis* Miller and Furnish .. .. 8    1107

*Neoglyphioceras subcirculare* ($P_{2b}$) Zone

Mudstone, with a few thin beds of limestone; *Caneyella membranacea*, *Dimorphoceras sp.*, *Sudeticeras stolbergi* .. .. .. 12    1119

Mudstone, calcareous with sporadic thin limestones; 2 ft bed with phosphatic nodules at 1123 ft; *Paraconularia sp.*, *Rugosochonetes sp.*, *Caneyella membranacea*, *Dunbarella sp.*, *Dimorphoceras sp.*, *Neoglyphioceras subcirculare* at 1124 to 1126 ft; *Sudeticeras spp.* (including *S. splendens* Bisat sp.) at 1135 ft .. 19    1138

*Mesoglyphioceras granosum* ($P_{2a}$) Zone

Mudstone, dark grey calcareous, with some thin-bedded limestone; *Asterocalamites scrobiculatus* (Schlotheim), *Paraconularia sp.*, *Caneyella membranacea*, *Dunbarella sp.*, *Kazakhoceras* cf. *hawkinsi* (Moore), *Mesoglyphioceras granosum*, *Sudeticeras sp.* .. 12    1150

---

[1] Only that part of the post-$D_1$ beds with a coral-brachiopod fauna (i.e. $D_2$ *pars*) is included here, the higher beds with a goniatite fauna being represented by their $P_1$–$P_2$ equivalents.

|  | Thick-ness ft | Depth ft |
|---|---|---|

Mudstone, dark grey calcareous, with a thin band of hard grey limestone in lowest 3 ft; *Paraconularia sp., C. membranacea, Dunbarella sp.,* orthocones, *M. granosum, Neoglyphioceras sp.* .. .. .. 12 1162

Mudstone, dark grey calcareous, with hard grey bioclastic and fragmental limestones; a little chert below 1172 ft; *Hyalostelia sp., Lingula sp.,* productoid spines, *Aviculopecten sp., C. membranacea, Dunbarella sp.* .. 24 1186

Limestone, with limestone pebbles and breccia and some mudstone fragments; some intraclasts of ? decomposed tuff; *Antiquatonia sp., Sudeticeras sp.* .. .. 10 1196

*Goniatites koboldi* (P$_{1d}$) Zone

Mudstone, lowest 5 ft with bands of brownish and grey limestone; *Antiquatonia sp., Chonetes sp.* [s.l.,] *Gigantoproductus sp., Epistroboceras sulcatum* (J. de C. Sowerby), orthocones, goniatites indet. 11 1207

Mudstone, dark grey calcareous, with bands of very fine-grained grey limestone in beds up to 1 ft; *?Posidonia becheri,* nautiloids, *Dimorphoceras sp., G. koboldi* Ruprecht group, *Sudeticeras sp.* .. .. .. 23 1230

*Goniatites sphaericostriatus* (P$_{1c}$) Zone

Mudstone, grey and greenish grey, also fine-grained grey limestones, some chert and brownish grey limestone; cf. *Sphenopteridium dissectum* (Goeppert), *Sphenopteris foliata* Stur, echinoid plates, *P. becheri* (abundant at 1233 and 1285 ft), *Goniatites bisati* Moore, *G. kajlovecensis* Patteisky, *G.* cf. *waldeckensis* Haubold, *G.*

*sphaericostriatus* Bisat, *Neoglyphioceras spirale* (Phillips) (at 1274 ft) .. .. 55 1285

Mudstone, with some limestone bands; *P. becheri, G. elegans* Bisat, *G.* aff. *falcatus,* striatoid goniatites indet. .. .. .. 17 1302

*Goniatites falcatus* (P$_{1b}$) Zone

Mudstone with limestone bands, some coarse shelly and crinoidal, more abundant than in higher beds, some chert; foraminifera, *Rugosochonetes sp.,* productoids including *Overtonia fimbriata; P. becheri, G. falcatus, G. spirifer* .. 35 1337

*Goniatites crenistria* (P$_{1a}$) Zone

Limestone, part dark grey argillaceous and fragmental, foraminiferal, and part brownish dolomite-mudstone .. .. .. 7 1344

Limestone, mainly grey fine-grained, with some brachiopod fragments, some mudstone; *Hyalostelia sp., P. becheri, Nomismoceras vittiger* (Phillips), *G. crenistria, G.* cf. *concentricus* Hodson and Moore .. .. .. 32 1376

Limestone, grey and grey-brown, with mudstone at top and dark grey mudstone with crinoid debris and pebbles of limestone and tuff; beds of tuffaceous limestone with pyrite at 1378 and 1384 ft .. .. 8 1384

Mudstone, with fine-grained grey limestone in beds up to 2 ft thick, sporadically crinoidal and sometimes shelly; some of the limestones pyritous, rarely cherty; mainly limestone 1409 to 1416 ft, tuffaceous debris at 1408 and 1415 ft, and in thin bands at 1426 and 1433 ft; 2 ft dark grey argillaceous limestone with

| | Thick- ness ft | Depth ft | | | Thick- ness ft | Depth ft |
|---|---|---|---|---|---|---|

pebbles of limestone and tuff at 1421 ft; *Dunbarella sp.*, *P. becheri*, *Rineceras sp.*, *Epistroboceras sulcatum*, and *G. crenistria* group .. **50 1434**

Limestone, pale grey coarsely bioclastic with worn shell and crinoid debris to 1436 ft; dark grey fine-grained with lapilli of greenish tuff to 1437 ft; black argillaceous with crinoid and shell debris, some in pyrite, and greenish tuff 'lapilli' to 1438 ft; grey-brown fine-grained to 1440 ft; dark argillaceous with crinoids, brachiopods, tuff 'lapilli' and limestone pebbles, thin layers of laminated tuffaceous mudstone near base to 1445 ft .. **11 1445**

Tuff and tuffaceous mudstone, banded, often pyritous, with banded and nodular chert; some slump structures .. .. .. **15 1460**

Limestone, platy and slightly tuffaceous .. .. .. **8 1468**

Tuff, bluish grey calcareous with thin tuffaceous mudstone at base, banded and nodular chert .. .. **7 1475**

Limestone, argillaceous and mudstone, with chert; bryozoa, *G. crenistria* group at 1478 ft .. .. .. **15 1490**

Upper *Beyrichoceras* (B₂) Zone

Mudstone and brownish grey or dark grey argillaceous limestone, shelly and crinoidal in parts, thin bands of tuff at 1493, 1508, 1512, 1536 and 1537 ft; *Dunbarella sp.*, *Beyrichoceras* aff. *delicatum*, *Bollandoceras micronotum*, *G. crenistria* group, *G.* aff. *hudsoni*, *G. struppus*, *Nomismoceras?* .. **52 1542**

Limestone, brownish grey or grey; cf. *Actinoconchus sp.*, *Chonetes sp.* (s.l.), *Pustula sp.*, *Beyrichoceras* cf. *delica-*

*tum*, *Prolecanites discoides*, *Pronorites cyclolobus* (Phillips), ostracods and a trilobite fragment .. .. **68 1610**

Limestone, dark brownish grey, with shale, tuffaceous and with slumping below 1632 ft; *Lingula sp.*, *Dunbarella sp.*, *Bollandites* cf. *phillipsi* (Bisat), *G. maximus* group possibly *G. maximus* var. *b* Bisat, *Prolecanites* cf. *discoides* .. .. .. **32 1642**

Limestone, brownish grey, argillaceous in places, with some thin calcareous mudstone bands; some chert from 1651 to 1664 ft; athyroid, spiriferoids, cf. *Parallelodon tenuistriatus* (McCoy), orthocone nautiloid, *Goniatites sp.* ?*maximus* group, *Nomismoceras sp.* .. **59 1701**

Lower part of Lower *Dibunophyllum* (D₁) Zone

Limestone grey or brownish grey fine-grained; calcilutite and chert nodules in places; nodular pebble bed at 1781 ft and crinoidal limestones with productoids and corals at 1853 ft; from 1701 to 1780 ft fossils as follows: bryozoa, *Antiquatonia?*, athyroid, *Chonetes hemisphaericus* von Semenow, *Chonetipustula carringtoniana* (Davidson), *Composita* cf. *ambigua* (J. Sowerby), *Krotovia spinulosa*, *Plicochonetes* aff. *elegans* (de Koninck), *P.* aff. *ventricosus* (Tornquist), *Rhipidomella?*, spiriferoids, *Ptychomphalus sp.*; 1781 ft, *Chaetetes sp.*, *Michelinia sp.*, *Avonia sp.*, *Chonetipustula carringtoniana*, *Plicochonetes sp.*, spiriferoid, ostracods; from 1781 to 1850 ft, *Antiquatonia?*, *P.* aff. *elegans*, ?*Coelonautilus konincki* (d'Orbigny);

J

| | Thick- ness ft | Depth ft |
|---|---|---|

from 1850 to 1852 ft, *Caninia amplexoides* (Wilmore), *Cladochonus sp.*, *L. pauciradiale*, *Syringopora?*, *Chonetipustula sp.*, chonetoid, orthotetoid, *Rhipidomella?*, *Paraparchites sp.*; from 1852 to 2027 ft, cf. *Permia*, *C.* cf. *carringtoniana*, *Lingula parallela* McCoy, *Megachonetes* cf. *dalmanianus* (de Koninck), *Orbiculoidea* cf. *tornacensis* Demanet, orthotetoid, *P.* aff. *elegans*, *Tornquistia* aff. *polita* (McCoy), orthocone nautiloid .. 326  2027

*Seminula* (S₂) Zone

Limestone, dark brownish grey dolomitic unfossiliferous, some calcilutites, chert quite common, some mudstone partings, rare beds with shell debris; 1-ft bed at 2028 ft with crinoid and shell debris and (algal) pellets; foraminifera, *Cryptophyllum?*, *Cyathaxonia rushiana* Vaughan, *Lithostrotion sociale*, *Rotiphyllum sp.*, *Rylstonia sp.*, cf. *R. benecompacta* Hudson and Platt [juv.], *Zaphrentites sp.* cf. *Z. lawstonensis* (Carruthers), chonetoids, *Composita?*, *Linoproductus corrugatohemisphericus*, *Overtonia fimbriata*, *Reticularia sp.*, *Bollandoceras?* .. 109  2136

Limestone, dark brownish grey, dolomitic cementstones with calcareous mudstone partings and chert; bands of calcilutite, some with pyrite; 1½ in of tuffaceous mudstone and limestone at 2300 ft, calcareous shale predominant and chert rare in top 143 ft; calcilutites and dolomite-mudstones more abundant in bottom 83 ft; *Cyathaxonia rushiana*, *Rotiphyllum sp.*,

| | Thick- ness ft | Depth ft |
|---|---|---|

bryozoa, *Antiquatonia?*, *Chonetipustula carringtoniana*, *C.* cf. *plicata* (Sarres), *?Cleiothyridina royssii* (Davidson), *Lingula elliptica* Phillips, *Linoproductus corrugatohemisphericus* [juv.], *O. fimbriata*, *Phricodothyris sp.*, *Pustula sp.*, *Semenewia sp.*, *Aviculopecten sp.*, 'Pseudamussium' *sp.*, *Phillibole* cf. *megalophthalma* (Weber), *P.* cf. *polleni* (H. Woodward), ostracods .. .. 226  2362

Limestone, dark grey fine-grained dolomitic, with shell and crinoid fragments, some chert and ?'lapilli' of light grey tuff .. .. 4  2366

Limestone, grey medium-grained, "conglomeratic", with shell and crinoid debris; pebbles of shale, pellet limestone and dark grey fine-grained limestone; many fragments, 'lapilli' or pebbles of blue and green tuff (up to 3 in diameter); abundant corals including *Caninia* cf. *amplexoides*, *Cyathaxonia rushiana*, *Diphyphyllum smithi* Hill, *Fasciculophyllum* cf. *densum* (Carruthers) [juv.], *Lithostrotion martini*, *L. minus* (McCoy), *Rotiphyllum sp.*, *Rylstonia dentata* Hudson and Platt subsp., chonetoid, *Davidsonina carbonaria* .. .. .. .. 5  2371

Limestone, dark grey fine-grained argillaceous, some calcilutites and shale; *Cyathaxonia rushiana*, *F. densum*, *Rotiphyllum sp.*, *Chonetipustula sp.*, *D. carbonaria*, *Linoproductus sp.* .. 20  2391

Breccia (?contemporaneous), of limestone and calcareous mudstone, passing down into mudstone with thin lime-

| | Thick-ness ft | Depth ft |
|---|---|---|
| stone bands, "both crumpled" .. .. .. | 1 | 2392 |
| Limestone, dolomitic, resting with high angle junction on fine-grained dolomitic limestone with pebbles of porcellanous limestone, recrystallised tuffaceous mudstone, pellet limestone, tuff and some chert. Fossils rare; *F. densum, Leiorhynchus sp.*, productoids, *Platyceras?* .. .. .. | 4 | 2396 |
| Limestone, dark grey argillaceous and dolomitic, with chert and rare shell and crinoid debris .. .. .. | 7 | 2403 |
| Limestone, fine- and coarse-grained dolomitic, with oblique mudstone partings (dip 60° to 70°), crinoid debris and limestone pebbles; *Cyathaxonia rushiana, F. densum, L.* cf. *martini* .. | 3 | 2406 |
| Limestone, dolomitic and crinoidal with chert, limestone conglomerate with mudstone parting (dip 70°) in middle; *Cyathaxonia sp., Rotiphyllum sp., Zaphrentites* cf. *ashfellensis* (Lewis), *D. carbonaria, Phricodothyris sp.*, productoids, rhynchonelloid .. .. .. | 2 | 2408 |
| Limestone, dark grey fine-grained, subangular limestone pebbles in lower 2 ft; chert and crinoid debris; *Z. enniskilleni, Lingula parallela* .. .. .. | 4 | 2412 |
| Limestone breccia, dark grey, with limestone and some tuff fragments (up to 4 in by 4 in); veined with calcite, baryte, sphalerite and fluorite; *L.* cf. *martini, Rotiphyllum sp., Rylstonia* aff. *dentata, Eomarginifera derbiensis, Pustula sp.* [juv.], *Rhipidomella michelini* .. | 5 | 2417 |
| Limestone, dark grey fine-grained, dolomitic slightly crinoidal, with some black chert; *Aulina sp., Cyathaxonia rushiana, F. densum* [juv.], *Michelinia sp., Lithostrotion sp., Rhopalolasma sp.* cf. *R. sympecta* Hudson, *Rotiphyllum nodosum* Smyth, *R.* aff. *rushianum* Smyth *non* Vaughan, *Antiquatonia?, Avonia sp., Chonetipustula* cf. *plicata, Echinoconchus eximius, Krotovia sp., Linoproductus sp., Plicochonetes sp.*, cf. *Productina transversistriatus* (Paeckelmann), rhynchonelloid, *Psephodus sp.* .. .. .. | 14 | 2431 |
| Limestone, dark grey fine-grained, dolomitic in upper part, thin mudstone partings throughout; rare chert; *Z. enniskilleni, L. corrugatohemisphericus* .. .. | 124 | 2555 |

Hudson and Cotton (ibid., p. 256) named the shales above 1302 ft the "Alport Shales" and these were underlain by the "Alport Limestones-with-Shales" extending to 1642 ft and the "Alport Limestones" to the base of the hole at 2555 ft. The term "Alport Shales" is here discontinued as the uppermost 45 ft (originally 26 ft) falls within the Namurian and the remaining terms are of local value only. The stratigraphy of the borehole is discussed on pp. 36–7 and it should be noted that fossils from the borehole have not been included in the lists in the appendices. I.P.S.

# PETROGRAPHY OF THE IGNEOUS ROCKS

## LAVAS

Most of the following descriptions apply to the Lower Miller's Dale Lava from which it has been possible to obtain reasonably fresh specimens. The less extensive outcrops of the Upper Lava have yielded satisfactory materials for slicing at one locality only.

## LOWER MILLER'S DALE LAVA

In all the specimens examined, the Lower Lava consists of olivine-basalt or basalt, of dark grey to grey-green colour, mottled black or red-brown (where hematitized) and containing sporadic pseudomorphous phenocrysts or glomeroporphyritic aggregates mainly 1 to 3 mm across. Vesicles or sub-vesicular patches are generally conspicuous, being filled with white carbonates and/or dark chlorite. The groundmass is generally fine-grained though crystals are usually visible macroscopically. Specimens from outcrop as well as from boreholes are invariably highly altered through chloritization, carbonation and hematitization, which have affected in particular primary ferromagnesian minerals. The feldspars, however, are least altered, and form in the groundmass a mesh of laths which are generally randomly arranged, though in places (particularly around vesicles and phenocrysts) they show fluxioning. Labradorite predominates in both groundmass and phenocrysts. In the least altered specimens, the essential ferromagnesian mineral is pale green clinopyroxene near augite in composition. Pseudomorphs after olivine are generally present. Opaque ores include ilmenite and magnetite. Evidence for the occurrence of composite lava flows is presented below.

*I.C.I. Borehole T/B/23A Great Rocks Dale* [0985 7614] (see p. 52)

(1) 30 to $63\frac{1}{2}$ ft: This consists predominantly of grey-green (speckled black) coarse-grained olivine-basalt (groundmass granularity averaging 0·6 by 0·1 mm), with calcite veining at 35 to 36 ft, and sporadic hematitization at 30 to 31 ft, 34 to 35 ft, 42 to 45 ft. The groundmass becomes a little finer grained with depth. A strongly pleochroic, fibrous clay mineral occurring in veinlets at 45 ft (with X = blue-green, Z = green; absorption X > Z), was identified by an X-ray powder photograph (NEX 801)[1] as a chlorite. A section (E 35229)[1] from 35 ft 8 in consists of pseudomorphous olivine phenocrysts, up to 3 mm in length, in glomeroporphyritic aggregates, and a groundmass mesh of randomly orientated labradorite laths (near $Ab_{46}An_{54}$) averaging 0·6 by 0·1 mm, which show some fluxioning around the phenocrysts. Accessory ilmenite forms randomly arranged laths, transgressing feldspars. Ferromagnesian minerals have been completely replaced by chlorite and carbonate. Feldspar laths show central alteration to a pinkish clay mineral, carbonate and leucoxene. Interstitial minerals include microgrowths of potash feldspar, apatite needles, xenomorphic quartz, chlorite and calcite. Chlorite is mainly olive-green, vermiform and radially fibrous, patchily stained by hematite. Subvesicles have chloritic rims and calcite cores. A further specimen (E 35230) of olivine-basalt at 45 ft 10 in is considerably less altered. The groundmass framework of labradorite (near $Ab_{43}An_{57}$) laths includes colourless to pale green, granular, allotriomorphic to hypidiomorphic fresh augite (averaging 0·08 mm across). Phenocrysts range up to 1·5 mm in length, are composed of carbonate, chlorites and chalcedony, and are pseudomorphous after probable olivine. Intergranular constituents include brown and green chlorites, chalcedony, quartz and rather sparse carbonate. There are plentiful laths of strongly transgressive accessory ilmenite. Specific gravities of three specimens at 36 ft 3 in, 38 ft 5 in and 45 ft 10 in were determined as 2·55, 2·54 and 2·63, respectively.

---

[1]NEX numbers refer to the X-ray powder photograph collection of the North of England Office of the Institute of Geological Sciences; E numbers refer to the Sliced Rock Collection of the Institute.

(2) $63\frac{1}{2}$ to 79 ft: A lithological change occurs at $63\frac{1}{2}$ ft, for the basalt is both finer grained, darker coloured and considerably more stained by hematite from this depth to 78 ft. Veining by calcite and vesicular development are also more pronounced (65 to 68 ft and 76 to 79 ft) than in the higher core described above. As a result of the hematitization, specific gravities are also a little higher— 2·79 and 2·76 for specimens at 76 ft 2 in and 77 ft 1 in respectively. A sectioned core at 66 ft (E 35231–2) is fine-grained, chloritized, hematitized and carbonated olivine-basalt, with pseudomorphous phenocrysts after probable olivine attaining 1·0 by 0·5 mm, scattered in a highly altered mesh of albitized plagioclase laths (averaging 0·4 by 0·08 mm), which are locally fluxioned around phenocrysts or subvesicles. Primary ferromagnesian minerals have been completely replaced. There are plentiful laths of partly leucoxenized ilmenite, with intergranular chlorite and carbonates. Olivine pseudomorphs are composed of patchy olive-green chlorite with hematitic veining. Vesicles (1 to 2 mm diameter) are filled with coarse calcite, green chlorite and chalcedony, fringed with opaque oxides and fluxional plagioclase laths. A stringer of the pleochroic clay mineral described above (at 45 ft) also occurs in this section. The fibres are arranged normal to the stringer walls. A section (E 35233) at 77 ft 4 in is closely similar.

The evidence that the two basalts described represent individual flows is supported by their distinct lithologies, and in particular by the intensive hematitization of the lower basalt which may indicate subaerial weathering prior to the later flow and development of vesicles.

*Sink-hole, north of Water Swallows* [0836 7557]

A specimen (E 32820) of the lava (see p. 52) is a green-grey (patchily stained purple) amygdaloidal olivine-basalt. It is somewhat coarse-grained with plagioclase laths forming the groundmass framework, averaging 0·6 by 0·1 mm, phenocrysts attaining 1·5 mm across. Glomeroporphyritic aggregates attain 3 mm across and other phenocrysts appear to represent pseudomorphous olivine crystals, replaced by carbonate and chlorite and veined by opaque oxides. Other coarse structures formed of iron oxide, chlorite and quartz are too irregular for identification, but may also represent replaced olivine crystals. Feldspars retain albite twinning but have been albitized and largely replaced by chlorite. There is abundant interstitial material (mainly olive-green chlorite, quartz and carbonate) and individual grains of chlorite may represent pseudomorphous pyroxene. There are plentiful opaque ilmenite laths and irregular hematite. Vesicles are commonly filled with calcite rimmed by chlorite and contain patches of chalcedony.

*Cave Dale, Castleton* [1478 8218]

Though the hand specimen (E 33466) is apparently fresh, dark green-grey, fine- to medium-grained, and charged with vesicles, in section the basalt is heavily carbonated and chloritized. The feldspar laths forming the rock framework (averaging 0·07 by 0·4 mm) are largely replaced by carbonate and chlorite. There is a brown mesostasis which includes leucoxene, chlorite and some potash feldspar. Accessory laths and needles of partially leucoxenized ilmenite are common. Vesicles average 1 mm diameter and contain radially fibrous chlorite in outer zones, with scalloped fringes lined with chalcedony and carbonate cores.

UPPER MILLER'S DALE LAVA

*Roadside exposure, Bole Hill* [1059 7545]

A specimen (E 32828) of relatively fresh olivine-basalt from this exposure is dense, dark grey and aphyric, with abundant, black, irregular micro-amygdaloidal patches up to 3 mm across. In section there are sparse chlorite pseudomorphs after olivine (up to 0·4 mm across) scattered in a mesh of fresh plagioclase laths, averaging 0·4 by 0·06 mm, later undulose plagioclase and intergranular allotriomorphic colourless to pale pink augite (0·1 to 0·2 mm across). The principal plagioclase is labradorite near $Ab_{42}An_{58}$. There are plentiful laths of opaque ore, principally magnetite. The fine-grained interstitial matrix is composed principally of chlorite, which commonly exhibits vermiform brownish zones set in a pale green, radial-fibrous base containing microglobules of chalcedony. Carbonate patchily replaces labradorite laths.

CRESSBROOK DALE LAVA

*Tansley Dale* [1728 7521]

A specimen (E 33474) is dark green-grey, vesicular, olivine-basalt, with amygdales showing preferred orientation and interconnection by strings of calcite and hematite. The basalt consists of likely phenocrysts after olivine (0·6 mm) and a plexus of fine (0·2 by 0·05 mm) albitized, plagioclase laths heavily replaced by chlorite, quartz and carbonate. Intra-feldspar matrix is occupied by olive-green chlorite, carbonate, hematite, quartz and opaque granules (pyrite, leucoxene, ilmenite-magnetite). There are plentiful laths of ilmenite which show some fluxioning. Primary ferromagnesian minerals may be represented by intra-feldspar chloritic grains and subhedra. Vesicles are common, ranging up to 4 mm diameter and are filled with microglobular green chlorite, fringed with darker chlorite, hematite or carbonate, and set in a coarse calcite cement.

A second specimen (E 33474A) is less vesicular and less altered, with glomeroporphyritic aggregates of carbonate and quartz, representing olivine phenocrysts (1 to 2 mm). Plagioclase laths (0·6 by 0·1 mm) are fluxioned in places and are composed of labradorite, near $Ab_{42}An_{58}$. Primary ferromagnesian minerals, however, have been completely altered to chlorite which, together with ilmenite and leucoxene, form much of the intra-feldspar matrix.

*Wardlow Mires No. 1 Borehole* [1850 7553] (see p. 93)

The lava is a dark green-grey basalt speckled with black chloritic patches. In places feldspar laths form microphenocrysts, and vesicles are prominent, especially towards the top of the flow, where they are filled with calcite and chlorite. Towards the base of the lava, there is a brecciated structure, with rounded masses of purple-grey basalt, marginally altered to green chlorite, engulfed in a complex of layered or irregular pumiceous fragments and cemented by dark basalt and chlorite. This coarse structure may, perhaps, represent a flow-breccia, though this interpretation awaits a complete description of the core. The high proportion of chlorite cement might perhaps suggest a degree of partial magmatic refusion of solidified and brecciated lava blocks prior to incorporation in the lava flow. The sharp contacts between pyroclastic (or pseudopyroclastic components) and chlorite cement, however, do not indicate that severe chemical or physico-chemical reaction between them has in fact occurred, and the injection of the chlorite cement took place within solidified and relatively non-reactive fragments.

A specimen (E 35260) of the least vesicular lava is dark grey, aphyric, speckled black with a streaky segregation of coarser chlorite. In section, the basalt is highly altered and consists of a mesh of albitized plagioclase laths (averaging 0·9 by 0·09 mm), skeletal opaque ores (partly leucoxenized), and much interstitial, olive-green, vermicular chlorite, some representing replaced primary ferromagnesian minerals, and a little carbonate.

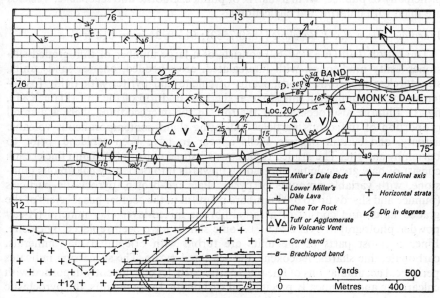

FIG. 14. *Sketch-map showing the structural relations of the Monk's Dale vents*

## PYROCLASTIC ROCKS

### DOVE HOLES TUFF

*Holderness Quarry* [0833 7821] (see p. 54)

The material (E 33477A) is an aggregate of calcite particles, with interstitial green chlorite particles, flakes and galls, charged with fine pyrite crystals. Two rounded limestone lapilli (E 33905–6) contain chlorite and pyrite particles, have vesicular chloritic skins, and show a degree of recrystallization.

### LITTON TUFF

*Temporary Section, Litton* [1676 7517] (see p. 75)

A volcanic bomb measures 12 cm by 8·5 cm maximum thickness and 11 cm maximum width, tapering with curved surfaces to a characteristic 'tail'. It consists of very vesicular purple-grey basalt, the vesicles being sparse on outer surfaces and showing little elongation on broken surfaces, where they are filled with white calcite. A thin section (E 33489) taken near the thickest part of the bomb, shows a mesh of feldspar lath swarms (laths averaging 0·1 by 0·02 mm) fluxioned around abundant vesicles (1 to 2 mm diameter) and pseudomorphous phenocrysts. The groundmass feldspar is principally labradorite near $Ab_{46}An_{54}$. The pseudomorphs (up to 1·5 mm length) appear to be after feldspar, and possibly olivine. The intra-feldspar matrix consists of chlorite, hematite, opaque dust (mainly leucoxene) and quartz. Vesicles are spheroidal to highly irregular, vary in size, and are filled with calcite, chlorite, fibrous chalcedony and a likely zeolite, probably analcime.

The tuffaceous matrix is highly weathered and consists of abundant lapilli, principally of pumice, set in a matrix of pumice, chlorite and leucoxene. Vesicles are filled with chlorite and chalcedony.

*Wardlow Mires No. 1 Borehole* [1850 7553]; 613 ft 3½ in to 614 ft 6 in

The Litton Tuff (E 35756–62) consists of a large number of thin graded silt-stones—silty mudstones, with intercalated, cross-bedded, coarser sandy units, structureless mudstone bands, seams of calcite and pyrite, with richly fossili-ferous microcoquinoidal limestone at the base and top of the tuff (Plate XXII, fig. 2). Fluorite also occurs near the top. In the microcoquina there is a small proportion of tuffaceous particles indicating slight contemporaneous volcanic activity, and certain organic shells appear to have been replaced, prior to burial, by chlorite. The graded tuffaceous siltstone units (6 to 20 mm thick) each consist of pale grey calcitic silt grading upwards into dark grey calcareous mud-stone, with variable amounts of pyrite. Pale red-brown, pyroclastic particles (pumice and shards up to 0·08 mm) are plentiful in the coarser basal layers of the graded units, and consist of a radial-fibrous mixed-layer clay mineral (X-ray powder photographs NEX 815–7). Some shards have been replaced by calcite. Finer topmost parts of the units are turbid with orientated clay minerals, carbonate, fine shards, leucoxene and scattered pyrite. The succession indicates periodical accumulations of fine pyroclastic particles concurrently deposited with calcarenite and chemically precipitated carbonate. In some units, the pyroclastic particles have been replaced by calcite.                                    R.K.H.

## REFERENCES

AGER, D. V. 1963. *Principles of Paleoecology*. New York.

ARNOLD-BEMROSE, H. H. 1894. On the microscopical structure of the Carboniferous dolerites and tuffs of Derbyshire. *Q. Jl geol. Soc. Lond.*, **50**, 603–44.

————— 1898. On a quartz-rock in the Carboniferous Limestone of Derbyshire. *Q. Jl geol. Soc. Lond.*, **54**, 169–83.

————— 1907. The toadstones of Derbyshire: their field-relations and petrography. *Q. Jl geol. Soc. Lond.*, **63**, 241–81.

BARNES, J. and HOLROYD, W. F. 1897. On the occurrence of a sea-beach at Castleton, Derbyshire, of Carboniferous age. *Trans. Manchr geol. Soc.*, **25**, 119–32.

BISAT, W. S. 1934. The goniatites of the *Beyrichoceras* zone in the North of England. *Proc. Yorks. geol. Soc.*, **22**, 280–309.

BLACK, M. 1933. The precipitation of calcium carbonate on the Great Bahama Bank. *Geol. Mag.*, **70**, 455–66.

BLACK, W. W. 1953. Critical sections in a Carboniferous reef-knoll. *Geol. Mag.*, **90**, 345–52.

————— 1954. Diagnostic characters of the Lower Carboniferous knoll-reefs in the north of England. *Trans. Leeds geol. Ass.*, **6**, 262–97.

BOND, G. 1950a. The Lower Carboniferous reef limestones of Northern England. *J. Geol.*, **58**, 313–29.

BOND, G. 1950b. The Lower Carboniferous reef limestone of Cracoe, Yorkshire. *Q. Jl geol. Soc. Lond.*, **105**, 157–88.

———— 1950c. The nomenclature of Lower Carboniferous 'Reef' limestones in the north of England. *Geol. Mag.*, **87**, 267–78.

BROADHURST, F. M. and SIMPSON, I. M. 1967. Sedimentary infillings of fossils and cavities in limestone at Treak Cliff, Derbyshire. *Geol. Mag.*, **104**, 443–8.

COPE, F. W. 1933. The Lower Carboniferous succession in the Wye Valley region of North Derbyshire. *J. Manchr geol. Ass.*, **1**, 125–45.

———— 1936. The *Cyrtina septosa* band in the Lower Carboniferous succession of North Derbyshire. *Summ. Prog. geol. Surv. Gt Br.*, for 1934, (2), 48–51.

———— 1937. Some features in the $D_1$–$D_2$ limestones of the Miller's Dale region, Derbyshire. *Proc. Yorks. geol. Soc.*, **23**, 178–95.

———— 1939. The mid-Viséan ($S_2$–$D_1$) succession in North Derbyshire and North-West England. *Proc. Yorks. geol. Soc.*, **24**, 60–6.

———— 1949. Woo Dale Borehole near Buxton, Derbyshire. *Q. Jl geol. Soc. Lond.*, **105**, iv.

DUNHAM, A. C. and KAYE, M. J. 1965. The petrology of the Little Whin Sill, County Durham. *Proc. Yorks. geol. Soc.*, **35**, 229–76.

DUNHAM, K. C. 1952. Fluorspar. 4th Edit. *Mem. geol. Surv. spec. Rep. Miner. Resour. Gt Br.*, **4**.

EARP, J. R., MAGRAW, D., POOLE, E. G., LAND, D. H. and WHITEMAN, A. J. 1961. Geology of the country around Clitheroe and Nelson. *Mem. geol. Surv. Gt Br.*

EDEN, R. A. 1954. in *Summ. Prog. geol. Surv. Gt Br.*, for 1953, 37.

———— ORME, G. R., MITCHELL, M. and SHIRLEY, J. 1964. A study of part of the margin of the Carboniferous Limestone 'massif' in the Pin Dale area, Derbyshire. *Bull. geol. Surv. Gt Br.*, No. 21, 73–118.

EVANS, W. B., WILSON, A. A., TAYLOR, B. J. and PRICE, D. 1968. Geology of the country between Macclesfield, Congleton, Crewe and Middlewich. *Mem. geol. Surv. Gt Br.*

FAREY, J. 1811. *A General View of the Agriculture and Minerals of Derbyshire*, **1**. London.

FEARNSIDES, W. G. and TEMPLEMAN, A. 1932. A boring through Edale Shales to Carboniferous Limestone and pillow lavas at Hope Cement Works near Castleton, Derbyshire. *Proc. Yorks. geol. Soc.*, **22**, 100–21.

FORD, T. D. 1952. New evidence on the correlation of the Lower Carboniferous reefs at Castleton, Derbyshire. *Geol. Mag.*, **89**, 346–56.

———— 1960. White Watson (1760–1835) and his Geological Sections. *Proc. Geol. Ass.*, **71**, 349–63.

———— 1964. A recent example of soil erosion on the Derbyshire limestone. *Mercian Geol.*, **1**, 31–3.

———— 1965. The palaeoecology of the goniatite bed at Cowlow Nick, Castleton, Derbyshire. *Palaeontology*, **8**, 186–91.

———— and MASON, M. H. 1967. Bibliography of the geology of the Peak District of Derbyshire (to 1965). *Mercian Geol.*, **2**, 133–244.

GEORGE, T. N. 1958. Lower Carboniferous palaeogeography of the British Isles. *Proc. Yorks. geol. Soc.*, **31**, 227–318.

GREEN, A. H., FOSTER, C. LE NEVE and DAKYNS, J. R. 1887. The geology of the Carboniferous Limestone, Yoredale Rocks, and Millstone Grit of North Derbyshire. 2nd edit. with additions by A. H. Green and A. Strahan. *Mem. geol. Surv. Gt Br.*

HIND, W. and HOWE, J. A. 1901. The geological succession and palaeontology of the beds between the Millstone Grit and the Limestone-Massif at Pendle Hill and their equivalents in certain other parts of Britain. *Q. Jl geol. Soc. Lond.*, **57**, 347–404.

HUDSON, R. G. S. and COTTON, G. 1945a. The Lower Carboniferous in a boring at Alport, Derbyshire. *Proc. Yorks. geol. Soc.*, **25**, 254–330.

———— 1945b. The Carboniferous rocks of the Edale Anticline, Derbyshire. *Q. Jl geol. Soc. Lond.*, **101**, 1–36.

———— and MITCHELL, G. H. 1937. The Carboniferous Geology of the Skipton Anticline. *Summ. Prog. geol. Surv. Gt Br.*, for 1935, (2), 1–45.

INSTITUTE OF GEOLOGICAL SCIENCES. 1968. Annual Report for 1967.

JACKSON, J. W. 1908. Carboniferous Fish remains in north Derbyshire. *Geol. Mag.*, **45**, 309–10.

———— 1925. The relation of the Edale Shales to the Carboniferous Limestone in North Derbyshire. *Geol. Mag.*, **62**, 267–74.

———— 1941. Description of a Carboniferous Limestone section with *Girvanella* in north Derbyshire. *J. Manchr geol. Ass.*, **1**, 239–46.

KENT, P. E. 1966. The structure of the concealed Carboniferous rocks of north-eastern England. *Proc. Yorks. geol. Soc.*, **35**, 323–52.

MAWE, J. 1802. *The Mineralogy of Derbyshire*. London.

MORRIS, T. O. 1929. The Carboniferous Limestone and Millstone Grit series of Stoney Middleton and Eyam, Derbyshire. *Proc. Sorby scient. Soc.*, **1**, 37–67.

NEWELL, N. D., RIGBY, J. K., FISCHER, A. G., WHITEMAN, A. J., HICKOX, J. E. and BRADLEY, J. S. 1953. *The Permian reef complex of the Guadalupe Mountains region, Texas and New Mexico*. San Francisco.

ORME, G. R. 1967. in *Geological Excursions in the Sheffield Region and the Peak District National Park*. Ed. R. Neves and C. Downie. University of Sheffield.

———— 1970a. The $D_2$–$P_1$ 'Reefs' and associated limestones of the Pin Dale–Bradwell Moor Area of Derbyshire. *C.r. VI Congr. Avanc. Étud. Stratigr. Géol. carb.*

———— 1970b. An unsual shallow water facies in the Viséan of Derbyshire. *C.r. VI Congr. Avanc. Étud. Stratigr. Géol. carb.*

PARKINSON, D. 1943. The age of the reef-limestones in the Lower Carboniferous of North Derbyshire. *Geol. Mag.*, **80**, 121–31.

———— 1947. The Lower Carboniferous of the Castleton district, Derbyshire. *Proc. Yorks. geol. Soc.*, **27**, 99–124.

———— 1952. Allometric growth in *Dielasma hastata* from Treak Cliff, Derbyshire. *Geol. Mag.*, **89**, 201–16.

———— 1953. The Carboniferous Limestone of Treak Cliff, Derbyshire with notes on the structure of the Castleton reef-belt. *Proc. Geol. Ass.*, **64**, 251–68.

———— 1965. Aspects of the Carboniferous Stratigraphy of the Castleton–Treak area of north Derbyshire. *Mercian Geol.*, **1**, 161–80.

PETTIJOHN, F. J. 1957. *Sedimentary Rocks*. New York. 2nd Edit.

PRENTICE, J. E. 1951. The Carboniferous Limestone of the Manifold valley region, north Staffordshire. *Q. Jl geol. Soc. Lond.*, **106**, 171–209.

RAYNER, DOROTHY H. 1953. The Lower Carboniferous rocks of the North of England: a review. *Proc. Yorks. geol. Soc.*, **28**, 231–315.

SADLER, HELEN E. 1964. The origin of the "Beach-Beds" in the Lower Carboniferous of Castleton, Derbyshire. *Geol. Mag.*, **101**, 360–72.

———and WYATT, R. J. 1966. A Lower Carboniferous S₂ inlier near Hartington, Derbyshire. *Proc. Geol. Ass.*, **77**, 55–64.

SARGENT, H. C. 1912. On the origin of certain clay-bands in the limestone of the Crich inlier. *Geol. Mag.*, **9**, 406–12.

——— 1921. The Lower Carboniferous chert formations of Derbyshire. *Geol. Mag.*, **58**, 265–78.

SCHNELLMANN, G. A. and WILLSON, J. D. 1947. Lead-Zinc Mineralization in North Derbyshire. *Trans. Instn Min. Metall.*, **485**, 1–14.

SHIRLEY, J. 1959. The Carboniferous Limestone of the Monyash-Wirksworth area, Derbyshire. *Q. Jl geol. Soc. Lond.*, **114**, 411–29.

——— and HORSFIELD, E. L. 1940. The Carboniferous Limestone of the Castleton–Bradwell area, North Derbyshire. *Q. Jl geol. Soc. Lond.*, **96**, 271–99.

——— ——— 1945. The structure and ore deposits of the Carboniferous Limestone of the Eyam District, Derbyshire. *Q. Jl geol. Soc. Lond.*, **100**, 289–310.

SIBLY, T. F. 1908. The faunal succession in the Carboniferous Limestone (Upper Avonian) of the Midland area (North Derbyshire and North Staffordshire). *Q. Jl geol. Soc. Lond.*, **64**, 34–82.

SMITH, E. G., RHYS, G. H. and EDEN, R. A. 1967. Geology of the country around Chesterfield, Matlock and Mansfield. *Mem. geol. Surv. Gt Br.*

SMITH, S. and YÜ, C. C. 1943. A revision of the coral genus *Aulina* Smith and descriptions of new species from Britain and China. *Q. Jl geol. Soc. Lond.*, **99**, 37–62.

STOCKDALE, P. B. 1922. Stylolites: their nature and origin. *Indiana Univ. Stud.*, **9** (55), 1–97.

TIDDEMAN, R. H. 1889. On Concurrent Faulting and Deposit in Carboniferous Times in Craven, Yorkshire, with a Note on Carboniferous Reefs. *Rep. Br. Ass. Advmt Sci.*, 600–3.

TRAILL, J. G. 1939. The geology and development of Mill Close Mine, Derbyshire. *Econ. Geol.*, **34**, 851–89.

VAUGHAN, A. 1905. The palaeontological sequence in the Carboniferous Limestone of the Bristol area. *Q. Jl geol. Soc. Lond.*, **61**, 181–307.

WATSON, W. 1811. *A delineation of the strata of Derbyshire*. Sheffield.

WELLER, J. M. 1960. *Stratigraphic principles and practice*. New York.

WHITEHURST, J. 1778. *An inquiry into the original state and formation of the earth*. London.

WOLFENDEN, E. B. 1958. Paleoecology of the Carboniferous reef complex and shelf limestones in north-west Derbyshire, England. *Bull. geol. Soc. Am.*, **69**, 871–98.

——— 1959. New sponges from Lower Carboniferous reefs of Derbyshire and Yorkshire, England. *J. Paleont.*, **33**, 566–8.

## Chapter III

# STRATIGRAPHICAL PALAEONTOLOGY OF THE CARBONIFEROUS LIMESTONE SERIES

## INTRODUCTION

THE MOST IMPORTANT FEATURE of the Carboniferous Limestone Series of the district is the development of reef-limestones at the northern edge of the outcrop, and from the time of Martin, J. and J. de C. Sowerby and McCoy, more than one hundred years ago, the area has been famous for its well-preserved fossils. Sibly (1908) was the first to apply Vaughan's (1905) coral-brachiopod zonal scheme to the Derbyshire limestones. It was only following Bisat's researches into Carboniferous goniatites (1924; 1934), however, that the importance of the apron-reef stratigraphy and its bearing on the correlation of the coral-brachiopod zones of the shelf facies with the goniatite zones of the basin facies was fully appreciated. Shirley and Horsfield (1940), Parkinson (1943; 1947; 1953) and Hudson and Cotton (1945a; 1945b) wrote important contributions on this topic and summaries of these and other papers dealing with the Carboniferous Limestone of north Derbyshire have already been given (pp. 10–1).

Eden and others (1964) described the Pin Dale area and most of the faunas listed by them are included in this account.

### TABLE 2

*Stratigraphical distribution of selected fossils from the Shelf Province of the Carboniferous Limestone Series (reef-limestone excluded)*

The full list of fossils is given in Appendix III where authors of specific names are also given.

The use of '?', 'cf.' or 'aff.' in this table respectively indicates doubt as to the identification of, similarity to, or departure from, the species named.

The key to the column numbers is as follows:

$P_2$    8. Eyam Limestones

$D_2$ { 7. Upper Monsal Dale Beds
{ 6. Lower Monsal Dale Beds

$D_1$ ⎰ 5. Miller's Dale Beds
4. Upper part of Chee Tor Rock
(above base of Laughman Tor coral band)
3. Lower part of Chee Tor Rock
(below base of Laughman Tor coral band)
2. Upper part of Woo Dale Beds

$S_2$    1. Lower part of Woo Dale Beds

128

| Genera and species | S₂ | D₁ | | | | D₂ | | P₂ |
|---|---|---|---|---|---|---|---|---|
| | 1 | 2 | 3 | 4 | 5 | 6 | 7 | 8 |
| *Lithostrotion arachnoideum* | × | | | | | | | |
| *Davidsonina carbonaria* | × | | | | | | | |
| *Daviesiella sp.* | × | | | | | | | |
| *Linoproductus corrugatohemisphericus* | × | | | | | | | |
| *Carcinophyllum vaughani* | × | | × | | | | | |
| *Lithostrotion* cf. *sociale* | × | × | × | × | | | | |
| *Athyris expansa* | × | × | × | × | cf. | × | | |
| *Lithostrotion martini* | × | × | × | × | × | × | × | |
| *L. portlocki* | aff. | | × | | × | × | × | |
| *Koninckophyllum* θ | | × | × | × | | | | |
| *Lithostrotion aranea* | | cf. | | cf. | | | | |
| *Dibunophyllum bourtonense* φ | | × | × | × | × | | | |
| *Lithostrotion pauciradiale* | | × | × | × | × | × | × | |
| *Palaeosmilia murchisoni* | | × | × | × | × | × | × | cf. |
| *Lithostrotion junceum* | | × | × | × | × | × | × | × |
| *Caninia* cf. *densa* | | | × | × | × | | | |
| *Delepinea comoides* | | | aff. | × | aff. | | | |
| *Gigantoproductus sp. maximus* group | | | × | × | × | | | |
| *G. dentifer* | | | × | | | × | × | |
| *Davidsonina septosa* | | | | | × | × | | |
| *Lithostrotion maccoyanum* | | | | aff. | cf. | × | | |
| *L. decipiens* | | | | ? | | × | × | |
| *Caninia benburbensis* | | | | × | × | × | × | |
| *Gigantoproductus edelburgensis* | | | | × | × | × | × | × |
| *Aulina furcata* | | | | | × | | | |
| *Gigantoproductus crassiventer* | | | | | × | × | × | |
| *Striatifera striata* | | | | | | × | | |
| *Caninia juddi* | | | | | | × | × | |
| *Diphyphyllum lateseptatum* | | | | | | × | × | |
| *Eomarginifera tissingtonensis cambriensis* | | | | | | × | × | |
| *Dibunophyllum bipartitum* subspp. | | | | | | × | × | × |
| 'Brachythyris' *planicostata* | | | | | | × | × | × |
| *Gigantoproductus giganteus* | | | | | | × | × | × |
| *Productus productus hispidus* | | | | | | × | × | × |
| *Pugilis pugilis* | | | | | | × | × | × |
| *Semiplanus sp. latissimus* group | | | | | | × | × | × |
| *Lonsdaleia duplicata duplicata* | | | | | | | × | |
| *Nemistium edmondsi* | | | | | | | × | |
| *Orionastraea placenta* | | | | | | | × | |
| *Aulophyllum fungites* subspp. | | | | | | | × | × |
| *Lonsdaleia floriformis floriformis* | | | | | | | × | × |
| *Palaeosmilia regia* | | | | | | | × | × |

The Carboniferous Limestone faunas collected during the present work are dealt with in two parts: first, the shelf facies in ascending stratigraphical order; and second, the marginal province with the reef-limestones. Full faunal lists for the former are given in Appendix III (p. 392) and for the latter in Table 5 (p. 142); the authors of specific names are included in these lists and not repeated in the text unless there is confusion over the names. Only the more important fossils are given in the text and the ranges of some of the stratigraphically significant fossils from the shelf limestones are given in Table 2 (p. 128). Some forms are illustrated on Plate XIV. The basin facies limestones found in the Alport and Edale boreholes are summarized on p. 113 from the accounts written by Hudson and Cotton (1945a; 1945b). The locality details are given in Appendix II (p. 384).

The type section for the Carboniferous Limestone of north Derbyshire is in the Wye valley, east of Buxton, and this was described by Cope (1933) with later additional information (1936; 1937; 1939).

The exposed Carboniferous Limestone ranges in age from the *Seminula* (S$_2$) Zone to the Upper *Posidonia* (P$_2$) Zone. In the S$_2$ to Upper *Dibunophyllum* (D$_2$) Zone part of the succession, the sequence of faunal assemblages is similar to that of the Avon Gorge (Vaughan 1905, pp. 195–9) and the Ashton Park Borehole (Kellaway 1967, pp. 64–8, table 1).

## SHELF PROVINCE

### WOO DALE BEDS

The best section of the Woo Dale Beds (p. 40 and Figs. 4, 5) is seen at the junction of Dam Dale and Hay Dale (loc. 3) where 165 ft of these beds are exposed. The lowest diagnostic fauna (loc. 3c) is 13 ft above the base of the section and includes *Carcinophyllum vaughani*, *Davidsonina carbonaria*, *Daviesiella sp.* and *Linoproductus corrugatohemisphericus*, an assemblage characteristic of the *Seminula* (S$_2$) Zone (Vaughan 1905, p. 195). *Daviesiella sp.* also occurs in the bed above (loc. 3d) but these are the only two records of this thick-shelled brachiopod from the present area. The genus occurs more extensively in the Wye valley (Cope 1939, p. 60, fig. 1; 1940, p. 207) where it ranges spasmodically through the undolomitized part of the S$_2$ beds and rarely up into the base of the Chee Tor Rock. The succeeding 129½ ft of beds (loc. 3e-n pars) yield few significant fossils. *Lithostrotion martini*, *Linoproductus sp.* hemisphaericus group and *Megachonetes sp.* papilionaceus group are the commonest forms with *Lithostrotion arachnoideum* and *L. cf. sociale* also present. The zonal position of this assemblage, however, cannot be expressed more precisely than S$_2$ D$_1$; but *Semiplanus sp. nov.* (of Ramsbottom *in* Fowler 1966, p. 76) occurs in two beds (locs. 3f, 3l). This latter species is considered to be characteristic of high S$_2$ beds in parts of the North-West Province and in Northumberland (see Ramsbottom *in* Day 1970, p. 171).

The top 11½ ft of the Woo Dale Beds of this section (loc. 3n pars–p) are marked by the incoming of the typical Lower *Dibunophyllum* (D$_1$) Zone fauna (Vaughan 1905, p. 198) with the corals *Dibunophyllum bourtonense*, *Koninckophyllum* θ and *Lithostrotion pauciradiale*. This sudden entry of the typical D$_1$ coral fauna at the base of the zone is one of the most consistent features of the Carboniferous Limestone of the South-West Province and all the faunas below this marked change are taken to be of S$_2$ age. In the Dam Dale–Hay Dale section gastropods are relatively common in the basal D$_1$ beds.

A fauna of D$_1$ age is also present in the top 14½ ft of the Woo Dale Beds at Hernstone Lane Head (loc. 4c, d) where the additional D$_1$ corals *Lithostrotion junceum* and *Palaeosmilia murchisoni* have been found.

The best section of Peak Forest Limestones, the lateral equivalent of the Woo Dale Beds, is seen near Hartle Plantation (loc. 8) and the record of *L. corrugato-hemisphericus* confirms the $S_2$ age of these beds. No $D_1$ fossils have been collected from this section but above it a small thickness of unexposed beds is thought to be present, below the base of the Chee Tor Rock, and this may be correlated with the high Woo Dale Beds of $D_1$ age in the southern part of the Peak Forest Anticline.

The distribution of lithologies and $D_1$ faunas in the top part of the Woo Dale Beds is shown in Fig. 5.

## CHEE TOR ROCK

The Chee Tor Rock (p. 43 and Fig. 6) has $D_1$ Zone faunas throughout. Corals are typically the commonest and most clearly diagnostic fossils, with the assemblage *Carcinophyllum vaughani*, *Dibunophyllum bourtonense*, *Koninckophyllum* θ, *Lithostrotion junceum*, *L. martini*, *L. pauciradiale* and *Palaeosmilia murchisoni* characteristic of these beds (Vaughan 1905, p. 198). The brachiopods present include *Athyris expansa*, *Davidsonina septosa*, *Delepinea comoides*, *Gigantoproductus maximus*, *Linoproductus sp. hemisphaericus* group and *Megachonetes sp. papilionaceus* group. The last two of these with *C. vaughani* and *L. martini* also occur in $S_2$. Gastropods have been commonly noted especially in the lowest 200 ft of the Chee Tor Rock.

The fossils usually occur in well-marked bands a few feet thick and much of the intervening limestone is poorly fossiliferous. The Upper and Lower *Davidsonina septosa* bands which occur near the top of the Chee Tor Rock are the best known of these bands.

There is no section showing the full sequence through the Chee Tor Rock but Fig. 6 gives a generalized succession. The more important fossil bands are shown on this section with some of the localities at which they occur. In the general account of the Carboniferous Limestone, the Chee Tor Rock has been divided into eight groups of beds for convenience of lithological description (p. 23) and these are used here.

(1) The lowest 40 ft of beds are well exposed in the Hay Dale, Dale Head and Hernstone Lane Head sections and a coral band is developed at the base of the Chee Tor Rock at these localities (locs. 3*q*, 2*f* and 4*e* respectively). The faunas are typical of the $D_1$ Zone and in addition to many of the corals listed above, *Caninia* cf. *densa* and *Lithostrotion* cf. *sociale* are present. The caninioid is the form noted by Hudson and Cotton (1945a, p. 306) which appears to be restricted to the $D_1$ Zone (Plate XIV, fig. 9). A small gigantoproductoid here named *Gigantoproductus? sp. nov.* [wrinkled concentric ornament] was found at two of these localities (loc. 3*q*, Plate XIV, fig. 2; and loc. 2*f*). Bryozoa are relatively common in this basal band and Cope (1939, p. 62, fig. 1) noted this level as a 'Polyzoan Bed' in the Wye valley.

(2) The succeeding group of beds, about 160 ft thick, again has a coral bed at the base and this is seen in Hay Dale (loc. 3*r*); a typical $D_1$ fauna is present. At Dale Head another coral band, 6 ft thick, is developed about 20 ft above the base of this group of beds (loc. 1*g*) but only a poor fauna is present. The same band is seen in Peter Dale (loc. 19) and here yields a rich fauna. *Gigantoproductus? sp. nov.* [wrinkled concentric ornament] is recorded from this horizon

and also from about 20 ft higher in the Dale Head section (loc. 1*h*). The latter is the highest record of this brachiopod which appears to be confined to the lower part of the Chee Tor Rock.

Near Wheston (locs. 22, 23) a minor band occurs about 40 ft higher but yields a poor fauna. This band is probably the same as the one seen on the east side of Hay Dale (loc. 26) where a richer fauna is noted, including *D.* aff. *comoides* [transverse form].

With the exception of scattered corals from the highest 20 to 30 ft, no fauna has been collected from the upper part of this group.

(3) An important coral band (about 10 ft thick) with typical $D_1$ fauna is recognized at Laughman Tor (loc. 14*b*) and bands at or near this horizon are seen on the east side of Middle Hill (loc. 16*b*) and south-west of Wheston (loc. 24). Scattered fossils are seen in the beds immediately above and below this horizon. Corals are also seen at approximately this level east of Hargatewall (loc. 18) and in Monk's Dale (locs. 20, 21) but it is not possible to make more accurate correlations. In the exposure of this band on the east side of Middle Hill (loc. 16*b*), *C.* cf. *densa* and *L.* cf. *sociale* are again present and this is the highest record of the latter species in the district. The locality south-west of Wheston (loc. 24) has yielded specimens of *Davidsonina septosa* and is the lowest occurrence of this fossil which characterizes the upper part of the Chee Tor Rock.

(4) There are about 80 ft of beds between the coral band of Laughman Tor and the base of the Lower *D. septosa* Band. The lower beds are seen on the east side of Middle Hill (loc. 16*c-e*) but only poor faunas have been recorded. On the north-west side of Middle Hill (loc. 17) corals have been collected about 10 ft above the base including *Lithostrotion* cf. *aranea* and *L.* aff. *maccoyanum* [more septa than typical]. A limited fauna is present at approximately the same level in Oxlow Rake (loc. 29). The uppermost 50 ft of these beds are well exposed in Duchy Quarry (loc. 12*a-i*). An algal band (loc. 12*d*) is present about 8 ft above the base of the section and from this a specimen of *D. septosa* [coated with algae] has been collected. This band is probably developed as a coral band south-west of Wheston (loc. 25) and south of Middle Hill (loc. 15) where the fauna includes *Davidsonina septosa* and *Delepinea* aff. *comoides* [transverse form]. *D. septosa* is also present in Duchy Quarry (loc. 12*i*) immediately below the Lower *D. septosa* Band while at Upperend Quarry it is recorded at this level (loc. 13*h*) and again about 15 ft lower (loc. 13*c*).

(5) The Lower *D. septosa* Band (1 to 4 ft thick) is a well-known marker band with a widespread distribution. Collections have been made from the band in Duchy Quarry (loc. 12*j*) and Tideswell Moor (loc. 27). The fauna is richer than that found in the Upper *D. septosa* Band and includes *Dibunophyllum bourtonense*, *Lithostrotion martini*, *P. murchisoni*, *Syringopora* spp., *Davidsonina septosa*, *Delepinea comoides*, *D.* aff. *comoides* [transverse form], *Gigantoproductus* sp., *Linoproductus* sp. *hemisphaericus* group, *M.* sp. *papilionaceus* group and *Straparollus* (*Euomphalus*) *sp.*

(6) *D. septosa* is again noted in the 35 to 40 ft of beds between the Lower and Upper *D. septosa* bands. In Upperend Quarry this species is present at two levels 5 ft (loc. 13*l*) and 13 ft (loc. 13*m*) above the Lower *D. septosa* Band.

(7) The Upper *D. septosa* Band (1½ to 5 ft thick) is also well marked and

widespread and is correlated with the "*Cyrtina septosa* Band" of the Wye valley (Cope 1936, p. 48). The best fauna from this band has been collected in Upperend Quarry (loc. 13*o*) where *Dibunophyllum bourtonense, Lithostrotion martini, Syringopora* cf. *geniculata, Davidsonina septosa, D. septosa transversa, Linoproductus sp.* *hemisphaericus* group, *M. sp. papilionaceus* group and gastropods are recorded.

Delepinea aff. *comoides* [transverse form] has been recorded in the Lower *D. septosa* Band and in two bands (locs. 15, 26) below this level but has not been found in the Upper *D. septosa* Band during the present survey. Cope (1936, p. 49), however, lists what is probably a similar form from the "*C. septosa* Band" in the Wye valley.

(8) The beds between the Upper *D. septosa* Band and the base of the Lower Miller's Dale Lava are 20 to 25 ft thick and except for the records of two more occurrences of *D. septosa* in Upperend Quarry (loc. 13*r, s*), the second in association with the subspecies *transversa*, few fossils are listed.

## MILLER'S DALE BEDS

The Miller's Dale Beds (p. 53 and Fig. 6) contain $D_1$ faunas similar to those of the underlying Chee Tor Rock. Few fossil bands are present and they are less well marked than many of those in the lower formation.

The limestones below the Dove Holes Tuff are seen in Dove Holes Dale Quarry (loc. 31*a, b*), Holderness Quarry (loc. 32*a-c*) and Smalldale (loc. 35*a-d*). Typical $D_1$ faunas are found at these localities and about 30 ft above the base of the Miller's Dale Beds (locs. 31*a*, 32*c*, 35*d*) there is a richly fossiliferous bed from which *Davidsonina septosa, D. septosa transversa* and *Delepinea* aff. *comoides* [transverse form, Plate XIV, fig. 1] have been collected. Jackson (1922, p. 467) recorded *Davidsonina septosa* and *Delepinea* aff. *comoides* from approximately the same horizon, 12 ft above the Lower Lava, in Miller's Dale. *Davidsonina septosa transversa* and *D. septosa* have been found together in the present collections both at this level, about 30 ft above the base of the Miller's Dale Beds, and also in and just above the Upper *D. septosa* Band at Upperend Quarry (loc. 13*o, s*). Mitchell and Stubblefield (1941, p. 212) recorded *D. septosa transversa* from beds with $B_2$ goniatites and *D. septosa* at Breedon Cloud. They suggested that *transversa* indicated "an horizon higher than that of the *C. septosa* Band of the North-West Province and Derbyshire". In the present collection, however, the transverse subspecies occurs as low as the Upper *D. septosa* Band. Parkinson (1965, p. 165) also discusses the records of *transversa*.

The beds above the Dove Holes Tuff are well exposed in Victory Quarry (loc. 30*a-j*) but few fossils have been collected from this locality. There is no evidence of the incoming of $D_2$ faunas at the top of the Miller's Dale Beds either here or in the old Peak Forest railway section near the Bull Ring (loc. 33*a*) where the highest beds are again seen.

The complete sequence of the Miller's Dale Beds is seen at Wall Cliff (loc. 28*k-p*) but only 63 ft of beds are present between the Lower and Upper lavas. *L.* cf. *maccoyanum* [septa withdrawn from a weak axis] has been recorded from 4 ft above the base of the section (loc. 28*k*). This level is probably the same as the coral bed with *L.* aff. *maccoyanum* immediately above the Lower Lava in the Wye valley (Cope 1933, p. 132) and the band with *L.* cf. *arachnoideum* and *L.* aff.

K

EXPLANATION OF PLATE XIV

FOSSILS FROM THE CARBONIFEROUS LIMESTONE

1. *Delepinea* aff. *comoides* (J. Sowerby) [transverse form]. About 30 ft above base of Miller's Dale Beds (D$_1$ Zone), Smalldale, quarry [0977 7700], 670 yd N. 17°W. of Rock Houses, loc. 35d, RS 175, × 1. See p. 133.

2. *Gigantoproductus?* sp. nov. [with wrinkled concentric ornament]. Coral band at base of Chee Tor Rock (D$_1$ Zone), section at junction of Hay Dale and Dam Dale [1187 7781], 535 yd E.30°S. of Loosehill Farm, loc. 3q, RS 407, × 1. See p. 131.

3. *Echinoconchus punctatus* (J. Sowerby). Back-reef limestone in Bee Low Limestones (D$_1$ Zone), quarry on south-west side of Snels Low [1125 8163], 1680 yd W.34°S. of Oxlow House, loc. 43a, RS 880, × 1. See p. 135.

4. *Echinoconchus* aff. *punctatus* [form with broad bands]. Apron-reef limestone (B$_2$ Zone), Tor Top Sink [0989 8126], Perryfoot, 115 yd S.8°W. of Whitelee, loc. 126a, RS 134, × 1. See p. 149.

5. *Productus productus* aff. *hispidus* Muir-Wood [form with two rows of spine-bases on flanks]. Back-reef limestone in Bee Low Limestones (D$_1$ Zone), Middle Hill, quarry at roadside [1208 8236], 500 yd W.17°S. of Oxlow House, loc. 44b, RS 778, × 1. See p. 135.

6. *Orionastraea placenta* (McCoy). 36 ft below *Orionastraea* Band, Upper Monsal Dale Beds (D$_2$ Zone), cliff section [2279 7560], Stoney Middleton, 900 yd E.27°N. of Highfields Farm, loc. 76a, RK 8854 (slide PL 317), × 1. See p. 139.

7. *Orionastraea* aff. *indivisa* Hudson [cf. Hudson 1926, pl. 8, fig. 3]. *Orionastraea* Band, Upper Monsal Dale Beds (D$_2$ Zone), cliff section [2279 7560], Stoney Middleton, 900 yd E.27°N. of Highfields Farm, loc. 76g, RK 8838 (slide PL 316), × 1. See p. 139.

8. *Aulina furcata* Smith. 11 ft above base of Miller's Dale Beds (D$_1$ Zone), Wall Cliff [1424 7738], 640 yd N.26°W. of Highfield House, loc. 28k, PT 2251 (slide PL 314), × 1. See p. 135.

9. *Caninia* cf. *densa* Lewis [Hudson and Cotton 1945a, p. 306]. Coral band at base of Chee Tor Rock (D$_1$ Zone), section at junction of Hay Dale and Dam Dale [1187 7781], 535 yd E.30°S. of Loosehill Farm, loc. 3q, RS 443 (slide PL 318), × 1. See p. 131.

10. *Dithyrocaris sp.* [mandible gnathal lobe]. Apron-reef limestone (B$_2$ Zone), Treak Cliff [1342 8304], loc. 136c, PJ 1909, × 3. See p. 150.

All specimens are in the Collection of the Institute of Geological Sciences, Leeds; registration numbers are given following the locality numbers.

FOSSILS FROM THE CARBONIFEROUS LIMESTONE

*maccoyanum* recorded by Shirley and Horsfield (1940, p. 276) 20 ft above the Lower Lava in Cave Dale and at Cop Round.

The rare coral *Aulina furcata* has been recorded (Mitchell 1965, p. 74; Clarke 1966, p. 226) from the lowest bed in the Wall Cliff section (loc. 28*k*, Plate XIV, fig. 8) and from about 10 ft higher in the sequence (loc. 28*m*). The type locality of this species is the high $B_2$ limestones of Siggate, Castleton (Smith 1925, p. 494; Smith and Yü 1943, p. 49), and a correlation between the horizon of these two localities is suggested. Clarke considers that this species characterizes a restricted horizon of uppermost $D_1$ (not below "the *D. septosa* band") and low $D_2$ age.

The higher beds with *A. furcata* (loc. 28*m*) also contain a brachiopod fauna including *Dielasma hastatum* and *Pugnax pugnus* [small form], species which suggest reef conditions (p. 24) and in the apron-reef belt would indicate a $B_2$ age. The possible link here between reef and shelf faunas is discussed further on p. 152.

## BEE LOW LIMESTONES

Detailed faunal correlations within the Bee Low Limestones (see also p. 25) are difficult and no additional information about the sequence of $D_1$ faunas is given by this formation.

A well-known coral band is developed in Pindale Quarry (loc. 47*f*) $32\frac{1}{2}$ ft below the top of the Bee Low Limestones and its fauna was discussed by Jackson (1941, p. 240) and Eden and others (1964, p. 78). In Cave Dale (loc. 45*b*) the coral *Lithostrotion* cf. *aranea* occurs $34\frac{1}{2}$ ft above the Lower Lava, a horizon that can be correlated with the beds yielding *L.* aff. *maccoyanum* in the Wye valley (Cope 1933, p. 132).

*Back-reef facies.* The Bee Low Limestones are well exposed in the areas to the south of the marginal reef belt and rich reef-brachiopod faunas are present at several localities (Plate XI).

The only goniatite record from these beds is that of *Bollandoceras sp. nov.*, a genus indicating a B Zone age, from Gautries Hill (loc. 39*c*). The associated reef fauna is poor and not diagnostic. In Eldon Hill Quarry, rich reef faunas are present in a bed with *D. septosa* (loc. 40*b*), probably the equivalent of the Upper *D. septosa* Band of the Chee Tor Rock, and at a second level about 60 ft lower (loc. 40*a*). Both these reef faunas indicate a $B_2$ age but no clear link is possible with the apron-reef faunas.

Abundant reef faunas have been collected from two other localities. By the roadside south-west of Snels Low, an old quarry (loc. 43*a*) has yielded *Acantho- plecta mesoloba*, *Dielasma hastatum*, *Echinoconchus punctatus* (Plate XIV, fig. 3), *Krotovia spinulosa*, *Ovatia sp. nov.*, *Pleuropugnoides pleurodon* and *Plicatifera plicatilis*. The *Ovatia sp. nov.* is the same form as found at Tor Top Sink, Perryfoot (loc. 126*a*) and Odin Fissure (loc. 135), localities which on goniatite evidence are of high $B_2$ age, and the assemblage from Snels Low is taken to be of this age. The old quarry at the roadside on the south side of Middle Hill (loc. 44*b*) has a similar fauna and in addition *Productus productus* aff. *hispidus*. This is a form of *hispidus* with only two instead of three rows of spine-bases on the flanks (Plate XIV, fig. 5) and it has been recorded in the apron-reef from the following $B_2$ localities; Perry Dale (loc. 125), Tor Top Sink (loc. 126*a*), Giant's Hole (loc. 130) and Castleton village (loc. 139).

The horizons of these two back-reef localities (locs. 43*a*, 44*b*) within the shelf succession are difficult to establish but they are about 100 to 150 ft below the top of the $D_1$ limestone, near the level of the Upper *D. septosa* Band, which is exposed a short distance to the east and they cannot be far below this level. If the correlations with the $B_2$ apron-reef faunas are reliable, then these $B_2$ beds can be equated with the limestones about 100 to 150 ft below the top of the $D_1$ Zone.

LOWER MONSAL DALE BEDS

With the exception of the Lower *Girvanella* Band no well-marked fossil bands occur in the Lower Monsal Dale Beds (p. 63 and Plate VI). The faunas are relatively poor, but indicate the Upper *Dibunophyllum* ($D_2$) Zone. Fossils include *Dibunophyllum bipartitum, Diphyphyllum lateseptatum, Lonsdaleia floriformis laticlavia, Gigantoproductus giganteus, Productus productus hispidus, Pugilis pugilis* and *Striatifera striata*, none of which occurs below the base of the Lower Monsal Dale Beds. Some elements of the $D_1$ fauna, including *Athyris expansa, Lithostrotion martini* and *Palaeosmilia murchisoni* persist but they are less common here than in the Bee Low Group.

The lower beds are exposed in Victory Quarry (loc. 30*k-y*), in Cressbrook Dale (loc. 54*a-k*) and in Earle's Quarry (loc. 70). The Lower *Girvanella* Band ($1\frac{3}{4}$ to 9 ft thick) is well developed at or near the base in Pin Dale (Jackson 1941, p. 239; Shirley and Horsfield 1940, p. 277; Eden and others 1964, p. 81). *Saccamminopsis sp.* also occurs in the basal beds and three beds with this fossil have been found in Cressbrook Dale (loc. 54*a, g, k*), the highest being 45 ft above the base of the Lower Monsal Dale Beds. These two fossils appear to be mutually exclusive for although they have both been collected from the old Peak Forest railway section, *Girvanella* at the base of the Lower Monsal Dale Beds (loc. 33*b*) and *Saccamminopsis sp.* from $26\frac{1}{2}$ ft above the base (loc. 33*d*), there is no record of them occurring together in the same bed.

At two localities on the south side of Bradwell Moor, *Lithostrotion maccoyanum* has been found (locs. 62*d*, 65) from near the base of these beds.

The higher beds of the Lower Monsal Dale Beds are seen in Cressbrook Dale (loc. 55*a-p*) and also in Earle's Quarry where the complete sequence (loc. 70*a-g*) between the Lower and Upper *Girvanella* bands is seen (Eden and others 1964, p. 81). Few fossils have been noted from this higher group of limestones.

Dark limestones of basin facies with zaphrentoids are seen near Bennetston Hall (locs. 51, 52), and at Barmoor Clough (loc. 50) the $P_1$ goniatite *Girtyoceras sp.* has been found at a horizon within 20 ft of the base of the Lower Monsal Dale Beds.

There is no development of back-reef limestones as seen in the Bee Low Limestone, but flat reefs occur close to the marginal belt. These can be seen in Earle's Quarry (loc. 70*b*) at a horizon which is correlated with the $P_{1b}$ beds of Upper Jack Bank Quarry (loc. 142; Eden and others, 1964, pp. 85, 90).

UPPER MONSAL DALE BEDS

The Upper Monsal Dale Beds (p. 75 and Plate VII) show a marked change in fauna from the beds below. Several fossil bands are present and some have rich coral faunas. The assemblages, particularly the ones from the *Lonsdaleia duplicata* and *Orionastraea* bands are similar to those from the classic $D_2$ localities of the South-West Province (Round Point, Rownham Hill, Wrington; Vaughan 1905, pp. 199–242; Kellaway 1967, p. 64). The corals found in the Upper Monsal Dale Beds include *Dibunophyllum bipartitum* subspp., *Diphyphyllum lateseptatum, Lithostrotion junceum, L. pauciradiale, L. portlocki, Lonsdaleia duplicata, L. floriformis, Nemistium edmondsi, Palaeosmilia murchisoni, P. regia* and *Orionastraea placenta*, and many of these are beautifully preserved. Brachiopods are also common, particularly *Gigantoproductus spp.*

The sequence of beds is set out on pp. 29–31. The limestones between the bands have relatively few fossils but the faunas of the bands are as follows.

(1) The Hob's House Coral Band (up to 5 ft thick), 50 to 60 ft above the base of the Upper Monsal Dale Beds, is only seen in Cressbrook Dale (locs. 55*v*, 83*a*) where *Caninia benburbensis, Clisiophyllum sp., Dibunophyllum bipartitum* subspp., *Diphyphyllum lateseptatum, L. martini, L. pauciradiale, L. portlocki, P. murchisoni* and brachiopod fragments have been collected.

(2) The Lower Shell Bed (6 to 7 ft thick) is 50 to 60 ft higher in the sequence and contains abundant specimens of *Gigantoproductus edelburgensis* in Middleton Dale (locs. 74*f*, 75*c*).

(3) The Black Bed (5 to 16 ft thick), 10 ft above the last band, is mainly restricted to Middleton Dale (locs. 73*d*, 74*j*, 75*g*) but is also present in Cressbrook Dale (loc. 83*l*). It has a limited fauna in which *L. floriformis floriformis* is the only important fossil. This coral also occurs in the beds immediately above the Black Bed especially at Shining Cliff (loc. 74*k*) and in Cressbrook Dale (loc. 83*m, o*). The occurrences of this subspecies are concentrated in this part of the sequence but it is also recorded from the base of the Upper Monsal Dale Beds in Cressbrook Dale (loc. 55*r*) at the horizon of the Upper *Girvanella* Band, from the *Lonsdaleia duplicata* Band in Intake Dale (loc. 88*b*) and from the Eyam Limestones. Only two specimens of *L. floriformis laticlavia* have been found, one in the Lower Monsal Dale Beds near Bennetston Hall (loc. 51*f*) and the other near the base of the Upper Monsal Dale Beds in Cressbrook Dale (loc. 55*u*).

(4) The Upper Shell Bed (1 to 4 ft thick) is about 25 ft above the Black Bed and as with the Lower Shell Bed can be traced only in Middleton Dale where *G. crassiventer* and *G. dentifer?* have been recorded (locs. 74*l*, 75*l*).

(5) No macrofossils have been collected from the White Bed (up to 7 ft thick) which is 30 to 35 ft above the Upper Shell Bed.

(6) The *Lonsdaleia duplicata* Band (up to 6 ft thick) is 5 to 30 ft above the White Bed horizon (see Plate VII and Fig. 10) and is well developed in the Bradwell Moor area (Shirley and Horsfield 1940, p. 277). The main elements of the fauna are given in Table 3. The *Dibunophyllum* aff. *muirheadi* Bed (Morris 1929, p. 42) occurs 3 ft above the White Bed at Shining Cliff (loc. 74*q*). It is of limited extent, and has a fauna which includes *Dibunophyllum bipartitum bipartitum* (? = *D.* aff. *muirheadi* of Morris), *D. bipartitum konincki, Lithostrotion junceum, L. pauciradiale, P. regia, G. edelburgensis* and *Spirifer spp*. This bed may be the equivalent of the *Lonsdaleia duplicata* Band but the correlation cannot be confirmed.

## TABLE 3

*Faunal distribution in the* L. duplicata *Band*

Locality numbers are set at the head of the table and details of these are given in the text and in Appendix II.

| Genera and species | 73p | 83r, q | 88b | 90d | 94a | 95 | 96b |
|---|---|---|---|---|---|---|---|
| Dibunophyllum bipartitum bipartitum .. | . | . | × | | | | |
| Diphyphyllum furcatum .. .. .. | . | . | . | . | × | | |
| D. lateseptatum .. .. .. .. | × | . | . | . | . | . | × |
| Lithostrotion junceum .. .. .. | × | × | . | × | | | |
| L. pauciradiale .. .. .. .. | × | . | . | × | | | |
| L. portlocki .. .. .. .. .. | × | × | . | . | × | | |
| Lonsdaleia duplicata duplicata .. .. | × | × | × | . | × | × | × |
| L. floriformis floriformis .. .. .. | . | . | × | | | | |
| Nemistium edmondsi .. .. .. | . | × | . | . | × | | |
| Palaeosmilia regia .. .. .. .. | . | × | . | . | . | . | × |
| Productus productus hispidus .. .. | . | . | . | . | . | . | × |
| Pugilis pugilis .. .. .. .. | . | . | . | . | . | . | × |

The rich coral faunas of the *L. duplicata* Band are similar to those of the *Orionastraea* Band (see Table 4). The differences between the faunas of the two bands may not be significant, except that *L. duplicata* is common at the lower level and has not been found in the upper band; on the other hand *Orionastraea placenta* is common at the higher level and has not been found at the lower level. Of the species found in the *L. duplicata* Band and not in the *Orionastraea* Band, *L. duplicata* has been found just above the *Orionastraea* Band both in Furness Quarry (loc. 73*u*) and at the southern end of Bradwell Dale (loc. 99*e*); *D. furcatum* is represented by a single record; *L. floriformis floriformis* is also found in the Eyam Limestones; while *N. edmondsi* has not been found above the *L. duplicata* Band, the only other record of the species in the shelf area being in Cressbrook Dale (loc. 83*e*) about 80 ft lower in the succession. *N. edmondsi* has also been found in the Castleton Borehole (see p. 112).

Intake Dale is one of the best localities where the relationships of the *L. duplicata* (loc. 88*b*) and *Orionastraea* (loc. 88*i*) bands have been seen in the same section. They are separated by 26 ft 8 in of limestones and abundant well-preserved specimens of both corals occur. The record here of *L. duplicata duplicata* and *L. floriformis floriformis* together in the *L. duplicata* Band is unique for this area.

(8) The *Orionastraea* Band (2 to 3 ft thick) is about 25 to 40 ft above the *L. duplicata* Band and 30 to 50 ft below the top of the Upper Monsal Dale Beds, and has been described by Shirley and Horsfield (1940, p. 277; 1945, p. 294). The important elements of the fauna are set out in Table 4.

## TABLE 4

*Faunal distribution in the* Orionastraea *Band*

Locality numbers are set at the head of the table and details of these are given in the text and Appendix II.

| Genera and species | 73r | 76g | 77c | 83u | 87i | 88i | 89 | 99c | 100j |
|---|---|---|---|---|---|---|---|---|---|
| *Aulophyllum fungites* | | | | | X | | | | X |
| *Caninia juddi* | | | | | X | | | | |
| *Clisiophyllum keyserlingi* | | | | | | | | | X |
| *Dibunophyllum bipartitum* | | X | | | X | | | | |
| *D. bipartitum bipartitum* | | | X | | | | | | X |
| *Diphyphyllum lateseptatum* | X | | X | X | | | | | X |
| *Lithostrotion decipiens* | | | | | | | | X | X |
| *L. junceum* | X | X | X | | X | | | X | X |
| *L. pauciradiale* | X | | | X | | | | | |
| *L. portlocki* | X | X | X | X | | X | X | | X |
| *Orionastraea* aff. *indivisa* | | X | | | | | | | |
| *O. placenta* | X | | X | | | X | | X | |
| *Palaeosmilia regia* | | | | | X | | | | |
| *Pugilis pugilis* | | | | | | X | | | |

*O. placenta* has been found at two horizons outside this band, one at Stoney Middleton (loc. 76a, Plate XIV, fig. 6) 10 ft below the White Bed and the other in Coombs Dale (loc. 77e) about 22 ft above the *Orionastraea* Band. The specimen of *O.* aff. *indivisa* from Stoney Middleton (loc. 76g, Plate XIV, fig. 7) is similar to the form figured by Hudson (1926, pl. 8, fig. 3). *O. indivisa* is recorded by Shirley and Horsfield (1945, p. 296) from Coombs Dale 25 ft below the *Orionastraea* Band and (p. 294) an atypical specimen from the horizon of this band at Castle Rock.

Beds with rich reef-brachiopod faunas are sometimes developed in the Upper Monsal Dale Beds. In Coombs Dale (loc. 77c, d) and at the south end of Bradwell Dale (loc. 99d) they occur high in these beds at and above the level of the *Orionastraea* Band. The brachiopod assemblages are comparable with those found in the overlying Eyam Limestones but are not so richly developed as in these. In the Upper Bench of Earle's Quarry, a 55-ft knoll-reef is seen with its base 26 ft 8 in above the top of the Upper *Girvanella* Band. The fauna of this knoll (loc. 98) has been described by Eden and others (1964, p. 89).

### EYAM LIMESTONES

The Eyam Limestones (Plate X) contain faunas similar to those of the Monsal Dale Beds. *Gigantoproductus spp.*, *Productus productus hispidus* and *Pugilis pugilis* occur while the corals include *Aulophyllum fungites* subspp., *Dibunophyllum bipartitum* subspp., *Lithostrotion junceum*, *Lonsdaleia floriformis floriformis*, *Palaeosmilia regia*, *Koninckophyllum interruptum* and *K. magnificum*. With the exception of the last two, which Hill (1938, p. 11) lists as occurring only above the Single Post Limestone of northern England (i.e. high in $D_2$), all these species are also present in the underlying beds. The assemblage is indicative of the $D_2$ Zone, and in the absence of goniatite evidence or of correlation with goniatite-bearing beds, would be referred to this coral-brachiopod zone. However, isolated exposures of limestone resting on $D_1$ near Brook House (locs. 121, 122) are correlated with the Eyam Limestones on lithological and faunal evidence. From one of these localities (loc. 122a) the goniatite *Sudeticeras sp.* has been collected and this dates the Eyam Limestones as of Upper *Posidonia* ($P_2$) age.

Dark limestones and shales of $P_2$ age, with *Rotiphyllum* cf. *costatum, R. densum, Posidonia corrugata* and *Sudeticeras sp.* are present at Watergrove Mine (loc. 123). Zaphrentoid corals are also recorded from black limestones at Hazlebadge Hall, Bradwell Dale, by Shirley and Horsfield (1940, pp. 286, 288) who correlated these limestones with the dark beds at Ashford and Cawdor Quarry, horizons which are known to be of $P_2$ age.

Two distinctive fossil bands can be traced in the Eyam Limestones, although fossils are common throughout. At Stoney Middleton (loc. 76s) a 1½-ft band is present about 35 ft above the base with *D. bipartitum craigianum* and *D. bipartitum konincki*. This is the *Aulophyllum* Band of Morris (1929, p. 54) but *A. fungites* has not been found here during the present work. The Eyam Dale Coral Band (3 ft thick) can be recognized about 60 ft above the base of the Eyam Limestones in Eyam Dale (loc. 110i). It is widespread in this area and the fauna includes the subspecies *bipartitum, craigianum* and *konincki* of *D. bipartitum* as well as *Spirifer spp.*

Knoll- and flat-reef limestones with rich brachiopod faunas are found in the Eyam Limestones. However, relatively few species are present although individuals may be common. Most of the species are also recorded from the apron-reef limestones of $B_2P_{1a}$ age developed at lower levels. The paucity of species compared with the rich reef faunas in the marginal province may be explained by the environment in the flat reefs of the Eyam Limestones being more restricted than that of the apron-reef front. As an assemblage the fauna is characteristic of the Eyam reef-limestones. The typical association is seen in the reef beds of Intake Dale (locs. 119, 120) where the following fauna has been collected: *Acanthoplecta mesoloba, Alitaria panderi, Antiquatonia antiquata, A. hindi, A. insculpta, Avonia davidsoni, A. thomasi, A. youngiana, Brachythyris integricosta, 'B.' planicostata, Buxtonia spp., Dielasma hastatum* [juvs.], *Echinoconchus punctatus, Eomarginifera lobata* aff. *laqueata, Girtyella? sacculus, Pleuropugnoides pleurodon, Pugnax pugnus* [small form], *Schizophoria resupinata* and *Spirifer (Fusella) triangularis.* In addition to these, *Antiquatonia sulcata, Krotovia spinulosa* (J. Sowerby), *Productus productus hispidus, Spirifer (Fusella) trigonalis, S. duplicicosta,* bivalves and more rarely gastropods have been recorded from other localities.

Of this fauna only *G.? sacculus* is confined to the Eyam Limestones and it is interesting to note that the original locality given for this species was "common in limestone, particularly near Eyem (sic) and Middleton" (Martin 1809, pl. 46, figs. 1, 2).

## MARGINAL PROVINCE

A map of the apron-reef outcrop, showing localities from which collections have been made, is given in Plate XI. Section and locality details are in the text (pp. 105–9) and in Appendix II (pp. 390–1). The fossils collected are listed in

Table 5 (pp. 142–8) but this list could almost certainly be extended by further collecting which would probably affect the specific ranges suggested.

Contributions on the geology of the apron-reef have been made by Jackson (1927; 1941), Bisat (1934), Shirley and Horsfield (1940), Ford (1952), Hudson and Cotton (1945a, b), Wolfenden (1958) and Parkinson (1943; 1947; 1953; 1965). The latest paper by Parkinson (1965) is an excellent summary of current views. A further review of the literature is unnecessary and the present account concentrates on the description and interpretation of the new collections that have been made. Acknowledgment must, however, be made of the value of earlier works which form the basis for this account. Wolfenden (1958, p. 871) studied primarily the palaeoecology and the lateral facies and faunal distribution in beds traced from the shelf province into the reef complex, whereas this account attempts to distinguish a sequence of faunal assemblages through the thickness of the reef beds. Wolfenden recognized two horizons at which algal reefs were developed as incipient barriers separating the shelf and back-reef areas from the fore reef and basin areas of deposition.

When goniatites are present in the reef faunas it is possible to establish the following sequence:

> Goniatites falcatus ($P_{1b}$) Zone
>
> Goniatites crenistria ($P_{1a}$) Zone
>
> Beyrichoceras ($B_2$) Zone

Subdivision of $B_2$ has been discussed by Hudson and Cotton (1945a, p. 301) who suggested a division into the two subzones, Beyrichoceras vesiculiferum (lower $B_2$) and B. delicatum (upper $B_2$) with five faunal assemblages, and by Parkinson (1947, p. 123). However, the stratigraphical relationships of the goniatite occurrences are difficult to establish and revision of the $B_2$ goniatite sequences is needed. It is not considered that the available information is adequate to sustain a subdivision of $B_2$ (fide Dr. W. H. C. Ramsbottom).

Where goniatites are not present, it is difficult to make faunal distinctions, especially between $B_2$ and $P_{1a}$ beds (Parkinson 1965, p. 162). However, by listing and comparing the faunas where the ages are indicated by goniatites it is possible to distinguish the following changes in the vertical distribution of faunas.

The $B_2$ beds contain the richest faunas (Table 5) with an abundance of species and individuals. $P_{1a}$ faunas, especially those inferred to be from low in the zone, are very similar to those from $B_2$, but the $P_{1b}$ horizons show a marked reduction in the number of species present though individuals are relatively abundant. The very extensive faunas from $B_2$ may reflect the greater area of these beds available for collection and the ease with which rich faunas can be obtained.

The species Dielasma hastatum, Pleuropugnoides pleurodon, Pugnax pugnus and Schizophoria resupinata are abundant in $B_2$ beds, less common in $P_{1a}$ and rare or absent in $P_{1b}$, although all are found in limited numbers in the reef development of Eyam Limestones. P. pugnus is represented by small forms in $B_2$ but in $P_{1b}$ only the larger form (Parkinson 1954, p. 563) has been recorded. Productus productus hispidus has not been found below the base of $P_{1b}$, but P. productus aff. hispidus [two instead of three rows of spine-bases on the flanks] is present in $B_2$.

**TABLE 5:** *List of fossils from Apron-Reef Limestones of the Marginal Province*

The locality numbers are referred to in the text (pp. 149–51) and full details listed in Appendix II. Localities have been arranged from west to east within each Zone. A query preceding a number in this table indicates that the identification of the fossil from that locality is doubtful. Goniatite identifications are by Dr. W. H. C. Ramsbottom.

| Genera and species | Localities | $B_2$ ZONE | | | | | | | | | | | $P_{1a}$ ZONE | | | | $P_{1b}$ ZONE | | | | | |
|---|---|---|---|---|---|---|---|---|---|---|---|---|---|---|---|---|---|---|---|---|---|---|
| | | 125 | 126a | 127 | 129 | 130 | 135 | 136 (a–w) | 137 | 139 | 143a | 147 | 126b | 141c | 143b | 143c | 140 | 141d | 142 | 144 | 145 | 146 |
| Algal nodules | | 125 | | | | | | | | | | | | | | | | | | | | |
| *Koninckopora inflata* (de Koninck) | | | | 127 | | | | 136c, e, g | | | | | 126b | | | | | | | | | |
| *Serpula* sp. | | | | | | | 135 | 136c | | | | | | | | | | | | | | |
| *Caninia* sp. *subibicina* McCoy group | | | | | | | 135 | 136e, g, l, m, r, s | 137 | | | | | | | | | | | | | |
| *Chaetetes depressus* (Fleming) | | | | | 129 | | | 136f, g | | | | | | | | | | | | | | |
| *C. radians* (Fischer) | | | | | | | | 136e | | | | | | | | | | | | | | |
| *Clisiophyllum keyserlingi* McCoy | | | | | | | | 136h | | | | | | | | | | | | | | |
| *Dibunophyllum bipartitum* (McCoy) | | | | | | | | | | | 143a | | | | | | | | | | | |
| *Heterophyllia* (*Heterophylloides*) cf. *ornata* McCoy | | | | | | | | 136h | | | | | | | | | | | | | | |
| *Koninckophyllum* sp. *dianthoides* (McCoy) group | | | | 127 | | | | | | | | | | | | | | | | | | |
| *Lithostrotion* cf. *aranea* (McCoy) | | | | 127 | | | | | | | | | | | | | | | | | | |
| *L. junceum* (Fleming) | | | | | | | 135 | 136f | | | | | | | | | | | | | | |
| *L. martini* Milne Edwards and Haime | | | | | | | 135 | 136h | | | | | | | | | | | | | | |
| *L.* aff. *martini* | | | | 127 | | | | 136i | | | | | | | | | | | | | | |
| *L. pauciradiale* (McCoy) | | | | 127 | | | | 136i | | | | | | | | | | | | | | |
| *L.* cf. *sociale* (Phillips) | | | | | | | | 136e | | | | | | | | | | | | | | |
| *Michelinia tenuisepta* (Phillips) [identified by Mr. D. E. White] | | | | 127 | | | | 136j | | | | | | | | | | | | | | |
| *Palaeosmilia murchisoni* Milne Edwards and Haime | | | | | | | 135 | 136j | | | | | | | | | | | | | | |
| *Rotiphyllum* sp. | | | | | | | | ?136e | | | | | | | | | | | | | | |
| *Zaphrentites* sp. *enniskilleni* (Milne Edwards and Haime) group | | | | | | | 135 | 136g | | | 143a | | | | | | | | | | | |
| *Zaphrentites?* | | | | | | | | | | | | | | | | | | | | | | |
| Bryozoa | | 125 | 126a | 127 | 129 | | 135 | 136c, d, f, g, k–m, q, t, v, w | | 139 | | 147 | | | | | 140 | | | 144 | | |
| *Fenestella* spp. | | 125 | 126a | 127 | 129 | 130 | 135 | 136a–c, e, g, j–m, p, r, t, v | 137 | | 143a | 147 | | | | | 140 | | 142 | 144 | | |
| *Fistulipora incrustans* (Phillips) | | 125 | | | | 130 | 135 | 136e, l | 137 | | | | | | | | | | | | | |
| *F.* sp. | | 125 | | | | | | | | | | | | | | | | | | | | |
| *Penniretepora* sp. | | 125 | 126a | 127 | 129 | 130 | 135 | 136c, f, g, l, m | | | | 147 | | | | | | | | | | |
| *Acanthoplecta mesoloba* (Phillips) | | | 126a | | | | 135 | 136d, e, j, l, m | | 139 | | | | | | | | | | | | |
| *Actinoconchus lamellosus* (Léveillé) | | | | | | | 135 | | | | | | | | | | | | | | | |
| *A.* cf. *lamellosus* | | | | | | | 135 | 136n, ?t | 137 | 139 | | | | | | | | 141d | | | | |
| *A.* sp. | | | | 127 | | | 135 | 136c, j–h, j–m | 137 | | ?143a | 147 | | | | 143c | | 141d | | | | |
| *Alitaria panderi* (Muir-Wood and Cooper) | | | | 127 | | | | 136f, v | | | | 147 | 126b | | | 143c | | | | | 145 | 146 |
| *A. triquetra* (Muir-Wood) | | | | 127 | | | | 136f, g, v | 137 | | 143a | 147 | 126b | | | | | | | 144 | 145 | |
| *A.* aff. *triquetra* | | | | | | | 135 | | 137 | | 143a | 147 | 126b | | | | | | | 144 | 145 | |
| *Antiquatonia antiquata* (J. Sowerby) | | 125 | 126a | 127 | | | 135 | 136b, e–g, o | | | 143a | 147 | 126b | | | | | | | 144 | 145 | |

## Table 5—continued

| Genera and species | Localities | 125 | 126a | 127 | 129 | 130 | 135 | 136 (a–w) | 137 | 139 | 143a | 147 | 126b | 141c | 143b | 143c | 140 | 141d | 142 | 144 | 145 | 146 |
|---|---|---|---|---|---|---|---|---|---|---|---|---|---|---|---|---|---|---|---|---|---|---|
| *Antiquatonia* cf. *antiquata* | | | 126a | | | | | | | | | | 126b | 141c | | | | | | | | |
| *A. hindi* (Muir-Wood) | | | | | | | | | | | | | | | | | | | | | 145 | |
| *A.* cf. *hindi* | | | | | | | | | | | | | | | | 143c | | | | | | |
| *A. hindi wettonensis* (Muir-Wood) | | | | | 129 | | | 136o | | 139 | | | | | 143b | | | | | | | |
| *A. hindi* cf. *wettonensis* | | | | 127 | | | | | | | | | | | | | | 141d | | | | |
| *A. insculpta* (Muir-Wood) | | | | 127 | | | | 136?j | | | | | ?126b | | | | 140 | 141d | | | | |
| *A.* cf. *insculpta* | | | | | | | | | | | | | | | | | 140 | 141d | | | | |
| *A.* aff. *insculpta* | | | | | | | 135 | 136h, k | | | | | | | | | | | | | | |
| *A. sulcata* (J. Sowerby) | | | 126a | 127 | | | | | | | | | | | | | | | | | | |
| *A. sp.* | | | | | | | | 136i, l, p, s, v | 137 | | | | | | | | | 141d | | | 145 | 146 |
| *Athyris expansa* (Phillips) | | | | | | | | 136h | | | | | | | | | | | | | | |
| *A.* cf. *obtusa* (McCoy) | | | | | | | 135 | 136d, v | | | | | | | | | | 141d | | | | |
| *Aulacophoria keyserlingiana* (de Koninck) | | | | | 129 | | 135 | 136e, g | | | | | 126b | | | | | | | | | |
| *Avonia davidsoni* (Jarosz) | | | | 127 | | | 135 | 136d, f–h, p | | | | | | | | | | | | 144 | 145 | 146 |
| *A.* aff. *davidsoni* | | | | 127 | | | 135 | 136f | | | | 147 | | | | | | | | | 145 | 146 |
| *A. youngiana* (Davidson) | | | | 127 | | | 135 | 136?a, b, f, j–m, v | 137 | 139 | | 147 | | | | | | | | ?144 | ?145 | |
| *A.* cf. *youngiana* | | | | | | | | 136b | | | | | | | | | ?140 | 141d | 142 | | | |
| *A. sp.* | | | | 127 | | | | 136n, ?o, ?t, ?w | | | ?143a | | ?126b | | | | | | | | | |
| *Brachythyris decora* (Phillips) | | | | 127 | | | 135 | 136d, n, r | 137 | | | 147 | | | | | | | | | | |
| *B. integricosta* (Phillips) | | | | 127 | | | | 136t | | | | | | | | | | | | | | |
| *B.* cf. *ovalis* (Phillips) | | | | | | | | 136f, t | | | | | | | | | | | | | | |
| *B. paucicostata* (McCoy) | | | 126a | | | | 135 | 136e, l, ?w | 137 | | | | 126b | 141c | | | ?141d | 142 | | | ?145 | 146 |
| *B. sp.* | | 125 | 126a | | | 130 | 135 | 136d, f, ?g, j, l | 137 | 139 | 143a | 147 | 126b | | | 143c | | | | | | |
| *'Camarotoechia' trilatera* (de Koninck) | | | | | | | | 136c–s, w | | | | | | | | | | | | | | |
| Chonetoids | | | | | | | 135 | 136v | 137 | | | | | | | | | | | | | |
| *Crurithyris sp.* [juv.] | | | | | | | 135 | 136t | | | | | | | | | | | | | | |
| *Derbyia sp.* | | | | | | | | 136s | | 139 | | | | | | | | | | | | |
| *Dictyoclostus pinguis* (Muir-Wood) | | | | | | | | 136e, ?o, p, s | | 139 | | | 126b | | | | | | | 144 | | |
| *D. semireticulatus* (Martin) | | | | | | | | | | | | | 126b | | | | | | | | | |
| *D.* cf. *semireticulatus* | | | 126a | 127 | | 130 | 135 | 136c, f–h, j, l, m, p, t | 137 | | | 147 | 126b | | | | | 141d | | | 145 | 146 |
| *Dielasma hastatum* (J. de C. Sowerby) | | | | | | | | | | | | | | | | 143c | | | | | | |
| *D. sp.* | | | | | | | | 136c, e, v | | | | | | | | | | 141d | | | | |
| *Echinoconchus* aff. *defensus* (I. Thomas) | | | | | 129 | | | 136c, e, v | | | | | 126b | | | | | 141d | | | | |
| *E. elegans* (McCoy) | | 125 | | | | | 135 | 136b, c, g, n, v | | | | | 126b | | | | | | | | | |
| *E.* cf. *eximius* (I. Thomas) | | | 126a | 127 | | 130 | 135 | 136a, b, e–g, i–k | | 139 | 143a | 147 | | | | | | 141d | | | 145 | 146 |
| *E. punctatus* (J. Sowerby) | | 125 | 126a | 127 | | | 135 | 136a, d, g, j | 137 | | | 147 | 126b | 141c | | | | 141d | | | 145 | 146 |
| *E.* cf. *punctatus* | | | | | | 130 | 135 | 136a–c, g, h | | 139 | | 147 | | | | | | 141d | | | | |
| *E.* aff. *punctatus* [broad bands] | | | | | | | 135 | 136t | | | | | | | 143b | | | | | | | |
| *E. subelegans* (I. Thomas) | | | | | | | 135 | 136q, r | | | | | | | | | | | | 144 | | |
| *E.* cf. *subelegans* | | | | | | | | | 137 | | | | | | | | | | | | | |
| *E. venustus* (I. Thomas) | | | | | | | | | | | | | | | | | | | | | | |
| *E. sp.* | | | | | | | | | | | | 147 | | | | | | | | | ?2145 | |
| *Eomarginifera derbiensis* (Muir-Wood) | | | | | | | | | | | | | | | | | | | | | | |
| *E. lobata* (J. Sowerby) *laqueata* (Muir-Wood) | | | | | | | | | | | | | | | | | | | | | | |

## Table 5—continued

| Genera and species | Localities | B₂ ZONE | | | | | | | | | | | P₁ₐ ZONE | | | | P₁ᵦ ZONE | | | | | |
|---|---|---|---|---|---|---|---|---|---|---|---|---|---|---|---|---|---|---|---|---|---|---|
| | | 125 | 126a | 127 | 129 | 130 | 135 | 136 (a–w) | 137 | 139 | 143a | 147 | 126b | 141c | 143b | 143c | 140 | 141d | 142 | 144 | 145 | 146 |
| Eomarginifera cf. pseudoplicatilis (Muir-Wood) | | | | | | | | | | | | 147 | | | | | 140 | | | | | |
| E. aff. setosa (Phillips) | | | | 127 | | | | 136v | | | | | | | | | | 141d | 142 | 144 | | |
| E. sp. | | | | | | | | 136d | | | | | | | | | | 141d | | 144 | | |
| Fluctuaria undata (Defrance) | | | | | | | 135 | 136f | | | | | | | | 143c | | | | | | |
| F. cf. undata | | | | | | | | 136c | | | | | | | | | | | | | | |
| F. sp. | | | | | | | | 136b, i | | | | | | | | | | | | | 145 | |
| Georgethyris cf. alexandri (George) | | | | | | | | 136e | 137 | | 143a | 147 | ?126b | | | | | | | | | |
| G. cf. elliptica (Phillips) | | | | | | | 135 | 136d, t | | | | | | | | | | | | | | |
| G. aff. lobata (Muir-Wood) | | | | | | | | | | | | | | | | | | | | | | |
| G. obtusa (J. Sowerby) | | | | | | | 135 | | | | | | | | | | | | | | | |
| Gigantoproductus sp. cf. bisati (Paeckelmann) | | | | | | | | | | | | | | 141c | | | | | | ?144 | | |
| G. edelburgensis (Phillips) | | | | 127 | | | ?135 | | | | | | | | | 143c | | | | | | |
| G. cf. giganteus (J. Sowerby) | | | | | | | | | | | | | | | | | | 141d | | | | |
| G. sp. cf. maximus (McCoy), group | | | | | | | | | | | | | | | | | | 141d | | 144 | | |
| G. sp. cf. moderatus (Schwetzow) | | | | | | 130 | | 136d | | | | | | | | | | | | | | |
| G. sp. | | 125 | | | | 130 | 135 | 136d | | | | | | | | | 140 | | | | | |
| Hustedia radialis (Phillips) | | | | | | | | | | 139 | 143a | 147 | 126b | | | | 140 | | | | | |
| Krotovia laxispina (Phillips) | | | 126a | | 129 | | | | | | | | | | | | | 141d | | | | |
| K. cf. laxispina | | | | | | | | 136c, e, f, p, v | 137 | 139 | 143a | 147 | 126b | | | | | | 142 | 144 | 145 | |
| K. spinulosa (J. Sowerby) | | | 126a | 127 | 129 | 130 | 135 | 136v | | | | | 126b | | | | | | | | | |
| K. cf. spinulosa | | | | 127 | | | | 136f, l | | | | 147 | | | | | | | | | | |
| K. aff. spinulosa | | | | | | | | 136k | | | | | | | | | | | | | | |
| K. spinulosa (Phillips non J. Sowerby) | | | | | | | 135 | 136e–g, s | | 139 | 143a | | 126b | | | | | | | | 145 | |
| K. sp. | | | | | | 130 | 135 | | | | | | | | | | | | | | | |
| Leptagonia analoga (Phillips) | | | 126a | | | | 135 | | | | | | | | | | | | | | | |
| L. sp. | | | | 127 | | | 135 | | | | | 147 | | | | | | | | 144 | | |
| Linoproductus cf. corrugatus (McCoy) | | | | 127 | | | 135 | 136n, o | | 139 | | | 126b | | | | | | 142 | 144 | | |
| L. sp. hemisphaericus (J. Sowerby) group | | | | | | | | 136a, k | 137 | | | | | | | | 140 | | | 144 | 145 | |
| L. sp. | | | | | | | | 136?e, f, s | | | | 147 | | | | | | | | 144 | | |
| Marginicinctus projectus (Muir-Wood) | | | 126a | | | | | 136b, d, u | 137 | 139 | | | 126b | | | | | | | | | |
| Marinia glabra (J. Sowerby) | | | | | | 130 | 135 | 136a, g, i | | | | | | | | | | 141d | | | | |
| M. sp. | | | | 127 | | | 135 | 136f, g, n | | | | | | | | | | | | | | |
| Martinothyris lineata (J. Sowerby) | | | | | | | 135 | 136a–f, ?j, k, l, n, p, s–u | 137 | 139 | 143a | 147 | 126b | | | | | | | | | |
| Megachonetes sp. | | | | | | | 135 | 136f | | | | | | | | | | | | | | |
| Orbiculoidea nitida (Phillips) | | | | | | | | 136e | | | | | | | | | | | | | | |
| O.? | | 125 | | | | | 135 | 136h, i, t, v | 137 | 139 | 143a | | 126b | | | | | 141d | 142 | | | |
| Orthotetoids | | 125 | | 127 | | | 135 | 136a–i, m, n, t, v | 137 | 139 | 143a | 147 | 126b | | | | | 141d | 142 | | | |
| Orthotetoids [with intercostal striae] | | | | | | | 135 | | | | | | | | | | | | | | | |
| Ovatia sp. nov. | | | 126a | | | | 135 | 136f | | | | | | | | | | | | | | |
| O. sp. [juv.] | | | | 127 | | | 135 | 136e | | | | | | | | | | | | | | |
| Overtonia fimbriata (J. de C. Sowerby) | | | | 127 | | | 135 | 136h, i, t, v | 137 | | | 147 | | | | | | | | | | |
| Phricodothyris periculosa George | | | | 127 | | | 135 | | | 139 | | | 126b | | | | | | | | | |
| P. verecunda? George | | | | | | | | | 137 | 139 | 143a | 147 | | | | | | | | | | |
| Pleuropugnoides pleurodon (Phillips) | | | | | | | 135 | | 137 | 139 | 143a | 147 | | | | | | | | 144 | 145 | |
| Plicatifera plicatilis (J. de C. Sowerby) | | | | | | | | 136d–k | 137 | | 143a | | | | | | | | | 144 | 145 | 146 |

Table 5—continued

| Genera and species | B₂ ZONE | | | | | | | | | | | P₁ₐ ZONE | | | | P₁ᵦ ZONE | | | | | |
|---|---|---|---|---|---|---|---|---|---|---|---|---|---|---|---|---|---|---|---|---|---|
| Localities | 125 | 126a | 127 | 129 | 130 | 135 | 136 (a-w) | 137 | 139 | 143a | 147 | 126b | 141c | 143b | 143c | 140 | 141d | 142 | 144 | 145 | 146 |
| *Plicatifera sp.* | | | | | | | 136 f, ?m, t, v, w | 137 | | 143a | 147 | 126b | | | | 140 | | ?142 | ?144 | | |
| *Plicochonetes buchianus* (de Koninck) | | | 127 | | | | | | | | | | | | | | | | | | |
| *P. cf. buchianus* | | | | | | | | | | | 147 | 126b | | | | | | | | | |
| *Proboscidella proboscidea* (de Verneuil) | | | | | | | 136f, g | | | 143a | | 126b | | | | | | | | | |
| *Productina margaritacea* (Phillips) | | | | | | | 136c | 137 | | | | | | | | | | | | | |
| *P. sp.* [juv.] | | | | | | | | | | | | | | | | | | | | | |
| *Productus productus* (Martin) | | | 127 | | | 135 | 136t | | | | | | 141c | | | | 141d | 142 | | | |
| *P. aff. productus* [broad form] | | | | | | | 136a, b, o, r, s | | | | | | 141c | | | | 141d | 142 | 144 | | |
| *P. productus hispidus* Muir-Wood | | | | | | | 136a | | | | | | ?141c | | 143c | | 141d | 142 | 144 | | |
| *P. productus aff. hispidus* [2 rows of spine bases on flanks] | 125 | 126a | | | 130 | | | | 139 | | | | | | | | | | | | |
| *P. productus aff. hispidus* [coarsely costate form] | | | | | | | | | | | | | | | | | 141d | 142 | | | |
| *P. sp.* | | | | | | | | | | | | 126b | | | | 140 | | | | | |
| *Pugnax acuminatus* (J. Sowerby) | | | | | | | | | | | | | | | | 140 | | | | | |
| *P. platylobus* (J. de C. Sowerby) | | | | | | 135 | 136b, e, f, v, w | | | 143a | 147 | | | | | 140 | | | | | |
| *P. cordiformis* (J. de C. Sowerby) | | | | | | 135 | 136b, d | | | | | | | | | | | | | | |
| *P. pseudopugnus* Parkinson | | | 127 | | | 135 | 136b, e, h, l, n, t, v | | | 143a | 147 | 126b | | | | 140 | | | | | |
| *P. pugnus* (Martin) [small form] | | | | | | 135 | | | | | | | | | | | | | | | |
| *P. pugnus* [large form] | | | | | | | 136b, c, f, g, i, l, m | 137 | | | | | | | | | | | 144 | 145 | 146 |
| *Pugnoides triplex* (McCoy) | | | | | | | 136s | | | | | | | | | | | | | | |
| *Pustula magnituberculata* I. Thomas | | | | | | | 136?a, g, s | | | | | | | | | | | | | | |
| *P. sp. rugata* (Phillips) group | | | | | 130 | 135 | | | | 143a | 147 | | | | | | 141d | | 144 | | |
| *P. sp.* | | | | | | 135 | | | 139 | | | | | | | | | | | | |
| *Reticularia mesoloba* (Phillips) | | | | | | 135 | 136k, l | | | | | | | | | | | | | | |
| *R. cf. mesoloba* | | | | | | | 136f | | 139 | | | | | | | | | | | | |
| *Rhipidomella michelini* (Léveillé) | | | | | | | 136f, h, v | 137 | | | 147 | 126b | | 143b | | | | | 144 | | |
| *R. michelini divaricata* (McCoy) | | | | | | | | | | | | ?126b | | | | | | | | | |
| *R. sp.* | | | | | | | 136c | | 139 | | | | | | | | | | | | |
| *Rugosochonetes sp.* | | | | | | | | | | | | | | | | | | | | | |
| *Schellwienella sp.* [with intercostal striae] | | | | | | | 136c | | | | | | | | | | | | | | |
| *S. sp.* | | | 127 | 129 | | 135 | 136b, c, e-h, j | 137 | | 143a | 147 | 126b | | | | | | | | | |
| *Schizophoria connivens* (Phillips) | | | | | | 135 | 136p | | | | | | | | | | | | | | |
| *S. palliata* (Demanet) | | | | | | 135 | 136d, e, i, k, m | 137 | | | | 126b | | 143b | | | | | | | |
| *S. aff. pinguis* Demanet | | | 127 | | | | | | | | | | | | | | | | | | |
| *S. resupinata* (Martin) | | | | 129 | | 135 | | 137 | 139 | | | 126b | | | | | | | | | |
| *S. resupinata gigantea* Demanet | | | 127 | | | 135 | 136d-n, p, q, s, t, v, w | 137 | 139 | 143a | 147 | 126b | 141c | | 143c | 140 | 141d | 142 | 144 | 145 | 146 |
| *S. sp.* | | | | | | 135 | 136v | | | | | | | | | | | 142 | | | |
| *Schuchertella sp.* | | | | | | | | | | | | | | | | | | | | | |
| *Sinuatella sinuata* (de Koninck) | | | 127 | | | 135 | 136d, e, p | | | | | | | | | | | | | | |
| Smooth spiriferoids | | | | | | | | | | | | 126b | | | | | | | | | |
| *Spirifer sp. bisulcatus* J. de C. Sowerby group | | 126a | 127 | | | | 136y | 137 | | 143a | 147 | 126b | 141c | | 143c | 140 | 141d | 142 | 144 | 145 | 146 |
| *S. duplicicosta* Phillips | | | | | | 135 | 136d, e, p | | | | | 126b | | | | | | | 144 | 145 | 146 |
| *S. cf. duplicicosta* | | 126a | | | | 135 | | | | | | | | | | | | | | | |
| *S. plicatosulcatus North* | | 126a | | | | 135 | | | | | | | | | | | | | | | |

Table 5—continued

| Genera and species | 125 | 126a | 127 | 129 | 130 | 135 | 136 (a-w) | 137 | 139 | 143a | 147 | 126b | 141c | 143b | 143c | 140 | 141d | 142 | 144 | 145 | 146 |
|---|---|---|---|---|---|---|---|---|---|---|---|---|---|---|---|---|---|---|---|---|---|
| Spirifer aff. plicatosulcatus | | | | | | | 136t | | | | | | | | | | | | | | |
| S. striatus (Martin) | | | 127 | | | | | | | 143a | | 126b | | | | | | | | | |
| S. (Fusella) cf. grandicostatus McCoy | | | 127 | | | | 136a-c, e-i | | | | | 126b | | | | | | | | | 146 |
| S. (F.) triangularis J. de C. Sowerby | | | | | | | 136g | | | | | | | | | | | | | | |
| S. (F.) trigonalis (Martin) | | | 127 | | | 135 | 136b, f-h, l | 137 | 139 | | 147 | 126b | | | | | 141d | | 144 | | |
| S. (F.) sp. | | | | | | | | | | | | | | | | | | | | | |
| S. sp. | | | | | | | | | | | | | | | | | | | | | |
| Spiriferellina insculpta (Phillips) | | 126a | 127 | | | 135 | 136s | 137 | | | | 126b | | | | | | | | | |
| S. octoplicata (J. de C. Sowerby) D (North) | | | | | 130 | | 136l | 137 | | | | | | | | | | | | | |
| S. cf. perplicata (North) | | | | | | | 136f | | | | | | | | | | | | | | |
| S. perplicata D (North) | | | | | | | 136c | | | | | | | | | | | | | | |
| S. sp. [juv.] | | | | | | 135 | | | | | | | | | | | | | | | |
| Streptorhynchus senilis (Phillips) | | | 127 | | | | 136t | 137 | | | | | | | | | | | | | |
| Striatifera striata (Phillips) | | | | | | | 136a | 137 | | | | | | | | | | | | | |
| Syringothyris cuspidata cuspidata (J. Sowerby) | | | | 129 | | | | | | | | | | | | | | | | | |
| S. cuspidata | 125 | 126a | 127 | | | 135 | 136n | | | | | 126b | 141c | 143b | | | | | | | |
| Tylothyris subconica (Martin) | | | | | | | 136?u | | | | | | | | | | | | | | |
| T. sp. | | | | | | ?135 | | | | | ?147 | 126b | | | | | | | | | |
| Bellerophon aff. tenuifascia J. de C. Sowerby | 125 | | 127 | | | | | | | | | | | | | | | | | | |
| B. sp. | | | | | 130 | | | | | | | | | | | | | | | | |
| Meekospira sp. | | | | | | | | | ?139 | | 147 | | | | | | | | | | |
| Mourlonia carinata (J. Sowerby) | | | | | 130 | 135 | | | | | | | | | | | | | | | |
| M. cf. naticoides (de Koninck) | | | | | 130 | | 136c | | | | | | | | | | | | | | |
| Naticopsis ampliata (Phillips) | | | | | | 135 | | | | | | | | | | | | | | | |
| N. cf. elliptica (Phillips) | | | | 129 | | | | | | | | | | | | | | | | | |
| N. elongata (Phillips) | | | | | | | | | | | | | | | | | | | | | |
| N. sp. | | | | | | | | | | | 147 | 126b | | | 143c | | | | 144 | | |
| Platyceras (Platyceras) cf. trilobum (Phillips) | | | | | | | | | | | | 126b | | | | | | | | | |
| P. (P.) vetustum (J. de C. Sowerby) | | | | | | | 136b, t, v | | | | | | | | | | | | | 145 | 146 |
| P. (P.) cf. vetustum | | | | | | 135 | | | | | 147 | 126b | | | | | 141d | | 144 | | |
| P. sp. [juv.] | | | | | | 135 | | | | | | | | | | | | | | | |
| Platyschisma glabratum (Phillips) | | | | | 130 | 135 | | | | | | | | | | | | | | | |
| P. ? | | | | | 130 | 135 | | | | | | | | | | | | | | | |
| Shansiella ? | | | | | | 135 | | | | | | | | | | | | | | | |
| Soleniscus ventricosus (de Koninck) | | | | | | | 136h, i | | | | | | | | | | | | | | |
| Straparella fallax (de Koninck) | | | 127 | | | | | | | | | | | | | | | | | | |
| S. ? | | | 127 | | | | | | | | | | | | | | | | | | |
| Straparollus (Euomphalus) catillus (Martin) | 125 | 126a | | | | | 136v | 137 | | | | 126b | | | | | | | | | |
| S. (Straparollus) mammulus de Koninck | | | | 129 | | 135 | 136c, w | 137 | | | | | | | | | | | ?144 | | |
| S. (S.) sp. | | | | | 130 | 135 | 136t, v | | | | | | | | | | | | | | |
| S. sp. | | 126a | 127 | | | 135 | 136?o | | | | | 126b | | | | | | | | | |
| Turbonitella biserialis (Phillips) | | 126a | 127 | | | | | 137 | | | 147 | 126b | | | | | | | | | |
| Acanthopecten nobilis (de Koninck) | | 126a | 127 | | | | | | | | | 126b | | | | | | | | | |
| A. stellaris (Phillips) | | | | | | | | | | | | | | | | | 141d | | | | |
| A. cf. stellaris | | | | | | | | | | | | | | | | | | | | | |
| A. sp. | | | | | | | | | | | | | | | | | | | | | |

Table 5—continued

| Genera and species | B₂ ZONE 125 | 126a | 127 | 129 | 130 | 135 | 136 (a–w) | 137 | 139 | 143a | 147 | P₁ₐ ZONE 126b | 141c | 143b | 143c | 140 | 141d | 142 | P₁ᵦ ZONE 144 | 145 | 146 |
|---|---|---|---|---|---|---|---|---|---|---|---|---|---|---|---|---|---|---|---|---|---|
| *Actinopteria persulcata* (McCoy) | | | | | | | 136p | | | | | | | | | | | | | | |
| *Annuliconcha sp.* [juv.] | | | | | | | 136t | | | | 147 | | | | | | | | | | |
| *Aviculopecten* aff. *pera* (McCoy) | | | 127 | | | | | | | | | | | | | | | | | | |
| *A. planoradiatus* McCoy | | | 127 | | 130 | | | | | | | | | | | | | | | | |
| *A. plicatus?* (J. de C. Sowerby) | | | | | | | 136c | 137 | | | | 126b 141c | | | | | | 142 | 144 | | |
| *A. sp.* | 125 | | | | | | 136b, e–g, i, n, o, t, v, w | | | | | | | | | | | | | | |
| *Aviculopinna mutica* (McCoy) | | | | | | | 136o, s | | | | | | | | | | | | | | |
| *Conocardium alaeforme* (J. de C. Sowerby) | | | 127 | | | | 136h, r, t | | | | | | | | | | | | | | |
| *C. rostratum* (Martin) | | | | | | | 136v | | 139 | 143a | | | | | | | | | | | |
| *Dunbarella* aff. *persimilis* (Jackson) | | | | | | | | | | | | | | | | 140 | | | | | |
| *Edmondia laminata* (Phillips) | | | | | 130 | | 136d | | | | 147 | | | | | | | | | | |
| *E. senilis* (Phillips) | | | | | | | | | | | | | | | | | | | | | |
| *E.* cf. *senilis* | | | | | | | | | | | | | | | | | | 142 | | | |
| *E. sulcata* (Phillips) | | 126a | 127 | | | | 136c, p | | | | | 126b | | | | | | 142 | | | |
| *E.* cf. *unionformis* (Phillips) | | 126a | 127 | | | | 136i | | | | | | | | | | 141d | 142 | | | |
| *E. sp.* | | | | | | | 136e | | | | | | | | | | | | | | |
| *Girtypecten tessellatus* (Phillips) | | | | | | 135 | 136g, r, s | | | | 147 | | | | | | | | | | |
| *Leiopteria hirundo* de Koninck | 125 | | | | | | | | | | | | | | | | | | | | |
| *L. laminosa* (Phillips) | | | | | 130 | 135 | 136d, p | | | | | 126b | | | | | | | | | |
| *L. lunulata* (Phillips) | | | | 129 | 130 | | 136d, p, ?s | | | | | 126b | | | | | | | | | |
| *L. sp.* | | 126a | 127 | | | | 136f | | | | | | | | | | | | | | |
| *Lithophaga lingualis* (Phillips) | | 126a | 127 | | | | 136f, t | | | | | | | | | | | | | | |
| *L.* cf. *lingualis* | | | | | | 135 | 136c, ?s | | | | | | | | | | | | | | |
| *L.?* | 125 | 126a | | 129 | | 135 | 136c, ?v | | | | | | | | | | 141d | | | | |
| *Palaeolima obliquiradiata* Hind | 125 | 126a | | | | | 136s | | | | 147 | ?126b | | | | | | | | | |
| *Parallelodon bistriatus* (Portlock) | | | | | | | 136m, o, s | | 139 | | | | | | | | | 142 | | | |
| *P. reticulatus* (McCoy) | | | | | | | | | | | | | | | | | | | | | |
| *P. sp.* | | | | | | | 136b, ?g | | | | | | | | | | | | 144 | | |
| Pectinoid | | | | | | | 136f, s | | | | | 126b | | | | | | | | | |
| *Posidoniella vetusta* (J. de C. Sowerby) | | 126a | | | | | 136e | 137 | | | 147 147 | | | | | | | | | | |
| *P. sp.* | | | | | | | 136g | | | | | | | | | | | | | | |
| *Pterinopectinella dumontiana* (de Koninck) | | | | | | | | 137 | | | | 126b | | | | | | | | | |
| *P. sp.* | 125 | 126a | 127 | | 130 | | 136o, r | | | | 147 | 126b | | | | | | | 144 | | |
| *Sanguinolites oblongus* Hind | | | | | | 135 | 136?b, c–f, ?i, ?t | | | | | | | | | | | | | | |
| *S. tricostatus* (Portlock) | | 126a | | | | 135 | 136s | | | | 147 | 126b 141c | | | | | | | | | |
| *S. sp.* | | | | | 130 | | | | | | | | | | | | | | | | |
| *Streblopteria hemisphaerica* (Phillips) | | | | | | 135 | | | | | | | | | | | | | | | |
| *S. laevigata* (McCoy) | | | | | | | | | | | | | | | | | | | | | |
| *S.* cf. *laevigata* | | | | | | | | | | | | | | | | | | | | | |
| *S. sp.* | | | | | | | | | | | | | | | | | | | | | |
| Orthocone nautiloids | | | | | | | | | | | | | | | | | | | | | |
| Nautiloid | | | | | | | | | | | | | | | | | | | | | |
| *Vestinautilus?* | | | | | | | | | | | | | | | | | | | | | |
| *Beyrichoceras rectangularum* Bisat | | | | | | | | | | | | | | | | | | | | | |

## Table 5—continued

| Genera and species | B₂ ZONE 125 | 126a | 127 | 129 | 130 | 135 | 136 (a–w) | 137 | 139 | 143a | 147 | P₁ₐ ZONE 126b | 141c | 143b | 143c | P₁ᵦ ZONE 140 | 141d | 142 | 144 | 145 | 146 |
|---|---|---|---|---|---|---|---|---|---|---|---|---|---|---|---|---|---|---|---|---|---|
| Beyrichoceras sp. [rectangularum or delicatum Bisat: cf. Bisat 1934, fig. 17, p. 288] | | 126a | | | | | | | | | | | | | | | | | | | |
| Beyrichoceratoid | | | | | | | | | | | | | | | | | | | | | |
| Bollandoceras cf. micronotum (Phillips) | | | | | | | | | | | | | | | | | | | | | |
| Goniatites crenistria Phillips [early form, cf. GSM 86644 from Blackhole, Malham] | | | | | | | | | | | | 126b | 141c | | | | | | | | |
| G. sp. crenistria group | | | | | | | | | | | | | | | | | | | | | |
| G. cf. hudsoni Bisat [cf. Bisat 1952, pl. 2, fig. 6] | | | | | | | | | | | | | | | | | | 142 | | | |
| G. striatus J. Sowerby | | | | | | | | | | | | | | | | | | | | | |
| G. cf. struppus Hodson and Moore | | | | | | 135 | 136n | | | | | | | | | | | | | | |
| G. sp. | | | | | 130 | | | | | | | | | | | | | | | | |
| Indeterminate goniatite | | | | | | | | | | | | | | | 143c | | | | | | |
| Brachymetopus maccoyi (Portlock) | | | | | | | | | | | | | | | | | | | | | |
| B. sp. | | | | | | | 136f, v | 137 | | | | | | | | | | | | | |
| Cummingella sp. | | | | | 130 | 135 | 136d, ?e, g, n, r, t, v, ?w | | | 143a | | | | | | 140 | | 142 | 144 | 145 | ?145 |
| Griffithides sp. | | | 127 | | | | | | | | 147 | 126b | | | | | | | | | |
| Weberides mucronatus (McCoy) | | | | | | 135 | 136a, c | | | | | 126b | | | | | | | 144 | | |
| W. sp. | | | | | | 135 | 136b | | | | | | | | | | | | | | |
| Trilobite fragments | 125 | | | | | | | | | | | | | | | | | | | | |
| Cyclus sp. | | | | | | | 136c | | | | | | | | | | | | | | |
| Dithyrocaris sp. [mandible gnathal lobe] | | | | | | | 136c | 137 | | | | | | | | | | | | | |
| D. ? [plate fragment] | | | | | | | | | | | | | | | | | | | | | |
| Cyprella chrysalidea de Koninck | | | | | | | | | | | | | | | | 140 | | | | | |
| Cypridina brevimentum Jones, Kirby and Brady | | | | | | | | | | 143a | | | | | | 140 | 141d | 142 | | | |
| C. cf. brevimentum | | | | | | | | | | 143a | | | | | | | 141d | | | | |
| C. ? | | | | | | | | | 139 | | | | | | | | | | | | |
| Cypridinella cf. cummingii Jones, Kirby and Brady | | | | | | | 136c–g, j–m, o, v | | | | | | | | | | | | | | |
| Ostracods | 125 | 126a | 127 | | 130 | 135 | 136c–g, j–m, o, v | | 139 | | | 126b | | | | | 141d | | 144 | | |

Species which are restricted, within the present area, to $B_2$ include *Acanthoplecta mesoloba*, '*Camarotoechia*' *trilatera*, *Echinoconchus* aff. *punctatus* [broad bands, Plate XIV, fig. 4], *Leptagonia analoga*, *Ovatia sp. nov.*, *Overtonia fimbriata*, *Plicatifera plicatilis*, *Proboscidella proboscidea*, *Pugnoides triplex* and *Sinuatella sinuata* (see Table 5). Although many of these are known from other horizons outside the Castleton reef-belt, their relative abundance, association with the remainder of the fauna and absence in the immediately overlying $P_{1b}$ beds suggest that they constitute a recognizable $B_2$ assemblage. Corals are almost entirely confined to $B_2$ beds.

*Striatifera striata* has been collected only from beds of $P_{1a}$ age and is probably the only fossil other than goniatites which can be used to distinguish $P_{1a}$ from $B_2$ faunas. Parkinson (1965, p. 171) notes this species as being of $P_{1a}$ to $P_{1b}$ age (or very high $D_1$) and the present collection supports the stratigraphical value of this fossil.

The ranges of gastropods, bivalves and other fossil groups are also given in Table 5, but apart from the fact that they are again more abundant in $B_2$, especially at Treak Cliff, little can be concluded.

The following list gives the significant elements of the faunas from each of the reef localities and the arguments for their suggested ages.

*Loc. 125, Perry Dale.* Only a small collection was made from this locality, but the presence of *Productus productus* aff. *hispidus* suggests a $B_2$ age.

*Loc. 126, Tor Top Sink, Perryfoot.* Collections have been made here from two horizons in the apron-reef. The lower horizon (loc. 126*a*) contains *Ovatia sp. nov.*, *P. productus* aff. *hispidus* and the goniatites *Beyrichoceras sp.* [*rectangularum* or *delicatum*, cf. Bisat 1934, fig. 17, p. 288] which indicates a high $B_2$ age. The upper horizon (loc. 126*b*) has a rich fauna of $P_{1a}$ age including *Goniatites sp. crenistria* group and *Striatifera striata*.

*Loc. 127, valley between Snels Low and Middle Hill.* *Leptagonia analoga*, *Overtonia fimbriata* and *Spirifer striatus* are present and the fauna is of $B_2$ age.

*Locs. 128a to 134 Peaks Hill, Giant's Hole.* This series of localities is from the complex area on the north side of Middle Hill (Parkinson 1943, p. 125; 1947, p. 107).

*Loc. 128a, west side of Peaks Hill.* The only specimen collected is the $B_2$ species *Goniatites wedberensis* Bisat. The locality is not listed in Table 5.

*Loc. 129, near west side of Peaks Hill.* A relatively limited reef fauna from here is probably of $B_2$ age.

*Loc. 130, Peaks Hill.* A $B_2$ fauna has been collected from this locality with *L. analoga, P. productus* aff. *hispidus*, *Goniatites crenistria* [early form, cf. GSM 86644 from Black Hole, Malham], and *G.* cf. *hudsoni* [cf. Bisat 1952, pl. 2, fig. 6]. The goniatite evidence suggests that this horizon is near the top of the $B_2$ Zone.

*Locs. 131–134, Giant's Hole.* No goniatites have been collected from these localities and the reef faunas, although of $B_2$ aspect, are too indefinite to be included in Table 5. *Palaeosmilia regia*, however, is recorded from locality 131 and the significance of this coral is discussed below (p. 152).

*Loc. 135, Odin Fissure.* A rich $B_2$ fauna is present and includes the goniatites *Beyrichoceras rectangularum, Bollandoceras* cf. *micronotum* and *Goniatites sp.* (Parkinson 1947, p. 110; 1953, p. 258; 1965, p. 162).

L

*Loc.* 136*a-w*, *Treak Cliff.* This is the classic collecting locality of the Castleton area. An extensive collection was made from a series of exposures across the width of the apron-reef, (*a*) being at the top of the Treak Cliff slope and (*w*) at the base of the slope near the limestone-shale junction (see inset map on Plate XI). Only a few poorly preserved goniatites were found, but the locality has long been famous for its isolated pockets of these fossils. The goniatite fauna listed by Parkinson (1947, p. 107) is of high $B_2$ age and there is no evidence of $P_{1a}$ being present. The richest faunas are found on the higher parts of the apron slope associated with the algal-reef limestones (Wolfenden 1958) which form the crest of Treak Cliff, and where presumably the widest range of ecological niches existed. Juvenile specimens are common at many localities and deformed shells are also found throughout the fauna. The shell damage is of the kind described by Brunton (1966, p. 355, pl. 60, figs. 4, 5) where the normal growth is distorted and the shell scarred to the anterior border. Brunton suggests that fishes may be responsible for crushing the shells. Fish remains are not here recorded from Treak Cliff but the mandible gnathal lobe (18 mm long, with six tooth-like projections on the crushing surface) of the crustacean *Dithyrocaris* has been found at locality 136*c* (Plate XIV, fig. 10) and this animal may have caused some of the shell damage.

Echinoderm remains, except for crinoid columnals, are very rare in the Castleton reefs and this contrasts with their abundance in the reef-limestones of the basin province at Clitheroe. Wolfenden (1958; 1959) records two species of sponges from the algal reef at Treak Cliff.

A specimen of the rare fossil '*Michelinia*' *balladoolensis* J. Smith (1911, p. 148, pl. 25, fig. 18) was collected from Treak Cliff [1341 8333] and presented to the Institute of Geological Sciences by the Rev. Fr. M.E.R. Le Morvan in 1965. The species, which is at present known only from three localities, is not a tabulate or rugose coral. Affinities either with the Conulariidae or with the Bryozoa are suggested. The type material comes from the Carboniferous Limestone (? $B_2$ reef) of Balladoole, Isle of Man, and 24 fragments of the species are present in the C. B. Salter Collection (I.G.S.) from the Hotwells Limestone ($D_2$) of Compton Martin, Mendip Hills.

*Loc.* 137, *Winnats.* An extensive $B_2$ fauna is recorded and *Ovatia sp. nov.*, *Overtonia fimbriata* and *Plicatifera plicatilis* are present.

*Loc.* 138, *Cow Low Nick.* Ford (1965, p. 186) listed a $B_2$ goniatite fauna from this locality which is placed in "lower $B_2$" by Parkinson (1965, p. 164) on stratigraphical grounds.

*Loc.* 139, *S.W. of Castleton church.* The reef fauna here includes *Echinoconchus* aff. *punctatus*, *L. analoga* and *P. productus* aff. *hispidus*, and is similar to the typical $B_2$ faunas from Treak Cliff.

*Loc.* 140, *ledge above entrance to Pindale Quarry.* Eden and others (1964, p. 90) correlated this locality with the $P_{1b}$ beds of Upper Jack Bank Quarry, but a specimen of *Pugnax pugnus* [small form] is present, and the limestones forming the floor of this ledge may be older than $P_{1b}$.

*Loc.* 141, *Lower Jack Bank Quarry.* Reef-limestone of both $P_{1a}$ and $P_{1b}$ age are developed (Eden and others 1964, p. 90). Poorly-preserved goniatite fragments are present at the lower level (loc. 141*c*) and one specimen shows "crenulate transverse ornament" diagnostic of the $P_{1a}$ species *G. crenistria*. *Striatifera* *striata* is also present. The upper horizon (loc. 141*d*) is equated both on fauna and geometry with the $P_{1b}$ beds of the apron slope.

Loc. 142, *Upper Jack Bank Quarry. Goniatites striatus*, figured from here by Bisat (1934, pl. 18, fig. 1), dates the beds as $P_{1b}$ and the fauna also includes *P. productus hispidus*.

Loc. 143, *Nunlow Quarry*. A small outcrop (loc. 143*a*) at the base of the quarry has a rich reef fauna of $B_2$ age including *L. analoga, Plicatifera plicatilis, Proboscidella proboscidea* and *Spirifer striatus*. The higher beds are difficult to relate and are not richly fossiliferous. However, locality 143*b* with *Striatifera striata* is thought to be of $P_{1a}$ age and the highest beds (loc. 143*c*) are of $P_{1a}$ or low $P_{1b}$ age.

Loc. 144, *Mich Low*. Parkinson (1947, p. 111) records *G. striatus* from Mich Low and the fauna with *Productus productus hispidus* and *Pugnax pugnus* [large form] is of $P_{1b}$ age.

Loc. 145, *near Bath Cottage. Brachymetopus maccoyi* is recorded here and the only other locality where it has been found is Mich Low. *Pugnax pugnus* [large form] is also present at both localities, which are of $P_{1b}$ age.

Loc. 146, *near Bath Cottage*. A small exposure about 100 yd west of the previous locality has a limited $P_{1b}$ fauna in which *Pugnax pugnus* [large form] is the only brachiopod of significance.

Loc. 147, *near New Bath Hotel*. A rich fauna has been collected which differs from that of the previous two localities and is of the $B_2$ type.

From a comparison of the apron-reef localities, the faunas which are suggested as typical of the three major reef horizons are as follows:

$P_{1b}$–Upper Jack Bank Quarry (loc. 142)

$P_{1a}$–the upper part of Tor Top Sink (loc. 126*b*)

$B_2$–Treak Cliff (loc. 136*a-w*)

A change occurs in the distribution of reef-limestones along the line of Dirtlow Rake in Pin Dale. To the west of this line almost the whole of the apron-reef limestones are of $B_2$ age and only the upper beds at Tor Top Sink (loc. 126*b*) are of $P_{1a}$ age. To the east of Pin Dale the apron slope is formed of $P_{1b}$ reef-limestones resting on $D_1$ shelf limestones. $B_2$ reef, which is only seen in isolated exposures at Nunlow Quarry (loc. 143*a*) and near the New Bath Hotel, (loc. 147) is not present over much of the eastern area.

This change of distribution is further emphasized by the presence of well-developed back-reef beds west of this line, while none is present to the east. Dirtlow Rake throws down about 75 ft to the east and whereas the top 50 ft of $D_1$ are seen in Pindale Quarry and show no development of apron-reef limestones, the $D_1$ shelf limestones seen to the west of Pin Dale may be more than 75 ft below the top of the zone and show extensive apron-reef development at several levels. It is possible, therefore, that no reef-limestones were formed corresponding with the highest beds of $D_1$, that the main mass of $B_2$ apron-reef correlates with limestones some way below the top of $D_1$ and that there is a faunal break between the $B_2$ and $P_{1b}$ reefs. The differences between $B_2$ and $P_{1b}$ faunas support this latter suggestion but the placing of the $P_{1a}$ faunas is not clear.

There has been a long-standing controversy about the relationship of the coral-brachiopod and goniatite zones, but the general equivalence of $D_1$ with

$B_2$ and lower $D_2$ with $P_{1a-b}$ has been accepted since the works of Hudson and Cotton (1945a, b) and Parkinson (1943; 1947); see also discussion of Shirley and Horsfield (1940, pp. 296–9).

The Pindale Quarry section shows the lateral passage of $P_{1b}$ apron-reef limestone into the shelf sequence at a horizon about 28 ft above the base of $D_2$ (Eden and others 1964, p. 90). The $B_2$ limestones are seen both on faunal and geometrical grounds to be the equivalent of beds high in the $D_1$ Bee Low Limestones in the Middle Hill and Snels Low areas (p. 136), and *Aulina furcata*, which occurs at Siggate in high $B_2$ beds and at Wall Cliff (loc. 28*k*, *m*) in the Miller's Dale Beds (high $D_1$) is a further link between $B_2$ and high $D_1$ zones. The position of $P_{1a}$ is therefore critical and is unfortunately the most difficult to relate.

Eden and others (1964, p. 90) considered that $P_{1a}$ beds "represent either the bottom few feet of $D_2$ or the top part of $D_1$". The presence in $P_{1a}$ of *Striatifera striata*, which has only been recorded from lower $D_2$ beds in the shelf limestones, suggests that $P_{1a}$ is probably not below the base of $D_2$. On this basis $P_{1a}$ correlates with part of the 28 ft of pre-$P_{1b}$ present at the base of $D_2$ in Pindale Quarry. Moreover, the occurrence of *Palaeosmilia regia* at Giant's Hole in beds which on goniatite evidence are high $B_2$ (Parkinson 1943, p. 126), agrees with $P_{1a}$ being also above the base of $D_2$. This coral is not known below the base of $D_2$ and, as shown by Parkinson (1947, fig. 1), $B_2$ probably overlaps with $D_2$. However, if Wolfenden's (1958, p. 881) record of *P. regia* "at all horizons in these ($B_2$) limestones" is accepted then this correlation cannot be correct. The placing of $P_{1a}$ in $D_2$ differs from that suggested by Bond (1950b, p. 179) for the Cracoe reefs where "Lower $P_{1a}$ is correlated with Upper $D_1$".

The boreholes at Castleton and Hope close to the apron-reef are included in the Marginal Province and details are given on pp. 111 and 112. In the Castleton Borehole, mudstones of $P_2$ age are found to 114 ft 6 in and limestones to the base of the hole at 203 ft 6 in. Rolled shell fragments are present from 139 ft 3 in to 203 ft 6 in and the lithology especially below 172 ft 6 in is typical of the Beach Beds. As at outcrop, the fauna consists of *Gigantoproductus* and *Spirifer* fragments and the only evidence of age is given by the coral band from 167 ft 3 in to 169 ft 7 in. The corals present include *Aulophyllum fungites*, *Diphyphyllum lateseptatum*, *Lithostrotion portlocki* and *Nemistium edmondsi* and suggest a position in the shelf sequence near the *L. duplicata* Band, high in the upper part of the $D_2$ Zone. In the Hope Cement Works Borehole *Lonsdaleia duplicata duplicata* and zaphrentoid corals are present from 299 ft to 310 ft 8 in and again this level is correlated with the *L. duplicata* Band.

## BASIN PROVINCE

The Alport and Edale boreholes proved parts of the Lower Carboniferous succession in the Basin Province and the faunas are summarized on pp. 113–9.

M.M.

## REFERENCES

Bisat, W. S. 1924. The Carboniferous goniatites of the North of England and their zones. *Proc. Yorks. geol. Soc.*, **20**, 40–124.

——— 1934. The goniatites of the *Beyrichoceras* Zone in the North of England. *Proc. Yorks. geol. Soc.*, **22**, 280–309.

BISAT, W. S. 1952. The goniatite succession at Cowdale Clough, Barnoldswick, Yorkshire. *Trans. Leeds geol. Ass.*, **6**, 155–81.

BOND, G. 1950. The Lower Carboniferous reef limestones of Cracoe, Yorkshire. *Q. Jl geol. Soc. Lond.*, **105** (for 1949), 157–88.

BRUNTON, C. H. C. 1966. Predation and shell damage in a Viséan brachiopod fauna. *Palaeontology*, **9**, (3), 355–9.

CLARKE, M. J. 1966. A new species of fasciculate *Aulina* from Ireland. *Scient. Proc. R. Dubl. Soc.*, (A), **2**, (14), 221–7.

COPE, F. W. 1933. The Lower Carboniferous succession in the Wye valley region of North Derbyshire. *J. Manchr geol. Ass.*, **1**, (3), 125–45.

———— 1936. The *Cyrtina septosa* Band in the Lower Carboniferous succession of North Derbyshire. *Summ. Prog. geol. Surv. Gt Br.*, for 1934, (2), 48–51.

———— 1937. Some features in the $D_1$–$D_2$ limestones of the Miller's Dale region, Derbyshire. *Proc. Yorks. geol. Soc.*, **23**, (3), 178–95.

———— 1939. The mid-Viséan ($S_2$–$D_1$) succession in North Derbyshire and North-West England. *Proc. Yorks. geol. Soc.*, **24**, (1), 60–6.

———— 1940. *Daviesiella llangollensis* (Davidson) and related forms: morphology, biology and distribution. *J. Manchr geol. Ass.*, **1**, (4), 199–231.

DAY, J. B. W. 1970. Geology of the country around Bewcastle. *Mem. geol. Surv. Gt Br.*

EDEN, R. A., ORME, G. R., MITCHELL, M. and SHIRLEY, J. 1964. A study of part of the margin of the Carboniferous Limestone 'massif' in the Pin Dale area, Derbyshire. *Bull. geol. Surv. Gt Br.*, No. 21, 73–118.

FORD, T. D. 1952. New evidence on the correlation of the Lower Carboniferous reefs at Castleton, Derbyshire. *Geol. Mag.*, **89**, 346–56.

———— 1965. The palaeoecology of the Goniatite Bed at Cowlow Nick, Castleton, Derbyshire. *Palaeontology*, **8**, (1), 186–91.

FOWLER, A. 1966. The stratigraphy of the North Tyne basin around Kielder and Falstone, Northumberland. *Bull. geol. Surv. Gt Br.*, No. 24, 57–104.

HILL, DOROTHY. 1938. A monograph on the Carboniferous Rugose corals of Scotland. *Palaeontogr. Soc.* [*Monogr.*], (1), 1–78.

HUDSON, R. G. S. 1926. On the Lower Carboniferous corals: *Orionastraea indivisa*, sp. n., and *Thysanophyllum praedictum*, sp. n. *Ann. Mag. nat. Hist.*, (9), **18**, 144–51.

———— and COTTON, G. 1945a. The Lower Carboniferous in a boring at Alport, Derbyshire. *Proc. Yorks. geol. Soc.*, **25**, (4), 254–330.

———— ———— 1945b. The Carboniferous rocks of the Edale Anticline, Derbyshire. *Q. Jl geol. Soc. Lond.*, **101**, 1–36.

JACKSON, J. W. 1922. On the occurrence of *Daviesiella llangollensis* (Dav.) in Derbyshire. *Geol. Mag.*, **59**, 461–8.

———— 1927. The succession below the Kinder Scout Grit in North Derbyshire. *J. Manchr geol. Ass.*, **1**, (1) 15–32.

———— 1941. Description of a Carboniferous Limestone section with *Girvanella* in North Derbyshire. *J. Manchr geol. Ass.*, **1**, (4), 239–46.

KELLAWAY, G. A. 1967. The Geological Survey Ashton Park Borehole and its bearing on the geology of the Bristol district. *Bull. geol. Surv. Gt Br.*, No. 27, 49–153.

MARTIN, W. 1809. *Petrificata Derbiensia; or, figures and descriptions of Petrifactions collected in Derbyshire*. Wigan.

MITCHELL, G. H. and STUBBLEFIELD, C. J. 1941. The Carboniferous Limestone of Breedon Cloud, Leicestershire, and the associated inliers. *Geol. Mag.*, **78**, 201–19.

MITCHELL, M. 1965. In *Summ. Prog. geol. Surv. Gt Br.*, for 1964, 74.

MORRIS, T. O. 1929. The Carboniferous Limestone and Millstone Grit Series of Stoney Middleton and Eyam, Derbyshire. *Proc. Sorby scient. Soc.*, **1**, 37–67.

PARKINSON, D. 1943. The age of the reef-limestones in the Lower Carboniferous of North Derbyshire. *Geol. Mag.*, **80**, 121–31.

———— 1947. The Lower Carboniferous of the Castleton district, Derbyshire. *Proc. Yorks. geol. Soc.*, **27**, (2), 99–124.

———— 1953. The Carboniferous Limestone of Treak Cliff, Derbyshire, with notes on the structure of the Castleton reef-belt. *Proc. Geol. Ass.*, **64**, (4), 251–68.

———— 1954. Quantitative studies of brachiopods from the Lower Carboniferous reef limestones of England. 2. *Pugnax pugnus* (Martin) and *P. pseudopugnus* n. sp. *J. Paleont.*, **28**, (5), 563–74.

———— 1965. Aspects of the Carboniferous stratigraphy of the Castleton–Treak area of North Derbyshire. *Mercian Geol.*, **1**, (2), 161–80.

SHIRLEY, J. and HORSFIELD, E. L. 1940. The Carboniferous Limestone of the Castleton–Bradwell area, North Derbyshire. *Q. Jl geol. Soc. Lond.*, **96**, 271–99.

———— ———— 1945. The structure and ore deposits of the Carboniferous Limestone of the Eyam district, Derbyshire. *Q. Jl geol. Soc. Lond.*, **100**, 289–308.

SIBLY, T. F. 1908. The faunal succession in the Carboniferous Limestone (Upper Avonian) of the Midland area (North Derbyshire and North Staffordshire). *Q. Jl geol. Soc. Lond.*, **64**, 34–82.

SMITH, J. 1911. Carboniferous Limestone rocks of the Isle of Man. *Trans. geol. Soc. Glasg.*, **14**, (2), 119–64.

SMITH, S. 1925. The genus *Aulina*. *Ann. Mag. nat. Hist.*, (9), **16**, 485–96.

———— and Yü, C. C. 1943. A revision of the coral genus *Aulina* Smith and descriptions of new species from Britain and China. *Q. Jl geol. Soc. Lond.*, **99**, 37–62.

VAUGHAN, A. 1905. The palaeontological sequence in the Carboniferous Limestone of the Bristol area. *Q. Jl geol. Soc. Lond.*, **61**, 181–307.

WOLFENDEN, E. B. 1958. Paleoecology of the Carboniferous reef complex and shelf limestones in north west Derbyshire, England. *Bull. geol. Soc. Am.*, **69**, 871–98.

———— 1959. New sponges from Lower Carboniferous reefs of Derbyshire and Yorkshire, England. *J. Paleont.*, **33**, (4), 566–8.

*Chapter IV*

# MILLSTONE GRIT SERIES (NAMURIAN)

## INTRODUCTION

THE MILLSTONE GRIT SERIES crops out over three-quarters of the Chapel en le Frith district. In the north it forms a broad tract including the plateau of Kinder Scout and adjacent moorlands. Farther south, relatively narrow outcrops lie between the Coal Measures of the Goyt Trough and the Carboniferous Limestone of the Derbyshire Dome. In the west it is exposed in the Todd Brook and Romiley anticlines, and in the east, the outcrop, narrowing southeastwards, separates the Carboniferous Limestone outcrop from the Coal Measures of the East Midlands Coalfield. Within the limestone area shales of early Millstone Grit age form an outlier at Wardlow Mires.

In the north much of the Millstone Grit Series, here some 4000 ft thick, is comparable with that of the Holmfirth and Glossop (Bromehead and others 1933) and the Huddersfield and Halifax (Wray and others 1930) districts. In the south-west, however, it shows many features in common with the Millstone Grit of the Staffordshire area (see for example Evans and others 1968). The south-eastern outcrops show the same southerly thinning as that noted around Sheffield (Eden and others 1957) and Chesterfield (Smith and others 1967).

**Previous Research.** Whitehurst (1778, p. 147) first introduced the term Millstone Grit in the Derwent valley, to the south of the present district, for a coarse sandstone (evidently the Ashover Grit) separated by shales from the Carboniferous Limestone. This description was followed and elaborated by Farey (1811) who recognized that these beds lie beneath the Coal Measures and are of generally similar character to them though lacking important seams of coal. Farey also noted the presence beneath the "Millstone Grit" of a lower series of sandstones which he termed the "Shale Grit" (ibid., p. 228) underlain by shales, the latter overlying the limestone. He also discussed the production of millstones, then as important in the present district as in the country to the south.

In 1864 Hull and Green described the Millstone Grit of the whole of the west Pennine area including Kinder Scout and the ground around the Goyt Trough. Their succession was:

> 1st Grit or Rough Rock
>
> 2nd Grit (not developed in the present district)
>
> 3rd Grit
>
> 4th Grit or Kinder Scout Grit
>
> Yoredale Beds, with a thick bed of sandstone, the Shale Grit,
>     at the top

These authors were apparently much influenced by the coarse, massive character of the Kinderscout Grit for they remark (p. 247) that "Were it not that the Kinder Scout Grit makes so natural and so generally received a base to the

155

Millstone-grit, the first of these groups [i.e. the Shale Grit] might well have been placed in that formation rather than among the Yoredale Rocks."

The sucession of Hull and Green was the basis of that used in two Geological Survey memoirs. The first of these (Hull and Green 1866) dealt with an area which included the country to the west of Chapel en le Frith, whereas Kinder Scout and the eastern outcrops fell within the region described by Green and others (1887). Although it was possible over the greater part of the area to correlate much of the Millstone Grit on a lithological basis (the only means then available), there were places where rapid lithological variation of the grits or faulting led inevitably to mis-correlation. Thus for example the Roaches Grit which was assigned to the 3rd Grit has since been shown to lie beneath it (Cope 1946, p. 142). Confusion also arose regarding the upper limit of the Millstone Grit due to the mistaken equation on lithological grounds of the Yard Coal of the New Mills area (in fact the Bassy Mine) with the Sand Rock Mine of Lancashire.

The unconformable nature of the junction between Millstone Grit shales ("Yoredale Beds") and Carboniferous Limestone was noted by Barnes and Holroyd as early as 1897. Anomalies in the relations between the two formations had been previously explained by faulting (see Old Series One-inch Sheet 81 N.E., published 1866).

The memoir on the northern part of the Derbyshire Coalfield by Gibson and Wedd, published in 1913, described the more important features of the Millstone Grit succession on the eastern edge of the district. The "Kinderscout Grit" of these authors has since been shown to be the Ashover Grit (Smith and others 1967, p. 68) and their "Longshaw Grit" has been proved to be the Rough Rock (Eden and others 1957, p. 18).

In a series of papers between 1923 and 1927 Jackson divided the lower part of the Millstone Grit in and around the Edale area into (in upward succession) Edale Shales, Mam Tor Sandstone, Shale Grit, Grindslow Shales and Kinderscout Grit. He recognized many of the faunal horizons, showed that the so-called Yoredale Beds were not the equivalent of those of north-west Yorkshire, and demonstrated the extent of the unconformity at the base of the Millstone Grit. Such advances in knowledge of the Millstone Grit were made possible by the pioneer work of Bisat (1924) on the Carboniferous goniatites of the north of England.

Knowledge of the relations between limestone and shale near the borders of the stable block (see p. 11) was further increased by Fearnsides and Templeman who, in 1932, described the succession in a borehole at Hope Cement Works. These authors recognized the presence of *Eumorphoceras bisulcatum* Girty in the higher part of the Edale Shales, though the fauna and zonal position of the lower beds have been re-interpreted in the present memoir (see pp. 200–1).

Stratigraphical work in more recent years has been limited to two important papers by Hudson and Cotton. These describe the successions in boreholes for oil, at Alport (1943) and Edale (1945) respectively. Both papers are quoted widely below.

In 1946 Cope gave a detailed description of contorted beds (see p. 161) in the Millstone Grit and Lower Coal Measures of the southern part of the Goyt Trough, partly within the present district.

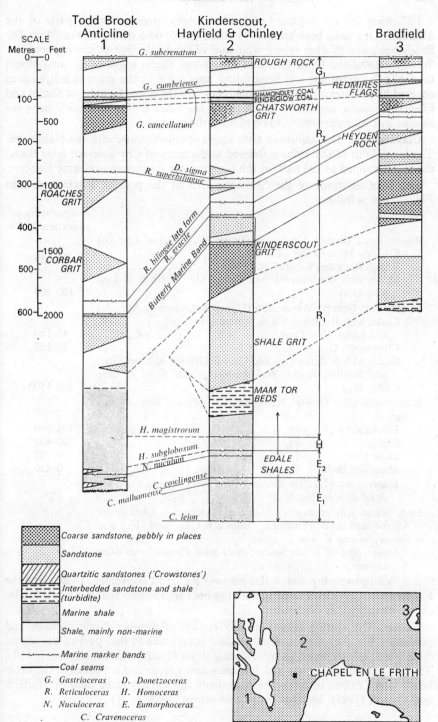

FIG. 15. *Generalized sections of the Millstone Grit Series*

Following the development of sedimentology in recent years, parts of the Millstone Grit have been studied by Allen (1960) who described the 'Mam Tor Sandstone', and Walker (1966) who studied the Shale Grit of the Alport area. Walker classed the Shale Grit and Grindslow Shales together as the "Alport Group"; this term has not been found applicable in the more extensive area covered by this memoir. Deep channels cut into the upper Grindslow Shales and filled with coarse sandstone which forms the basal Kinderscout Grit are considered by Collinson (1970)[1] to be of fluviatile origin.

**Classification.** The Millstone Grit Series is synonymous with the Namurian. The lower limit of the series is defined at the entry of *Cravenoceras leion* Bisat; the top is defined as the base of the *Gastrioceras subcrenatum* Marine Band.

The chief divisions of the Millstone Grit in the present district (see also Fig. 15) are as follows:

|  | Approximate thickness in feet |
|---|---|
| Rough Rock Group: Yeadonian (Lower *Gastrioceras* Age, $G_1$) | |
| Six-Inch Mine or Pot Clay Coal ⎫ $G_{1b}$ .. .. ..  Rough Rock and Rough Rock Flags ⎭ .. .. .. | 50–150 |
| Shales with *Gastrioceras cancellatum* Marine Band at base ($G_{1a}$–$G_{1b}$) .. .. .. .. .. .. .. | 100–280 |
| Middle Grit Group: Marsdenian (Upper *Reticuloceras* Age, $R_2$) | |
| Shales with Redmires Flags, Simmondley Coal ⎫  and Ringinglow Coal ⎬ $R_{2c}$ ..  Chatsworth Grit ⎭ .. | 55–140  25–150 |
| Shales with Roaches Grit and Corbar Grit (or Ashover Grit) and Heyden Rock; *Reticuloceras gracile* Band at base ($R_{2a}$–$R_{2c}$) .. .. .. .. .. .. .. | 300–1500 |
| Kinderscout Grit Group: Kinderscoutian (Lower *Reticuloceras* Age, $R_1$) | |
| Kinderscout Grit ⎫ .. .. .. .. .. .. | 100–500 |
| Shales ⎪ .. .. .. .. .. | 100–400 |
| Shale Grit ⎬ $R_{1c}$ .. .. .. .. .. | 0–700 |
| Mam Tor Beds ⎭ .. .. .. .. .. | 0–450 |
| Upper part of Edale Shales with *Homoceras magistrorum* Band at base ($R_{1a}$–$R_{1c}$) .. .. .. .. .. | 150 |
| Beds below Kinderscout Grit Group: Alportian, Chokierian, Arnsbergian and Pendleian (*Homoceras* $H_2$ and $H_1$, and *Eumorphoceras* $E_2$ and $E_1$ ages) | |
| Lower part of Edale Shales; beds with *Cravenoceras leion* at base .. .. .. .. .. .. .. .. | 700 |

For descriptive purposes in this memoir it is most convenient to describe the Kinderscout Grit Group and underlying beds together so that the Edale Shales can be dealt with as one unit.                                               I.P.S.

**Faunas.** The marine faunas usually consist of non-benthonic bivalves and goniatites, but they are restricted in their occurrence to that part of the faunal cycle which is interpreted as representing times of maximum salinity (see p. 164). The goniatite stages, zones and subzones now recognized are given in the table below which differs in certain respects from the zonal scheme used by Hudson and Cotton (1945); see also Evans and others (1968, p. 21).

---

[1] COLLINSON, J. D. 1970. Deep channels, massive beds and turbidity current genesis in the central Pennine basin. *Proc. Yorks. geol. Soc.*, **37**, 495–519.

| Stage | Zone | Subzone |
|---|---|---|
| Yeadonian ($G_1$) | *Gastrioceras cumbriense* ($G_{1b}$) | —— |
| | *G. cancellatum* ($G_{1a}$) | *G. crencellatum* / *G. cancellatum* / *G. branneroides* |
| Marsdenian ($R_2$) | *Reticuloceras superbilingue* ($R_{2c}$) | *Donetzoceras sigma* / *R. superbilingue* |
| | *R. bilingue* ($R_{2b}$) | *R. metabilingue* / *R. bilingue* |
| | *R. gracile* ($R_{2a}$) | —— |
| Kinderscoutian ($R_1$) | *R. reticulatum* ($R_{1c}$) | *R. coreticulatum* / *R. reticulatum* |
| | *R. nodosum* ($R_{1b}$) | *R. nodosum* / *R. dubium* |
| | *R. circumplicatile* ($R_{1a}$) | *R. todmordenense* / *R. circumplicatile* |
| Alportian ($H_2$) | *Homoceratoides prereticulatus* ($H_{2c}$) | *Ht. prereticulatus* / *H. eostriolatum* |
| | *Homoceras undulatum* ($H_{2b}$) | —— |
| | *Hudsonoceras proteus* ($H_{2a}$) | —— |
| Chokierian ($H_1$) | *Homoceras beyrichianum* ($H_{1b}$) | —— |
| | *H. subglobosum* ($H_{1a}$) | —— |
| Arnsbergian ($E_2$) | *Nuculoceras nuculum* ($E_{2c}$) | *N. nuculum* / *N. stellarum* |
| | *Cravenoceratoides nitidus* ($E_{2b}$) | *Ct. nititoides* / *Ct. nitidus* / *Ct. edalensis* |
| | *Eumorphoceras bisulcatum* ($E_{2a}$) | *E. bisulcatum* / *C. cowlingense* |
| Pendleian ($E_1$) | *Cravenoceras malhamense* ($E_{1c}$) | —— |
| | *E. pseudobilingue* ($E_{1b}$) | —— |
| | *C. leion* ($E_{1a}$) | *E. hudsoni* / *E. stubblefieldi* / *E. tornquisti* |

In $E_1$, $E_{2a}$ and $R_{1c}$ to $G_1$ the goniatite-bearing horizons are very distinct and are usually separated by unfossiliferous beds. But from $E_{2b}$ to $R_{1b}$ there is only a small total thickness of unfossiliferous beds, the goniatite horizons are closer together and the succession is almost wholly marine, even between the goniatite horizons, where mollusc spat is common (Ramsbottom, Rhys and Smith 1962, p. 115). Most of the faunas have been described in greater detail from the Clitheroe district by Ramsbottom *in* Earp and others (1961, pp. 186–95).

Benthonic faunas of brachiopods, crinoids, trilobites, etc. are known only from a few horizons, as for example in the basal $E_1$ beds, in places at the horizon of *C. malhamense*, at the *Ct. nititoides* horizon, and in the Butterly Marine Band where the fauna consists of benthonic bivalves and gastropods as well as brachiopods.

Non-marine faunas are restricted in this district to $R_2$ and $G_1$. It is worth noting that the record of *Carbonicola* in the unusual lithology of the Heyden Rock constitutes the earliest record of the genus in the Pennine area. W.H.C.R.

**Lithology.** The sandstones form the most striking feature of the Millstone Grit, being much better exposed than the shales, and forming much bolder features; they are, however, less important so far as thickness is concerned. Following the account of the Millstone Grit sandstones of the Bradford area by Dunham (*in* Stephens and others 1953) it is now customary to avoid the use of the term 'grit' in the lithological descriptions of these rocks, as it implies an angularity of grain which is not always present; it is retained for historical reasons in the stratigraphical names.

The coarser sandstones of 'fluvial-grit' lithofacies (see p. 162) are highly feldspathic, commonly pebbly, cross-bedded, and in places conglomeratic. The pebbles are usually quartz although feldspar occurs rarely. The medium- to coarse-grained sandstones (such as the Shale Grit and parts of the Roaches Grit) show a different development with massive sandstone beds, rarely cross-bedded, but with sole-marks, particularly large load-casts. Thin partings of flaggy sandstone, siltstone or shale separate the thicker sandstone beds.

Fine-grained sandstones and siltstones are developed at some horizons, the best examples being the Rough Rock Flags and Redmires Flags, though both are of local development only. The finer-grained sandstones are frequently micaceous, thin-bedded and flaggy. A development of alternating thin sandstones and shales which forms the Mam Tor Beds has been described as of 'turbidite' lithofacies (see p. 173).

Quartzitic sandstones are present in the lower part of the succession in the south-west of the area. These may be conveniently termed 'crowstones', the name given to similar beds in the Staffordshire area. The use of the term 'crowstone' has been recently criticized by Holdsworth (1964) largely on account of its misuse by Hudson and Cotton (1943). There seems, however, no need to discard it for the thick quartzitic sandstones of Staffordshire, for which it was originally used by Hull and Green (1866, pp. 9, 67), and for similar beds in the present district.

In this account the argillaceous rocks are usually described as 'shale', the term 'mudstone' being restricted to rocks which can be seen to be poorly or non-fissile. A variety of types, many of which grade into each other, is present: the most frequent is a grey shale of moderate fissility, but micaceous, silty or sandy shales also occur. Shales which are otherwise free of silt-grade material may contain abundant mica flakes. Marine bands, rarely exceeding a few feet in thickness (in the upper, predominantly non-marine beds), show in many cases a lithology different from that of the non-marine shales, being either more calcareous and blocky, or darker and more regularly fissile.

Ironstone bands and nodules are of frequent occurrence in the grey shales. Limestone is present as bands or lenses in the lower, marine part of the sequence and as bullions (concretions), associated with the marine bands in the upper part. Pyrite also occurs in the marine shales, and sometimes also in the associated unfossiliferous shales. Frequently shales of marine or quasi-marine type present a yellowish 'sulphurous' appearance when weathered. Hartley (1957, pp. 19–23) showed that similar 'sulphurous' weathering in shales of Viséan, Namurian and Lower Coal Measures ages in Yorkshire was due to the presence of the secondary mineral jarosite, resulting from the decomposition of pyrite.

Coals are present at four horizons in the upper part of the Millstone Grit, though the lowest (that below the Butterly Marine Band) is known from one locality only. They are usually thin, the greatest thickness known in the present district being 30 in for the Simmondley Coal. Despite their locally pyritous nature the three main coal horizons have all been worked in places, though often in conjunction with the underlying seatearth. Seatearths or ganisters are usually present beneath the coals, or may mark their horizon when the coal is absent. Many of the thicker grits have a small thickness of ganister or ganister-like sandstone at the top and in one place (see p. 234) ganister is present within the Chatsworth Grit, perhaps marking a split into two cyclothems.

A remarkable feature of the Millstone Grit succession in the area of the Goyt Trough is the presence of thin contorted bands in the shales, usually at constant horizons. These beds were first described in detail by Cope (1946, pp. 139–76) who found that they were closely associated with, and usually just above marine bands. Cope noted eight such beds in the part of the Millstone Grit succession above the *Reticuloceras bilingue* Band. In the ground lying south of Whaley Bridge within the present district he noted contorted bands above the *Gastrioceras cancellatum* and above the *G. cumbriense* bands. Following an analysis of the structural features (shear-planes and small folds) present in the bands, Cope concluded that they were of tectonic origin and resulted from bedding-plane slip in the least competent beds in the succession during folding.

Additional contorted beds have been found during the present work in the Edale Shales, particularly near the junction with the Carboniferous Limestone. Similar beds are also present in the Coal Measures (see p. 267).

**Sedimentation.** The Millstone Grit was laid down in two contrasting areas: (*a*) a 'negative' area of thick sedimentation extending from the Alport valley through Kinder Scout to the area now occupied by the Todd Brook Anticline and from thence into North Staffordshire. In this basin of deposition the overall thickness of the Millstone Grit varies from about 3500 ft at Alport and 4000 ft at Kinder Scout to perhaps 5200 ft[1] in the Todd Brook Anticline. Owing to lack of information regarding the thickness of the lower beds, it is not possible to define the northern and western limits of this basin, but it probably extended north-westwards into the Rossendale and Bradford areas. In this area of sedimentation evidence from the Alport and Edale boreholes shows a conformable relationship between Namurian and Viséan rocks; (*b*) a 'positive' area of thin sedimentation over the Derbyshire Dome, coinciding broadly with the present outcrop of the Carboniferous Limestone. Although Namurian sediments are rarely present over the area, thinning in these beds can be observed in many places as the Dome is approached. The basal unconformity (see below) of the Namurian provides additional evidence of the 'positive' influence of the limestone massif. Outcrops of Namurian on the massif at Great Hucklow and in the outlier at Wardlow Mires show a strongly condensed succession. Round the edges of the Derbyshire Dome there is, in addition to the thinning noted above, a general tendency for thinning to the south, largely accounted for by the southerly disappearance of many of the 'grits'.

The pattern of sedimentation shows a marked vertical variation as follows:

(*a*) The Edale Shales show uniform conditions with the quiet deposition of marine muds and only minor incursions of coarser sediment. These beds show

---

[1] This figure is based partly on the estimate of 1800 ft for the total average thickness of the $E_1$ and $E_2$ stages in the Macclesfield district (Evans and others 1968, p. 19).

cyclic sedimentation, with a repetition of faunal phases (see p. 164) as noted by Ramsbottom, Rhys and Smith (1962, pp. 114–7) in the Ashover area. These cycles, being of the type ABCDCBA, however, differ in arrangement from the sedimentary cycles of the higher beds (ABCDABCD); (b) The Mam Tor Beds and Shale Grit represent, respectively, the earlier and later stages of the advance into the area of a fan of 'turbidite' deposits on the frontal slopes of a delta-fan coming from the north; (c) The main part (see below for beds of different facies) of the Kinderscout Grit is of the 'fluvial-grit' type of Trotter (1952, pp. 78–9) and represents the incursion from the north of the main mass of the delta-fan. The top of this part of the Kinderscout Grit is in many places ganister-like, and in one place shows a thin coal. Thus, by this time, the basin had been largely filled, allowing the establishment of more uniform conditions of cyclic sedimentation; (d) The beds above the Butterly Marine Band, where the Kinderscout Grit is split, or above the base of the *Reticuloceras gracile* Band, where the grit is united, show the characteristic Millstone Grit cyclic sedimentation.

The sedimentary cycles or cyclothems are similar to those described in the Chesterfield area (Smith and others 1967, pp. 5, 80–1), and may be summarized thus:

> Coal
>
> Seatearth
>
> Siltstone or sandstone
>
> Grey mudstone becoming silty upwards
>
> Dark mudstone, frequently marine at or near the base

The cycles are often incomplete with one or more elements missing and others expanded. In some cases (e.g. the Chatsworth Grit) the sandstone element is coarse and massive and of 'fluvial-grit' facies. The $R_2$ and $G_1$ stages are made up of about 8 to 10 cycles and these vary in thickness from as little as 15 ft (beneath *G. cancellatum* Band) to 320 ft (beneath *G. cumbriense* Band in the Goyt Trough).

**The Basal Unconformity.** The importance of this feature was first recognized by Barnes and Holroyd (1897, pp. 125, 129). Its widespread nature was confirmed and its vertical extent determined by Jackson (1925, 1927). Hudson and Cotton (1945, p. 14), referring to the area south of Mam Tor, stated that *Reticuloceras* occurred in a pocket of shale on the limestone surface. No precise locality was given for this occurrence, despite its importance. It is difficult to reconcile the record with the presence of $E_{2c}$ faunas high in the landslip at Mam Tor and well above the limestone in a stream north of Perryfoot. Perhaps the shale from which Hudson obtained the *Reticuloceras* was not *in situ;* however, if correct, it would indicate a greater amount of overlap than anywhere else in the district.

The present work confirms the presence of an angular unconformity at the Namurian–Viséan junction at most places where it is seen at surface within the present district. Whether the record of *Reticuloceras* quoted above is accepted or not, it is evident that the salient of Viséan near Mam Tor stood out as a headland or shoal and as an area of non-deposition during $E_1$ and part of $E_2$ times (and possibly longer) and was progressively buried, producing an unconformity with large overstep. Where differences in the relief are not too pronounced, as at Windy Knoll and around Brook House near Buxton, the Namurian shales, instead of being steeply banked against the limestone, occupy erosional hollows on its upper surface.

Simpson and Broadhurst have demonstrated (1969) the presence of a boulder bed at the base of the shales overlying the apron-reef at Treak Cliff. This bed has resulted from the downhill movement of large masses of limestone on the steeply-inclined frontal slope of the apron-reef and their incorporation into the mudstones. The concept, adopted here, of the gradual burial of the apron-reef during early Namurian times suggests an $E_1$ age for the boulder bed at Treak Cliff in view of its position on the reef slope. Minor features associated with the unconformity include 'Neptunian dykes' of shale and breccia in the limestone at Treak Cliff, Castleton (Ford 1952, pp. 352-3) and of shale in the limestone at Windy Knoll, near Mam Tor, and the local concentration of uraniferous phosphate on the uppermost surface of the apron-reef limestone at Castleton (Peacock and Taylor 1966, pp. 19-32). Similar concentrations of phosphate have been found associated with breaks in sedimentation in other places. For example, Dietz, Emery and Shepard (1942, pp. 815-47) have described the phosphatization of the walls of steep escarpments and submarine canyons off the coast of southern California, all in areas of non-deposition. This probably presents a fairly close parallel to the conditions obtaining prior to the deposition of the Edale Shales in the reef area.

Where shale of Viséan ($P_2$) age is present beneath the Namurian, as in the Castleton Borehole (see p. 111) and in the Wardlow Mires outlier, there is no evidence of an unconformity between the two and the stratigraphical break is at the base of the $P_2$ shale. The intra-Carboniferous movements involved form part of those distinguished by Hudson and Mitchell (1937, pp. 28-31) as 'Sudetian' in the area of the Skipton Anticline (see also p. 322); they do not, however, coincide precisely with any of the three phases of movement noted by these authors.

# GENERAL STRATIGRAPHY

## KINDERSCOUT GRIT GROUP AND UNDERLYING BEDS
### (Pendleian, $E_1$ to Kinderscoutian, $R_1$)

These strata are described together, as the uppermost part of the Edale Shales is included in the Kinderscout Grit Group, and it is convenient to describe these shales under one heading. The lower subdivision extends upwards from the base of the *Cravenoceras leion* Zone to the base of the *Homoceras magistrorum* Band (see Plate XV) and the Kinderscout Grit Group, from thence to the base of the *Reticuloceras gracile* Band (see Plate XVII).

### EDALE SHALES

The Lower Carboniferous is succeeded by a group of marine shales which is best exposed in the Edale valley (see Fig. 16). These have been called the Edale Shales by Jackson (1923, p. 307) and the name can be appropriately used throughout the present district though their top, marked by the incoming of sandy beds of $R_1$ age (Mam Tor Beds, Shale Grit or Kinderscout Grit), is diachronous, even in the type area. The zonal succession is set out on p. 159 and the stratigraphy is described under these headings below. It should be noted that the present zonal boundaries do not everywhere coincide with those used by Hudson and Cotton (1943, 1945). These authors also used the term 'Edale Shale' in a more restricted sense in the Alport and Ashop valleys.

The Edale Shales are about 800 ft thick in the type area where the detailed succession is well exposed in stream sections except for the lowest beds which were proved in the Edale Borehole (Hudson and Cotton 1945). In the Alport valley a thicker succession totalling 1220 ft has been found in surface exposures and in the Alport Borehole (Hudson and Cotton 1943).

The chief lithological types within the Edale Shales are as follows:

(a) Soft grey shale, poorly fossiliferous or unfossiliferous and often showing 'sulphurous' weathering, resulting from the alteration of pyrite. These shales contain ironstone bands and nodules in places, but limestone bands or nodules are rare or absent.

(b) Dark shales, usually calcareous and with an abundant fauna. Thin limestone bands, up to a few inches thick, are present in places, and bullions up to 3 ft diameter occur at some horizons (particularly in H and $R_1$). At one horizon (see pp. 189–91) a thin bed of dolomitic siltstone and dolomite is present.

(c) Silty shales occur in part of $E_1$. These contain interbedded thin quartzitic siltstones in the Alport Borehole where they have been given the name "Alport Crowstones" (Hudson and Cotton 1943, p. 147) but, as Holdsworth (1964) has pointed out, the use of the term 'crowstone' is here inappropriate as these rocks are quite unlike the typical 'crowstones' of the Staffordshire area (see for example Evans and others 1968, pp. 26–33).

(d) Hard quartzitic sandstones, the typical 'crowstones' of Staffordshire, are present in the south-west of the district.

(e) Potassium bentonite has been found to occur as thin bands of yellow, blue or orange clay or pyritous clay in the $E_{2a}$ and $E_{2b}$ zones in Edale by Trewin (1968, pp. 73–91). These are interpreted as altered bands of volcanic ash.

Following the discovery of shales of high radioactivity in the upper part of the $E_{1a}$ Zone in boreholes in the Castleton and Ashover areas (Ponsford 1955, pp. 31–3; Cosgrove in Ramsbottom, Rhys and Smith 1962, p. 164) a radiometric survey was made by Mr. J. Taylor on the Wardlow Mires outlier. This showed that a similar horizon showing high radioactivity was present close above the base of $E_1$. At Ashover the high radioactivity was found to be linked with the presence of phosphatic fish remains and collophane which had absorbed the uranium. The horizon is that of the "Barren Beds" ($E_{1a}$) of the Edale Borehole and the "Alport Crowstones" of the Alport Borehole, neither of which are exposed; however, the stratigraphical relationships of these beds have been since proved in the Wardlow Mires boreholes (see pp. 208–9).

The Edale Shales show the type of cyclic sedimentation described by Ramsbottom, Rhys and Smith in the Ashover area (1962, p. 114). The ideal cycle consists of a number of faunal phases: (11) Fish phase, (10) *Planolites* phase, (9) *Lingula* phase, (8) Mollusc spat phase, (7) *Anthracoceras* or *Dimorphoceras* phase, (6) Thick-shelled goniatite phase, (5) *Anthracoceras* or *Dimorphoceras* phase, (4) Mollusc spat phase, (3) *Lingula* phase, (2) *Planolites* phase, (1) Fish phase. These phases were considered to reflect variations in salinity. They bring out the otherwise indistinct cyclic nature of the shales of E, H and R age.

The extensive fossil collections of Hudson and Cotton in the Edale and Alport boreholes and neighbouring surface exposures (part of which have been donated to the Geological Survey) have been widely referred to in the present account, though some of the identifications have been revised. During the present work fossil collecting has been mainly aimed at filling such gaps as were found to exist in the Hudson and Cotton collections.

Evidence of the nature of the Namurian succession in the area of the 'block' is sparse, though the Wardlow Mires boreholes and the Hucklow Edge No. 1 Borehole show strongly condensed sequences at the base. Such condensed sequences appear to be restricted to slightly down-warped areas on the limestone surface.

Such clastic material as entered the area during the deposition of the Edale Shales has been considered by Holdsworth (1963) to have been introduced by northward-flowing turbidity currents bringing material from the south Midland landmass known, among other names, as the "Midland Barrier" (Trueman 1947, p. lxxx). The parent rocks were considered to have been greatly leached by weathering. The coarser true 'crowstones' were considered by Holdsworth to have been formed nearer the source of supply than the finer-grained and distal "Alport Crowstones".

*Cravenoceras leion* Zone ($E_{1a}$). The beds of this zone are known within the present district from boreholes only, as in most places they are overlapped by higher beds in the vicinity of the limestone massif. In the Castleton area their small outcrop is concealed beneath drift, while in the Wardlow Mires outlier they are present, though in condensed form.

Revision of the zonal boundaries (see p. 188) has caused more beds to fall within this zone than had previously been the case. The thickness varies considerably, from 436 ft in the Alport Borehole to 2 ft 1 inch in the Hucklow Edge No. 1 Borehole. The zone may be subdivided as follows:

(a) A lower and more uniformly developed part is made up of the beds with *Cravenoceras leion* Bisat at the base. These are shales with some thin limestones in places. In the Edale Borehole these beds are 73 ft thick with the *C. leion* fauna in the lowest 20 ft; a 6-ft bed with limestone pebbles occurs near the base, indicative of the contemporaneous erosion of the nearby limestone massif. Elsewhere, these beds are 45 ft in the Alport Borehole, 10 ft 7 inches in the Castleton Borehole and only 7 inches in the Hucklow Edge No. 1 Borehole.

(b) An upper poorly fossiliferous variable group of shales including the "Barren Beds" of the Edale Borehole (175 ft thick) and the "Alport Crowstones" of the Alport Borehole (391 ft thick). These beds are silty shales, with thin quartzitic siltstones. In the vicinity of the limestone massif they thin greatly and tend to lose their silty character; in the Hucklow Edge No. 1 Borehole they are only 1 ft 6 in thick. Plant and fish debris are the most important organic remains, though a sparse goniatite-bivalve fauna does occur (see p. 206).

I.P.S.

The fauna of this zone in the Alport Borehole was discussed by Bisat (1950) who recognized representatives of all three subzonal faunas, though *Eumorphoceras tornquisti* (Woltersdorf) itself was not seen in the lowest of them. The middle fauna (*E. stubblefieldi* Subzone) includes *Lyrogoniatites tonksi* (Moore), *Eumorphoceras bisati* Horn (= *E. pseudobilingue* A Bisat) and *C. leion*. Yates (1962, p. 414) lists *E. medusa* Yates (recorded by Bisat 1950 as *E.* aff. *hudsoni* Gill) and *E. medusa* cf. *sinuosum* Yates in the upper fauna of the zone in the Alport Borehole. In the other boreholes circumstances have precluded detailed

M

discrimination of horizons in $E_{1a}$, though the *E. tornquisti* Subzone is known in the Edale Borehole.

An example of the rare fish tooth *Edestus* (*Edestodus*) was collected near the top of the zone in the Castleton Borehole. Among the bivalves, *Posidonia corrugata* (Etheridge) occurs throughout the zone, *Caneyella membranacea* (McCoy) and *Obliquipecten costatus* Yates near the base only, and *P. trapezoedra* (Ruprecht) in the upper part only. W.H.C.R.

*Eumorphoceras pseudobilingue* Zone ($E_{1b}$). Beds of this age are known mainly from boreholes, though they also appear in a small outcrop in the centre of the Edale Anticline (see Plate XVI A); they also crop out at Castleton though here are beneath the drift. They consist of shales, silty in places and with some thin limestones. The latter are best developed at Edale where the uppermost 29 ft of the zone, here 34 ft in all, are exposed. The zone is subject to only minor variations in thickness in the area bordering the limestone massif, being $33\frac{1}{2}$ ft and 27 ft in the Castleton and Hope Cement Works boreholes respectively. At Mam Tor, however, it must be subject to even more rapid overlap than $E_{1d}$ (see p. 201). In the thick succession proved in the Alport Borehole these beds, 130 ft thick, are shales, silty in part and with scattered thin limestones. In contrast to this, in the condensed sequence proved by the Hucklow Edge No. 1 Borehole, the zone is only 4 ft in thickness. I.P.S.

The principal goniatite horizon contains *E. pseudobilingue* (Bisat) and the reported occurrence of *Cravenoceras* in this zone in the Alport Borehole (Hudson and Cotton 1943, p. 168) has not been confirmed. Yates (1962, pp. 414–5) comments on the absence in the Alport Borehole of *E. angustum* Moore which occurs in Ireland immediately below the horizon of *E. pseudobilingue* associated with a *Kazakhoceras?* with nodose venter. This last-named was found just below *E. pseudobilingue* in the Castleton Borehole, and a similar form occurs in the Hope Borehole. *E. pseudobilingue* C Bisat characterized by early dying away of the transverse ribs was found just above the horizon of *E. pseudobilingue* s.s. in the Castleton Borehole in a position similar to that recorded by Yates (1962, p. 380) and is also known to the south in the Calow Borehole (see Smith and others 1967, p. 65). *P. corrugata* and *P. trapezoedra* are the commonest bivalves.

Above the *E. pseudobilingue* C horizon bivalves and Dimorphoceratids occur up to the overlying *C. malhamense* Zone. W.H.C.R.

*Cravenoceras malhamense* Zone ($E_{1c}$). The beds of this zone are known at outcrop only around the core of the Edale Anticline. They were also proved in the Castleton, Hope and Alport boreholes.

In the Edale valley the zone consists of some 50 ft of shale with about 20 ft of shales with thin limestones and siltstones in the middle. The uppermost part of the Castleton Borehole showed $14\frac{1}{2}$ ft of beds of this zone, though it is not certain whether the whole thickness was present. The Hope Cement Works Borehole proved a thick development; the lower beds with the *C. malhamense* fauna were dark calcareous shales with some thin limestone bands 16 ft thick, and these were overlain by 70 ft of dark shales with fish debris. The Alport Borehole showed 27 ft of shale with thin calcareous siltstone bands on 30 ft of shale with thin limestones with the *C. malhamense* fauna. The development in the Hope Cement Works Borehole is exceptional in that the zone is thicker here than in the Alport Borehole. In the Hucklow Edge No. 1 Borehole the zone is represented by 1 ft 4 in of mudstone and in the Wardlow Mires No. 1 Borehole by 10 in of mudstone; neither yielded a diagnostic fauna. I.P.S.

The characteristic goniatite fauna includes *C. malhamense* (Bisat) and *Kazakhoceras scaliger* (Schmidt), together with indeterminate Dimorphoceratids. Characteristic species other than goniatites are *Caneyella membranacea* and *Posidonia corrugata*. A nautiloid, *Rayonnoceras sp.*, was collected from the Castleton Borehole, and the Alport Borehole yielded *Reticycloceras sp.*, *Actinopteria sp.*, fragmentary orthotetoids and productoids and the crustacean *Ceratiocaris*. W.H.C.R.

*Eumorphoceras bisulcatum* Zone ($E_{2a}$). This zone is known from surface exposures in the Edale valley and in the Wardlow Mires outlier and from the Hope Cement Works and Alport boreholes. Along much of the limestone–shale junction it rests unconformably on the limestone, as between Eyam and Bradwell and at Perryfoot. Along the western edge of the massif between Dove Holes and Buxton (One-inch Sheet 111) beds of $E_{2a}$ or later $E_2$ age rest on the limestone. In the basin the thickness varies from about 80 ft in the Edale valley to 116 ft in the Alport Borehole. The sections at both localities show a development of beds with thin limestones in the lower part of the zone. The *C. cowlingense* Subzone at the base is some 40 to 45 ft in thickness at Edale and 68 ft in the Alport Borehole. The higher shales are somewhat silty in the Edale valley and pyritous with thin calcareous siltstones in the Alport Borehole. In the condensed sequence over the limestone massif a typical section of the zone is provided by Hucklow Edge No. 1 Borehole which proved a thickness of 20 ft 4 in with the *C. cowlingense* Band at the base and the *E. bisulcatum* Band at 6 ft 9 in above the base.

Recent work by Trewin (1968, pp. 76–8) has shown the presence of five thin bands of potassium bentonite in the upper part of the *E. bisulcatum* Subzone at Edale; these are mostly 1 cm or less in thickness though one band, near the top of the subzone, reaches 4 cm (see also p. 164). The bands have also been recognized in north Staffordshire. I.P.S.

Three levels in the zone yield a goniatite fauna. In the *C. cowlingense* Subzone at the base (which was referred to as $E_{1d}$ by Hudson and Cotton 1943) *C. cowlingense* Bisat occurs with poorly preserved *Eumorphoceras bisulcatum*. The main faunal horizon at the base of the succeeding subzone contains *E. bisulcatum* of which both Yates's subspecies *erinense* and *ferrimontanum* are found, *Kazakhoceras scaliger*, rare *Cravenoceras sp.*, *Chaenocardiola footii* (Baily), *Dunbarella sp.* and crinoid debris. This fauna agrees closely with that found at the same horizon at Slieve Anierin, Co. Leitrim (Yates 1962) and elsewhere. Higher in $E_{2a}$ in the Alport Borehole there is another horizon containing *E. bisulcatum* s.l. at a depth of 367 to 371 ft (Hudson and Cotton 1943, p. 163), but the specimens are badly preserved and the band has not been recognized elsewhere. W.H.C.R.

*Cravenoceratoides nitidus* Zone ($E_{2b}$). Beds of this zone are exposed in the Edale valley, at Mam Tor, and were proved in the Hope Cement Works and Alport boreholes.

The thickness is about 60 ft in the Edale valley and in the Hope Cement Works Borehole and 65 ft in the Alport Borehole. In the Edale valley the succession may be summarized:

ft

(c)  Shale, ferruginous in places    ..    ..    ..    ..    ..    ..    ..    12

(b)  Shale, with thin cherty limestones and silicified mudstones containing the
      *Cravenoceratoides nititoides* (Bisat) fauna including brachiopods and tri-
      lobites, mainly in a thin band near the middle    ..    ..    ..    ..    3

(a)  Shale, with several goniatite horizons (see below); *Cravenoceratoides eda-
      lensis* (Bisat) at base  ..    ..    ..    ..    ..    ..    .. about 40–45

Near Mam Tor beds of this zone are considered by Hudson and Cotton (1945,
p. 13) to rest directly on the Carboniferous Limestone. The Hope Cement
Works Borehole yielded *Cravenoceratoides nititoides* near the top of 51 ft of
shales referred to the $E_{2b}$ Zone. The Alport Borehole succession is generally
similar to that in the Edale valley.

Trewin (ibid., p. 78, 80) has recognized the presence of a thin band of potassium bentonite near the base of the zone at Edale; the band has also been found
in north Staffordshire.                                                    I.P.S.

The basal faunal band of $E_{2b}$, that of *Ct. edalensis*, was first described (Bisat
1932) from the River Noe, near Skinner's Hall. The immediately succeeding
faunas are not well known in the Edale valley though *C. subplicatum* Bisat and
*Ct. bisati* Hudson were recorded in the Noe by Hudson and Cotton (1945, p. 9).
The succession given by Hudson (1944, p. 241) is now subject to correction
through increased knowledge of the succession elsewhere (summarized by Yates
1962, pp. 420–2). No examples of *Ct. nitidus* (Phillips) or its associated fauna
are known from the Edale valley or from the Alport Borehole. Records of
*C.* cf. *nitidus* by Hudson (1944) and Hudson and Cotton (1945) in the Edale
valley and of the *Ct. nitidus* band in the Alport Borehole (Hudson and Cotton
1943) are now known to refer to *Ct. nititoides*. The next highest faunal band
collected during the present work was that of *C. holmesi* Bisat which in the
River Noe is found some 9 to 10 ft below the thin cherty limestone containing
the *Ct. nititoides* fauna, which also includes crinoid debris, *Productus hibernicus*
Muir-Wood, *Eumorphoceras rostratum* Yates, *Weberides sp.* and other fossils.
This is one of the most persistent faunal bands in $E_2$ (see Ramsbottom, Rhys
and Smith 1962, p. 131). It should be noted that the fauna reported by Hudson
and Cotton (1945) to contain *Ct. nititoides* has been re-determined as the lowest
of the three *N. nuculum* horizons (ibid., p. 130).

Three sections show the relationship between the *Ct. nititoides* band and the
overlying band containing *N. stellarum* (Bisat) and its associated fauna some
6 to 12 ft above (see below).                                          W.H.C.R.

*Nuculoceras nuculum* Zone ($E_{2c}$). As re-defined by Holdsworth (1965, p. 229)
this zone now includes (a) the *N. stellarum* Subzone, formerly included in $E_{2b}$,
and (b) the *N. nuculum* Subzone (corresponding to the former limits of $E_{2c}$).

Beds of this zone are well exposed in the Edale valley and are also seen in
smaller exposures at Mam Tor (though here in slipped material), at Perryfoot
and at Dove Holes. The only complete section is provided by the Alport Borehole. They also appear in the axial region of the Todd Brook Anticline, near
Saltersford.

In the Edale valley the zone consists of 45 to 60 ft of shale, with the *N. nuculum*
fauna (see below) known mainly from a band 15 ft from the base. The Alport
Borehole shows 89 ft of shale with thin bands of hard calcareous siltstone; the

*N. stellarum* fauna occurs in a thin band at the base and the *N. nuculum* fauna in three bands, respectively 42 ft, 64 ft, and 73 ft above the base.

At Dove Holes the railway-cutting section, near the limestone boundary, shows 32 ft of beds of this zone beneath beds of $H_1$ age. *N. nuculum* Bisat and *E. bisulcatum* occur in a band about 14 ft from the top. This suggests that the section finished just above the horizon of the middle band with *N. nuculum* proved in the Alport Borehole. If this is the case, the zone has, compared with the Edale valley, expanded a little in the vicinity of the boundary of the limestone massif.

The development of beds ascribed to $E_{2c}$ in the Todd Brook Anticline resembles that of the Staffordshire region (see Evans and others 1968, p. 33) in that thick quartzitic sandstones of 'crowstone' type occur throughout a thickness of some 200 ft of shales. These overlie about 50 ft of shale with, at one place, *N. nuculum*. I.P.S.

The *Nuculoceras stellarum* fauna at the base of the zone is also characterized by the presence of *Posidoniella* aff. *vetusta* (J. de C. Sowerby). The relationship between this fauna and the underlying *Cravenoceratoides nititoides* fauna is seen at the important Skinner's Hall section in the River Noe and in the outcrop near the confluence of the Noe and Crowden Brook. The *N. stellarum* fauna was also recognized in the Alport Borehole (Hudson and Cotton 1945, pp. 161–2).

Three horizons in the Alport Borehole are now known to contain *N. nuculum* and these correspond to the three such levels known elsewhere. The lowest of these seems always to have a poorer fauna than the two higher. Besides *N. nuculum*, the fauna of the zone includes *Eumorphoceras bisulcatum* [poorly preserved examples of two forms, one with less than 30 ribs per whorl and one with about 39 ribs per whorl], *Ct. fragilis* (Bisat) and bivalves. This zone marks the highest occurrence of *Posidonia corrugata*. W.H.C.R.

*Homoceras subglobosum* Zone ($H_{1a}$). This is represented in the Edale valley by about 30 ft of shale with beds of bullions at some levels. *H. subglobosum* Bisat occurs in a band with bullions at the base and, at least in places, at two higher horizons. In the Alport Borehole the zone is only 10 ft in thickness; it is thus the most variable part of H. I.P.S.

In addition to *H. subglobosum*, this zone includes *Aviculopecten sp.* (= *Pterinopecten rhythmica* Demanet 1943 *non* Jackson) and *Posidoniella* aff. *variabilis* Hind. Near the top of the zone in the Alport Borehole a band contains *Homoceras sp. nov.* with widely spaced trough-like striae and a well-developed lingua (Ramsbottom, Rhys and Smith 1962, p. 128, pl. 6, fig. 11). W.H.C.R.

*Homoceras beyrichianum* Zone ($H_{1b}$). In the Edale valley these beds show similar lithological characters to $H_{1a}$, though there is a lesser development of bullions. The shales, some 50 ft thick, are sparsely fossiliferous, with *H. beyrichianum* (Haug). In the Alport Borehole the zone comprised 57 ft of mainly unfossiliferous shale, with *H. beyrichianum* in the lowest 6 ft. I.P.S.

The fauna was confirmed in an exposure at Fulwood Holmes only. Here, in addition to *H. beyrichianum*, it included *Aviculopecten sp.*, *Reticycloceras sp.* and *Posidoniella sp.* The record from the Edale valley is on the authority of Hudson and Cotton (1945, pp. 6–7). Fossil collecting from the Alport Borehole was not detailed enough to reveal more than one horizon containing *H. beyrichianum*, but in the Ashover boreholes three such horizons were recognized

(Ramsbottom, Rhys and Smith 1962, p. 128) which contain variants similar to those found in the Alport Borehole. It is therefore possible that these three horizons are actually present in this borehole.                               W.H.C.R.

*Hudsonoceras proteus* ($H_{2a}$), *Homoceras undulatum* ($H_{2b}$) and *Homoceratoides prereticulatus* ($H_{2c}$) zones. In much of the Edale valley and in the Alport Borehole these zones comprise some 30 ft of shale with bands of bullions. At Upper Booth they expand to 40 ft.                               I.P.S.

At the base is the persistent band containing *Hd. proteus* (Brown) which also includes *Homoceras smithii* (Brown) towards the top. Higher up are successive goniatite horizons with *Homoceras undulatum* (Brown), *H. eostriolatum* Bisat and *Ht. prereticulatus* Bisat (see Hodson 1957). *Caneyella semisulcata* (Hind) and *Posidoniella sp.* are the characteristic bivalves.                               W.H.C.R.

*Reticuloceras circumplicatile* Zone ($R_{1a}$) (formerly called *R. inconstans* Zone). In the Edale valley this zone maintains a rather uniform thickness of 70 to 80 ft and consists of shale with some silty bands and thin ironstones. In the Alport and Ashop valleys it is thinner, Hudson and Cotton (1943, p. 164) recording some 45 ft of beds which now fall within $R_{1a}$.                               I.P.S.

The lower limit of the zone is defined at the base of the band containing *Homoceras magistrorum* Hodson. Slightly higher, *H. henkei* Schmidt and *Homoceratoides varicatus* Schmidt occur. Specimens of *Reticuloceras* in this zone fall into two main groups: (*a*) Evolute forms with strong umbilical plications such as *R. circumplicatile* (Foord) and *R. paucicrenulatum* Bisat and Hudson and (*b*) Relatively involute forms with fine transverse ornament such as *R. todmordenense* Bisat, *R. pulchellum* (Foord) and *R. subreticulatum* (Foord). Members of these two groups are often found together; *R. circumplicatile* and *R. pulchellum* occur in the lower part of the zone; *R. paucicrenulatum* and *R. todmordenense* near the top. *R. adpressum* Bisat and Hudson is found at the top of the zone. The associated fauna includes species of *Caneyella* and *Posidonia*, with *Dunbarella* only in the upper part.                               W.H.C.R.

*Reticuloceras nodosum* Zone ($R_{1b}$) (formerly called *R. eoreticulatum* Zone). This zone is some 40 to 70 ft thick in the Edale valley and is, except for being slightly more silty, lithologically similar to $R_{1a}$. In the Alport and Ashop valleys it is about 80 ft (Hudson and Cotton 1943, p. 164). At least 56 ft of the zone are exposed in the landslip-scar at Mam Tor.

At the western end of the Edale valley the zone is locally thin, *R. nodosum* Bisat and Hudson occurring only 34 ft below the base of the Mam Tor Beds; this is accounted for by a downward extension of the Mam Tor facies, possibly into the higher beds of $R_{1b}$. This contrasts with the tendency, noted below in $R_{1c}$, immediately to the south (see p. 171).                               I.P.S.

*R. dubium* Bisat and Hudson, an almost non-crenulate finely ornamented derivative of *R. adpressum*, characterizes the lowest beds of the zone. Higher beds yield members of the *R. nodosum* group in which the shell is usually moderately evolute with strong transverse striae which regularly bifurcate. *Hudsonoceras ornatum* (Foord and Crick) and species of *Homoceratoides* and *Homoceras* also occur in the higher part of the zone. The type specimens of *R. nodosum*, *R. stubblefieldi* Bisat and Hudson and *H. spiraloides* Bisat and Hudson are from the present district.                               W.H.C.R.

*Reticuloceras reticulatum* Zone ($R_{1c}$). Owing to the diachronous character of the base of the Mam Tor Beds the part of this zone falling within the Edale Shales varies considerably in thickness. Over much of the Edale valley a mere 10 to 20 ft of shale of $R_{1c}$ age underlie the Mam Tor Beds. This shale is silty with some thin sandstones, especially towards the top, and forms a transition to the overlying beds. A similar development is present in the Alport and Ashop valleys. At the western end of the Edale valley all but 15 ft of the zone falls within the Mam Tor Beds while only half a mile to the south, in Whitemoor Sitch, 80 ft of $R_{1c}$ are included in the Edale Shales. Similar conditions obtain near Hope, where some 70 ft of $R_{1c}$ shales underlie the Mam Tor Beds. A small faulted inlier is present near Hathersage.

In the south-east of the district, at Eyam, the general regional thinning in that direction is more than compensated for by the lateral passage of Mam Tor Beds into shale; thus about 135 ft of shale of $R_{1c}$ age here underlie the Shale Grit.                                                                                          I.P.S.

The main palaeontological feature of the zone is a band at the base with *Reticuloceras reticulatum* (Phillips), *R. davisi* (Foord and Crick), *Homoceras striolatum* (Phillips) and *Homoceratoides sp.* In places this horizon contains a benthonic fauna including crinoid debris, brachiopods and gastropods (Hudson and Cotton 1943, p. 153).                                                           W.H.C.R.

## MAM TOR BEDS

At the type locality (see Plate XVI B) the Mam Tor Beds consist of an alternating sequence of sandstones, siltstones and shales, at least 450 ft thick, which were originally described by Jackson (1927, pp. 17–8) as 'Mam Tor Sandstones'. This name was found inappropriate during the present work when these beds were examined over a larger area as the sandstone element is often much less developed than at Mam Tor. For this reason the term 'Mam Tor Beds' has been introduced (Gaunt 1960, p. 32) as being less misleading than, though synonymous with, the older term.

The stratigraphical limits of the Mam Tor Beds vary in some degree owing to the diachronous nature of both base (see also p. 170) and top. The base is marked either by the incoming of sandstone beds of some importance (say over one foot in thickness and of reasonably frequent occurrence) or, in places where the sandstones are less developed, by a general upward increase in the arenaceous content of the shales. The upper limit, also, is diachronous as where thick sandstone bands of massive character develop about this horizon they are placed in the Shale Grit.

The Mam Tor Beds are present in the Edale, Alport, Ashop, Derwent, Perryfoot and Hope valleys and at Mam Tor. Their thickness varies from about 200 ft in the northern and eastern exposures (northern side of Edale valley, Ashop and Alport valleys and near Hathersage) to 450 ft at Mam Tor. In a south-westward direction they pass laterally into the lowest part of the Shale Grit. This is attributed to the deposition of the Mam Tor Beds of normal facies in a shallow-water environment adjacent to the massif where the winnowing action of currents removed the clay fraction. The mode of disappearance of the Mam Tor Beds to the south-east, however, is rather different; while the uppermost beds pass laterally into a sandstone which joins the Shale Grit, in a similar manner to that described above, the lower beds lose their identity near Brough owing to a general decrease in the arenaceous content, and pass laterally into shales which are no more than the normal thin sandy phase transitional to the Shale Grit.

The type section at Mam Tor has been the subject of a detailed study by Allen (1960, pp. 193–208), on which the following description is largely based. The exposed section of these beds, given by Allen as 350 ft[1] in thickness, is made up of a large number of cyclic units, the whole showing certain similarities to flysch deposits. The overall sedimentary pattern has affinities with the Edale Shales at the base, there being a marked shale component; towards the top the sandstone component is more important, indicating an approach to Shale Grit conditions.

The cyclic unit, which usually varies from 3 to 8 ft in thickness, is made up as follows:

4. Shale and mudstone
3. Laminated siltstone
2. Laminated sandstone
1. Massive sandstone

The massive sandstones usually have a little feldspar (1·0 to 11·9 per cent), which is either orthoclase, microcline or albite (Allen 1960, p. 197). There are some mica- or siderite-rich bands. The laminated sandstones show similar mineralogy but are, as would be expected, commonly more micaceous; they show low-angle cross-bedding and ripple-drift bedding (Sorby 1859) in many cases. The base of each sandstone is sharp and frequently shows sole-marks, the several types of which are summarized below. Graded bedding is another conspicuous feature. The massive sandstones often contain beds of shale conglomerate or breccia. The conglomerate fragments are the smaller (up to about 5 cm) but Allen (1960, p. 197) records fragments up to 75 cm diameter in the breccias. So far as the width of the section at Mam Tor (about 100 yd) is concerned, the sandstones show little lateral variation.

The siltstones, apart from their smaller grain-size, are generally similar to the laminated sandstones. The shales and mudstones, in the lower part of the Mam Tor Beds, closely resemble the Edale Shales, but tend to be more sandy in the higher part. The shales yield, in places and especially towards the base, a poor goniatite–bivalve fauna indicative of the $R_{1c}$ Zone. The chief forms are *Reticuloceras* cf. *reticulatum*, *Homoceras sp.*, *Homoceratoides sp.* and Dimorphoceratids, together with *Caneyella* and *Dunbarella*. Carbonized wood debris is of fairly common occurrence in both sandstones and shales, and larger plant stems are sometimes present in the sandstones.

The undersides of the sandstones bear a variety of sole-markings, all described and several figured by Allen who also lists the literature on this subject. The most important of these structures are: (a) Flute casts; these are ridges with an abruptly raised end towards the source of the current which produced them, and tapering off in the opposite direction. Their orientation has been shown by Allen to be nearly due south, indicating currents flowing from the north.

---

[1]This appears to be under-estimated, the present estimate being 400 ft for the beds exposed in the face, with a further 50 ft of unexposed strata above.

(b) Groove casts; these are linear structures attributed to the grooving of the underlying surface by objects, such as drifting wood fragments, dragged over it by the current. (c) Load-casts are unorientated lobate structures formed by the unequal loading of mud by coarser sediment. (d) Trail casts and burrows are cylindrical casts of marks left by organisms on the mud surface before the overlying sandstone was deposited.

Penecontemporaneous deformational (slump) structures, though not so typical of the Mam Tor Beds as sole-marks, are present locally. They include convolute bedding, corrugated bedding, crumpled bedding (Allen 1960, p. 202) and slump-balls. Some of the latter structures reach a diameter of 50 cm.

The present work, extending over a wide area, has shown the Mam Tor Beds to be much more variable than is apparent at Mam Tor. Only a mile to the east, at Brockett Booth, sandstones are much less frequent and are much thinner. Elsewhere relatively thick sandstones of mappable proportions are locally present.

The conditions of deposition have been discussed in detail by Allen, who states "that intermittent turbidity currents . . . rapidly deposited the sandstones in a deep marine environment where clays were normally being slowly formed" (op. cit., p. 205). This is considered to have been at the distal end of the slopes fronting a large delta, lying to the north.

## SHALE GRIT

The Shale Grit consists of a thick sequence of sandstones, often massive, with shale bands. The name (Farey 1811, p. 228) apparently derives from the presence of the shale bands or from the common occurrence of shale pellets in the sandstones. The Shale Grit forms a broad spread of moorland flanking Kinder Scout but showing nearly everywhere a less craggy topography than that formed by the Kinderscout Grit. In places, however, the lower or middle parts of the Shale Grit stand out as strong escarpments.

As in the case of the Mam Tor Beds, the limits of the Shale Grit are somewhat diachronous. The lower limit has already been discussed above. The practice here followed in the fixing of the upper limit has been to include in the Shale Grit all sandstones below the thick shale sequence ('Grindslow Shales', see p. 175) separating the Shale and Kinderscout grits. If the general relations of the shale partings and sandstones within the Shale Grit are considered, two types of development are apparent in the Chapel en le Frith district: (a) A regular development in the north around Kinder Scout extends eastwards to include the upper valley of the Derwent, Hathersage and Offerton Moor. Here the thickness varies from 700 ft near Hayfield and 600 ft in the Ashop valley and in the western end of the Edale valley to about 300 ft near Hathersage. The sandstones tend to be some 50 to 60 ft in thickness (though they are thicker where the shale partings fail locally) and the shale partings up to about 10 ft. Individual shale bands can be traced laterally for considerable distances. The lowest part of the Shale Grit shows a lesser development of massive sandstones than the overlying beds, their place being taken by alternating beds of 'Mam Tor' type. The middle part shows a marked development of thick, massive sandstones, which are in places coarse and pebbly; a thin ganister-like sandstone, with *Stigmaria* locally, is present at the top of this subdivision in the middle part of the Ashop valley.

The highest beds are to some extent transitional to the 'Grindslow Shales' with two or more sandstones very regularly developed in the ground north-east of Kinder Scout. (b) A marginal development, in the south-western and south-eastern outcrops, heralds the southern disappearance of the Shale Grit. Here the different sandstones are irregular, contrasting strongly with the regular development described above. The shales, for example, may coalesce, giving thick beds, up to 100 ft at Bagshaw and 60 ft at Abney; these beds are, however, of small lateral extent. In general, in the south-western outcrops the Shale Grit thins rapidly to the south and disappears a little north of Dove Holes. In the south-eastern outcrops, in addition to the southerly thinning there is a tendency for the shale partings to disappear towards and to some extent over the limestone massif giving a single sandstone at Eyam—the Eyam Edge Sandstone of Morris (1929, pp. 41, 58)—which is about 200 ft thick. This, however, dies out just south of the present district near Curbar.

Near the northern boundary of the district the Charlestown Borehole (see p. 211) was put down to a depth of 586 ft below the base of the Kinderscout Grit and failed to prove any Shale Grit. Since the intervening shales are only some 300 ft thick in outcrops a mile away, at least the uppermost 286 ft of the Shale Grit must have failed here. It is considered that the whole of the Shale Grit may be absent at this locality. The Shale Grit is normally developed, except for some increase in the shale partings, in the nearest outcrops.

The Shale Grit is also absent in the Todd Brook Anticline; it thus fails in a westerly direction in the Goyt Trough in addition to the southerly direction already noted around the Derbyshire Dome. The thick development at Hayfield, however, suggests the presence here of a concealed tongue of Shale Grit extending westwards into the Goyt Trough.

The sandstones of the Shale Grit are of medium to coarse grain, usually massive, with common included shale pellets and large load-casts on the under surfaces of beds. Ripple-marks have also been observed in many places. Very coarse sandstones, nearer in lithology to the Kinderscout Grit, occur locally near the top of the middle part of the Shale Grit. In a detailed consideration of the sedimentology of the Shale Grit in the Alport, Ashop and Derwent valleys, Walker (1966, p. 99) states that the coarser sandstones are in individual beds up to 10 ft thick and show scouring at their bases with downcutting to the extent of 3 to 4 ft. Groups of these sandstones may total 50 to 100 ft in thickness. The bases are sharp and in places show sole-marks, especially load-casts (up to 18 in diameter) and less commonly flute casts (up to 3 ft by 1 ft). He further recognizes beds of alternating type, comparable with the Mam Tor Beds; these show a variety of sole-markings. The argillaceous beds were divided by Walker (ibid., pp. 101–4) into silty mudstones, pebbly mudstones and thinly laminated black mudstones. The silty mudstones commonly contain thin interbedded sandstones, the pebbly mudstones frequently show slump structures, and the black mudstones recall the lithology of the fossiliferous marine strata in the Edale Shales but are usually barren.

Walker (ibid., pp. 95–101) attributed the deposition of all the arenaceous beds in the Shale Grit to the action of turbidity currents. The beds of 'Mam Tor' type, more frequent in the lower part of the Shale Grit, were interpreted as of distal origin (see p. 173) while the massive sandstones of less obvious 'turbidite' origin and more frequent in the upper part of the Shale Grit, were considered to be proximal, the delta front having advanced farther into the area. The thicker sandstones are regarded by Walker as 'turbidites' mainly on account of their sharp bases with sole-marks and lack of cross-bedding. This mode of origin accords well with the uniformity of the bedding and shale partings in the 'regular development' of the Shale Grit. The thick sandstone bands of 50 to 100 ft are considered by Walker to be of composite origin, the mud fraction having been winnowed out by current action.

### SHALES BETWEEN SHALE GRIT AND KINDERSCOUT GRIT

The Shale Grit is separated from the Kinderscout Grit by a series of shales and sandy shales with thin sandstones and siltstones and a few thicker sandstone bands. These beds, about 220 ft thick around Grindslow Knoll and Grinds Brook, have been called the "Grindslow Shales" by Jackson (1927, pp. 16–7), who noted their variability in thickness elsewhere. Farther east they are 120 to 150 ft thick, but thin somewhat to the south-east to 100 ft at Abney. The maximum development is north of the Kinder Scout plateau where 350 to 400 ft of shale are present; from here rapid thinning to the south-west occurs, only 30 ft being present in the lower reaches of the Kinder valley, near Hayfield. Still farther south the beds appear to expand, but this is partly due to lateral passage of the Lower Kinderscout Grit into shales. With the disappearance of the Shale Grit the group cannot be recognized beyond Dove Holes.

The lithology of the shales in the Alport, Ashop and Derwent valleys has been considered in detail by Walker (ibid., pp. 101–4) and the following types recognized: mudstones (not falling in other groups), paper laminated mudstones (rare), sandy and silty mudstones with horizontal burrows, unbedded sandy and silty mudstones. The last two types are said by Walker to be restricted to the Grindslow Shales. The associated sandstones (ibid., pp. 104–5) are divided into parallel-bedded silty sandstones and parallel-bedded carbonaceous sandstones, the former being restricted to the Grindslow Shales as a whole and the latter restricted to their upper part.

### KINDERSCOUT GRIT

The Kinderscout Grit is made up of one or more leaves of massive sandstone, typically coarse and commonly pebbly. The lithology contrasts strongly with that of the Shale Grit, but shows features in common with that of higher beds, particularly the Chatsworth Grit.

The Kinderscout Grit gives rise to many of the most striking topographical features in the district, particularly the plateau of Kinder Scout; here wind-eroded stack-like outcrops are frequent, the most striking being Crowden Tower and Noe Stool. Derwent Edge shows similar features such as the Wheel Stones and the Salt Cellar. Strong scarp features are also typical, good examples being Kinder Downfall (see Plate XVIII A), The Edge, Derwent Edge and Bamford Edge.

Except in the south-west and south-east of the district it has been possible to subdivide the Kinderscout Grit as follows:

(c) Upper Kinderscout Grit

(b) Shales with Butterly Marine Band

(a) Lower Kinderscout Grit

This follows the usage in the Holmfirth and Glossop district (Bromehead and others 1933, p. 15). The two divisions of the Kinderscout Grit are not, however, named separately on the map. The recognition of the Butterly Marine Band as far south as Hayfield and Strines (near Bradfield) represents a considerable extension southwards of the known occurrences of this band.

(a) The Lower Kinderscout Grit is coarse, feldspathic, massive and frequently pebbly. It is 300 ft thick on the west side of Kinder Scout and 250 ft at Hayfield. It thins rapidly northwards from Hayfield, being only 74½ ft in the Charlestown Borehole. Traced southwards, it splits into a number of leaves which then die out, the upper ones failing first; the whole of the Lower Kinderscout Grit thus fails by the time Chapel en le Frith is reached.

To the east of the Derwent valley the Lower Kinderscout Grit is most fully developed around the head of Howden Dean where it consists of a 150-ft thick lower leaf and an 80-ft thick upper leaf, separated by about 120 ft of shales and siltstones. The lower leaf, a coarse pebbly massive sandstone where it forms the main escarpment north of Howden Dean, splits southwards into two beds, both of which thin and die out below Derwent Edge. The upper leaf expands southwards to between 120 and 180 ft of coarse pebbly massive sandstone where it forms the main escarpments of Derwent Edge and Bamford Edge. Large-scale current-bedding foresetting towards the south-west is conspicuous and results in anomalous oblique features running across and down the fronts of these edges. The Lower Kinderscout Grit thins appreciably southwards along Bamford Edge. To the north of Hathersage it unites locally with the Upper Kinderscout Grit. East and south of Hathersage a number of individual beds again become discernible, though they cannot be correlated with any certainty with the Upper and Lower Kinderscout grits.

The top surface of the Lower Kinderscout Grit is often ganister-like and sometimes bears *Stigmaria*.

(b) The shales with the Butterly Marine Band vary on the western outcrops from 87 ft (in the Charlestown Borehole) to 33 ft. Some of this variation may be due to channelling at the base of the Upper Kinderscout Grit. The best section was provided by the Charlestown Borehole: this showed a thin coal and seatearth at the base, overlain by the Butterly Marine Band, here showing the remarkable thickness of 27 ft. The band has the typical benthonic mollusc and brachiopod fauna listed by Bromehead and others (1933, p. 145) and, in addition, a phase with *Reticuloceras sp. nov.* near the base. The upper part of the band showed two non-marine interdigitations with *Cochlichnus kochi* (Ludwig) and fish debris (see below). A thinner development of the band was found at Chunal and south-east of Hayfield and it probably dies out a short distance to the south of the latter place.

Near South Head, east of Chinley, the lateral passage of the upper part of the Lower Kinderscout Grit into shales results in an overall thickening of the overlying shales to some 350 ft. A dark carbonaceous mudstone towards the top of these shales may represent the horizon of the thin coal below the Butterly Marine Band in the Charlestown Borehole.

On the eastern outcrops these shales are some 70 ft thick. About three-quarters of a mile north of Foulstone Delph the Butterly Marine Band was found near the base as an 11-in band with *Reticuloceras sp.* and fish debris, the benthonic fauna being apparently absent here. At Bole Hill the shales disappear where the Upper and Lower Kinderscout grits unite. I.P.S.

The present district sees a transition in the fauna of the Butterly Marine Band. In the Holmfirth and Glossop (86) district to the north (Bromehead and others 1933, pp. 145–6) the fauna comprises largely benthonic forms such as *Lingula mytilloides* J. de C. Sowerby, *Orbiculoidea nitida* (Phillips), *Retispira undata* (Etheridge jun.), *Sanguinolites ovalis* Hind and *S. tricostatus* (Portlock) together with a few nectonic species such as *Aviculopecten dorlodoti* Delépine and extremely rare goniatites (only two specimens are known from the Glossop district despite intensive collecting). This type of fauna persists in the western part of the present district but goniatites are more common. In the Charlestown Borehole, where the band is some 27 ft thick, the basal 4 ft has a typical benthonic fauna without goniatites. From 4 to 5 ft above the base *Lingula* and *S. ovalis* occur with *Reticuloceras sp. nov.* Higher beds contain mainly *Lingula* and bivalves, but at two levels there are apparently non-marine interdigitations, the lower of which contains the sineoid trail *Cochlichnus kochi* and the upper fish debris only. Around Chunal the band is thinner and contains a benthonic fauna as well as *Reticuloceras*. To the east (see pp. 223–4) the fauna consists of *Reticuloceras sp. nov.* and fish debris only. *Anthracoceras* has not been collected anywhere at this horizon. W.H.C.R.

(c) The Upper Kinderscout Grit consists of sandstones, usually of medium grain, with some coarser bands and in places bands with shale pellets. The thickness varies in the western outcrops from about 60 to 150 ft, and in the eastern from 60 to 90 ft. At Chinley, borehole evidence (see p. 221) suggests the lateral passage of the top 40 ft or so of the Upper Kinderscout Grit into silty mudstones and siltstones: otherwise it is much less variable than the Lower Kinderscout Grit. On the main western outcrop the Upper Kinderscout Grit continues south of Chapel en le Frith to the southern edge of the district. On the eastern outcrops it joins with the Lower Kinderscout Grit at Bole Hill, as has already been noted.

The top of the Upper Kinderscout Grit is often ganister-like, bearing rootlets and *Stigmaria*. A thin seatearth, and in the north-east a thin coal, overlain by a small thickness of shale, is usually present between it and the *Reticuloceras gracile* Marine Band.

In the area where the Kinderscout Grit has not been subdivided it is developed as follows: (1) In the Todd Brook Anticline 180 ft of sandstone with subsidiary shale partings are present in the northern part; farther south it thickens to about 320 ft and a lower leaf appears, some 50 ft below the upper and 70 ft thick. No correlation of these leaves with Upper and Lower Kinderscout Grit is implied. (2) On the eastern side of the area, east and south of Hathersage, after the junction of Upper and Lower Kinderscout grits already noted, a shale parting appears again in approximately the position of the shale with the Butterly Marine Band. This splits the Kinderscout Grit, some 200 ft thick in all, into two approximately equal leaves between Hathersage and Grindleford. To the south of the latter place, thinning occurs and the Kinderscout Grit eventually dies out between Bakewell and Rowsley.

The conditions of deposition of the Kinderscout Grit were considered by Trotter (1952, pp. 78–9) who classed it as of "fluvial-grit facies". This was made up of massive spreads of grit "of the valley-spread type and possibly of the piedmont type". The Kinderscout Grit is now generally accepted as of fluviatile origin (see, for example, Holdsworth 1963, p. 135; Collinson and Walker 1967, pp. 80, 87–8) and the orientation of the current-bedding indicates a northern or north-easterly source. The petrography also falls into line with this conclusion, the sandstones being coarse arkoses of the type described farther north in the west and north Yorkshire area by Gilligan (1920, pp. 253–75) and shown by him to have been derived from the denudation of an area of crystalline rocks lying to the north-east.

After the deposition of the Lower Kinderscout Grit it is evident that more regular conditions favouring cyclic sedimentation were established (Reading 1964, p. 345). These conditions continued during the deposition of most of the higher part of the Millstone Grit in the north Derbyshire area.

## MIDDLE GRIT GROUP
### (Marsdenian, R$_2$)

The Middle Grit Group extends from the base of the *Reticuloceras gracile* Marine Band to the base of the *Gastrioceras cancellatum* Marine Band (see Plate XIX). In contrast to most of the Millstone Grit so far described, this succession shows clear-cut widely traceable cyclothems, there being six or seven present over most of the district and eight where the *Reticuloceras superbilingue* and *Donetzoceras sigma* bands are so split as to lie in different cyclothems. Several of these cyclothems are, however, incomplete, and in certain parts of the group there is considerable lateral variation.

### BEDS BETWEEN KINDERSCOUT GRIT AND CHATSWORTH GRIT

These beds show great variation in thickness. In the western outcrops they range from 550 ft near Glossop to 1500 ft south of Kettleshulme. This southerly thickening is not, however, so acute near the limestone massif, the thickness near Dove Holes being only about 900 ft. The eastern outcrops show the southerly thinning usual on that flank of the Pennines, from some 400 ft between Bradfield and Stanage to 250 ft south of Hathersage. The thickness variations are at least in part accounted for by the incoming or disappearance of sandstones.

*Western area.* The lowest 80 to 100 ft of these beds are the least variable and show a fairly uniform development of four marine bands whose fauna is discussed on p. 183:

(a) The *Reticuloceras gracile* Marine Band, usually from 4 to 5 ft thick, is a persistent feature at the base. It is typically grey platy shale with *R. gracile* Bisat and other fossils.

(b) The *R. bilingue* early form Marine Band, from about 2½ to 7 ft thick, consists usually of dark calcareous shale, locally pyritous. It lies from 14 to 30 ft above the *R. gracile* Band.

(c) The *R. bilingue* Marine Band lies approximately half way between the bands with *R. bilingue* early form and *R. bilingue* late form. Its thickness varies from 7 in to 7 ft. The thinner developments are usually made up of dark calcareous shale, but in the thicker ones the marine fauna extends upwards into the overlying grey shales.

(d) The *R. bilingue* late form Marine Band, lying at the top of the group, is usually some 3 to 5 ft thick. The lithology varies from dark 'sulphurous' shale to dark slightly silty shale.

The following exceptional thicknesses in this district of the three lowest of these bands were proved in a borehole at Whitehough: *R. gracile* 18 ft 10 in, *R. bilingue* early form 14 ft 9 in and *R. bilingue* 14 ft 8 in.

The relationships of the above beds are illustrated in Plate XIX. Between the *R. bilingue* late form horizon and the *R. superbilingue* horizon the great variability contrasts strongly with the regularity of the beds below. The thickness of this group of strata varies from as little as 50 ft at Chunal to 1000 ft in the Todd Brook Anticline. The belt of thick sedimentation between these horizons extends southwards and also south-westwards into the Macclesfield area (Evans and others 1968, p. 57). The most important feature associated with this thickening is the incoming of the Corbar and Roaches grits, which together are equivalent to the Ashover Grit of the country to the east (Smith and others 1967, p. 60). The Corbar Grit is the least extensively developed, though it is present in the south-western part of the district where the Roaches Grit fails. It is in general fine-grained and flaggy and reaches a thickness of 300 ft in the Todd Brook Anticline, south of Kettleshulme (Plate XIX), dying out northwards. At Corbar Hill, in the Buxton (111) district to the south, it lies close beneath the Roaches Grit, giving rise to a composite sandstone group 600 ft thick. To the north-east of Corbar Hill the grit rapidly dies out, but its place is taken by a group of silty or sandy shales with thin siltstones or sandstones. The Roaches Grit, separated from the Corbar Grit by some 250 ft of shale in the Todd Brook Anticline, is more extensively developed. It makes its appearance at Chinley and thickens southwards. It is very variable in development with a marked tendency for different leaves to split or to unite. The lithology varies from coarse and massive to fine-grained and flaggy. Some beds bear large solemarks, mainly load-casts. The coarser massive developments recall the 'fluvial grit' facies (see p. 162), while the finer are comparable to the Shale Grit.

The *Reticuloceras superbilingue* Marine Band is separated from the Roaches Grit, in the present area, by from 30 to 200 ft of shale; typically, in the southern part of the Goyt Trough, the marine band lies directly on the grit (see Cope 1946, p. 142). At Chunal the marine band consists of 2 ft 4 in of dark shale. In the vicinity of Combs Moss, however, it is usually represented by only a few inches of hard calcareous flags.

The *R. superbilingue* Marine Band is overlain by shales some 300 to 350 ft thick. Apart from the local occurrence of the *Donetzoceras sigma* Band these beds call for little comment, except to note the presence in them of several impersistent sandstones north of Chinley.

In the western outcrops the *Donetzoceras sigma* Marine Band is known only near Hayfield where it is 4 in thick. Judging from the eastern outcrops (see below) and from the position of the band at Oakenclough, in the Macclesfield (110) district (Evans and others 1968, p. 63), it would appear to lie about 40 to 50 ft above the *R. superbilingue* Band, though this interval cannot be directly determined in the present area.

*Eastern area.* The succession here is thinner than in the west, the thickness being 400 ft in the north around Strines Moor and about 300 ft below Millstone Edge. The general succession is the same as in the west, though the *R. bilingue* early form Marine Band appears to be absent and in the north-east of the district the Heyden Rock is present beneath the *R. bilingue* late form Marine Band (see Eden and others 1957, p. 19).

The Heyden Rock is present between Bradfield Moors and Moscar House, but then dies out rapidly southwards. It is usually medium-grained in texture, but locally becomes coarser and pebbly. The thickness is about 80 ft at Bradfield. The Heyden Rock is more extensively developed in the Holmfirth and Glossop district where it was mistakenly correlated with the Rivelin Grit (Edwards 1932, pp. 180–1), a mistake rectified by Davies (1941, pp. 241–4). In the present district, a temporary section on Strines Moor (see p. 232), showed the unusual feature of a 3-in band of sandstone with abundant casts and moulds of *Carbonicola* just below an 11-ft shale parting towards the top of the Heyden Rock. This is the lowest record of *Carbonicola* in the Pennine province.

The shales above the *R. bilingue* late form Marine Band, some 200 ft thick, contain a better development of the *R. superbilingue* and *D. sigma* marine bands than in the western outcrops. The *R. superbilingue* Marine Band lies about 100 ft above the *R. bilingue* late form Band, and in the south-eastern corner of the area overlies the Ashover Grit which comes in to the south and is fully developed in the Chesterfield district (Smith and others 1967, p. 60). The *R. superbilingue* and *D. sigma* bands are about 30 to 50 ft apart.

The *R. superbilingue* Marine Band is usually hard and platy shale or dark 'sulphurous' shale from $1\frac{1}{2}$ to 2 ft in thickness. Near Sugworth it is possible that the band expands, the fauna extending over as much as 30 ft of shale. A thick development of the band was present near Sheffield in the Hallam Head Borehole (Davies 1941, pp. 241–2) where the fauna ranged through at least 16 ft and the bottom was possibly not proved.

The *Donetzoceras sigma* Band is typically composed of two leaves, each about 3 in thick and 5 ft apart. The upper seems always to contain the *D. sigma* fauna while the lower commonly contains *Lingula* only. At Jacob Plantation, Sugworth, the *D. sigma* fauna ranges over 8 ft of shale; as in the case of the *R. superbilingue* Band this may be compared with the Hallam Head Borehole where the fauna was present "at certain horizons" (Davies 1941, p. 242) over 15 ft of shale.

Near Grindleford both *R. superbilingue* Bisat and *D. sigma* (Wright) are present in some abundance in both bands.

**Source of the Roaches, Corbar and Ashover grits.** The northward disappearance of the Roaches and Corbar grits in the west and the Ashover Grit in the east contrasts strongly with the behaviour of other grits, such as the Kinderscout Grit, which die out towards the south. The Roaches Grit is coarse and pebbly at the type locality (Challinor 1921, p. 81) but becomes finer grained in general northwards; in the Chesterfield district Smith and others (1967, p. 68) also noted that the Ashover Grit is coarsest in the south and becomes finer grained northwards. These facts alone suggest that the two grits were derived from the south, and this is further supported by the northward orientation of the foreset-bedding[1]. The outcrops of the Roaches Grit show both a coarse development comparable with the 'fluvial-grit' facies and a finer-grained facies with shale partings which is in many respects comparable to the Shale Grit. These can be attributed respectively to the upper surface and frontal slopes of a delta advancing into the area from the south.

---

[1] See Stephens, E. A. 1952, pp. 224–5, though the Ashover Grit is here mistakenly called the 'Upper Kinderscout Grit' (Smith and others 1967, p. 71).

## CHATSWORTH GRIT

With the exception of the Rough Rock, the Chatsworth Grit forms the most constant sandstone horizon in the Millstone Grit of the region; moreover, the sandstone, or its equivalent the Huddersfield White Rock, occurs over most of the south Yorkshire and Derbyshire area. For these reasons it has been possible to abandon such local names as Rivelin Grit, current in the Sheffield area (Eden and others 1957, pp. 14, 18), or Shining Tor Grit, put forward by Cope (1946, p. 142) in the Goyt Trough, in favour of the more widely used term 'Chatsworth Grit'.

The Chatsworth Grit shows marked variations both in lithology and thickness. Two main lithological types may be recognized:

A fine-grained lithofacies occurs north of a line from Kettleshulme, through Chinley Churn to Moscar. The Chatsworth Grit is here a coarse- to medium-grained sandstone, yellow-brown in colour, and up to about 80 ft thick. Near the southern limit of this facies it frequently thins to as little as 25 ft (Bradfield Dale) and 43 ft (Grove Mill Borehole, New Mills) and at Kettleshulme it is apparently split, thin and ill defined, though here there is some complication due to faulting. Near Hayfield it is split into two leaves and to the west of Chunal, although the split is not apparent, its position is marked by a ganister. In the north-east of the area, near Bradfield, the Chatsworth Grit is also split.

Over the remainder of its outcrop the Chatsworth Grit is very coarse, pebbly, massive and feldspathic; it is usually cross-bedded in a general southerly direction, though in the east this changes to south-west or west. These lithological characters have much in common with those of the Kinderscout Grit. The thickness is usually about 100 ft, but reaches 150 ft locally in the Goyt Trough, where the grit shows marked red staining in most exposures and the associated mudstone partings near the base are also deeply stained (see p. 235). The eastern exposures show no red staining, the colour of the rock being yellow-brown, though otherwise the lithology is similar to that of the reddened rock. An impersistent lower leaf, some 40 to 60 ft thick, is present beneath the main bed around Hathersage.

The coarse facies of the Chatsworth Grit forms marked features, such as Combs Moss and Eccles Pike on the west and Stanage Edge, Higger Tor, Froggatt Edge and Curbar Edge on the east. Around Hathersage millstones were, until recent years, produced from the Chatsworth Grit.

The top of the Chatsworth Grit is usually either ganister-like or a true ganister, and is overlain by the Ringinglow Coal or its seatearth.

The red coloration of the Chatsworth Grit in the western outcrops is a striking feature largely restricted to the coarse facies in which it is almost ubiquitous. Associated finer sediments such as flaggy sandstones and siltstones at the base are deeply stained, and associated mudstone partings even more so. Thin sections (see p. 260) show the reddening to be subsequent to both sedimentation and diagenesis as the hematite lies outside secondary quartz overgrowths. It seems likely that the coarse sandstones were particularly favourable to the transmission of reddening solutions during or just before the Permo-Trias.

N

BEDS BETWEEN CHATSWORTH GRIT AND *Gastrioceras cancellatum*
MARINE BAND

These beds, which include two complete cyclothems above the Ringinglow Coal, vary somewhat in the present district, the average thickness on the east being about 120 ft, while on the west it is only some 60 ft.

*Western area.* These beds are usually about 55 to 60 ft thick, but on the western side of Eccles Pike they reach 100 ft and represent a westward extension of the thick sedimentation of the eastern area (see below).

At the base, the Ringinglow Coal and its seatearth normally lie on top of the Chatsworth Grit. The coal varies between 10 and 19 inches in most places south of New Mills, but on Wild Moor, near the southern edge of the district, it thickens to 23 in. To the north of New Mills the coal is frequently thin or absent. It is usually underlain by 1½ to 2 ft of seatearth, but locally, west of Whaley Bridge, it rests on 8 in of ganister. Near Hayfield (see p. 236) a 3-ft ganister occurs a little below the horizon of the Ringinglow Coal and is separated from it by a few feet of sandstone.

In the north, at Simmondley, the beds between the Ringinglow and Simmondley coals are about 40 ft thick and consist entirely of shale. They are not well exposed between Rowarth and Chinley but to the south they show rather constant characters in the Goyt Trough and Todd Brook Anticline. Here the thickness is usually about 25 ft, the upper part, and in many places the lower, being sandstone, though this fails around the nose of the Todd Brook Anticline. In Fernilee No. 1 Borehole the top 17 ft were sandstone and yielded *Naiadites sp.* in a 1-ft mudstone parting near their base.

The Simmondley Coal reaches its greatest thickness, 24 to 30 in, at the type locality near Glossop where it has also been called the 'Two Sheds'. Farther south information regarding the seam is scanty as far as Whaley Bridge, but it is better known south of that place. In Fernilee No. 1 Borehole the coal was 24 in thick but included an 8-in mudstone parting. Around Errwood the thickness was 15 in; the coal was of better quality here and was worked under the name 'Little Mine'.

A variable thickness of shales separates the Simmondley Coal from the *G. cancellatum* Marine Band. These beds are probably only a few feet in thickness near Glossop. In the Goyt Trough, south of Whaley Bridge, they are usually about 20 ft, but they thin again southwards to under 9 ft near Errwood. Between Whaley Bridge and Eccles Pike a sandstone in the position of the Redmires Flags is present in this group of strata. The Fernilee No. 1 Borehole proved the unusual thickness of 32¼ ft, the lowest 2½ ft being dark carbonaceous mudstone with *Curvirimula belgica* (Hind); this was overlain by 7 ft 8 in of mudstone with *Planolites sp.*, *Anthracoceras sp.*, a productoid and *Sanguinolites sp.*; the uppermost beds were shales, locally pyritous. This marine band has not been located in surface exposures in the vicinity, and its presence seems to be related to the local thickening of this part of the succession. A *Lingula* band is present at this horizon at several places on the eastern side of the district and in the adjacent Sheffield area.

*Eastern area.* The thickness of these beds varies from 120 ft at Bradfield to 140 ft near Brown Edge. The Ringinglow Coal at the base varies from 12 in to 24 inches in thickness. It is succeeded by 90 to 100 ft of shales. The *Lingula* band above the coal is exposed in Oaking Clough; the band is better developed in the adjacent Sheffield area (Eden and others 1957, pp. 20–1). The Simmondley Coal is not known, owing to lack of exposures; it is, however, sporadically developed in the adjacent Sheffield area, for example in the Rod Moor No. 3 Borehole (ibid., p. 20) where it was 21 in thick. Sandy mudstones and sandstones between the two coals have been called the "Brown Edge Flags" by Pulfrey (1934, p. 259), but the name is invalidated by the fact that the sandstone of Brown Edge has proved to be Redmires Flags. The Redmires Flags make up the remainder of these beds; they are medium-grained flaggy sandstones usually 20 to 35 ft thick, but absent near Moscar.                              I.P.S.

*Fauna of the Middle Grit Group.* Two sorts of *Reticuloceras* occur in the *Reticuloceras gracile* Zone: evolute forms with strong umbilical plications and involute, finely ornamented forms. Both types show similar lattice-like ornament when adult and both occur in the *R. gracile* Marine Band. The involute forms are *R. gracile* and the evolute forms are either *R. gracile* early form or *R. gracile* late form—the names not necessarily implying stratigraphical position. The *R. gracile* Marine Band also contains forms of *Anthracoceras* with strongly bilobate growth lines, *Caneyella rugata* (Jackson) and *Dunbarella speciosa* (Jackson). With the exception of *Lingula* at one locality no benthonic fossils are known in the *R. gracile* Marine Band. Some of the original type specimens of *R. gracile* late form came from Raddlepit Rushes within the present district (Bisat 1924, p. 117).

The three principal marine horizons of the *R. bilingue* Zone contain in upward succession *R. bilingue* early form, *R. bilingue* (= mut.β Bisat) and *R. bilingue* late form. The first two of these are evolute, and the last is relatively involute and finely ornamented. All show a greater development of the lingua than species found in the *R. gracile* Zone and reticulate ornament tends to be restricted to the lingua. The *R. bilingue* early form horizon has not been found in the eastern part of the district. Forms of *Hudsonoceras ornatum* occur in both the *R. bilingue* early form and *R. bilingue* late form horizons, but they have not been found in collections from the *R. bilingue* level. The lower marine band in Whitehall Works Borehole (see p. 228) near Chinley contains a fragment with straight crenulate transverse striae resembling *Gastrioceras*. The *R. bilingue* horizon in the same borehole contains a *Reticuloceras* with closely-set subcrenulate striae on the flanks, which is similar to a specimen recorded from the Macclesfield district (Evans and others 1968, table 6). The marine fauna of the zone is entirely devoid of benthonic species in this district, except for a thin 2-in band containing *Lingula mytilloides* within the *R. bilingue* Marine Band in the Whitehall Works Borehole. Records of non-marine faunas from the *R. bilingue* Zone are infrequent. *Curvirimula?* (and plants) occurs above the *R. bilingue* Band in the Whitehall Works Borehole, and in the Heyden Rock on Strines Moor a thin layer with moulds of *Carbonicola sp.* was found, unusually in a sandstone matrix.

The typical fauna of the *R. superbilingue* Marine Band comprises *Caneyella rugata*, *Dunbarella sp.*, *Gastrioceras spp.* [undescribed species with crenulate transverse striae], *Homoceratoides fortelirifer* Ramsbottom [type locality at Pears House Clough within the present district] (see Ramsbottom 1958, pp. 29–31) and *R. superbilingue* [abundant]. *Donetzoceras sigma* occurs rarely as does *Homoceratoides* aff. *divaricatus* (Hind). The fauna tends to be more varied in the east, where the band is thick, than in the west where it is thin. In the *D. sigma* Band, *D. sigma* is the only abundant goniatite, but *R. superbilingue* and *Gastrioceras sp.* occur rarely together with *Caneyella sp.* and *Dunbarella sp.* At one locality near Mitchell Field, an example of the gastropod *Pseudozygopleura* was collected. In the stream between Strines and Dale Dike reservoirs *Lingula* occurs some 5 to 6 ft below the *D. sigma* Band.                         W.H.C.R.

## ROUGH ROCK GROUP
### (Yeadonian, G₁)

These beds comprise shales below and the Rough Rock above, overlain by a seatearth and sometimes a thin coal (see Plate XIX). The top is defined by the base of the *Gastrioceras subcrenatum* Marine Band. A thin lower and a thick upper cyclothem are usually present and begin respectively with the *G. cancellatum* and *G. cumbriense* marine bands; there is some indication that the upper cyclothem splits locally.

#### SHALES BELOW THE ROUGH ROCK

*Western area.* These are in most places from 180 to 220 ft thick, though they attain 260 to 280 ft around New Mills and Fernilee. They consist for the greater part of shales, though some thin sandstones and siltstones are present in the upper part; locally thin sandstones at the top join the Rough Rock. The most important stratigraphical features are the presence of the *Gastrioceras cancellatum* and *G. cumbriense* marine bands respectively at and near the base of the shales.

The *G. cancellatum* Marine Band consists of dark calcareous shales, usually from 1½ to 3 ft thick. In most places some 20 to 26 ft of shale separate the *G. cancellatum* and *G. cumbriense* bands. Around the Goyt Trough these shales frequently contain a thick contorted band overlying the *G. cancellatum* horizon and a thinner contorted band a few feet below the *G. cumbriense* Band. To the north-west, around Rowarth, the shales expand to about 80 ft. In the Fernilee No. 1 Borehole, *Planolites sp.*, *Cochlichnus kochi* and fish debris occur between the *G. cancellatum* and *G. cumbriense* marine bands.

The *G. cumbriense* Marine Band (see also pp. 239–42) usually consists of 5 to 14 in of dark calcareous shale, though the Fernilee No. 1 Borehole showed the unusual thickness of 2 ft 11 in. A thin *Lingula* band has been observed close to the base of the band near Errwood.

The shales overlying the *G. cumbriense* Band are usually poorly exposed. What must be nearly the full thickness, 220 ft, was proved in Fernilee No. 2 Borehole. This showed shales with a few thin siltstones and flaggy sandstones in the upper part. The top 90 ft yielded *Naiadites sp.* [juv.] at three levels. The uppermost of these, about 28 ft below the Rough Rock, recalls the occurrence of non-marine bivalves (*Carbonicola*) in a similar position beneath the Rough Rock at Langsett (Bromehead and others 1933, p. 163). The uppermost beds show a gradual passage upwards where the Rough Rock Flags (see below) occur and a more abrupt change where the Rough Rock only is present.

*Eastern area.* Here only 100 to 130 ft of shales occur between the *G. cancellatum* Band and the Rough Rock. However, the interval between the *G. cancell-*

*atum* and *G. cumbriense* bands is greater than in the west, being about 70 ft. The general features of the marine bands are as described for the western outcrops, the only additional feature of note being the local development of a band 2 ft thick with sporadic *Lingula* beneath the *G. cumbriense* Band at Holes Clough, Bradfield. This is in the position of the *Lingula* band noted above near Errwood.

## ROUGH ROCK

The Rough Rock consists usually of a single massive coarse sandstone, often current-bedded and in places with small pebbles. Finer-grained, flaggy beds are frequently developed at the base and the most distinctive of these, though of local development only, have been separately shown on the Six-inch maps (though not on the One-inch) as 'Rough Rock Flags'.

The thickness of the Rough Rock is somewhat variable (see Plate XIX). In the north-western outcrops it is 50 to 60 ft between Glossop and Chinley, but this increases westwards to 85 ft at Broadbottom, where it is well exposed in the deep gorge of the River Etherow, and 163 ft in a borehole at Mellor. In the southern part of the Goyt Trough a lower leaf of up to 40 ft of flaggy sandstones develops in places and this is separated from the coarse upper leaf, about 100 to 110 ft thick, by a few feet of shale. Near Kettleshulme the Rough Rock thins, appears to split, and is only some 20 to 30 ft in thickness.

On Mellor Moor the Rough Rock is split into two equal leaves and there is some indication from a borehole at Capstone (see p. 244) that the base of this parting may represent the horizon of the Sand Rock Mine of Lancashire. The horizon is known from the Romiley Dyeworks Borehole (Taylor and others 1963, p. 152) where it is separated from the Six-Inch Mine by only 11 ft of measures, including the seatearth of the latter.

The main bed of the Rough Rock is underlain in places by up to 30 ft of Rough Rock Flags. These beds are best developed on Chinley Churn where they are regularly bedded fine-grained micaceous flaggy sandstones. At this locality they were formerly quarried and even mined as roofing flags. Less distinctive flaggy beds are also present near the base of the Rough Rock and as thin leaves which split off it, for example, at Rowarth.

On the eastern side of the district outcrops of Rough Rock are restricted. The main characters are as already described and the thickness is 50 to 60 ft. The only noteworthy feature is the splitting into an upper coarse-grained leaf and a lower fine-grained leaf in the vicinity of Brown Edge and Stanage.

A regional study of the Rough Rock by Shackleton (1962), which includes the present district, described its petrography and sedimentary structures. The rock is an arkose, containing about 20 per cent feldspar (mainly microcline-perthite with a little oligoclase-andesine) with small quantities of muscovite, biotite, clay minerals and rarely calcite. Heavy minerals include tourmaline, rutile, zircon, garnet and apatite.

### BEDS ABOVE ROUGH ROCK

The Rough Rock is overlain in western outcrops by a seatearth some 2½ ft thick which locally bears a thin coal, the Six-Inch Mine. To the west of Mellor the coal occurs as a single thin seam of about 6 in; to the east, however, a split seam (for example, coal 5 in, dirt 3 in, coal 5 in) is present. The Shaw Farm

Borehole, near Brook Bottom, proved the thickest section of the coal in the district: coal 11½ in, dirt 7 in, coal 5½ in. To the south of Brook Bottom the Six-Inch Mine has not been found.

To the south of Whaley Bridge about 2 ft of seatearth rest on the Rough Rock. At Fernilee 2¼ ft of unexposed measures, probably shale, lie between the seatearth and the *G. subcrenatum* Marine Band, but in the Fernilee No. 2 Borehole the marine band rests directly on the seatearth.

On the eastern side of the district, around Ughill, the seatearth is well developed as the Pot Clay of the Sheffield area (Eden and others 1957, pp. 36–7[1]). It is about 3½ to 6 ft thick and is overlain by the Pot Clay Coal up to 2 in thick. A small thickness of shale separates the coal and marine band. Workings for the Pot Clay at Sugworth show 10 in of shale with a thin *Carbonicola* band towards the base; this was first noted by Eagar (1953, p. 174).                               I.P.S.

*Fauna of the Rough Rock Group.* Faunally the *G. cancellatum* Marine Band is composite and contains three faunas characterized by, in ascending order: *G. branneroides* Bisat; *G. cancellatum* Bisat and *R. superbilingue; Agastrioceras carinatum* (Frech) and *G. crencellatum* Bisat. The lowest of these has not been recognized in collections made in the district. *Anthracoceras sp., Caneyella multirugata* (Jackson) and *Dunbarella elegans* (Jackson) are the other fossils typically found in this marine band; *Homoceratoides* occurs rarely.

In the *G. cumbriense* Marine Band, *G. cumbriense* Bisat and *G. crenulatum* Bisat are the typical goniatites, but *Homoceratoides* aff. *divaricatus* is commonly found. *Aviculopecten* aff. *losseni* (von Koenen) is characteristic of this level, but is not always found. Other fossils include *Caneyella multirugata, Dunbarella sp.* and *Anthracoceras.* A thin band with *Lingula* occurs about 7 in below the *G. cumbriense* Band near Errwood.                               W.H.C.R.

In the Fernilee No. 2 Borehole the shales below the Rough Rock contained sporadic, small *Naiadites sp.* cf. *productus* (Brown) and *Cochlichnus spp.* at several horizons. The widespread musselband occurring immediately below the *G. subcrenatum* Marine Band in the adjacent parts of the Lancashire and Yorkshire coalfield (Eagar 1953) has been found at Bradfield Moors; the fauna included '*Anthraconaia' lenisulcata* (Trueman) and *Geisina arcuata* (Bean). M.A.C.

## DETAILS

### KINDERSCOUT GRIT GROUP AND UNDERLYING BEDS

#### EDALE SHALES

*Edale valley* (see Fig. 16). Beds of the *Cravenoceras leion* Zone ($E_{1a}$) do not crop out at Edale and they are known only from the Edale Borehole [1078 8493], 665 yd E. 11°S. of Highfield. The lowest beds, lying between 305 and 325 ft, are described by Hudson and Cotton (1945, pp. 4, 12) as shales with pebbles of limestone and "limy shale" between 309 and 315 ft. The pebbles are evidence of the strong erosion of the Carboniferous Limestone of the upturned edge of the block south of Mam Tor (see p. 113). The fauna, with names amended where necessary, consisted of *Caneyella membranacea, Posidoniella sp., Eumorphoceras sp.* and *Orthoceras sp.* between 315 and 325 ft, and *C. membranacea,* cf. *C. semisulcata, Cravenoceras leion, Eumorphoceras tornquisti, E.* cf. *E. tornquisti* between 305 and 315 ft. From 252 to 305 ft the borehole

---

[1]These authors included the Pot Clay in the Coal Measures for descriptive purposes.

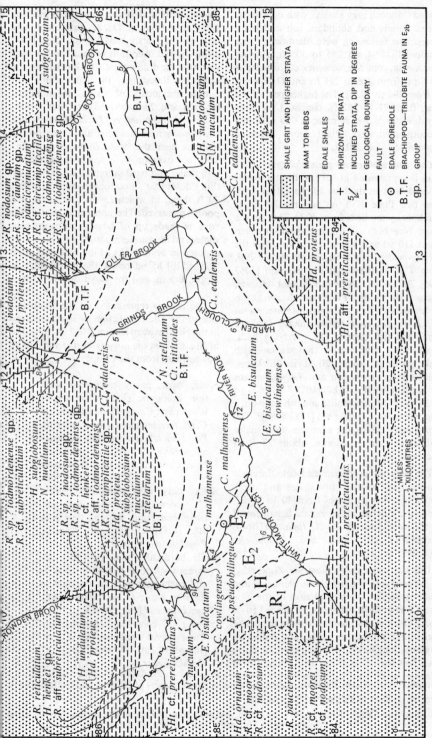

FIG. 16. *Sketch-map of the outcrop of the Edale Shales in the Edale valley*

proved "smooth grey shales" with occasional marine fossils and abundant fish and plant debris. The following were identified: "rare posidoniellids" from 265 to 305 ft, *Pseudamussium sp.* and *Sanguinolites* cf. *angustatus* (Phillips) from 286 to 291 ft, and *Orthoceras sp.* from 296 to 302 ft. The higher part of the zone, from 77 to 252 ft, consisted of "grey silty mudstones, fine micaceous shales, or fine siliceous siltstones". Plant debris occurred between 152 and 252 ft.

The *Eumorphoceras pseudobilingue* Zone ($E_{1b}$) is represented at surface by the lowest exposed beds in the core of the Edale Anticline (see Plate XVI A) and in the Edale Borehole by the beds from 40 to 77 ft. In the River Noe beds of $E_{1b}$ age occur between about 110 yd and 420 yd upstream from the railway-bridge. The lowest strata (Edale 79)[1], some 4 ft in thickness, are exposed [1063 8495] 700 yd N. 38°W. of Manor House[2], where they yielded *Posidonia* cf. *corrugata*, *Posidoniella* cf. *laevis* (Brown), *Anthracoceras sp.*, *Eumorphoceras pseudobilingue* [old-age form] and fish debris. The overlying beds, thin-bedded argillaceous limestones with shale partings, about 20 ft in thickness, yielded *P.* cf. *corrugata* and *Dimorphoceras sp.* at a locality (Edale 80) [1073 8491] 620 yd N. 33°W. of Manor House and fish debris in a nearby exposure (Edale 83) [1083 8483][3]. The uppermost beds of the zone are dark shales with some thin ironstones, about 23 ft in thickness.

The beds of this zone in the Edale Borehole consisted of "shales, silty shales and occasional siliceous siltstones from 40 to 77 ft" yielding *Posidonia* cf. *corrugata* [small form] *P.* cf. *membranacea*, *P.* cf. *radiata* (Hind), *Posidoniella* cf. *laevis* and *Posidoniella variabilis*.     I.P.S.

The *Cravenoceras malhamense* Zone ($E_{1c}$) is exposed on the northern flank of the Edale Anticline along the banks of the Noe east of Highfield, where it is nearly 50 ft thick. The lowest 15 to 20 ft consist mainly of shales and mudstones with thin dolomitic siltstones and limestones, and contain fairly abundant fossils, as noted from outcrops in the southern flank of the anticline by Hudson and Cotton (ibid., p. 11). An outcrop of these beds in the north bank of the Noe [1047 8506] showed over 15 ft of thin-bedded shale with thin hard dolomitic silty limestones and dolomitic limestones, the latter containing minute fragments of shells, foraminifera, bryozoa and ?ostracods. These beds yielded *Caneyella membranacea, Posidonia corrugata, Cravenoceras malhamense,* and in carbonate bullions near the base, *Dimorphoceras sp.* They are overlain by a sequence of alternating shales and hard dolomitic siltstones with thin dolomitic limestones, approximately 20 ft thick, in which fossils are sparse and poorly preserved. The uppermost part of the zone is made up of largely unfossiliferous shales and mudstones up to 10 ft thick, poorly exposed in both banks of the Noe at a point [1042 8510] 85 yd below the road-bridge to Upper Booth, and for 40 yd downstream.

                                        G.D.G.

The zone is represented on the south-east side of the Edale Anticline by shale outcrops in the Noe between a point [1092 8479] 110 yd upstream from the railway-bridge and 210 yd downstream from the bridge:

|  | ft |
|---|---|
| Shale, dark, partly calcareous with thin dolomitic shaly siltstones in lower part; *Caneyella membranacea, C.* aff. *membranacea* [elongate form], *Posidonia* aff. *corrugata, Posidoniella* aff. *vetusta, Cravenoceras malhamense, C. sp.* at 5 to 10 ft from base about | 35 |
| Shale, dark with thin silty bands; *C. membranacea, Myalina* aff. *compressa* Hind, *M. sp., Posidonia* aff. *corrugata, P.* aff. *vetusta, Pseudamussium sp.* .. .. .. | 4 |
| Shale, hard calcareous, with thin calcareous siltstone bands up to 4 in and carbonate bullions up to 9 in thick at base; *C. membranacea, C. membranacea* [elongate form], *Posidonia* aff. *corrugata, Hyolithes sp., Cravenoceras malhamense, C. sp.* [smooth form], *Dimorphoceras sp., Kazakhoceras sp.* .. .. .. | 6 |

[1] Numbers preceded by the word 'Edale' indicate locality numbers of Hudson and Cotton (1945).
[2] Upstream limit of Edale 79: [1058 8497].
[3] Hudson and Cotton (ibid., p. 4) state that the lowest 30 ft of beds at the surface are the same as the uppermost 30 ft of the borehole. The latter figure is approximately correct if taken from the point (37 ft) at which coring commenced.

A. River Noe, Barber Booth: Edale Shales of the $E_{1b}$ Zone

(L 207)

B. Mam Tor: Landslip-scar in Mam Tor Beds

PLATE XVI

(L 212)

All the fossil records are from Hudson and Cotton (ibid., p. 11).

Excavations for the sump for the Edale Borehole in shale and siltstone yielded *Posidonia membranacea, Posidoniella* cf. *laevis* and *P. variabilis* from this zone. From Hudson and Cotton's description of $E_{1b}$ (ibid., p. 10) it is clear that these authors refer all the beds in the borehole down to 40 ft to $E_{1c}$, although only the lowest 3 ft of these was cored.

The *Eumorphoceras bisulcatum* Zone ($E_{2a}$) is represented by some 80 ft of shale with carbonate bullions and thin dolomitic siltstone bands cropping out in the banks of the Noe over a distance of a mile between the most easterly exposures of the *C. malhamense* Zone near Barber Booth and the most westerly exposure with *Cravenoceratoides edalensis* [1285 8512], (see p. 190). These beds yielded fish debris only, at a point 215 yd downstream from the railway-bridge. The upper part of the zone is exposed in the river farther downstream. At a point [1140 8475] about ¼ mile N.E. of Manor House, the following section was measured:

|  | ft | in |
|---|---|---|
| Shale, grey 'sulphurous', platy in lower part: 3-in ironstone at 3 ft from base: rare goniatites    .. | 10 | 0 |
| Shale, platy decalcified    ..    .. |  | 3 |
| Shale, pale grey 'sulphurous'    .. | 1 | 4 |
| Ironstone    ..    ..    ..4 in to |  | 5 |
| Shale, soft grey 'sulphurous'    .. |  | 9 |
| Shale, grey ..    ..    ..    .. | 3 | 6 |
| Shale, soft ferruginous    ..    .. |  | 2 |
| Ironstone    ..    ..    ..0 in to |  | 6 |
| Shale, pale grey    ..    ..    .. | 3 | 0 |
| Shale, decalcified ferruginous    .. |  | 3 |
| Ironstone    ..    ..    ..    .. |  | 2 |
| Shale, grey ..    ..    ..    .. |  | 10 |

The fossils recorded by Hudson and Cotton (ibid., p. 10) from this section (Edale 111) are: *Posidonia sp., Pseudamussium jacksoni* Demanet, *P. sp., Cravenoceras cowlingense* group, *C. sp.* [smooth form], *Dimorphoceras sp., Kazakhoceras* cf. *scaliger, Orthoceras sp.,* crinoid columnals, and (specified as occurring in the lower part of the section) *Posidonia* cf. *corrugata* and *Eumorphoceras bisulcatum.* A similar fauna including *E. bisulcatum* was also recorded from the confluence of the Noe with a small stream (Edale 112) [1163 8478], nearly ½ mile E.N.E. of Manor House.

I.P.S.

The total thickness of the zone exposed in the River Noe near Upper Booth is not much more than 80 ft, but a few feet at the top may be faulted out, as suggested by Hudson and Cotton (ibid., p. 10). Outcrops occur in the Noe from 40 yd below the confluence with Crowden Brook [1020 8522] to 20 yd downstream (Edale 74) [1037 8512] from the road-bridge to Upper Booth, where Hudson and Cotton (ibid., p. 11) record *C. cowlingense* in the basal beds of the zone. At the road-bridge [1035 8513] to Upper Booth thin bands of cross-bedded fine sandstone, partly decalcified, partly dolomitic, and up to 1 in thick, are present in interbedded shales and dolomitic siltstones in the northern bank. An exposure on the south bank of the Noe [1024 8520] consists of 18 ft of shale with ironstone bands and containing *Anthracoceras* or *Dimorphoceras sp.,* overlying 6½ ft of hard shaly mudstone with *P. corrugata elongata* Yates, *Eumorphoceras bisulcatum* and *Kazakhoceras scaliger.* The lowest occurrence of *E. bisulcatum* is estimated to be 40 to 45 ft above the base of the zone at this locality. Farther east, Hudson and Cotton (ibid., pp. 10–11) record a fauna from shale with thin ironstone bands on the north bank of the Noe (Edale 124) [1237 8519] 200 yd S. 30° E. of Edale station and a similar outcrop (Edale 123a) on the south bank by the footbridge 60 yd upstream, which they refer to as "$E_{1d}$"; recent collecting has shown the fossils to be mainly fragments of old-age *Eumorphoceras.* Though determination of horizon is not possible, it seems probable that these outcrops are of $E_{2a}$ age as defined on p. 167.

The *Cravenoceratoides nitidus* Zone ($E_{2b}$) has a total thickness of about 60 ft, but it is difficult to make a reliable estimate due to faulting and poor exposure. It consists of shales and mudstones with subsidiary ferruginous siltstones, thin ironstone bands, carbonate bullion horizons and a distinctive bed of dolomitic siltstone and dolomite containing a brachiopod-trilobite fauna in the upper part of the zone. Most of the zone, including the lowest beds with the *Cravenoceratoides edalensis* fauna are probably cut out by faulting in the vicinity of Upper Booth. The following section of the upper part of the zone and the basal part of the *Nuculoceras nuculum* Zone occurs on the east bank of Crowden Brook [1022 8526],

10 to 20 yd above its confluence with the
Noe:

|  | ft | in |
|---|---|---|
| Mudstone .. .. .. .. | 5 | 6 |
| Siltstone, ferruginous, passing locally into ironstone .. 3 in to | | 5 |
| Mudstone, with thin ferruginous silty bands .. .. .. | 9 | 0 |
| Siltstone; *Posidoniella* aff. *vetusta*, *Nuculoceras stellarum* .. .. | | 9 |
| *Not exposed* .. .. .. about | 2 | 0 |
| Mudstone .. .. .. .. | 5 | 0 |
| Mudstone, black .. .. .. | 1 | 6 |
| Mudstone, hard black silty .. | 1 | 0 |
| Dolomite, pale grey fine-grained hard splintery and partly laminated, with ferruginous silty partings and fragmented shell debris .. .. .. .. | | 4 |
| Mudstone, hard black silty, with thin lenticular dolomite partings; *P.* aff. *vetusta*, nautiloid cf. *Tylonautilus, Cravenoceras?, Eumorphoceras* cf. *rostratum* and *Weberides sp.* mainly concentrated in a band 14 in above the base .. | 2 | 6 |

*Productus hibernicus, Rugosochonetes sp.*
and *E. rostratum* have been identified in
Hudson and Cotton's collections from this
exposure (Edale 45), which confirms the
position of the brachiopod-trilobite-*Ct. nititoides* fauna as being 12 ft below the base of
the succeeding zone, as represented by
*N. stellarum.*

Beds of $E_{2b}$ age crop out extensively in the
vicinity of Grindsbrook Booth. The lowest
horizon identified is on the east bank of
Grinds Brook [1237 8589], just north-north-
east of Edale church, where 3 ft of shale
contains poorly preserved *Cravenoceratoides
sp.*, possibly *Ct. edalensis* or *Ct. bisati.*
Exposures on both banks 20 yd farther
upstream have yielded *Posidonia corrugata,
Anthracoceras* or *Dimorphoceras sp., Cravenoceratoides sp.* and fish fragments. The
dolomite horizon has not been identified,
but the following section occurs on the east
bank of Grinds Brook by the bridge [1233
8604]:

|  | ft |
|---|---|
| Shale, soft .. .. .. .. | 3 |
| Shale, hard flaggy rather silty, with large calcareous bullions at the base; | |

|  | ft |
|---|---|
| athyroids indet., *Orbiculoidea nitida, Productus hibernicus, Rugoso-chonetes sp., Posidonia* aff. *vetusta, Eumorphoceras sp., Kazakhoceras sp.*, and crinoid columnals .. .. | 2 |
| Shale, with soft black bands, thin hard ferruginous siltstone bands and small carbonate bullions .. .. | 11 |

The abundance of *P.* aff. *vetusta* indicates
proximity to the *N. stellarum* band at the
base of the overlying *N. nuculum* Zone.

The $E_{2b}$ Zone strata crop out sporadically
in the banks of the Noe between its confluence with Grinds Brook and the small
reservoir south-east of Nether Booth. Black
iron-stained shale with *Ct. edalensis* crops out
in the south bank [1285 8512]. This outcrop is
either the type locality, or is very near to the
type locality for *Ct. edalensis.* To the east,
the upper part of the zone and the basal part
of the succeeding $E_{2c}$ Zone are present in
exposures along the river near Skinner's
Hall, between the footbridge [1295 8516]
leading from Edale to Hollins Cross and the
junction with Oller Brook [1326 8535], a
composite section being:

|  | ft |
|---|---|
| Shale, ferruginous; *P. corrugata, P.* aff. *vetusta* and *N. stellarum* in upper 1 ft .. .. .. .. | 10 |
| Shale, flaggy, with 2-in band of hard dolomitic siltstone at base contain-ing *Lingula mytilloides, Productus hibernicus, Rugosochonetes sp., Euchondria* aff. *levicula* Newell, *Epistroboceras?, Cravenoceratoides nititoides, Dimorphoceras sp., Eumorphoceras* cf. *rostratum, Weberides sp.* and crinoid debris | 2 |
| Shale, soft, with sporadic carbonate bullions containing *Posidonia cor-rugata* and *Cravenoceras sp.* .. | 8 |
| Shale, hard thin-bedded, with *P. corrugata, Posidoniella variabilis, Cravenoceras* cf. *holmesi* and ortho-cone indet. .. .. .. .. | 5 |

Still farther east, shale with *Ct. edalensis*
is again exposed in both banks of a bend in
the Noe [1332 8531] 1130 yd E.2°S. of
Edale station, and outcrops, probably of
upper $E_{2b}$ and lower $E_{2c}$ strata, occur for
another 300 yd downstream to the north-east.

Beds of this zone have been proved in the river banks south-east of Nether Booth and over 30 ft of strata, including the brachiopod-trilobite horizon, are exposed in the south bank of the small reservoir [1457 8580] some 500 yd S.E. of Nether Booth, as indicated by Hudson and Cotton (ibid., pp. 9–10). This outcrop (Edale 181) was inaccessible during the present survey.                         G.D.G.

The $E_{2b}$ Zone, although probably present in Whitemoor Sitch, is poorly exposed; it is represented, and this somewhat doubtfully, by an outcrop [1075 8455], where *Posidonia corrugata* was collected.                   I.P.S.

The *Nuculoceras nuculum* Zone ($E_{2c}$) is about 60 ft thick near Upper Booth, but possibly less than 50 ft in Grinds Brook, and it may be even thinner farther east. It consists of a fairly uniform sequence of shale and mudstone with a few thin, in places slightly dolomitic, siltstone beds and rare carbonate bullion horizons. The exposure near the confluence of Crowden Brook with the River Noe, containing the basal *N. stellarum* band in the uppermost 1 ft, has already been noted on pp. 189–90. This fossil has also been recorded from a number of localities nearby (Edale 42, 43, 46) (Hudson and Cotton ibid, pp. 8–9). The occurrence of this band in the Skinner's Hall section has also been mentioned.

*N. nuculum*, which is characteristic of the middle and upper parts of the zone, is present in a number of localities near Upper Booth. It has been identified in collections made by Hudson and Cotton from a locality in Crowden Brook (Edale 49) [1025 8540] and is recorded by them together with *Cravenoceratoides fragilis*, from an outcrop a little higher upstream (Edale 51) [1025 8548] where the following succession, typical of the lithology of the zone, is exposed:

|  | ft |
|---|---|
| Shale, with carbonate bullions 3 ft from top  ..  ..  ..  .. | 9 |
| Mudstone, black  ..  .. 1½ ft to | 2 |
| Limestone, dolomitic ..  .. ½ ft to | 2½ |
| Shale, black  ..  ..  ..  .. | 2 |

G.D.G.

*N. nuculum* has also been found during the re-examination of collections made by Hudson and Cotton from the north bank of the Noe (Edale 32) [1007 8535]. Together with *Ct. fragilis*, it had been recorded by these

authors from a shale outcrop in a small stream a short distance to the south-west (Edale 39) [0996 8519].          G.D.G., I.P.S.

Farther east, Hudson and Cotton record cf. *N. nuculum* from the base and *Homoceras subglobosum* from the upper part of a section (Edale 148) in Grinds Brook [1221 8616], ¼ mile N.15°W. of Edale church, where rather fissile shale, 15 ft thick, is exposed in the east bank.

In the south bank of the Noe [1365 8542], 1010 yd E.32°S. of Ollerbrook Booth, the following section is present, partially repeated by faulting:

|  | ft | in |
|---|---|---|
| Shale ..  ..  ..  ..  .. | 9 | 0 |
| Shale, silty with dolomitic siltstone lenses  ..  ..  .. 0 to | 5 | |
| Siltstone, dolomitic; *Glabrocingulum sp.*, *Actinopteria persulcata* (Mc-Coy)  ..  ..  .. 4 in to | 5 | |
| Mudstone  ..  ..  ..  .. | 5 | 6 |
| Shale, silty ferruginous  ..  .. | 2 | 9 |

A short distance to the east [1370 8545], an exposure, complicated by faulting and slipping, is probably in the same strata; it has yielded *A. persulcata*, *Leiopteria longirostris* Hind, *P. corrugata* and *Posidoniella sp.* Hudson and Cotton (ibid., pp. 8, 9) record *N. nuculum* from the base and *H. subglobosum* s.l. from the top of this exposure (Edale 176).

The *Homoceras subglobosum* ($H_{2a}$) and *Homoceras beyrichianum* ($H_{2b}$) zones consist of mudstones, partly shaly, over 70 ft thick, with sporadic carbonate bullion horizons and with fossils mainly in the *H. subglobosum* Zone. The basal strata are exposed in the east bank of Crowden Brook [1027 8552] at Upper Booth:

|  | ft |
|---|---|
| Shale, with black bands; a line of carbonate bullions at the base containing *H. subglobosum*  ..  .. | 17 |
| Mudstone, black, with thin shaly bands  ..  ..  ..  .. | 9 |

As previously stated, Hudson and Cotton recorded *H. subglobosum* from a locality (Edale 148) in the east bank of Grinds Brook [1221 8616] at Grindsbrook Booth. *Posidoniella* cf. *variabilis* was found 3 ft above the base of this outcrop during the present survey. Hudson and Cotton also record *H. subglobosum* s.l. from the top of an outcrop (Edale 176) on the south bank of the Noe [1370 8545] (see p. 191). The following section of beds of $H_{1a}$ age is present in the River Noe [1475 8597], 710 yd S.15°E. of Clough Farm and for a further 110 yd downstream:

|  | ft |
|---|---|
| Shale, with thin ironstone nodules .. | 4 |
| Mudstone, black, with ferruginous shale bands .. .. .. .. | 2 |
| *Not exposed* .. .. .. about | 2 |
| Shale .. .. .. .. .. | 2 |
| *Not exposed* .. .. .. about | 2 |
| Shale with large carbonate bullions; *Aviculopecten sp.*, cf. *H. subglobosum* .. .. .. .. | 2 |
| Shale .. .. .. .. .. | 2 |
| *Not exposed* .. .. .. .. | 2 |
| Shale, with black mudstone bands and large carbonate bullions at base .. | 2 |
| Shale .. .. .. .. .. | 5 |
| Shale, ferruginous, with large carbonate bullions .. .. .. | 0¾ |
| Mudstone, soft .. .. .. | 1 |
| Shale .. .. .. .. .. | 2½ |
| Mudstone, with soft black bands .. | 1 |
| Shale, ferruginous, with thin nodular ironstone bands and large carbonate bullions containing *H. subglobosum* | 0¾ |
| Shale .. .. .. .. .. | 6 |
| *Not exposed* .. .. .. about | 4 |
| Shale, black .. .. .. .. | 2 |
| Shale, silty ferruginous, with ironstone lenses and carbonate bullions containing *H. subglobosum* .. | 0¼ |
| Shale .. .. .. .. .. | 2 |

About 70 yd downstream [1477 8611] the following section is exposed in the north bank:

|  | ft |
|---|---|
| Shale, black in lower part, with 2-in silty ironstone at base .. .. | 3¼ |
| Shale, black towards base, with rare disseminated plant debris and very thin lenticular coal partings near top; *H. subglobosum* and *Homoceras sp.* about 2½ ft above base .. | 6 |
| Shale, black, with interbedded ferruginous silty shale and thin nodular ironstone bands .. .. .. | 2¼ |

Some of the *Homoceras sp.* recorded in this section resemble the *Homoceras sp. nov.*

figured by Ramsbottom, Rhys and Smith (1962, pl. 6, fig. 11) from between the *H. beyrichianum* and *H. subglobosum* faunas.

No outcrops containing *H. beyrichianum* were proved in the Edale valley during the present work.

The *Hudsonoceras proteus* ($H_{2a}$), *Homoceras undulatum* ($H_{2b}$) and *Homoceratoides prereticulatus* ($H_{2c}$) zones are represented by over 30 ft of shale and mudstone, with sporadic bullion horizons. A good section can be seen in a series of outcrops in Crowden Brook for a distance of over 50 yd downstream from a point [1025 8563], 640 yd N. 3°E. of Highfield. Hudson and Cotton record *Homoceras smithii* and *Hudsonoceras proteus* from this section (Edale 57). The higher beds containing the *Homoceras undulatum* and *Homoceratoides prereticulatus* horizons are apparently not exposed in Crowden Brook.

The lowest fauna is present in Grinds Brook [1222 8626] near Grindsbrook Booth, in the following section:

|  | ft |
|---|---|
| Shale, black .. .. .. .. | 3 |
| Carbonate bullions with *Hd. proteus* | — |
| Shale .. .. .. .. .. | 3 |
| Carbonate bullions with *H. smithii* .. | — |
| Shale, with rare thin bands of sandy clay; *Caneyella semisulcata* near base .. .. .. .. .. | 10 |

The bullion band with *Hd. proteus* was also proved about 80 yd upstream (Edale 152) by Hudson and Cotton (ibid., pp. 6–7). The higher $H_2$ faunas have not been proved in Grinds Brook and the whole zone is virtually unexposed in the banks of the Noe south-east of Nether Booth.     G.D.G.

An important section in $H_2$ is that [0957 8554] on the left bank of the Noe at the junction with Grain Clough:

|  | ft |
|---|---|
| Shale, pale grey 'sulphurous' with ferruginous bands at 7 ft above base | 17½ |
| Shale, soft black, with bullion at base; *Anthracoceras* or *Dimorphoceras sp.*, *Homoceras undulatum* .. .. | 2 |
| Shale, grey 'sulphurous' .. .. | 2½ |
| Shale, soft black; *Caneyella semisulcata*, *Anthracoceras* or *Dimorphoceras sp.*, *Homoceras* cf. *smithii* | 1½ |

ft

Shale, grey 'sulphurous' .. .. 7

Shale, soft black; *C. semisulcata, H.*
    *smithii, Hudsonoceras proteus* .. $1\frac{1}{2}$

Beds a few feet higher in the succession are exposed in a meander [0953 8551] about 50 yd upstream from the last section. Here (Edale 14) Hudson and Cotton (ibid., pp. 6–7) record *Homoceratoides* cf. *prereticulatus.*

In Whitemoor Sitch a $H_{2c}$ fauna was obtained by Hudson and Cotton (ibid., pp. 6–7) from the right bank of the stream (Edale 100) [1045 8428]: *Dunbarella* cf. *mosensis* (de Koninck), *Homoceras sp.* cf. *H. diadema* (Beyrich) group, *Homoceratoides prereticulatus.* $H_2$ faunas are also present in two places in Harden Clough. The lower of these [1244 8446] occurred in a 6-in band of dark, partly decalcified shale, which yielded *Homoceras smithii* and *Hudsonoceras proteus.* Some 30 ft higher in the succession and 60 yd farther upstream [1246 8440] the top of the *Homoceras* stage is marked by a band yielding *Posidonia sp., Anthracoceras* or *Dimorphoceras sp., Homoceras sp., Homoceratoides* aff. *prereticulatus.* I.P.S.

The *Reticuloceras circumplicatile* Zone ($R_{1a}$) consists mainly of shale with sporadic thin ferruginous siltstone and ironstone bands, about 80 ft thick. *Homoceras magistrorum* has not been proved in the Edale valley. In Crowden Brook the lowest exposure of the zone [1025 8566] above Upper Booth consists of shale, 17 ft thick, with carbonate bullions near the base, containing *Caneyella sp., Dunbarella sp. nov., Anthracoceras* or *Dimorphoceras sp., Homoceras sp., Reticuloceras sp.* of *R. circumplicatile* group and *R.* aff. *todmordenense.* In the west bank, about 50 yd upstream, but apparently little higher in the succession, 6 ft of shale with a 4-in ironstone band near the top contain *Caneyella sp., Dunbarella sp., Posidonia sp.,* an orthocone nautiloid, *Homoceras* cf. *henkei, Reticuloceras* cf. *pulchellum* and *Reticuloceras sp.* Higher beds are seen in the east bank a few yards farther north [1022 8572] and consist of 12 ft of shale with silty ironstone bands containing an orthocone nautiloid, *Homoceras sp.* of *H. eostriolatum* group and *Reticuloceras sp.*

A good section of lower $R_{1a}$ strata is present in the east bank of Grinds Brook [1214 8635] 25 yd E. of Grindslow House:

ft

Shale, soft .. .. .. .. 5

Shale, hard splintery fissile .. .. 4

Shale, soft black fissile .. .. 7

Shale, with thin ironstone bands and very thin pale brown clay bands; *Dunbarella sp., Posidoniella sp., Homoceras sp.* and *Reticuloceras sp.* of *R. ? todmordenense* group in a band 4 ft above base .. .. 6

Siltstone, ferruginous, with ironstone nodules; *Posidoniella sp., Reticuloceras sp.* .. .. .. .. $0\frac{1}{4}$

Shale, with thin ferruginous siltstone bands; *Dunbarella sp., Anthracoceras* or *Dimorphoceras sp., Homoceratoides sp., R.* cf. *pulchellum, R.* cf. *subreticulatum* and *Cypridina sp.* in a band 3 ft above base .. .. 6

Shale, soft black at base, becoming hard with thin ferruginous siltstone bands towards the top .. .. 15

Upper $R_{1a}$ and possible lower $R_{1b}$ beds are almost continuously exposed for 50 yd upstream from a point [1215 8639] 60 yd N.W. of Grindslow House; they consist of over 30 ft of soft shale containing rare *Dunbarella sp.* and with sporadic ferruginous siltstone nodules and ironstone bands up to 3 in thick yielding *Dunbarella* cf. *speciosa, Posidonia ?, Promytilus ?* and *Reticuloceras sp.*

Three outcrops of $R_{1a}$ age have been proved in Oller Brook. In the west bank [1285 8611] just north of Ollerbrook Booth 9 ft of shale contain *Dunbarella sp.* and *Reticuloceras sp.* of *R. ? todmordenense* group in a band 2 ft above the base. Some 30 yd upstream [1285 8614], 4 ft of shale yield *Dunbarella sp., R.* cf. *todmordenense* and *R.* cf. *circumplicatile* from a band 1 ft above the base, and an exposure a farther 15 yd upstream on the east bank [1285 8615] shows 4 ft of shale with *Dunbarella sp.* and *Reticuloceras* cf. *paucicrenulatum.*

No fauna of $R_1$ age has been proved from the River Noe south-east of Nether Booth, where the banks are largely obscured by head and landslip. The only fossiliferous beds exposed are on the south bank [1487 8612], where indeterminate goniatites occur in shale with carbonate bullions. G.D.G.

In the upper reaches of the River Noe, $R_{1a}$ faunas are present at several places, the zone being here about 75 ft in thickness. The lowest exposure (Edale 9) [0939 8566], above Lee House, yielded *Dunbarella* cf. *rhythmica* (Jackson), *Anthracoceras* or *Dimorphoceras sp.*, *Homoceras sp.* and *Reticuloceras* aff. *subreticulatum* [crenulate flanks]. Some 40 yd farther upstream (Edale 8) Hudson and Cotton (ibid., p. 5) record *Dunbarella* cf. *carbonaria* (Hind), *D. rhythmica*, *D. sp.*, *Posidoniella minor* (Brown), *P. minor* Hind non Brown, *Dimorphoceras sp.*, *Homoceras henkei* group, *H. sp.* cf. *H. striolatum* group.

Faunas here referred to the $R_{1a}$ Zone were found by Hudson and Cotton (ibid., p. 5) in Whitemoor Sitch, in shales with large bullions, about 70 ft thick: (a) (Edale 100a) [1039 8424], nearly due south of the mouth of Cowburn Tunnel; *Dunbarella sp.*, *Dimorphoceras splendidum?* (Brown), *Homoceras sp.*, *Reticuloceras inconstans* (Phillips) group, *R.* cf. *pulchellum*: (b) (Edale 100b) 40 yd upstream from the last locality: *Homoceras sp.*, *Reticuloceras adpressum*, *R. paucicrenulatum*.                                    I.P.S.

The *Reticuloceras nodosum* ($R_{1b}$) Zone, about 40 ft in thickness, consists of shale and mudstone, with distinct thin siltstone beds and with locally abundant groups of thin ironstones. An outcrop in the west bank of Crowden Brook [1021 8575], above Upper Booth, contains a low $R_{1b}$ or possibly high $R_{1a}$ fauna:

|                                                                                                           | ft | in |
|-----------------------------------------------------------------------------------------------------------|----|----|
| Shale, soft black        ..        ..        ..                                                          | 2  | 0  |
| Siltstone, sandy, hard blocky ferruginous    ..      ..      .. 4 in to                                  |    | 8  |
| Shale with soft black bands and ironstone bands $\frac{1}{2}$ in to 2 in thick                          | 6  | 0  |
| Shale, black in lower part ..        ..                                                                  | 7  | 6  |
| Clay, pale brown sandy    ..        ..                                                                   |    | 0$\frac{1}{2}$ |
| Shale, black, with ferruginous nodules; *Homoceratoides divaricatus* and *Reticuloceras spp.* ( *?nodosum* and *todmordenense* groups)        .. | 1  | 0  |
| Shale with rare disseminated plant debris near base ..        ..        ..                              | 5  | 6  |
| Shale with abundant ferruginous siltstone bands $\frac{1}{4}$ to $\frac{3}{4}$ in thick        ..      | 3  | 0  |

In Grinds Brook [1212 8647], near Grindslow House, *R. nodosum* has been found in ironstone nodules contained in 8 ft of shale with abundant very thin ironstone bands, probably from near the top of the zone.

The widest range of $R_{1b}$ faunas has been proved in Oller Brook. In the east bank [1281 8626], 760 yd N.E. of Edale church, 8 ft of mudstone with shaly bands contain *Dunbarella speciosa* and *Reticuloceras sp.* ( *? dubium* group) in the lowest 2 ft, probably the basal horizon of the zone. In the east bank 30 yd upstream [1281 8629], 8 ft of shale, hard and platy near top, contain *Caneyella squamula* (Brown), *D. speciosa*, *Anthracoceras* or *Dimorphoceras sp.*, *Homoceras?* and *Reticuloceras sp.* in the lower 4 ft. About 15 yd farther upstream on the same side [1281 8630], the following section is exposed:

|                                                                                                                               | ft |
|-------------------------------------------------------------------------------------------------------------------------------|----|
| Shale with a thin ferruginous siltstone band in the lower part        ..        ..                                          | 4  |
| Shale    ..        ..        ..        ..        ..                                                                          | 1$\frac{1}{2}$ |
| Mudstone, soft black; *Posidoniella sp.*, *Anthracoceras* or *Dimorphoceras sp.*, *Ht. divaricatus* and *Reticuloceras sp.* (*nodosum* group)        ..        ..        .. | 1  |
| Mudstone, black, alternating with thin bands of hard ferruginous shale        ..                                            | 1$\frac{1}{2}$ |
| Shale    ..        ..        ..        ..        ..                                                                          | 2  |
|                                                                                                                               | G.D.G. |

In the upper reaches of the Noe, an $R_{1b}$ fauna was obtained by Hudson and Cotton (ibid., p. 5) from a point (Edale 6) [0932 8572], $\frac{1}{4}$ mile W.N.W. of Lee House: *Coelonautilus sp.* and *Reticuloceras inconstans* group. The zone is about 65 ft thick in this vicinity.

The following section was measured on the hillside [1000 8470] near The Orchard:

|                                                                                                                        | ft |
|------------------------------------------------------------------------------------------------------------------------|----|
| Base of Mam Tor Beds        ..        ..                                                                              | —  |
| Shale, with thin sandstone bands in top 6 ft        ..        ..        ..        ..                                 | 18 |
| Shale, soft ferruginous and decalcified                                                                               | 0$\frac{3}{4}$ |
| Mudstone, grey    ..        ..        ..                                                                              | 13 |
| Shale, dark; *Caneyella* cf. *rugata*, *Dunbarella rhythmica*, orthocone nautiloids, *Dimorphoceras sp.*, *Homoceras* cf. *striolatum*, *Hudsonoceras ornatum*, *Reticuloceras* cf. *moorei* Bisat and Hudson, *R.* cf. *nodosum* and mollusc spat; a 15-in band of grey 'sulphurous' mudstone with *Dunbarella sp.* at 6 in from base    ..        ..        ..        .. | 3  |
| Mudstone, grey; *Dunbarella sp.* at 3 in from top        ..        ..        ..        ..                            | 6  |

|  | ft |
|---|---|
| *Not exposed* .. .. .. .. | 5 |
| Mudstone, grey .. .. .. | 10 |

The fauna is indicative of the $R_{1b}$ Zone though there is room for some beds of $R_{1c}$ age at the top of the section.

In Whitemoor Sitch the $R_{1b}$ Zone consists of some 70 ft of shales with some bands of bullions. A fauna indicative of the zone was collected [1029 8412] 400 yd S.S.W. of the Cowburn Tunnel entrance: *Caneyella* cf. *squamula, Dunbarella rhythmica, Posidonia sp. nov., Promytilus foynesianus* (Baily), *Homoceras sp., Reticuloceras* cf. *moorei, R.* cf. *nodosum*, mollusc spat. Another locality nearby, 25 yd downstream, yielded *Homoceras sp.* and *R. nodosum* group.       I.P.S.

Beds of the *Reticuloceras reticulatum* ($R_{1c}$) Zone within the Edale Shales appear to be less than 30 ft thick on the northern side of the Edale valley, and consist of shales and mudstones becoming increasingly silty towards the top, and with fairly abundant thin ironstone and ferruginous siltstone bands. A composite section from Crowden Brook [1021 8578] some ¼ mile N. of Upper Booth and for a further few yards upstream is:

|  | ft | in |
|---|---|---|
| Base of Mam Tor Beds .. .. | — | — |
| Siltstone, flaggy .. .. .. | 1 | 0 |
| Shale, silty, with micaceous laminae and rare thin silty ironstone bands | 9 | 6 |
| Ironstone .. .. .. .. | | 1 |
| Shale, hard slightly silty, with ironstone nodules near base .. .. | 1 | 0 |
| Ironstone .. .. .. .. | | 1½ |
| Shale, hard slightly silty, with sporadic thin ironstone bands .. .. | 4 | 6 |
| Siltstone, sandy ferruginous .. | | 8 |

Similar sections can be seen in Grinds Brook [1208 8649], in Oller Brook [1281 8634] and in Lady Booth Brook [1415 8622] immediately south of the small dam. The following section in a landslip-scar on the south side of the Noe [1500 8604], 750 yd S.E. of Clough Farm is in the highest beds of the Edale Shales:

|  | ft | in |
|---|---|---|
| Base of Mam Tor Beds .. .. | — | — |
| Shale, silty, with thin ironstone bands in upper part .. .. | 4 | 0 |
| Carbonate bullions with indeterminate goniatites .. .. | — | — |

|  | ft | in |
|---|---|---|
| Shale, silty .. .. .. .. | 1 | 6 |
| Ironstone, silty .. .. .. | | 4 |
| Shale, silty .. .. .. .. | 2 | 9 |
| Shale, silty, with sporadic thin ironstone bands .. .. .. | 3 | 3 |
| Shale .. .. .. .. .. | 5 | 0 |

G.D.G.

In the upper reaches of the Noe an $R_{1c}$ fauna with *Reticuloceras davisi* group and *R. reticulatum* was found in about 3 ft of shales with thin sandstones lying about 15 ft below the base of the Mam Tor Beds [0925 8576], above Lee House. The whole of $R_{1c}$ within the Edale Shales is here more silty than the beds below, and thin sandstones are quite common, particularly as the Mam Tor Beds are approached; the thickness of these beds is about 15 ft. In Whitemoor Sitch, $R_{1c}$ faunas were not found though there is room for some 80 ft of beds of this zone within the Edale Shales. The expansion is explained by diachronism of the base of the Mam Tor Beds.       I.P.S.

*Alport and Ashop valleys.* The Alport Borehole [1360 9105], described by Hudson and Cotton (1943), provides the most complete single section of the Edale Shales in the district. The uppermost 129 ft or so of beds in the borehole also crop out nearby. A summary of the record with revised zonal boundaries and re-determination of certain fossils, is given below:

|  | Thickness | Depth |
|---|---|---|
|  | ft | ft |
| Lower *Reticuloceras* zones ($R_1$) | | |
| No core .. .. .. | 100 | 100 |
| Calcareous shale; *H. henkei* and *R. circumplicatile* group 108 ft to base .. .. | 10 | 110 |
| *Homoceratoides prereticulatus, Homoceras undulatum* and *Hudsonoceras proteus* zones ($H_2$) | | |
| Shale with hard calcareous bands and bullions; *Ht.* aff. *prereticulatus* 110–112 ft; *H. eostriolatum* 116–127 ft; *H.* cf. *undulatum, H.* cf. *smithii* and *Hd. proteus* 127–139 ft .. .. .. | 29 | 139 |

|  | Thick-<br>ness<br>ft | Depth<br><br>ft |
|---|---|---|

*Homoceras beyrichianum* and *Homoceras subglobosum* zones ($H_1$)
Shales with hard calcareous bands and bullions; *H. beyrichianum* 190–196 ft; *Homoceras sp. nov.* 196–199 ft; *H. subglobosum* 200–206 ft (and in bullions at 200 ft)    67    206

*Nuculoceras nuculum* Zone ($E_{2c}$)
Shale and mudstone with hard calcareous siltstone bands; *E. bisulcatum* and *N. nuculum* 216–222 ft; *Cravenoceratoides fragilis, E. bisulcatum* and *N. nuculum* 228–231 ft; *Ct. fragilis* group, *E. bisulcatum* and *N. nuculum* [originally recorded as *Ct. nititoides,* now equated with lowest *N. nuculum* horizon (see Ramsbottom, Rhys and Smith 1962, p. 130)] 250–253 ft; *P. aff. vetusta* and *N. stellarum* 294–295 ft    89    295

*Cravenoceratoides nitidus* Zone ($E_{2b}$)
Shale and mudstone, with numerous thin calcareous siltstone and cherty limestone beds to 306 ft; productoids, chonetoids, athyroids, *Ct. nititoides, E. rostratum* and trilobites 301–306 ft. Shales with calcareous siltstones to 360 ft; *Cravenoceras sp.* 312–314 ft and *Ct. edalensis* 354–360 ft    65    360

*Eumorphoceras bisulcatum* Zone ($E_{2a}$)
Platy pyritous shale and mudstone with calcareous siltstone bands to 449 ft, argillaceous limestones with thin shale partings to 463 ft; calcareous mudstones with thin limestones at top to 476 ft; *E. bisulcatum* 367–

371 ft; *E. bisulcatum* and *Kazakhoceras scaliger* 399–408 ft; *Chaenocardiola footii, Cravenoceras cowlingense* and *E. bisulcatum* s.l. 453–476 ft    ..    ..    ..    116    476

*Cravenoceras malhamense* Zone ($E_{1c}$)
Silty shale with calcareous siltstone bands to 503 ft; calcareous mudstone with thin silty limestone bands to 533 ft; orthotetoid and productoid fragments, *Actinopteria sp., Reticycloceras sp., C. malhamense, K. scaliger* and *Ceratiocaris sp.* 503-533 ft    ..    ..    ..    ..    57    533

*Eumorphoceras pseudobilingue* Zone ($E_{1b}$)
Silty shale at top, passing down into calcareous shale and mudstone with thin silty limestone bands; *E. pseudobilingue* 594–599 ft; *E. sp.* 660–662 ft[1] ..    ..    ..    130    663

*Cravenoceras leion* Zone ($E_{1a}$)
Hard, calcareous, slightly silty and locally pyritous shale with abundant thin quartzose siltstone bands and sporadic thin silty limestone bands; bivalves, orthocone nautiloids, rare goniatites, mainly dimorphoceratids and plant debris to 1054 ft. [The "Alport Crowstones" of Hudson and Cotton (ibid., p. 168). Indications of faulting in core suggested that up to 75 ft of strata may have been repeated]. Shale and mudstone with calcareous and silty bands and thin silty limestone bands in the lower part, and containing *Eumorphoceras medusa* 1064 ft; *C. leion* and *Lyrogoniatites* cf. *tonksi*

---

[1]*Eumorphoceras* specimens from this level are poorly preserved, but may well belong to the $E_{1a}$ Zone. But see Trewin, *Palaeontology,* **13** (1970), p. 44. W.H.C.R.

| | Thickness ft | Depth ft |
|---|---|---|

1071 ft; *Cravenoceras sp.* and *Eumorphoceras bisati* 1072–1075 ft; *Cravenoceras sp., E.* cf. *E.* form A Moore, *L. tonksi* and *Sudeticeras sp.* 1075–1078 ft; *C. leion* early form and *Girtyoceras* cf. *waitei* Moore 1098–1099 ft. [The faunal list for the lower part of the $E_{1a}$ zone and the revised depth of the base of the Namurian in the borehole are based on Bisat (1950, p. 14)] .. .. 436  1099 Viséan (see p. 115).

The lowest Edale Shales at outcrop in the Alport and Ashop valleys are two small inliers of $H_2$ beds. In the more northerly, an abandoned river bank [1401 9029], 1010 yd S. 32°E. of Alport Castles Farm, has yielded *Homoceras sp.* in weathered shale 4 ft above the base. Hudson and Cotton (1943, p. 159) record *H. smithii* and *Hd. proteus* and 4 ft higher, *H.* cf. *undulatum* from this locality (Alport 16)[1]. Nearly half a mile to the south, the southerly inlier is exposed in the east bank of the Alport [1412 8969] east-north-east of Hayridge Farm:

| | ft |
|---|---|
| Shale, black with thin hard ferruginous bands; *Aviculopecten sp.*, orthocone nautiloid, *H.* cf. *henkei*, cf. *Homoceratoides varicatus* and *Reticuloceras sp.* in top 2 ft; *Posidonia sp.* and *Homoceras sp.* (possibly the *H. magistrorum* band) in the bottom 2 ft .. .. .. | 12 |
| Shale, with thin hard ferruginous bands; *Ht. prereticulatus* 2 ft above base .. .. .. .. .. | 16 |

The exact relationships in this exposure are somewhat complicated by contortion and minor faulting. It would appear to be the locality (Alport 13) from which Hudson and Cotton recorded *H.* cf. *eostriolatum*. Their record of *H. eostriolatum* in bullions on the opposite side of the river (Alport 13a) has not been confirmed, probably due to the presence

of a wall across part of the outcrop, but *Caneyella sp., Dunbarella rhythmica* and *Anthracoceras* or *Dimorphoceras sp.*, suggestive of an $R_{1a}$ age, have been obtained from a nearby outcrop of shale [1408 8965]. A few yards downstream [1411 8963], *Aviculopecten sp., Caneyella sp., Dunbarella sp., H.* cf. *henkei* and *Reticuloceras sp. (circumplicatile* group) occur in shale. It is apparent that a change in the course of the river has altered the exact arrangement of outcrops since Hudson and Cotton examined them, and the lower horizons of $H_2$ age have been largely obscured. In the east bank [1415 8962] an exposure of shale contains *Ht.* cf. *divaricatus*, suggesting an upper $R_{1a}$ age. Shale, about 27 ft thick, now largely weathered, forms a roadside bank (Alport 1) [1438 8951], near Gillethey Farm; the section (Bisat and Hudson 1943, p. 390) is as follows:

| | ft |
|---|---|
| Shale .. .. .. .. .. | — |
| Bullion bed; *R.* aff. *pulchellum* and variants, *R. sp. nov.* [crenulate form] | — |
| Shale .. .. .. .. .. | 6 |
| Bullion bed; *R.* aff. *pulchellum* and *R. sp.* .. .. .. .. | — |
| Shale .. .. .. .. .. | 4 |
| Shale with *H.* cf. *henkei, Homoceratoides sp., R.* cf. *coronatum* (Foord) and *R. sp. nov.* [non-crenulate form] | $0\frac{1}{4}$ |
| Shale .. .. .. .. .. | 4 |
| Shale, with *H. henkei angustum* Schmidt, *H. sp. nov., Ht. varicatus* group, *R. sp.* .. .. .. | $0\frac{1}{4}$ |
| Shale .. .. .. .. .. | 5 |

Shale, about 45 ft thick, crops out between 100 yd and 200 yd south of this exposure in the east bank of the Ashop [1438 8938]. Hudson and Cotton (1943, pp. 155–6) obtained from this exposure (Alport 3) an $R_{1a}$ fauna consisting of *H. henkei, Ht.* cf. *varicatus* and *R. sp.* cf. *R. inconstans* group from 6 ft above the base and a possible low $R_{1b}$ fauna from near the top of the exposed succession.

Adjacent to the more northerly inlier of $H_2$ strata, in the west bank of the Alport [1405 9012], $\frac{1}{3}$ mile N.N.E. of Hayridge Farm, shale 10 ft thick contains *Caneyella sp.* [juv.], *Dunbarella sp.* and *Anthracoceras* or *Dimorphoceras sp.* in a band $4\frac{1}{2}$ ft above the base and *R.* cf. *circumplicatile* and *R.* cf.

---

[1]Numbers preceded by the word 'Alport' indicate locality numbers of Hudson and Cotton (1943).

o

*pulchellum* in bullions 6 ft above the base. On the northern side of this inlier, Hudson and Cotton record an upper $R_{1a}$ fauna. This outcrop (Alport 17) [1402 9035], $\frac{1}{2}$ mile S.E. of Alport Castles Farm, has been extended by river action, but the older exposure is now partly obscured by alluvium, and the section, complicated by faulting, but suggestive of an $R_{1b}$ age, is:

|  | ft |
|---|---|
| Shale, black; *Ht. divaricatus* and *Hudsonoceras ornatum* 7 ft above base; |  |
| *Reticuloceras sp.* 9$\frac{1}{2}$ ft above base .. | 15 |
| Shale; *Promytilus sp.* in upper 1 ft .. | 18 |

About $\frac{1}{4}$ mile farther south on the west bank [1408 8998], and intermittently exposed for a further 170 yd downstream, are shales with soft black bands, rare ironstone bands and bullions, from two of which (Alport 14, 15) Hudson and Cotton obtained *Hd. ornatum*. An almost fully exposed sequence of upper $R_{1b}$ and $R_{1c}$ strata up to the base of the Mam Tor Beds is present in the lower reaches of Swint Clough (Alport 20–22) for a distance of about 200 yd upstream from a point [1347 9106] some 50 yd west of Alport Castles Farm. Bisat and Hudson (1943, p. 389) recorded:

|  | ft |
|---|---|
| Shale .. .. .. .. .. | 13 |
| Shale; *R. reticulatum* s.s. .. .. | 0$\frac{1}{2}$ |
| Shale, sparsely fossiliferous .. .. | 12 |
| Shale; *Hudsonoceras ornatum, R.* aff. *moorei, Reticuloceras* cf. *regularum* Bisat and Hudson, *R.* cf. *stubblefieldi* .. .. .. .. .. | 1 |
| Shale, sparsely fossiliferous .. .. | 15 |
| Bullion band; *Homoceras striolatum* early form, *R. sp* .. .. .. | — |
| Shale .. .. .. .. .. | 13 |
| Shale; *Dunbarella sp.* .. .. .. | 0$\frac{1}{2}$ |
| Shale .. .. .. .. .. | 8 |
| Bullion band; *Homoceras spiraloides* Bisat and Hudson, *H. striolatum* early form, *R. nodosum* .. .. | — |
| Shale .. .. .. .. .. | 6 |

In the Ashop valley, about 35 ft of shale with thin ironstone bands in the upper part underlies Mam Tor Beds [1410 8942], 170 yd S.S.W. of the junction of the Alport and Ashop rivers, and the same strata are exposed in a number of other outcrops in the south bank of the river downstream as far as Rowlee Bridge. An exposure (Alport 8), 120 yd N. of this bridge on the north bank [1495 8918], showing 27 ft of rather silty shale with thin silty ironstone bands, and containing *Caneyella rugata, Dunbarella sp.*, indeterminate goniatites and mollusc spat from 12 ft above base, is recorded by Hudson and Cotton (1943, pp. 153–4) as being basal Mam Tor Beds, but in the absence of distinct sandstones, is here assigned to the Edale Shales.

*Fulwood Holmes.* Edale Shales occur along the River Noe north of Hope as a narrow inlier near Fulwood Holmes. The lowest exposed strata are from near the base of the *H. subglobosum* Zone; they crop out in the east bank of the river [1666 8537], west-south-west of Harrop Farm:

|  | ft |
|---|---|
| Shale .. .. .. .. .. | 4 |
| Shale, with ferruginous bands: a bullion band containing *Posidoniella sp., Homoceras subglobosum* and *H. sp.* at the top .. .. .. | 9$\frac{1}{2}$ |
| Shale, with large bullions containing *H. subglobosum* .. .. .. | 1$\frac{1}{2}$ |
| Shale with *Posidoniella sp.* .. .. | 3$\frac{1}{2}$ |

*H. subglobosum* also occurs in shales with calcareous mudstone bands about 60 yd downstream in the same bank [1662 8537], and also, together with *Aviculopecten sp.*, in similar strata on the opposite bank farther upstream [1662 8548]. About 100 yd farther upstream on the west bank [1659 8554], 11 ft of shale with calcareous bullions contain *Aviculopecten sp., Posidoniella sp., Reticycloceras sp.* and *H. beyrichianum.* Shale with ferruginous bands and calcareous bullions containing poorly preserved *Homoceras sp.* is exposed in the east bank [1662 8526] and may represent either the *H. beyrichianum* Zone or be of lowest $H_2$ age.

The *Hudsonoceras proteum–Homoceras smithii* and *Homoceras undulatum* horizons have not been proved, but are probably present in river bank exposures 120 to 180 yd S. of Normans Farm and also 200 yd W. to W.N.W. of Fulwood Holmes. The highest $H_2$ horizon, represented by *Aviculopecten sp., Anthracoceras* or *Dimorphoceras sp., Ht. prereticulatus* and fish debris, occurs in shales and bullions at the base of an outcrop in the east bank [1670 8500]. At 150 yd farther south, in the east bank of the river [1667 8487], 10 ft of shale with thin black mudstone bands contains *D. rhythmica* and *Reticuloceras sp.* of *R. circumplicatile* group in the lower beds.

The best outcrops of $R_{1b}$ strata are adjacent to the railway-bridge over the river near Normans Farm. In 7 ft of shale on the west bank [1663 8569] south-west of the bridge, *Caneyella?*, *D.* cf. *rhythmica*, *Posidonia sp.*, *Homoceras?* and *Reticuloceras sp.* (with $R_{1b}$ affinities) occur, and in similar strata on both banks immediately north-east of the bridge, *Hd. ornatum* and *Reticuloceras sp.* (*nodosum* group) are present in the lowest beds.

The presence of $R_{1c}$ faunas has not been proved, but in the river banks [1667 8585] at Normans Farm, are good exposures of silty shales and mudstones becoming increasingly silty upwards, and with thin ironstone bands concentrated into certain parts of the succession. This sequence is very similar to the strata of $R_{1c}$ age exposed in Crowden Brook and Swint Clough.                     G.D.G.

*Hope valley.* In the western part of the Hope valley the Edale Shales form a broad tract on the northern side. Much of the outcrop is concealed beneath head, and exposures are for the most part confined to the banks of Peakshole Water and smaller streams.

The Castleton Borehole [1410 8293] proved the following section of the lowest part of the Edale Shales, resting on shales of $P_2$ age:

| | Thickness ft in | Depth ft in |
|---|---|---|
| Head .. .. .. | 15 0 | 15 0 |
| *Cravenoceras malhamense* Zone ($E_{1c}$) | | |
| Mudstone (little recovered) | 1 6 | 16 6 |
| Mudstone, grey, soft and fissile .. .. .. | 4 2 | 20 8 |

Mudstone, hard silty calcareous to 22 ft, hard brownish below; *Caneyella membranacea*, *C.* cf. *membranacea*, *Posidonia* cf. *corrugata* and dimorphoceratid indet. [bilobate growth-lines] at 20 ft 9 in to 23 ft 4 in; *C. membranacea* and

| | Thickness ft in | Depth ft in |
|---|---|---|
| *Cravenoceras malhamense* at 23 ft 9 in; *C. membranacea*, *P.* cf. *corrugata* and *Rayonnoceras sp.* at 24 ft 1 in to 24 ft 10 in .. .. | 4 2 | 24 10 |
| Mudstone, hard slightly silty; *Caneyella membranacea*, *C.* cf. *membranacea*, *P.* cf. *corrugata*, *Cravenoceras malhamense*, *C.* cf. *malhamense*, dimorphoceratid indet. [bilobate growth-lines], fish debris indet. | 4 8 | 29 6 |
| *Eumorphoceras pseudobilingue* Zone ($E_{1b}$) | | |
| Mudstone, soft grey; pyritous with fucoid markings below 35 ft; scales of *Elonichthys sp.*, *Rhadinichthys sp.*, palaeoniscids indet., *Rhabdoderma?* .. | 20 6 | 50 0 |
| Mudstone, grey silty, slightly calcareous; *P.* cf. *corrugata*, *P.?* and fish debris in phosphatic nodules 50 ft to 50 ft 9 in; sponge spicules, *P.* cf. *corrugata*, dimorphoceratid [bilobate growth-lines], *Eumorphoceras pseudobilingue*, *E. pseudobilingue* C, *Rhadinichthys sp.* 51 ft to 54 ft 10 in .. | 4 10 | 54 10 |
| Mudstone, grey, with guilielmites (cores broken) to 58 ft, silty below; *C.* aff. *membranacea*, *P.* aff. *corrugata*, goniatite indet., fish indet. 56 ft to 60 ft 3 in; *P.* cf. *corrugata*, *Kazakhoceras?* [with nodose venter] 61 ft to 62 ft 3 in | 8 2 | 63 0 |
| *Cravenoceras leion* Zone ($E_{1a}$) | | |

Mudstone, grey slightly bituminous with fish debris; slightly more

| | Thickness | | Depth | |
|---|---|---|---|---|
| | ft | in | ft | in |

silty below 72 ft; calcareous and silty below 86 ft 6 in; *Edestodus sp. nov.* at 65 ft; *P. cf. corrugata, P. sp.*, crinoid columnals, *Rhadinichthys sp.* [scale], indet. palaeoniscid scale at 68 ft 6 in to 88 ft 6 in .. .. 26 0 89 0

Mudstone, calcareous and bituminous; *P. cf. corrugata, P. sp.*, dimorphoceratid, indet., goniatites and palaeoniscid scales 89 ft to 94 ft 9 in; *P. sp.* and *Cravenoceras sp.* 95 ft 6 in; *P. cf. corrugata, P. sp., Cravenoceras leion, C. cf. leion*, dimorphoceratid indet. [bilobate growth-lines], *Eumorphoceras sp.* 97 ft 6 in to 98 ft 7 in .. .. 9 7 98 7

Siltstone .. .. .. 11 99 6

Mudstone, dark; *Caneyella membranacea, Cravenoceras cf. leion* .. 1 99 7

Viséan (see p. 111). .. — — — —

The thinning of the lower zones adjacent to the limestone massif is very apparent in this section, $E_{1a}$ and $E_{1b}$ being together only 70 ft thick, as compared with 285 ft at Edale.

The Hope Cement Works Borehole (Fearnsides and Templeman 1932, pp. 100–5) provides the most complete section of the lower part of the Edale Shales in the Hope area. The site was at Salter Barn[1] (approximate ref. 1678 8228), and the hole, after passing through 16 ft of head, proved beds referred to the Edale Shales to 280 ft overlying Lower Carboniferous rocks (see p. 113). There was a $2\frac{1}{2}$-ft bed of limestone at $98\frac{1}{2}$ ft and several thinner calcareous bands were noted. The fossils from the borehole, collected by A. Templeman, have been re-examined with the following results:

*Nuculoceras nuculum* Zone ($E_{2c}$)

16 to about 26 ft: Mudstone, no core;

these beds have been placed in $E_{2c}$ on account of the position of the underlying *Ct. nititoides* fauna which bears a fairly constant relation to the *N. stellarum* band.

*Cravenoceratoides nitidus* Zone ($E_{2b}$)

About 26 to 77 ft: Mudstone, no core to 38 ft; mudstone, calcareous and platy shale; *Cravenoceratoides nititoides* at $38\frac{1}{2}$ ft; *Leiopteria longirostris*, orthocone nautiloid at 39 ft; *Mitorthoceras?* at 43 ft; *Posidonia corrugata gigantea, Cravenoceras* cf. *holmesi, Cravenoceratoides* cf. *nitidus, Eumorphoceras sp.* at 50 ft; *P. corrugata, Streblochondria?* at 55 ft; *P. corrugata, P. corrugata elongata* Yates, orthocone nautiloid, *Anthracoceras glabrum* Bisat at 58 ft; *P. corrugata elongata, Anthracoceras?* at 75 ft; *A.* cf. *glabrum* at 76 ft; *Posidoniella* aff. *vetusta* group at 77 ft.

*Eumorphoceras bisulcatum* Zone ($E_{2a}$)

77 to 125 ft: Mudstone, as above to 115 ft; platy shale below; *Eumorphoceras sp.*, fish debris at 86 ft; *Anthracoceras?* at 88 ft; *Posidoniella* aff. *vetusta* group, orthocone nautiloid at 92 ft; *Leiopteria longirostris, Posidoniella* aff. *vetusta* group, goniatite indet. at 101 ft 6 in; *Eumorphoceras bisulcatum, Kazakhoceras scaliger* at 111 to 114 ft; fish debris at 125 ft.

*Cravenoceras malhamense* Zone ($E_{1c}$)

125 to 211 ft: Mudstone with a 3-ft calcareous bed at 168 ft and "an earthy bituminous calcareous group" with thin limestones between 195 and 210 ft; fish debris at 180 ft; *Anthracoceras* or *Dimorphoceras sp.* at 195 ft; *A.* or *D. sp.* at 202 ft; *Cravenoceras malhamense, D. sp., Kazakhoceras scaliger* at $203\frac{3}{4}$ to 209 ft; *Posidonia sp.* at 211 ft.

*Eumorphoceras pseudobilingue* Zone ($E_{1b}$)

211 to about 238 ft: Shale with 9 ft of "hard, rather sandy ferruginous mudstone" at 237 ft; fish debris at 220 ft; *P. corrugata, Anthracoceras* or *Dimorphoceras sp.* at 230 ft; *Kazakhoceras?* at 237 ft.

*Cravenoceras leion* Zone ($E_{1a}$)

About 238 to 280 ft: Mudstone, hard calcareous and bituminous below 276 ft; *Posidonia?* at 274 ft; fish debris at 275 ft; *Obliquipecten costatus, Catastroboceras* cf. *quadratum* (Fleming), *Cravenoceras leion, Eumorphoceras sp.* at 277 to 280 ft. These

---

[1]Demolished, now part of cement works.

beds are underlain by shales of $P_2$ age (see p. 112).

The borehole showed little in the way of silty beds at the horizon of the 'Crowstones' in the $E_{1a}$ Zone of the Alport Borehole. The flaggy limestones developed in the $E_{1b}$ Zone in both the Edale and Alport boreholes were not present, though those at the *Cravenoceras malhamense* horizon ($E_{1c}$) were.

The lowest beds of the Edale Shales are seldom exposed, though an excavation at Messrs. Earle's Hope Cement Works [1658 8221] showed dark shale dipping steeply north-eastwards, and some 15 yd away from the Carboniferous Limestone. The exposure must lie close to the position of the trench at the entrance to Nun Low Quarry, which Jackson (1925, pp. 271–2) described as showing shale resting on "an irregular and channelled surface of limestone". Jackson further noted that "pockets and fissures of the limestone are also filled with shale".

I.P.S.

To the south-east of Hope, strata of $R_{1b}$ age were proved in temporary excavations at the sewerage works. Bullions containing *Reticuloceras* cf. *nodosum* and *R. stubblefieldi* occurred in shale at a depth of 12 to 15 ft [1764 8317]. A few yards farther west *Posidonia obliquata* (Brown) and *Reticuloceras sp.*, found in shale at a depth of 12 to 15 ft [1759 8319], are either of high $R_{1b}$ or low $R_{1c}$ age.

No surface outcrops containing $R_{1b}$ or earlier faunas have been proved in the vicinity of Hope. In the west bank of the Noe [1751 8344], 300 yd E. of Hope church, shale 2 ft thick contains *Caneyella squamula, Dunbarella rhythmica, ? P. obliquata* and *Anthracoceras* or *Dimorphoceras sp.* in a ferruginous horizon near the top of the outcrop, and may be of high $R_{1b}$ or $R_{1c}$ age. At a point 80 yd upstream on the east bank [1747 8350], shale 5 ft thick contains *C. squamula, D. rhythmica, Anthracoceras* or *Dimorphoceras sp., Homoceras sp., Homoceratoides sp.* [juv.] and *Reticuloceras sp.*, indicating a probable $R_{1c}$ age.

A number of outcrops of silty shale with ironstone bands, typical of the highest Edale Shales, are present farther upstream as far as Killhill, and similar strata are exposed in the south bank of the Noe a few yards east of the confluence with Peakshole Water.

G.D.G.

In Peakshole Water the following section of beds of probable $R_{1c}$ age was measured on the right bank [1666 8215], ¼ mile S.E. of Marsh Farm:

|  | ft |
|---|---|
| Shale, soft dark, ferruginous in places; *Dunbarella sp.* [juv.], *Posidonia sp.* and *Homoceratoides divaricatus* 3 ft from base; *Caneyella sp., Dunbarella sp., Homoceras striolatum* and *Reticuloceras sp.* 9 ft from base .. .. | 15 |
| Shale, ferruginous, with bullions 2 ft to | 3¼ |
| Shale, soft black .. .. seen | 12 |

Another section, 25 yd downstream and on the left bank, yielded *Anthracoceras* or *Dimorphoceras sp.* and *Hudsonoceras ornatum.*

At The Folly, a number of shallow boreholes, put down by Messrs. Earle to prove shale reserves for the Cement Works, have penetrated the upper part of the Edale Shales. One of these (No. S5) [1717 8295] passed through Mam Tor Beds to 82¾ ft, and Edale Shales to 160 ft: the latter consisted of silty and finely micaceous mudstones with sporadic thin sandstone bands (one 2 ft thick occurred at 105½ ft). The borehole yielded the following fossils: 113 ft 5 in to 123 ft 9 in, *Caneyella* cf. *rugata, Dunbarella rhythmica, Posidonia obliquata, Promytilus sp., Anthracoceras* or *Dimorphoceras sp., Reticuloceras sp.;* 125 ft to 145 ft 5 in, *Caneyella* cf. *squamula, P. obliquata, Anthracoceras* or *Dimorphoceras sp., Reticuloceras* cf. *reticulatum;* 146 ft to 152 ft, *Caneyella sp., Dunbarella sp., Posidonia sp. nov.*, palaeoniscid scales; 153 ft 6 in to 156 ft 5 in, *C. squamula, D. rhythmica, Anthracoceras* or *Dimorphoceras sp., Reticuloceras spp.* [includes *? R. nodosum* group]. The faunas indicate an $R_{1c}$ age for most of the beds passed through, though $R_{1b}$ may have been reached at the base.

*Mam Tor.* In the vicinity of Mam Tor the unconformity between Millstone Grit and Carboniferous Limestone is at its most acute. The outcrop of the Carboniferous Limestone here forms a marked salient into the Millstone Grit (see also p. 162). As the junction of the two formations is traced northwards from near the Castleton Borehole (see p. 199), where *C. leion* and shales of $P_2$ age are present, progressively higher horizons in the shale are found to rest on the Carboniferous Limestone. The mapping shows the surface of the limestone to be very irregular.

An exposure in the upper part of Odin Sitch [1329 8351], showed the following section some 3 ft above the Carboniferous Limestone:

|  | ft |
|---|---|
| Shale, dark grey .. .. .. | — |
| Mudstone, hard calcareous; fossiliferous crinoidal limestone debris[1] .. | 5 |
| Limestone, dark, thin-bedded towards top; *?Posidonia corrugata* .. .. | 3 |

The dark limestone is probably that stated by Hudson and Cotton (1945, p. 13) to contain in addition *Posidoniella sp.*, *Syncyclonema carboniferum* Hind and *Eumorphoceras bisulcatum;* it was considered that "the age of the fauna was certainly $E_{2b}$". It should be noted that this is one of the few localities at or near the unconformity which shows derived material in the Edale Shales.

The only section of Edale Shales *in situ*[2] in the landslip-scar at Mam Tor is [1299 8346] some 300 yd S.E. of the summit. The section[3] commences some 30 ft below the base of the Mam Tor Beds, though a small fault downthrowing to the north may cut out about 10 ft of shale:

|  | ft | in |
|---|---|---|
| Shale, silty with thin sandstone bands, becoming more numerous towards top .. .. about | 7 | 0 |
| Shale, weathered, with poorly preserved fossils; *Dunbarella rhythmica*, orthocone nautiloid indet., *Dimorphoceras sp.*, mollusc spat | 5 | 0 |
| Ironstone .. .. .. 2 in to | | 7 |
| Shale; *Posidonia sp.* [elongate form], *Reticuloceras* cf. *moorei*, *R. sp.* (*R. nodosum* group) .. .. | 5 | 0 |
| Shale, black: *Posidonia* cf. *obliquata*, *Homoceratoides* aff. *divaricatus*, *Reticuloceras sp.* .. .. .. | | 6 |
| Ironstone; *Dimorphoceras sp.*, *Hudsonoceras ornatum*, *Reticuloceras sp.* .. .. .. .. 2 in to | | 7 |

|  | ft | in |
|---|---|---|
| Shale; *Caneyella squamula* [elongate form], *Dunbarella rhythmica*, *Dimorphoceras sp.*, *Homoceras sp.*, *Homoceratoides sp.* [juv.] [*divaricatus* group] .. .. .. | 11 | 6 |
| Shale, 'sulphurous'; *Dimorphoceras sp.*, *Reticuloceras* cf. *nodosum* .. | 2 | 6 |
| Ironstone; *Caneyella sp.*, *Homoceras sp.*, *R.* cf. *nodosum* .. 1 in to | | 7 |
| Shale, 'sulphurous'; *Dunbarella sp.*, orthocone nautiloid, *Dimorphoceras sp.*, *Reticuloceras sp.* (*nodosum* group) .. .. .. | 3 | 6 |
| Shale, dark platy .. .. .. | | 10 |
| Ironstone; *Caneyella sp.*, *D. rhythmica* .. .. .. 0 to | | 3 |
| Shale, 'sulphurous'; indet. goniatites, fish fragments .. .. | 2 | 6 |
| Ironstone .. .. .. 0 to | | 2 |
| Shale, platy 'sulphurous' .. | 9 | 9 |
| Shale, pale 'sulphurous' in top 2 ft 8 in; darker below with ferruginous patches; *C. squamula*, *D. rhythmica*, *Dimorphoceras sp.*, *Reticuloceras sp.* .. .. .. | 3 | 6 |
| Ironstone; *Hd. ornatum*, *Reticuloceras sp.* .. .. .. 2 in to | | 4 |
| Mudstone, ferruginous at top; *Dunbarella sp.*, *Dimorphoceras sp.*, *Homoceras sp.*, *Reticuloceras sp.* | 1 | 9 |
| Shale, soft pale 'sulphurous' .. | 5 | 0 |
| Shale, platy; *Dunbarella sp.* .. | | 2 |
| Shale, grey 'sulphurous' .. .. | | 10 |

The whole of the section, with the possible exception of the lowest 6 ft which contains no diagnostic fauna, is referred to the $R_{1b}$ Zone; the base of $R_{1c}$ probably lies just above the topmost beds. The base of the section is estimated to lie about 330 ft above the Carboniferous Limestone, giving a total thickness for the Edale Shales of some 400 ft. Near the limestone boundary the Edale Shales appear to have thinned further. The presence of $E_{2b}$ faunas as noted above in Odin Sitch, in an area where the general tendency is for thinning towards the limestone massif, suggests a figure of as little as 300 ft for the Edale Shales here.

---

[1]Fossils include *Lithostrotion* cf. *martini* Milne Edwards and Haime, *Syringopora?*, *Martinia?*, productoid fragments, *Rugosochonetes sp.*

[2]This was described by Jackson (1927, pp. 23–4) as the upper of three slipped masses within the landslip. The main features of the fauna were recognized.

[3]Including information from Messrs. W. E. Graham and J. Pattison.

The presence of lower faunas than the above within the Mam Tor landslip was recognized by Jackson (ibid., p. 24) and confirmed during the present work in two places:

2. [1305 8349]. Probably the "middle block" of slipped material of Jackson. This shows approximately 44 ft of shale with *Reticuloceras circumplicatile* at the top, *Homoceras* cf. *magistrorum* at 12 ft from the top and *Hudsonoceras proteus* at the base. This has apparently moved very little.

1. [1315 8345]. Probably at or near the position of the "lower block" of Jackson. This yielded only indeterminate dimorphoceratids and *Posidonia* cf. *corrugata* during the present work, though Jackson records *Eumorphoceras bisulcatum* and *Nuculoceras nuculum*. The present stratigraphical position of this locality is about 180 ft above the Carboniferous Limestone.

Between the Blue John Mine [1317 8319] and Windy Knoll [126 830] the shale occupies an irregular erosional hollow on the limestone surface. A small inlier of limestone occurs within the shale east of the road junction [129 831]; the sides of the inlier are steep and it protrudes through the shale, forming a small hill.

Hudson and Cotton (1945, p. 14) record *Reticuloceras* from a pocket of shale on the limestone surface, presumably in the vicinity of Windy Knoll. This is unfortunately not accurately located and the record has not been confirmed during the present work. The occurrence of *Reticuloceras* in this position is difficult to reconcile with the stratigraphy, and it is suggested that the shale may not have been *in situ*.

*Perryfoot valley.* Between Sparrow Pit and Windy Knoll the outcrop of the Edale Shales occupies a broad valley between the escarpment of Rushup Edge and the Carboniferous Limestone. Much of the ground is obscured by head, though streams cut through this in places. The junction of shale and limestone is marked by a line of swallow-holes.

Hudson and Cotton (ibid., p. 13) described the faunas from a series of tips from shafts along the line of an old sough driven northwards from the limestone boundary near Peakshill. Their results were:

(a) Northern tip [1175 8307]: *Homoceras*

*sp.* and *Homoceratoides varicatus* [or *Ht. prereticulatus*].

(b) Tip [1173 8303] 50 yd S.23°W. of (a): *Homoceras henkei, H. henkei angustum, H. striolatum* group, *Reticuloceras inconstans* and *R. pulchellum*.

(c) Tip [1172 8295] 85 yd S.8°W. of (b): *Homoceras sp.* cf. *H. eostriolatum* towards *H. henkei*.

These faunas indicate a low $R_1$ horizon and Hudson and Cotton were of the opinion that the beds resting on the limestone here were of high $H_2$ age, the total thickness of Edale Shales being here some 150 to 200 ft. However, this figure seems most unlikely in view of the present estimate of 400 ft at Mam Tor (see p. 202), where the thickness would be expected to be less than at the present locality. There is no evidence that any of the shafts reached Carboniferous Limestone.

The limestone-shale junction is exposed in several sink-holes in the valley bottom, one [1027 8161], north-north-east of Perryfoot showing the following section:

|  | ft |
|---|---|
| Head .. .. .. .. .. | 5 |
| Shale, soft grey .. .. .. | 10 |
| Shale, dark hard platy .. .. | 18 |
| Shale, soft grey .. .. .. | 1 |
| *Not exposed* .. .. .. .. | 3 |
| Shale, grey .. .. .. .. | 2 |
| *Not exposed* .. .. .. .. | 4 |
| Shale, grey, with blocks of dark limestone up to 6 in diameter (dip at base roughly follows unconformity) | 3½ |
| *Unconformity* .. .. .. .. | — |
| Limestone, massive fossiliferous (fore reef .. .. .. .. about | 10 |

Near Tor Top, Perryfoot [approximately 0985 8130], Jackson (1925, p. 270) recorded "black shales and limestones . . . dipping N.W. off massive limestones . . . Higher in the shales are thinly-bedded black limestones containing crinoid-stems, goniatites (especially *Eumorphoceras bisulcatum*), nautiloids, etc. These are probably not more than 75 feet above the junction."

Higher beds than the above are exposed in a stream [0986 8170], north of Perryfoot, where hard dark calcareous flags yielded cf. *Cravenoceras darwenense* Moore, *Eumorphoceras sp., Nuculoceras nuculum, N.* cf.

*tenuistriatum* Demanet (E$_{2c}$ Zone). Some 160 yd farther upstream an exposure showed 10 ft of soft grey 'sulphurous' shale with a 2½-in band containing unidentifiable goniatites at 4½ ft from base.

To the west of Perryfoot, records in Coalpithole Mine show 216 ft of shale on limestone in No. 8 Shaft [0957 8119] and 424 ft of shale on limestone in No. 10 Shaft [0915 8129]. The latter starts some 75 ft below the base of the Shale Grit giving an approximate thickness of 500 ft for the shale at this locality.

*Sparrow Pit and Barmoor Clough.* At Sparrow Pit some 350 ft of shale separate the Shale Grit and Carboniferous Limestone. The shale is mostly ill exposed. At Barmoor Clough [approximate site 078 796], close beneath the Shale Grit, Jackson (1927, p. 31) recorded shale with "*Reticuloceras reticulatum, Homoceras striolatum* and *Pterinopecten rhythmicus*".

*Dove Holes.* The most complete section in this area is provided by the railway-cuttings north of Dove Holes, starting at the tunnel entrance [0761 7933] and with the bottom of the section [0770 7901], 240 yd to the south:

| | ft | in |
|---|---|---|
| To base of Shale Grit .. .. | 100 | 0 |
| Shale, dark grey weathered; *Caneyella sp., Dunbarella rhythmica, Posidonia sp., 'Stroboceras' sp., Dimorphoceras, Homoceras* cf. *striolatum, Hudsonoceras ornatum, Reticuloceras sp.* (*R. nodosum* group), mollusc spat .. .. | 14 | 0 |
| Shale, soft black decalcified; *Dunbarella sp., Reticuloceras sp.* 4 in to | | 5 |
| Shale, blue grey, weathered; *Caneyella sp., D. rhythmica, R.* cf. *nodosum, R.* cf. *subreticulatum*; 2½-ft bed of bullions, at 2½ ft from base, with *R.* cf. *nodosum* and *R.* cf. *pulchellum* .. .. .. | 18 | 0 |
| Ironstone .. .. .. 2 in to | | 4 |
| Shale, dark grey, poorly exposed .. | 12 | 0 |
| *Not exposed* .. .. .. about | 25 | 0 |
| Shale, dark grey; *Caneyella sp.* [juv.], *Promytilus sp.* .. .. | 33 | 0 |
| Bullion; *Homoceras subglobosum* (base of H$_{1a}$ Zone) .. up to | | 6 |
| Shale, grey .. .. .. .. | 11 | 0 |
| Shale, hard ferruginous; *Posidoniella sp.* .. .. .. .. | | 6 |

| | ft | in |
|---|---|---|
| Shale, grey, very weathered .. | 1 | 6 |
| Shale, dark decalcified; *Posidonia corrugata, Posidoniella variabilis, Dimorphoceras sp., Eumorphoceras bisulcatum, Nuculoceras nuculum* .. .. .. .. | | 4 |
| Shale, hard ferruginous; *Eumorphoceras bisulcatum* .. .. .. | | 6 |
| Shale, dark grey .. .. .. | 2 | 6 |
| Ironstone; *P. corrugata* and goniatites indet. .. .. .. | | 6 |
| Shale, grey .. .. .. .. | 12 | 0 |
| Ironstone band, impersistent; *P. corrugata* .. .. .. .. | | 8 |
| Shale .. .. .. .. .. | 3 | 0 |
| To Carboniferous Limestone about | 90 | 0 |

This gives a total thickness of about 325 ft for the Edale Shales at this locality.

Near Dove Holes the section of the main line railway tunnel shows some 300 ft of shale to be present, though these beds thicken southwards to 500 ft between Turner Lodge and Brook House.

From Dove Holes to a point some ½ mile N. of Brook House, the limestone-shale junction appears to be regular; farther south the shale rests on a strongly eroded limestone surface, filling well-marked valleys, respectively 500 yd N. and 200 yd S. of Brook House. A small outlier of shale is present on the level upper surface of the limestone at a point 700 yd N.E. of Brook House.

The shales are exposed in several places in a stream just south of Dove Holes; an exposure [0730 7772] here showed dark decalcified shale 1 ft 6 in, grey 'sulphurous' shale about 4 ft, soft decalcified shale with *Posidoniella sp., Anthracoceras* or *Dimorphoceras sp.* and *Cravenoceratoides* or *Cravenoceras sp.* 7 in, hard dark grey shale 1 ft 3 in, obscured 3 ft, on soft grey shale 2 ft. The horizon is probably near the top of E$_2$. Exposures of the shales are fairly continuous for some 320 yd upstream from the above locality, but no fossils were found.

Beds near the junction with the limestone are well seen in Nun Brook. An exposure [0663 7485], south-west of Brook House, yielded *Posidonia corrugata, Posidoniella variabilis, Spirorbis sp.* on carbonized plant fragment, *Anthracoceras sp., Dimorphoceras?*; this probably indicates an E$_2$ horizon.

Downstream, the following section is exposed [0655 7469]:

|  | ft | in |
|---|---|---|
| Shale, grey 'sulphurous' with a 3-in dark decalcified band in middle .. | 4 | 0 |
| Shale, dark decalcified with a 10-in bullion .. .. .. .. | 1 | 9 |
| Shale, contorted and sheared, base clear-cut .. .. .. .. | 6 | 0 |
| Shale, soft 'mushy' .. .. 1 in to | | 2 |
| Shale, grey, with 1-in band at base with *Posidonia corrugata, Posidoniella variabilis, Anthracoceras sp.* and an orthocone nautiloid .. | 2 | 1 |
| Shale, hard platy .. .. .. | 1 | 6 |
| Shale, 'sulphurous' .. .. .. | 3 | 0 |

This fauna is considered to be probably $E_2$ and most likely one of the *Anthracoceras* bands in $E_{2b}$.

*Todd Brook valley.* Beds underlying the Kinderscout Grit are present along the axis of the Todd Brook Anticline between Kettleshulme and Eaves Farm, near the southern boundary of the district. To the north of Green Booth the Edale Shales are, with few exceptions, obscured by boulder clay. Between this locality and Redmoor Farm, however, shales with thick sandstones of 'crowstone' type are exposed in a strongly-folded belt. Typical sections are provided by a quarry [9819 7651] on the southern side of Fox Hill:

|  | ft | in |
|---|---|---|
| Shale, grey .. .. .. about | 6 | 0 |
| Sandstone, irregularly bedded massive siliceous .. .. .. | 12 | 0 |
| Shale with lenses of sandstone up to 6 in thick .. .. .. .. | 3 | 0 |
| Sandstone, siliceous, in beds up to 6 in thick with shale partings .. | 1 | 9 |

A larger working [9815 7590] ¼ mile S.W. of Saltersford Hall showed:

|  | ft | in |
|---|---|---|
| Sandstone, siliceous, in beds up to 2 ft thick .. .. .. .. | 5 | 6 |
| Shale parting, poorly exposed, about | | 4 |
| Sandstone, coarse massive siliceous | 4 | 4 |
| Shale, with thin siltstone bands 0 to | | 5 |
| Sandstone, fine-grained flaggy siliceous .. .. .. .. | | 6 |
| Sandstone, pale violet siliceous .. | 4 | 4 |
| Shale .. .. .. .. 0 to | | 1 |

|  | ft | in |
|---|---|---|
| Sandstone, hard siliceous, with a 3-in shale parting in middle .. | 7 | 1 |
| Shale, pale-weathered grey .. | 4 | 0 |

Veins of baryte are present in the sandstones in the latter exposure (see p. 317).

The only evidence of horizon of the shales in this area is provided by an exposure [9813 7636], on the left bank of Todd Brook, near Saltersford Hall: dark decalcified shale with *Posidonia corrugata, Dimorphoceras sp., Eumorphoceras bisulcatum* s.l. and *Nuculoceras nuculum* 6 ft 6 in, on dark decalcified shale with bands of siliceous sandstone up to 3 in thick 3 ft 3 in, on soft grey shale 3 ft. This exposure, indicative of an $E_{2c}$ horizon, cannot, owing to folding and possible faulting, be accurately related to the beds with 'crowstones' of Fox Hill, though it appears certain that the ages of the two cannot differ greatly. It seems likely that most of the exposed 'crowstones' are slightly younger in age than the *N. nuculum* band, but do not extend far up in the succession, and are all referred to the $E_2$ stage.      I.P.S.

*Derwent valley.* The highest beds of the Edale Shales crop out near Hathersage and also near the submerged village of Derwent, in Ladybower Reservoir. In the west bank of Hood Brook [2286 8121], 250 yd N.E. of Nether Hall, 16 ft of rather silty shale with small ferruginous nodules, containing *Caneyella?, Dunbarella sp. Posidoniella sp., Homoceras sp.* and *Reticuloceras* cf. *reticulatum* are exposed. Similar fossiliferous strata are present in the banks of the same stream about 10 to 20 yd above its confluence with the Derwent. The intervening exposures along the stream are of typical basal Mam Tor Beds, and it appears that the two exposures mentioned are parts of a small upfaulted mass of Edale Shales, the remainder of which is obscured by Drift.

In 1959 the level of Ladybower Reservoir was sufficiently low to expose the remains of Derwent village and parts of the original channel of the river, including a number of outcrops of lower and basal Mam Tor Beds. These beds were seen, in the cores of 'valley bulge' folds [1819 8858], to be underlain by dark silty shales with thin ironstone bands identical in lithology to the highest Edale Shales to the west and south.      G.D.G.

*Bradwell and Great Hucklow.* With the disappearance of the Mam Tor Beds and some lateral passage of the Shale Grit into the underlying shales in this area, the Edale Shales south of Bradwell may include a small thickness of beds higher than in the type area, though with the general south-easterly thinning, this rapidly loses its significance. These beds consist of shales, some 550 ft in thickness and somewhat silty, and with thin sandstone bands towards the top. The greater part is ill exposed.

The unconformity at the junction with the Carboniferous Limestone is in places very irregular. At Bradwell Hills this is particularly so, and appears to be due to the original depositional form of the flat reef in the Eyam Group, further modified by pre- or early Namurian erosion. Thus, about 500 yd S. of St. Barnabas's Church, Bradwell, a narrow valley in the limestone appears to be filled by shale. The records of shale with *Eumorphoceras bisulcatum* noted by Jackson (1925, pp. 273–4) in this vicinity have been confirmed just west of the small brook 500 yd S. of St. Barnabas's Church. The shales at this locality [1757 8063], and for a distance of 90 yd upstream, yield *Aviculopecten sp., Chaenocardiola?, Posidonia corrugata, P. corrugata gigantea, Dimorphoceras sp.* and *Eumorphoceras bisulcatum erinense* Yates, indicative of an $E_{2a}$ horizon.

From the tips of Wortley Mine [1774 8105], near Bradwell church, Jackson (1927, p. 31) recorded a fauna, considered by him to be the same horizon as that from Bradwell Hills, which included *E. bisulcatum* and *Kazakhoceras scaliger* (recorded as "goniatite with lattice ornament").

To the south and south-east of Bradwell Hills the unconformable junction with the Carboniferous Limestone is much more regular than in the Castleton–Bradwell area, and in most places the mapping suggests that there is little or no angular discordance between the two formations.

A number of shafts, to workings on the Hucklow Edge Vein, have penetrated the shales south of Hucklow Edge. The greatest thickness was passed through at Silence Mine [1880 7799] which, starting about 75 ft below the base of the Shale Grit, proved 328 ft of shale before reaching limestone.

A sink-hole [1868 7746], ½ mile N.W. of Foolow, shows beds low in the shale sequence: dark shale 6½ ft, not exposed 9 ft, on grey to dark grey thin-bedded cherty limestone (Eyam Limestones) 9 ft.

The Hucklow Edge No. 1 Borehole [1941 7781], west of Bretton, proved a nearly complete section of the Edale Shales. This is summarized below:

| | Thickness | | Depth | |
|---|---|---|---|---|
| | ft | in | ft | in |
| *Reticuloceras circumplicatile* Zone ($R_{1a}$) Mudstone with some hard calcareous bands; *R.* cf. *todmordenense* at 27 ft 6 in; *Homoceras henkei* at 32 ft 4½ in; *H. magistrorum* band 3 ft 4 in at 38 ft 4 in    ..    .. | 38 | 4 | 38 | 4 |
| *Homoceratoides prereticulatus, Homoceras undulatum* and *Hudsonoceras proteus* zones ($H_2$) Mudstone; *Homoceratoides prereticulatus* band 7 in at 39 ft 8 in and *Hudsonoceras proteus* band 1 ft 1 in at 50 ft    ..    ..    .. | 11 | 8 | 50 | 0 |
| *Homoceras beyrichianum* and *H. subglobosum* zones ($H_1$) Mudstone with a 5-in silty band at 51 ft 4 in; fucoid markings at 54 ft 3 in; thin pyrite lens at 70 ft 9 in; probable *H. beyrichianum* band 1 ft 7 in at 64 ft; several horizons with *H. subglobosum* between 67 ft 1 in and 77 ft 1 in    .. | 27 | 1 | 77 | 1 |
| *Nuculoceras nuculum* ($E_{2c}$) and *Cravenoceratoides nitidus* ($E_{2b}$) zones Mudstone with a 5-in argillaceous limestone at 112 ft; 3-in band with sphaerosiderite at 113 ft 3 in; 19-in hard calcareous band at base; upper *N. nucu-* | | | | |

*lum* band 7 in at 87 ft 1 in; middle *N. nuculum* band 1 ft 4 in at 99 ft 2 in; lower *N. nuculum* band 2 ft 6 in at 111 ft 9 in; *Cravenoceratoides nititoides* band 9 in at 132 ft 3 in; *Ct. nitidus* band 4 in at at 135 ft 5 in; *Ct. edalensis* in 10 in at 150 ft 4 in; *Cravenoceras subplicatum* in 1 ft 0½ in at 151 ft 5 in .. .. 74 4 151 5

*Eumorphoceras bisulcatum* (E$_{2a}$) Zone

Mudstone with a 2½-in fine-grained limestone at 164 ft 2½ in; *E. bisulcatum* band 5 ft at 165 ft; *Cravenoceras cowlingense* band 1 ft 9 in at 171 ft 9 in .. .. 20 4 171 9

*Cravenoceras malhamense* (E$_{1c}$) Zone

Mudstone; *C. malhamense* band at 173 ft 1 in .. .. .. 1 4 173 1

*Eumorphoceras pseudobilingue* (E$_{1b}$) Zone

Mudstone; *E. pseudobilingue* band 3 in at 177 ft 1 in .. .. 4 0 177 1

*Cravenoceras leion* (E$_{1a}$) Zone

Mudstone with a 6-in fine-grained limestone at 178 ft; *C. leion* band 7 in at 179 ft 2 in .. 2 1 179 2

Viséan (see Appendix I, p. 370) .. .. .. — — — —

The great condensation of the E$_1$ strata is a striking feature of this section (see Plate XV).

The Hucklow Edge No. 2 Borehole [2025 7760] (see Appendix I, pp. 371–2) proved the following succession:

|  | Thickness |  | Depth |  |
|---|---|---|---|---|
|  | ft | in | ft | in |
| No core .. .. .. | 207 | 9 | 207 | 9 |

*Cravenoceratoides nitidus* (E$_{2b}$) Zone

Mudstone, with a 1-ft limestone band at 219 ft; *Ct. nititoides* band 1 ft 2 in at 212 ft 6 in; *Ct. edalensis* band 7 in at 236 ft .. .. 35 10 243 7

*Eumorphoceras bisulcatum* (E$_{2a}$) Zone

Mudstone with thin limestone bands, 1½ in at 251 ft 2½ in, 5 in at 252 ft 9 in and 3 in at 253 ft 3 in; *Leiopteria longirostris* at 243 ft 11 in; *E. bisulcatum* band 4 ft 9 in at 251 ft 11 in .. .. .. 21 11 265 6

Viséan (see p. 101, etc.) — — — —

This borehole shows a similar development of the pre-Namurian unconformity to the Bradwell Hills area and Mill Lane, Eyam.

*Eyam, Stoney Middleton and Calver.* To the north of Eyam the full thickness of shale was passed through at Ladywash Shaft [2189 7754] between the base of the Shale Grit at 312 ft and the Carboniferous Limestone at 699 ft, a thickness of 387 ft. The Edale Shales have been met with in places in the mine workings. The recent driving of a crosscut between Stoke Sough and the Old Edge Vein provided evidence of the horizon of the lowest shales underground. Material from a few feet above the Carboniferous Limestone [2253 7683], 1050 yd S. 42°E. of Ladywash Shaft, yielded *P. corrugata, P. corrugata elongata, Cravenoceras* cf. *cowlingense, E. bisulcatum,* crinoid debris and fish fragments. This indicates the main *E. bisulcatum* horizon in E$_{2a}$. Material from some 160 yd farther down the crosscut [2262 7693] and about 15 ft above the horizon of the last locality yielded the following fauna: *P. corrugata, P. corrugata gigantea,* orthocone nautiloids and *Cravenoceratoides edalensis.* The fauna is indicative of the *Ct. edalensis* band at the base of E$_{2b}$.

Several sections in the shales occur in a stream north of Eyam. The lowest of these, about 170 ft above the Carboniferous Limestone [2145 7695], east of Eyam View, showed 11 ft of dark shale with a 1-ft bed yielding *Anthracoceras* or *Dimorphoceras sp.* at 2½ ft from the base. The fauna gives no indication of horizon. A higher section [2142 7719] in the same stream is as follows:

|  | ft |
|---|---|
| Base of Shale Grit .. .. .. | — |
| Sandstone and siltstone, interbedded | 6 |
| Mudstone, silty .. .. .. | 7 |
| *Not exposed* .. .. .. .. | 12 |
| Sandstone .. .. .. .. | 8 |
| Mudstone, silty and siltstone .. .. | 17 |
| Shale, dark pyritous in part .. .. | 4 |
| Mudstone, silty .. .. .. | 4 |
| *Not exposed* .. .. .. about | 20 |
| Shale, dark .. .. .. .. | 25 |
| *Not exposed* .. .. .. about | 25 |
| Shale, dark .. .. .. .. | 6 |
| Shale, dark, with marine fossils in top 1 ft and lowest 3 ft 6 in; *Caneyella sp.*, *Dunbarella sp.*, orthocone nautiloid, *Dimorphoceras sp.*, *Homoceras sp.*, *Reticuloceras* cf. *reticulatum, Cypridina sp.* .. .. .. | 7⅓ |

The fossiliferous horizon is in high $R_1$ and the section corresponds to that in the Edale valley where the *R. reticulatum* band lies close beneath the base of the Mam Tor Beds. Much of the beds between the *R. reticulatum* band and Shale Grit in the above section is, therefore, the lateral equivalent of the Mam Tor Beds.

To the south-east of Eyam, shale occupies an erosional valley in the $P_2$ flat-reef limestones near Cliffstile Mine [2246 7595]. Nearby, in Mill Lane [2236 7616] 1 ft 9 in of shale with thin bands of dark limestone yielded *Obliquipecten sp.*, *P. corrugata*, '*Cyrtoceras*' *rugosum* (Phillips *non* Fleming), '*Stroboceras*' *sp.*, *Anthracoceras* or *Dimorphoceras sp.*, *Eumorphoceras sp.* and *Coleolus?*. From this locality Jackson (1927, p. 31) recorded *Posidonomya sp.*, *Pseudamussium* cf. *praetenuis* (von Koenen), *A. glabrum* and *E. bisulcatum*; the record of *E. bisulcatum* affords clearer evidence of horizon ($E_2$) than does the more recently collected fauna.

The shales thin in an easterly direction, only some 250 to 300 ft being present at Stoney Middleton. Between the latter place and Coombs Dale, shales occupy a broad depression eroded in the dip-slope of the Carboniferous Limestone. Part of the shale

outcrop is obscured by boulder clay (see p. 342). The best exposure lies in a landslip-scar [2296 7525] south-west of Stoney Middleton church where 2 ft of dark decalcified shale is seen. A little farther to the south-east, Morris (1929, p. 58) records "*Posidonomya membranacea*" and a few goniatites near Trinkey Lane, about ¼ mile S.E. of Stoney Middleton.

*Wardlow Mires outlier.* Shale occupies a shallow depression some three quarters of a mile in length, with Wardlow Mires on its western edge. The presence of shales of $P_2$ age has already been noted (see p. 105), though in view of their normal absence adjacent to the shelf area, the practice followed on the map has been to limit their outcrop as far as the scattered surface provings allow. Forefield Shaft [1882 7577] at Watergrove Mine passed through 30 ft of shale before entering limestone; some at least of this shale is $P_2$ in age.

The only surface evidence of horizon of the higher beds ($E_2$) was obtained by Shirley and Horsfield[1] (1945, p. 293) from a trench crossing the outlier. The collections made were presented to the Geological Survey and details are as follows:

1. [1858 7547]: *P. corrugata*, *Cravenoceras sp.*, *E. bisulcatum* s.s.
2. [1847 7549]: *P. corrugata*, smooth pectinoid indet., *Cravenoceras sp.*, dimorphoceratid indet., *E. bisulcatum*.
3. [1839 7550]: *P. corrugata*, *E.* cf. *bisulcatum*.

An attempt to supplement the above information was made during the present work by digging at several points. The only locality yielding fossils was a bank [1816 7568] 70 yd N.E. of Wardlow Mires, where *P. corrugata*, coiled nautiloid (*Catastroboceras?*) and indet. goniatites were found.

The Wardlow Mires No. 1 Borehole [1850 7553] (see also p. 378), near the centre of the shale outlier, proved a very condensed sequence of Namurian shales with the following generalized section:

| | Thickness | | Depth | |
|---|---|---|---|---|
| | ft | in | ft | in |
| *Eumorphoceras bisulcatum* ($E_{2a}$) Zone | | | | |
| Mudstone; *E. bisulcatum* | | | | |

---

[1] Chief elements of faunas including *E. bisulcatum* listed, but no precise localities given.

| | Thickness | | Depth | |
|---|---|---|---|---|
| | ft | in | ft | in |
| horizon at 10 ft 1 in; *Cravenoceras cowlingense* horizon at 20 ft 9 in | 20 | 9 | 20 | 9 |
| *Cravenoceras malhamense* ($E_{1c}$) Zone Mudstone     .. | | | 10 | 21 7 |
| *Eumorphoceras pseudobilingue* ($E_{1b}$) Zone Mudstone; *E. pseudobilingue* band at 29 ft 8½ in  ..     ..     .. | 8 | 1½ | 29 | 8½ |

| | Thickness | | Depth | |
|---|---|---|---|---|
| | ft | in | ft | in |
| *Cravenoceras leion* ($E_{1a}$) Zone Mudstone; *?C. leion* band 4 ft 3 in at base     .. | 38 | 0½ | 67 | 9 |
| Viséan shales (see p. 105) | — | — | — | — |

The Wardlow Mires No. 2 Borehole [1825 7586] (see Appendix I, pp. 378–82) proved a comparable section, the *C. leion* band being confirmed in this case.     I.P.S.

## MAM TOR BEDS

*Alport and Ashop valleys.* An inlier of Mam Tor Beds is present in the Ashop valley between the Snake Inn and the western end of Ladybower Reservoir, extending up the Alport valley to approximately a mile north of Alport Castles Farm. Their thickness in the Ashop valley is over 200 ft but they thin northwards to 123 ft in Swint Clough, where nearly continuous exposures of thin-bedded sandstones and silty shales are present. In the south bank of Swint Clough [1318 9106], 9 ft of shale and siltstone contain *Dunbarella sp.*, *Posidonia obliquata*, *Anthracoceras* or *Dimorphoceras sp.*, *Homoceras?* and *Reticuloceras?*. This outcrop (Alport 24), noted by Hudson and Cotton (1943, p. 154), is from the upper part of the Mam Tor Beds.

Good exposures of thin-bedded sandstone and silty shale occur in the banks of the Ashop, especially near the Snake Inn, south of Wood Houses and opposite the junction with the River Alport. An 18-in mass of sandstone, probably a slump-ball, is enclosed in silty shale on the west bank of Lady Clough [1079 9091]. In the south bank of the Ashop [1498 8905] 520 yd S.W. of Rowlee Farm, the basal Mam Tor Beds consist of silty shales and sandy siltstones with thin sandy ironstone bands 23 ft, with *Caneyella?* and *R.* cf. *reticulatum* at 8½ ft above, and *Anthracoceras* or *Dimorphoceras sp.* and *Homoceras?* at 13½ ft above the base of the section. Farther east [1541 8896], from about the middle of the Mam Tor Beds, *Caneyella sp.*, *Dunbarella sp.*, *Posidonia sp.* and *Anthracoceras* or *Dimorphoceras sp.* are present in thin silty shale alternating with siltstone and

sandstone bands 32 ft. Hudson and Cotton (1943, p. 154) record *R.* cf. *reticulatum* s.s. from a locality (Alport 30) which lies near this outcrop.

*Derwent valley (northern area).* Mam Tor Beds occupy the lower slopes of the Derwent valley between Marebottom Farm and Riding House, where, including strata now submerged beneath the reservoirs, they are at least 330 ft thick. A borehole immediately south of Howden Dam [1693 9245] which commenced in strata approximately 80 ft below the top of the Mam Tor Beds, penetrated thin-bedded sandstones, siltstones and shales to at least 252 ft; these contained rare *Dunbarella sp.* and *Anthracoceras* or *Dimorphoceras sp.* in the lower part; beneath were strata of Edale Shales type, with a poor goniatite-bivalve fauna, extending to the bottom of the hole at 284 ft 6 in. Thin-bedded sandstones, siltstones and silty shales are well exposed along the lower reaches of Abbey Brook, Ouzelden Clough and Mill Brook. Rare indeterminate goniatites and fish debris are present in shaly beds 18–22 ft above the base of a small cliff [1716 8974] on the western side of the outflow from Derwent Dam. In the west bank of the Derwent [1720 8963], north of Ladybower Reservoir, *Anthracoceras* or *Dimorphoceras sp.*, *Homoceras sp.*, *Reticuloceras sp.* and acanthodian spines occur in shales 5 to 10½ ft and 26 to 40 ft above the base of an alternating sequence of shale, siltstone and sandstone, 40 ft thick. Clayey, sandy and ferruginous nodules, the latter containing traces of oolitic structure, are present in this outcrop, and plant debris is fairly common. Stobbs (*in* Hind 1905, pp. 231–4) has recorded a fauna of bivalves, *Orthoceras*, goniatites and fish fragments from shales alternating with "grits" 30 to 60 ft below ground surface in the foundation trench of Derwent Dam. In 1959, lowering of the water level in Ladybower Reservoir due to drought revealed many outcrops of Mam Tor Beds, including typical basal shales with thin silty ironstone bands and rare thin ferruginous silty sandstones at the confluence of Mill Brook and the original course of the Derwent.

*Edale, Mam Tor and Rushup Edge.* On the northern slopes of the Edale valley the Mam Tor Beds vary from 225 ft to over 300 ft in thickness. The basal strata, consisting of shales and silty shales, locally ferruginous, with soft micaceous sandstones and thin bands and nodules of hard silty ironstone, can be seen overlying Edale Shales in Crowden Brook [1021 8578] and Grinds Brook [1208 8649], and in the River Noe [1665 8586] north of the railway-bridge near Normans Farm. Alternating thin-bedded sandstones, siltstones and silty shales are exposed in many places in Crowden Brook, Grinds Brook, Oller Brook and Lady Booth Brook and in the Noe between Clough Farm and Normans Farm. Unidentifiable fragments of goniatites and pectinoids have been found in the lower part of the Mam Tor Beds in a landslip-scar [1500 8604], 750 yd S.E. of Clough Farm (see p. 195) and in the banks of the Noe [1667 8599] ½ mile S.E. of Edale End.                                    G.D.G.

In the western part of the Edale valley the Mam Tor Beds maintain their usual lithological character, their thickness varying from 200 ft near Lee House to 350 ft near Chapel Gate [105 840]. This southward thickening is partly due to the lateral passage of the lowest beds of the Shale Grit into Mam Tor Beds west of the Cowburn Tunnel entrance and partly due to an overall thickening. To the north-west of Lee House a sandstone up to 20 ft thick is present near the top.

Along the south side of the Edale valley the Mam Tor Beds thicken to about 400 ft near Barker Bank and Lose Hill. In Harden Clough a shale parting "about 50 feet above the base" (Hudson and Cotton 1945, p. 5) yielded *Posidoniella sp.*, *Dunbarella sp.*, *Dimorphoceras sp.*, rare *Reticuloceras sp.* and *Homoceras sp.*                          I.P.S.

On Lose Hill, a short distance below the top of the Mam Tor Beds, a prominent feature indicates an abnormally thick bed of sandstone which either thins out or is laterally replaced by normal strata farther west, as seen in the Back Tor landslip-scar.                                             G.D.G.

At the type locality of Mam Tor these beds are at least 450 ft thick. Some 400 ft of alternating sandstone and shale are exposed in the landslip-scar on the eastern side of the hill (see Plate XVI B) and there are a further 50 ft of unexposed beds between the top of the scar and the summit, which must be just below the base of the Shale Grit. This section has been described in detail by Allen (1960, pp. 193–208).

Around Brockett Booth the Mam Tor Beds appear to contain little sandstone. Lateral passage of the uppermost beds into Shale Grit is described on p. 171.

To the west of Mam Tor the group can be traced below Rushup Edge as far as Rushup, though with some passage of the highest beds into the base of the Shale Grit. To the south-west of Rushup, however, the amount of sandstone within the group increases and the whole of the Mam Tor Beds passes into the basal beds of the Shale Grit before No. 10 Shaft [0915 8129] of Coalpithole Mine is reached.                                          I.P.S.

*Hope Valley and Derwent valley (southern area).* In the Hope Valley, the Mam Tor Beds are over 300 ft thick. They form a marked hill at The Folly, about ⅓ mile S. of Hope. Several boreholes have been drilled here by Messrs. Earle to prove shale reserves; a typical hole (No. S5) [1717 8295] (see also p. 201) passed through Mam Tor Beds of normal type to 82 ft 9 in; an 8 ft 3 in sandstone occurred at 36 ft 3 in and a 5 ft 9 in sandstone at the base. On the eastern side of the hill, the lower beds, here containing fewer and thinner sandstones than in the country to the north, are seen to overlie the Edale Shales in Messrs. Earle's shale quarries near Eccles House. The succession in the eastern pit [1740 8286] is siltstone with rare thin soft micaceous sandstones and hard ferruginous sandstones 14 ft, on silty shale with siltstone bands 11 ft. In the western pit [1736 8281] alternating thin beds of sandstone, soft silty and micaceous, with siltstone bands 5 ft, overlie silty shale and siltstone, with rare thin micaceous sandstones, 27 ft. Silty and sandy ferruginous nodules, plant fragments and thin carbonaceous streaks are also present.

G.D.G., I.P.S.

The lower, mainly argillaceous Mam Tor Beds, are also exposed in the River Derwent, north-west of Bamford station. The upper part of the Mam Tor Beds in the Hope valley is thin-bedded sandstones, siltstones and silty shales similar to the succession farther north. Good exposures occur in the banks of the Noe, east of Hope and along the River Derwent for 500 yd downstream from its confluence with the Noe. They are strongly contorted and with slumped bedding in the bank of the Derwent [1997 8415] north-west of Bamford.

Near Hathersage the Mam Tor Beds are 200 to 250 ft thick. The basal strata, silty shales with some thin silty ironstone bands

and nodules, but with very few thin sandstones, are exposed in the stream [2284 8108] 190 yd E. of Nether Hall. Thin-bedded sandstones and silty shales at higher horizons in the Mam Tor Beds are exposed intermittently in the stream from Brookfield Manor to Hathersage.                G.D.G.

Between Brough and Shatton Moor the relations of the Mam Tor Beds to the Shale Grit are complex, with thicker sandstones making their appearance in the upper part of the group. One of these sandstones forms a marked scarp and dip-slope near Elmore Hill Farm. About ⅓ mile to the south several sandstones come in at a slightly higher horizon and thicken eastwards to join the lowest part of the Shale Grit. To the south of Brough it has not been found practicable to separate Mam Tor Beds, which here lose their identity.

## SHALE GRIT

*Glossop and Hayfield.* To the north of Kinder Scout the Shale Grit forms a broad spread of moorland including Black Moor, Hurst Moor and Ramsley Moor. Between Leygatehead and Ramsley Moor the higher ground is formed by the upper part of the Shale Grit. Lower beds, however, are exposed (up to about 350 ft) in an inlier in Bray Clough [056 917] and in a faulted inlier at Span Clough. Much of the area is obscured by peat, and stream sections show a normal development of massive sandstones with shale partings, the latter often silty and with thin sandstone bands.

A borehole at Messrs. Olive and Partington's works at Charlestown, Glossop [0297 9331] (here called the Charlestown Borehole) showed that at least the upper part of the Shale Grit was absent at this locality, the base of the Kinderscout Grit being at 275 ft and the underlying beds, to the bottom of the hole at 851½ ft, silty mudstones and siltstones with thin sandstone bands.

Between Kinder Reservoir and Oaken Clough the Shale Grit occupies a westward sloping tract, in many places approximating to a dip-slope. On the west the grit forms the sharp feature of Kinder Bank. Here several shale partings, each up to 30 ft in thickness, are present, though the intervening sandstones are very massive and have been quarried near Bowden Bridge and farther

north [0506 8802]. The latter section showed 100 ft of very coarse yellow-brown sandstone with pebbles up to half an inch and with thin leaves of silty mudstone in places. The available evidence suggests a thickness of 700 ft for the Shale Grit here.

*South Head and Brown Knoll.* Around the head of the Edale valley the Shale Grit forms steep slopes in which shale slacks and accompanying spring-lines run regularly and for considerable distances. The thickness of the Shale Grit is about 600 ft in the slopes north of Lee House and about 650 ft on, and west of, Horsehill Tor. In this area individual shale bands do not exceed 20 ft in thickness.

Between Horsehill Tor and South Head the Shale Grit forms a broad spread of upland, deeply dissected, and well exposed in Roych Clough and to a lesser degree in smaller cloughs near Upper Fold. An eastward extension of the outcrop forms Rushup Edge. Between Chinley and Chapel Milton, the Forge Works No. 2 Borehole [0478 8202] probably passed through the Shale Grit (sandstones and shales with some thicker sandstone beds) between 480 and 730 ft. The higher beds in this borehole are difficult to correlate.

*Bagshaw and Barmoor Clough.* In the ground to the south of the Chapel en le Frith–Castleton road the Shale Grit dips westwards, and its broad outcrop approximates to a dip-slope. This area shows a marked contrast with that around Kinder Scout in the considerable irregularity of individual beds of sandstone and shale. There is an overall tendency for the group to thin to the south. Shale bands are mostly of normal thickness, though at Bagshaw shales about 100 ft thick are present towards the middle of the Shale Grit; these disappear northwards between Mag Low and Breck Edge by interdigitation with sandstone. To the south also the thick shale band is replaced by sandstone, though the horizon of its top and bottom may be traced for a further ¾ mile southwards as minor shale partings. Near Coalpithole Mine No. 10 Shaft the lateral passage of the Mam Tor Beds into the base of the Shale Grit has already been noted (see p. 210). Between here and Barmoor Clough the lower part of the Shale Grit forms a well-marked escarpment. The southerly thinning is particularly evident near Barmoor Clough where several sandstones are observed to wedge out to the south-west, the intervening shale partings first joining and then disappearing in the same direction.

The sides of Barmoor Clough show well-bedded sandstones with shale partings which are overlain by coarse massive sandstones, at least 35 ft in thickness, in the approach cutting [0722 7975] to the railway-tunnel north-east of Paradise. To the south of Barmoor Clough the Shale Grit thins rapidly and it disappears on the northern outskirts of Dove Holes.          I.P.S.

*Alport and Ashop valleys.* In the upper Ashop and Alport valleys and on the adjacent moors the Shale Grit varies from 400 to over 600 ft thick near the summit of the Snake Pass. The following sequence, typical of the lower part of the Shale Grit, is exposed in the lower reaches of Ashop Clough [1061 9069] and for a further 300 yd upstream:

|  | ft |
| --- | --- |
| Sandstone with silty shale partings .. | 14 |
| Sandstone with thin siltstone bands .. | 2 |
| Siltstone with thin sandstone bands .. | 2¾ |
| Sandstone, massive    ..       ..       .. | 3½ |
| Siltstone with thin sandstone bands .. | 8¾ |
| Shale, silty, with thin sandstone bands | 2¼ |
| Shale, in part silty, with thin sandstone bands in middle part       ..       .. | 14½ |
| Shale, silty, with thin sandstone bands | 6 |
| Sandstone, massive    ..       ..       .. | 4 |

|  | ft |
| --- | --- |
| Shale, silty, and siltstone with abundant thin sandstone bands ..     .. | 14 |
| Sandstone, flaggy, with many thin siltstone bands       ..       ..       .. | 7¾ |
| Sandstone, massive   ..       ..       .. | 5 |

Beds of similar lithology are exposed in Swint Clough where some 413 ft of the lower and middle parts of the Shale Grit can be seen. The middle part of the Shale Grit is also well exposed in the Alport Castles landslip-scar [142 915], where the following section was measured:

|  | ft |
| --- | --- |
| Sandstone, medium- to coarse-grained, massive      ..       ..       ..       .. | 22 |
| Shale, silty with thin sandstone bands | 5 |
| Sandstone, massive    ..       ..       .. | 15 |
| Sandstone, with thin silty shale and siltstone bands       ..       ..       .. | 39 |
| Sandstone, massive    ..       ..       .. | 19 |
| Shale, thin-bedded silty, alternating with siltstone and sandstone       .. | 5 |
| Sandstone, massive    ..       ..       .. | 19 |
| Sandstone, with thin silty shale and siltstone bands       ..       ..       .. | 18 |
| Sandstone, massive    ..       ..       .. | 15 |
| Shale, silty, and siltstone with thin sandstone bands    ..       ..       .. | 10 |
| Shale, thin-bedded silty, alternating with sandstone       ..       ..       .. | 14 |
| Shale, silty, and siltstone with thin sandstone bands    ..       ..       .. | 26 |
| Sandstone with thin silty shale bands | 8 |
| Shale, silty, and siltstoe with thin sandstone bands    ..       ..       .. | 11 |
| Sandstone    ..       ..       ..       .. | 6 |

Similar strata outcrop in the Cowms Rocks landslip-scar [124 905].

On the moors west of Lady Clough the thick massive sandstones of the middle part of the Shale Grit are succeeded by silty shales and siltstones up to 40 ft thick, exposed in the upper reaches of Upper Gate Clough, and overlain by a flaggy fine-grained sandstone with siltstone partings, about 30 ft thick [0936 9175] and in small exposures for another 250 yd downstream.

A similar sequence, consisting of a thin sandstone overlying silty shale, forms outliers resting on the middle part of the Shale Grit on the moors west of Nether Reddale Clough, on the northern part of Hope Woodlands 1½ miles N. of Alport Castles Farm and on Birchinlee Pasture. A fourth outlier is present on Rowlee Pasture and extends eastwards to Lockerbrook Heights, being exposed at the top of Bellhagg Tor [1594 8951], where 39 ft of fine- to medium-grained well-bedded or flaggy sandstone, with thin siltstone bands in the upper part overlies 19 ft of silty shale, in turn resting on the middle part of the Shale Grit. Features on Rowlee Pasture ½ mile N. of Rowlee Farm indicate the presence of a second higher sandstone over a small area of moorland, but there are no outcrops.

In Fair Brook the upper part of the Shale Grit succession consists of 60 ft of medium-grained massive sandstone, becoming coarser towards the top, resting on 45 ft of silty shale with rare thin sandstone bands on massive fine-grained sandstone 26 ft, on silty shale 40 to 45 ft, overlying the middle part of the Shale Grit. Numerous small exposures of these upper beds can be found in the tributaries of Ashop Clough and Fair Brook. In Blackden Brook [from 1223 8842 to 1198 8834] this succession, thinning eastwards, is estimated as follows:

|  | ft |
|---|---|
| Sandstone, fine- to medium-grained, mainly well-bedded but in part massive .. .. .. .. | 48 |
| Shale, in part silty .. .. .. | 12 |
| Shale, with very thin sandstone bands | 5 |
| Sandstone, fine-grained .. .. | 6 |
| Shale, with thin sandstone bands .. | 7 |
| Shale .. .. .. .. .. | 10 |

Features and small exposures in the streams suggest that a similar succession continues across Ashop Moor, but in Jaggers Clough [from 1453 8775 to 1420 8774] the two sandstones are seen to have thickened out again, the succession being estimated:

|  | ft |
|---|---|
| Sandstone, mainly massive medium- to fine-grained .. .. .. | 58 |
| Beds not exposed, believed to be shales about | 25 |
| Sandstone, mainly well-bedded medium-grained .. .. .. | 45 |
| Shale, in part silty, with rare thin sandstones .. .. .. .. | 12 |

On Crook Hill features suggest that two sandstone bands in the upper shaly part of the Shale Grit are each approximately 30 ft thick, and old pits on these features contain fragments of fine-grained flaggy sandstone.

*Upper part of Derwent valley.* In the Derwent valley north of Crook Hill and on the adjacent moors, the Shale Grit varies from 350 ft to over 450 ft in thickness. Sandstones, predominantly medium-grained and well bedded, with subsidiary thin beds and partings of siltstone and silty shale, can be seen in many small outcrops in the lower Westend valley and adjacent tributaries north of Birchinlee Pasture, and in Ouzelden Clough and its tributaries. Near the top of Ouzelden Clough [1510 9108], there is a continuous outcrop of 51 ft of medium- and coarse-grained massive sandstone, pebbly in places. Numerous exposures of sandstone are present in Abbey Brook. Farther south, the best exposures are in Mill Brook. Traced southwards the sandstones become better bedded, partly thin-bedded and flaggy, even the higher sandstones in the middle part of the Shale Grit losing their massive character, coarseness of grain and pebble content. Shaly partings and thin bands become thicker and increasingly numerous, and micaceous horizons in the sandstones become more pronounced. In a 40-ft outcrop of sandstone in a stream bank [2055 8645] east-south-east of Ladybower House, only the top 20 ft is massive and coarse-grained, the remainder being flaggy, with thin siltstone bands.

Along the eastern flank of the upper Derwent valley the upper part of the Shale Grit consists of up to 150 ft of shales and siltstones with two thin sandstones, as on the moors to the west. The higher sandstone caps an outlier of these beds on Nether Hey. In Abbey Brook [1948 9264] 25 ft out of an estimated total of 40 ft of silty shale are seen directly overlying the main sandstone sequence of the Shale Grit. The succeeding 6 ft of sandstone, fine-grained and flaggy, form a small waterfall 250 yd upstream [1968 9256]. In a stream which joins Abbey Brook at this point, this sandstone is seen to be succeeded by silty shale 30 ft, this in turn being overlain by the higher sandstone, medium-grained well bedded, 40 ft. This higher sandstone forms a gorge at the confluence [1979 9244] of Abbey Brook and a tributary stream. Farther south, in Far Deep Clough [1873 9047], flaggy sandstone 12 ft, is seen overlying silty shale 8 ft, approximately

P

10 ft above the top of the main sandstone sequence of the Shale Grit. This sandstone maintains its thickness of 15 to 20 ft as far south as Riding House, beyond which it cannot be traced. The silty shales between the lower and higher sandstones are rarely exposed, but appear to decrease in thickness from nearly 40 ft on Abbey Bank to less than 15 ft at Riding House. The higher sandstone, 30 to 50 ft thick south of Abbey Brook, forms locally prominent features, such as Abbey Bank and Pike Low, with small outcrops in streams; old pits show mainly fine- to medium-grained well-bedded or flaggy sandstone. In the upper reaches of Far Deep Clough and Dovestone Clough nearly continuous outcrops indicate that it has thickened to about 70 ft; farther south it becomes increasingly flaggy and poorly exposed.

*Edale valley.* On the northern slopes of the Edale valley, the Shale Grit is 400 to 475 ft thick. Numerous outcrops of the main sandstone sequence occur in Crowden Brook, Grinds Brook, Oller Brook and Lady Booth Brook, consisting largely of medium-grained, well-bedded sandstones with subsidiary siltstone and silty shale bands and partings. The slopes between the streams are poorly exposed, and only one major shale horizon, at about the middle of the succession, can be traced continuously. The following section from Grinds Brook [1182 8702] is typical of the lower part of the Shale Grit:

|  | ft |
|---|---|
| Sandstone .. .. .. .. | 12 |
| Shale, in part silty .. .. .. | 3 |
| Sandstone .. .. .. .. | 2 |
| *Not exposed* .. .. .. .. | 6 |
| Sandstone .. .. .. .. | 15 |
| Shale, silty, with thin sandstone bands | 7 |
| Sandstone .. .. .. .. | 6 |
| Shale, silty .. .. .. .. | 3 |
| Sandstone, with thin silty shale bands | 8 |
| Alternating thin bands of silty shale and sandstone .. .. .. | 3 |
| Shale, in part silty .. .. .. | 3 |
| Sandstone .. .. .. .. | 2 |

Somewhat higher strata are exposed in a quarry [1517 8723] ¾ mile N.W. of Edale End Farm:

|  | ft |
|---|---|
| Sandstone, medium-grained massive | 10 |

|  | ft |
|---|---|
| Sandstone, fine- and medium-grained, with thin siltstone bands .. .. | 7 |
| Sandstone, medium-grained massive | 25 |

The upper 25 ft of the main sandstone sequence are medium- to coarse-grained and massive in a series of exposures complicated by folding in the upper part of Grinds Brook [1151 8718].

The shales succeeding the middle part of the Shale Grit are approximately 30 ft thick in Grindsbrook Clough, but thicken to over 50 ft in Oller Brook. Both the upper and lower sandstones which can be traced within the upper part of the Shale Grit, here totalling nearly 200 ft, are split by well-developed shale bands in Oller Brook, where they are both sporadically exposed. In Grindsbrook Clough they are largely obscured, but the upper sandstone appears to be at least 70 ft thick.

*Hope valley.* Approximately 230 ft of Shale Grit are present but poorly exposed on Lose Hill. In the Back Tor landslip-scar, 81 ft of the lower part of the Shale Grit can be seen resting on Mam Tor Beds. The exposed sequence consists of sandstones, thin-bedded and flaggy near the base, passing upwards into thicker units, with thin bands and partings of siltstones and silty shales decreasing in abundance up the succession.

On the steep slope under Hope Brink, west of Win Hill, four shale bands in the lower and middle parts of the Shale Grit are indicated by prominent spring lines. These horizons can be traced by means of localized features and individual springs along the south side of Win Hill, where the total thickness is over 400 ft. Exposures are few, small and confined to flaggy or well-bedded sandstones. Two distinct sandstone features can be traced in the unexposed shales forming the upper part of the Shale Grit around Win Hill. The higher bed appears to be approximately 50 ft thick, and outcrops as fine- to medium-grained sandstone 22 ft in an old quarry [1796 8523] ½ mile W. of Win Hill summit.

*Bamford and Hathersage.* On the eastern slopes of the Derwent valley between Ladybower House and Hathersage the Shale Grit thins southwards from over 350 ft to slightly under 300 ft in thickness. The bands of argillaceous strata, particularly one near the base and one in the middle of the succession, increase considerably in thickness southwards from Brookfield Manor. The sandstones are, almost without exception, fine-grained, well-bedded or flaggy, with abundant thin micaceous bands and partings. Exposures in Dale Bottom, east of Hathersage indicate that the interbedded siltstones and silty shales account for at least 40 per cent of the thickness of the Shale Grit. The middle part of the Shale Grit is exposed [2375 8174] along Dale Bottom:

|  | ft |
|---|---|
| Siltstone .. .. .. .. | 30 |
| Sandstone, in part flaggy .. .. | 9 |
| Siltstone, with thin flaggy sandstone bands .. .. .. .. | 16 |
| *Not exposed* .. .. .. .. | 6 |
| Sandstone, in part flaggy .. .. | 9 |
| Siltstone, thin-bedded, alternating with sandstone .. .. .. | 9 |
| Sandstone .. .. .. .. | 8 |
| Sandstone with siltstone bands .. | 6 |
| Siltstone .. .. .. .. | 7 |
| Siltstone with thin sandstone bands .. | 12 |
| Sandstone with thin siltstone bands .. | 14 |
| Sandstone .. .. .. .. | 2 |
| Sandstone with thin siltstone bands .. | 6 |
| Siltstone with sandstone bands .. | 17 |

The shales and siltstones of the upper part of the Shale Grit are not exposed south of Ladybower House, and there is no trace of the lower of the two sandstones known farther north and west. An outcrop [2074 8664] of medium-grained fairly massive sandstone 4 ft, in the stream some ¼ mile E.N.E. of Ladybower House, is part of the upper sandstone which forms a slight feature between this point and Greens House, 1½ miles N. of Hathersage. It cannot be traced farther south, unless it equates with an isolated lenticular sandstone containing thin siltstone bands, which occurs within shaly strata at least 40 ft above the main sandstone sequence and is exposed at Kimber Court [2384 8247] in the lane on the north-east side of the house.                                    G.D.G.

*Brough, Great Hucklow and Eyam.* To the south of Brough the Shale Grit forms the complex features of Bradwell Edge, Durham Edge and Hucklow Edge. A sandstone, present beneath it in the Mam Tor Beds at Shatton (see also p. 211), when followed southwards is seen to join the base of the Shale Grit in a manner closely resembling

that occurring near Sparrow Pit (p. 210).
                                    I.P.S.

In the high ground between Bamford and Abney the Shale Grit thins south-westward from over 350 ft to under 300 ft. It consists of an upper and lower group of sandstones containing thin interbedded shales and siltstones, separated by a thick succession of shales and siltstones with minor sandstones, and succeeded by a thin shaly sequence with a thin sandstone at the top. The lower part of the succession is flaggy sandstones with interbedded siltstones and silty shales in numerous small exposures along Abney Clough and Highlow Brook. The best outcrops of the upper sandstone group are in two minor streams between Banktop and Offerton Hall, and in Dunge Brook. The top of the highest sandstone is exposed in Dunge Brook [2132 8037] as 13 ft of sandstone, in part flaggy.                                    G.D.G.

Between Shatton and Abney the Shale Grit forms high ground, deeply dissected by Overdale Brook, which provides a typical section of the lower beds for some 500 yd below its source at Silver Well [1872 8031]. Here the normal features of these beds, sandstones with numerous regular shale partings, are seen. On the western side of Bleak Knoll, at least 300 ft of Shale Grit are present. South of Abney the Shale Grit has fewer shale partings and these are limited to the lower part of the formation, much of the middle part consisting of a single bed of sandstone giving rise to the uniform dip-slope north-west of Abney, at Abney Grange and near Camphill Farm, with a narrow tongue extending westwards along the scarp to Bretton. A return to the normal development of sandstones with shale partings is apparent in the highest beds on Abney Low and Bretton Moor.

The full thickness of Shale Grit was passed through in Ladywash Shaft [2189 7754], where 219 ft of interbedded 'gritstone' and 'shale' were recorded. To the south-east of this locality the grit continues to give rise to a marked escarpment though the shale partings disappear, giving a much more massive aspect to the subdivision than is usually the case. A typical section in a quarry [2251 7651], near Riley House, showed coarse flaggy sandstone with many siltstone partings 18 ft, on massive coarse sandstone with sporadic siltstone partings 25 ft.

Between Eyam and the Derwent valley the Shale Grit forms a well-marked escarpment. At Calver the grit appears to fail, though it seems to be present farther to the east at

Curbar, from whence it can be traced, though thinning, in a southerly direction beyond the border of the district.

## BEDS BETWEEN SHALE GRIT AND KINDERSCOUT GRIT

*Kinder Scout.* These beds reach their maximum thickness of 350 to 400 ft on Black Ashop Moor, north of Kinder Scout. The beds are well exposed in the vicinity of Ashop Head and in the southern tributaries of Ashop Clough, as mudstones, silty mudstones and siltstones with thin sandstone bands. This lithology is typical.

Below Kinder Downfall the group is mainly obscured by landslip; its thickness here is about 300 ft. It thins southwards around the Kinder Scout plateau, being only 175 ft near Kinder Low, 150 ft at The Cloughs and 120 ft in Crowden Brook. I.P.S.

On Blackden and Ashop moors these beds are slightly over 250 ft thick and contain local lenticular sandstones in their upper part. The succession in the upper reaches of Blackden Brook [1198 8834 to 1171 8833] is:

|                                                                          | ft  |
|--------------------------------------------------------------------------|-----|
| Shale, silty, and siltstone with thin sandstone bands  ..   ..   ..      | 30  |
| Sandstone, fine- to medium-grained, well-bedded ..   ..   ..   ..        | 28  |
| Shale, silty, and siltstone with rare thin sandstone bands  .. about     | 100 |
| Sandstone, fine-grained   ..   ..                                        | 9   |
| Shale, silty, with rare thin sandstone bands   ..   ..   ..   ..          | 78  |

There is a marked thinning eastwards, less than 200 ft of siltstone and silty shale with no distinct sandstones being present in Jaggers Clough, but between this valley and Lady Booth Brook an impersistent sandstone, estimated to be up to 40 ft thick, is developed about half-way up the sequence. About 220 ft of beds are present in the upper reaches of Grinds Brook, where nearly 100 ft of silty shales passing up into siltstones are seen resting on Shale Grit [1098 8729]; there are continuous sections upstream for 200 yd. A number of sandstone bands, generally less than 1 ft thick, are of two main types: either soft, lenticular, micaceous and flaggy, or hard, strongly cemented and nodular. Some 40 ft of siltstones with thin sandstone bands are seen immediately below the Lower

Kinderscout Grit farther upstream [1069 8729]. A sandstone feature is present around the eastern and southern slopes of Grindslow Knoll.                                   G.D.G.

*Western outcrops.* To the east of Chunal some 350 ft of shale, siltstone and thin sandstone are present at this horizon. A section in the upper part of Whitethorn Clough [0466 9085] showed silty mudstone with thin sandstones 14 ft, soft shale, 'sulphurous' at base, silty and 'blocky' above 22 ft, on silty shale with thin flaggy sandstones and siltstones 10½ ft. In this vicinity lenticular sandstones are present near both the base and top of the group.

The Charlestown Borehole (see also pp. 363–4) passed through the Kinderscout Grit into shale, part of which must belong to the present beds, but with the failure of the Shale Grit the subdivision loses its identity and these shales are best considered as a whole.

To the south of Hollingworth Clough a rapid thinning of the beds between the Kinderscout Grit and Shale Grit takes place: from at least 300 ft in Hollingworth Clough to some 50 ft on Middle Moor. In the Kinder valley only some 30 ft of shale are present, but between there and the slopes [052 861] to the west of Coldwell Clough, lateral passage of the uppermost part of the Shale Grit into shale results in a substantial thickening of the subdivision. From the last locality the beds continue southwards, though displaced by faults, as far as Upper Fold. South of this point they can be followed as a shallow depression, often drift-filled, to Chapel Milton. South of Chapel en le Frith the outcrop expands greatly owing partly to a flattening of the dip, partly to the inclusion within the subdivision of shale which is the lateral equivalent of the lower part of the Kinderscout Grit (see p. 219), and partly to thickening; in all some 350 ft of shale are present here. At Martinside and near Cow Low, lateral passage of the upper part of these beds into Kinderscout Grit takes place, and the group thins in consequence. Near

Dove Holes it loses its identity with the southerly disappearance of the Shale Grit.

I.P.S.

*Eastern outcrops.* The succession on Win Hill, estimated to be just over 150 ft thick, is not exposed, but old pits in a narrow feature around the upper slopes contain fragments of fine-grained flaggy sandstone. On Crook Hill a vestigial feature on the north-eastern slope suggests the local development of a similar sandstone.

East of the Derwent valley and north of Ladybower House, the beds between the Shale Grit and Kinderscout Grit vary from 100 ft to 200 ft thick, for although in general the strata thin out to the south, the succession is expanded locally owing to wedging out of successive units of the lower part of the Lower Kinderscout Grit with consequent addition of the equivalent and intervening strata to the sequence below. Between Abbey Bank and Pike Low two thin impersistent sandstones form features in the otherwise concave slopes of these beds and old pits along the outcrops contain loose fragments of fine-grained flaggy sandstone. A similar impersistent sandstone feature is present on the slopes east of Riding House. Exposures of the argillaceous strata are rare. Sporadic outcrops of silty shales, siltstones and very thin sandstones, the latter either hard and nodular or soft and micaceous, occur in a stream which joins the upper part of Abbey Brook near the shooting cabin, and a thin ripple-marked sandstone occurs in the stream [1968 9198] near the top of these beds. In Dovestone Clough [1921 8980] 8 ft of silty shale, with very thin hard sandstone bands, crop out a few feet above the top of the Shale Grit.

Under Bamford Edge there appear to be about 200 ft of beds between the Shale Grit and Kinderscout Grit. An impersistent coarse massive sandstone, at least 20 ft thick, is present in the upper part of these beds, forming crags on the steep slopes of Priddock Wood [2065 8642]. Outcrops of siltstone and silty shale with rare thin sandstone bands occur in a number of small exposures in the upper reaches of the stream east of Bamford Filters, at approximately 400 yd S.W. of Bole Hill. Siltstones with thin sandstone bands, 11 ft thick, crop out [2373 8348] east of North Lees. A number of small outcrops of the strata underlying the Kinderscout Grit, mainly siltstone and silty shale, laminated and micaceous, with abundant thin sandy bands and partings, are present in a stream [238 835] east-north-east of North Lees. Outcrops of micaceous siltstone and silty shale with very thin hard flaggy sandstone bands occur in Dale Bottom between 200 yd and 470 yd W. of Mitchell Field.

The argillaceous beds between the highest sandstone on Offerton Moor, which is considered to be Kinderscout Grit, and the Shale Grit below, decrease in thickness southward across the moor from 150 ft to less than 100 ft. In Dunge Brook [2132 8037] siltstones with thin flaggy sandstones 9 ft, overlie the top of the Shale Grit. Some 85 yd farther upstream [2125 8039], and a few feet below the base of the Kinderscout Grit, 6 ft of silty shale are exposed.          G.D.G.

On the western side of Shatton Moor from 100 to 120 ft of shale are present at this horizon. The top is irregular owing to lateral passage into the lowest part of the Kinderscout Grit.

Ladywash Shaft (see p. 372), commencing about 20 ft below the base of the Kinderscout Grit, passed through made ground to 29 ft and continued in shale to 93 ft before reaching sandstones referred to the Shale Grit. The beds thin eastwards from this locality and, with the disappearance of the Shale Grit at Calver Sough, cannot be recognized beyond this point.

## KINDERSCOUT GRIT

*Kinder Scout.* At the type locality the Lower Kinderscout Grit, about 300 ft thick, forms an elevated plateau; the Upper Kinderscout Grit is absent. Much of the outcrop is obscured by peat. The grit is typically a very massive coarse-grained feldspathic sandstone, cross-bedded and pebbly in places, characters well shown at Kinder Downfall [0815 8883] (see Plate XVIII A) where 265 ft of it are exposed. The section below Crowden Tower [0959 8706], shows 240 ft of the same beds overlain by a shale band some 20 ft thick, and the latter by 40 ft of coarse sandstone forming the top of the escarpment. Elsewhere shale partings occur infrequently near the base of the grit on the northern edge of the plateau [091 899] and on its westward extension. Shale bands, including the one seen at Crowden Tower, are also present along the south-western edges of the plateau as at Edale Head.                    I.P.S.

Under Seal Edge the lower and upper leaves of the Lower Kinderscout Grit appear to be united, but in the upper reaches of Blackden Brook [1170 8834], the lower leaf is present as coarse-grained sandstone 34 ft thick, succeeded by 33 ft of siltstone with thin sandstone bands. The same leaf forms a rather ill-defined feature under Blackden Edge, becoming more distinct eastwards, where [1397 8855], ⅓ mile W.N.W. of Crookstone Knoll, 17 ft of coarse-grained pebbly sandstone are exposed. Ripple-marks are seen in siltstones and thin sandstone bands in a stream running down Blackden Edge [1290 8819], between the lower and upper leaves of the Lower Kinderscout Grit. The lower leaf and overlying beds are seen in Jaggers Clough [1410 8774]:

|  | ft |
|---|---|
| Siltstone and silty shale .. .. | 11 |
| Sandstone, massive .. .. .. | 1 |
| Sandstone, flaggy .. .. .. | 1 |
| Siltstone .. .. .. .. | 3 |
| Sandstone, with thin siltstone bands and partings .. .. .. | 11 |

The lower leaf is apparently not present between the upper reaches of Lady Booth Clough and Golden Clough, but it can be seen at the base of Nether Tor, where 12 ft of sandstone are separated from the upper leaf by 17 ft of siltstone with thin sandstone bands. In a stream [1100 8756] west of Upper Tor, this succession is reduced to 6 ft of sandstone overlying 12 ft of siltstone, but it thickens in the upper reaches of Grindsbrook Clough, where in a crag [1075 8735], 17 ft of sandstone are exposed. The succeeding argillaceous beds are 15 to 25 ft thick in this locality, but are nowhere exposed. The lower leaf forms a local feature south-west of Grindslow Knoll, with small exposures of coarse-grained pebbly sandstone, mainly massive, but this feature cannot be traced north-eastwards around the head of Crowden Brook.

The upper leaf of the Lower Kinderscout Grit is about 80 ft at the eastern end of the Kinder Scout plateau, thickening westwards to about 150 ft south of Seal Edge. Abundant crags and residual blocks, mainly of massive coarse-grained pebbly sandstone, are present on Seal Edge. A continuous exposure of 72 ft of sandstone occurs along Blackden Brook [1166 8836] westwards for 100 yd. It is mainly massive and coarse-grained, and locally pebbly, but some medium-grained well-bedded and even flaggy varieties can be seen. False-bedding is present in the craggy outcrops 100 yd south of this point.

In exposures along Jaggers Clough, westwards for 190 yd from a point [1406 8778] ¾ mile N. of Rowland Cote, 53 ft of mainly massive coarse-grained and pebbly sandstones are succeeded by 17 ft of medium to coarse-grained sandstone, mainly well-bedded or flaggy.

In addition to crags and residual blocks on top of the plateau, the southern edge of Kinder Scout north of the Edale valley shows a number of cliff-like outcrops in which the massive coarse-grained pebbly nature of the sandstones can be seen. Up to 42 ft of sandstone crops out at the western end of Ringing Roger, up to 65 ft on Nether Tor, and up to 75 ft on Upper Tor. Where Grinds Brook cuts down through the edge of the plateau [1066 8744] and for 100 yd upstream, the succession is:

|  | ft |
|---|---|
| Sandstone, well-bedded medium- to coarse-grained .. .. .. | 35 |
| Sandstone, false-bedded coarse-grained pebbly .. .. .. | 45 |
| Sandstone, flaggy fine- to medium-grained, with micaceous silty partings .. .. .. .. | 4 |
| Sandstone, false-bedded coarse-grained pebbly .. .. .. | 30 |

Numerous crags and residual blocks of sandstone are present on Grindslow Knoll. Towards the centre of the plateau, streams have cut through the peat to expose abundant loose blocks and small exposures of sandstone; these are mainly massive or well-bedded, medium- to coarse-grained, but include some thin-bedded flaggy sandstones.                    G.D.G.

*Glossop to Chapel en le Frith.* In this area the Kinderscout Grit contains a thick bed of

A. KINDER DOWNFALL: ESCARPMENT OF LOWER KINDERSCOUT GRIT

B. CHINLEY CHURN: ESCARPMENTS OF SANDSTONES IN UPPER PART OF
MILLSTONE GRIT SERIES AND LOWER PART OF LOWER COAL MEASURES

(L 231)

shale with the Butterly Marine Band, separating the Lower Kinderscout Grit from the thinner Upper Kinderscout Grit. The three separate components are described separately below. In all but the southern part of this area (see below) the Upper Kinderscout Grit is closely overlain by the *Reticuloceras gracile* Marine Band. Both sandstone leaves give rise to marked scarp and dip-slope features, although where the shale underlying the Lower Kinderscout Grit thins, as on Kinder Bank, Hayfield, the lower scarp disappears and a composite feature of Kinderscout Grit and uppermost Shale Grit results.

The **Lower Kinderscout Grit,** some 250 ft thick, reaches its thickest development in the Hayfield area. A typical section is provided by the old quarries [0432 8658] south of the Sportsman's Inn, where some 160 ft of yellow-brown massive coarse-grained sandstone show large-scale cross-bedding, with some channelling of individual beds. Farther north, Hollingworth Clough provides a good section of massive cross-bedded coarse and frequently pebbly sandstone. From this locality the grit thins northwards, being some 150 ft thick east of Chunal. The Charlestown Borehole (see pp. 363–4) proved only 74½ ft of Lower Kinderscout Grit, with the base at 275 ft; the upper 12 ft were coarse-grained but contained argillaceous partings; they were underlain by a 2½-ft bed of silty mudstone resting on 1 ft of silty seatearth with rootlets at 216 ft. The underlying sandstone was coarse and massive with pebbly bands though the lowest 8½ ft formed a passage to the underlying shale.

To the south of Hayfield, also, the Lower Kinderscout Grit varies greatly. Between the River Kinder and Mount Famine, its development continues as in the north. To the south and south-east, however, the grit emerges southwards from a much-faulted area as two distinct leaves. The uppermost of these again splits southwards into two parts, both of which die out between Andrews Farm and a point 700 yd south of it. The lower leaf maintains its character southwards as a coarse massive pebbly rock and gives rise to the strong scarp of South Head. It is exposed in the following section [0566 8212] in the railway-cutting south-east of New Smithy [top of section at 0534 8232]:

| | ft |
|---|---:|
| Sandstone, yellow-brown well-bedded, with a few flaggy partings  ..    .. | 12 |
| Sandstone, massive coarse feldspathic | 40 |
| Shale, silty, with thin sandstone bands | 9 |
| Sandstone, yellow-brown well-bedded medium-grained  ..    ..    .. | 6 |
| Sandstone, yellow-brown massive coarse  ..    ..    ..    .. | 20 |

The total thickness here is about 110 ft.

The last exposure of any importance of the depleted Lower Kinderscout Grit is in the railway-cutting [0557 8164] south of Chapel Milton, where sandy shale overlies 27 ft of massive feldspathic sandstone. Farther south the small feature at Chapel en le Frith suggests that the whole bed degenerates and passes into shale about a quarter of a mile south-west of the town.

A complete section of the **shales with the Butterly Marine Band,** 87 ft 3 in thick, is provided by the Charlestown Borehole:

| | Thickness | | Depth | |
|---|---:|---:|---:|---:|
| | ft | in | ft | in |
| Base of Upper Kinderscout Grit  ..    .. | — | — | 113 | 3 |
| Mudstone, silty, and siltstone with sandstone bands ..    ..    .. | 25 | 9 | 139 | 0 |
| Sandstone, argillaceous, with partings of silty mudstone and siltstone in lower part ..    .. | 26 | 5 | 165 | 5 |
| Mudstone, pale grey with thin bands and occasional lenses and nodules of ironstone    .. | 4 | 7 | 170 | 0 |

Butterly Marine Band
Mudstone, lithology as above; *Sanguinolites sp.* and bivalve fragment indet. at 170 ft to 170 ft 6 in; palaeoniscid scale 182 ft; coiled shell indet. 184 ft; *Lingula mytilloides, Retispira?* [juv.], *Rhabdoderma sp.* [scales] at 185 ft 3 in; *L. mytilloides, Polidevcia* aff. *attenuata* (Fleming), *San-*

| | Thick-ness | | Depth | |
|---|---|---|---|---|
| | ft | in | ft | in |

guinolites sp. [juv.]
186 ft to 187 ft 9 in;
cf. *Cochlichnus kochi*,
*Planolites ?*, cf. *Cypri-*
*dina sp.* 188 ft 4 in to
188 ft 6 in; *L. mytil-*
*loides, Sanguinolites*
*ovalis* 189 ft to 190 ft
2 in; *L. mytilloides*,
turreted gastropod
indet., *Sanguinolites*
*ovalis, Reticuloceras*
*sp. nov.*, mollusc spat
191 ft 6 in to 192 ft
9 in; *L. mytilloides,*
*Orbiculoidea nitida,*
*Retispira undata,*
*Aviculopecten* aff.
*dorlodoti, Edmondia*
*sp., Polidevcia* aff.
*attenuata, Sanguino-*
*lites ovalis, S.* cf. *tri-*
*costatus, Schizodus*
*antiquus* Hind, *Ephip-*
*pioceras?* 193 ft 3 in
to 197 ft 3 in    ..   27   3    197   3

Mudstone, dark platy
micaceous, with some
irregular pyrite    ..   1   11½   199   2½

Coal    ..    ..    ..       3½   199   6

Seatearth, banded silty
with rootlets ..    ..   1   0   200   6

This is a much fuller development of the Butterly Marine Band than anywhere at outcrop in the district. Non-marine inter-digitations are present at 182 ft and between 188 ft 4 in and 188 ft 6 in. Other aspects of the fauna have already been discussed on p. 177.

A section[1] on Chunal Moor [0376 9151] is:

| | ft |
|---|---|
| Base of Upper Kinderscout Grit    .. | — |
| *Not exposed*    ..    ..    .. about | 10 |
| Mudstone, sandy    ..    ..    .. | 5 |
| Mudstone, grey silty, darker below   .. | 14 |
| Mudstone, dark grey; *Reticuloceras* | |
| *sp. nov.*; fish debris ..    ..    .. | 1 |

| | ft |
|---|---|
| *Not exposed*    ..    ..    .. about | 5 |
| Lower Kinderscout Grit    ..    .. | — |

Another section [0361 9054] in a stream east-north-east of the Grouse Inn showed a greater thickness of shale:

| | ft |
|---|---|
| Sandstone, coarse, thin-bedded in lower part    ..    ..    ..    .. | 10 |
| Mudstone, silty    ..    ..    .. | 20 |
| Sandstone, thin-bedded    .. about | 3 |
| Mudstone, sandy    ..    ..    .. | 4½ |
| Mudstone, very silty at top, less so below    ..    ..    ..    .. | 20½ |
| Mudstone, dark shaly with *Lingula mytilloides, Aviculopecten* aff. *dor-lodoti, Sanguinolites ovalis, Reti-culoceras sp.*, mollusc spat ..    .. | 5 |
| Mudstone, dark silty; plant debris   .. | 0½ |
| Mudstone, grey with rootlets and lenses of ganister-like sandstone   .. | 2¾ |
| Lower Kinderscout Grit    ..    .. | — |

This section, although farther away from the Charlestown Borehole than that quoted above, shows a closer similarity to it. This is due to the northward thickening of the 3-ft sandstone present within the shale near the Grouse Inn and its junction with the main bed of the Upper Kinderscout Grit. Rapid variations in the thickness of the shale to the south are probably due to the same cause.

At Little Hayfield the beds between Upper and Lower Kinderscout grits, poorly exposed, are about 60 ft in thickness.

The southernmost exposure [0458 8603] showing the Butterly Marine Band is ¼ mile N. of The Heys[1]:

| | ft |
|---|---|
| Upper Kinderscout Grit    ..    .. | — |
| *Not exposed*    ..    ..    ..    .. | 30 |
| Mudstone, dark silty, becoming less silty below    ..    ..    .. | 9 |
| Mudstone, dark fossiliferous, especially at 3 in from top and at base; *Reticuloceras sp. nov.*, mollusc spat | 2⅔ |
| Mudstone, dark; goniatite fragment and rare plant debris    ..    .. | 1¼ |
| Mudstone, dark silty and sandy, part tending towards seatearth, about | 1 |
| Lower Kinderscout Grit    ..    .. | — |

---

[1]Measured by Mr. C. G. Godwin.

To the south of Mount Famine the lateral passage of part of the Lower Kinderscout Grit into shale results in a considerable thickening of the shales between the Upper and Lower Kinderscout Grit, which here reach 350 ft. In the region of this passage the horizon of the coal of the Charlestown Borehole is perhaps represented [0529 8435] north-north-west of Andrews Farm by 2½ ft of soft black micaceous silty mudstone resting on black micaceous siltstone. Lower beds in these shales are exposed in the stream, from a point 200 yd E. of Andrews Farm, downstream for a third of a mile as silty mudstones and siltstones with thin sandstone bands. To the south, the shale group, though thinning, can be traced through Chapel Milton and Chapel en le Frith. Beyond the latter place, with the disappearance of the Lower Kinderscout Grit, it cannot be recognized, and has been included with the beds between Kinderscout Grit and Shale Grit.

In the Charlestown Borehole the **Upper Kinderscout Grit** is represented by at least 62 ft 7 in of beds, with the base at 113 ft 3 in; the lithology is mainly fine- to medium-grained sandstone, micaceous and flaggy in places and with carbonaceous partings. Coring commenced 10 in below the top of the sandstone, so details of the overlying beds are lacking.

The Upper Kinderscout Grit, some 90 ft thick, forms a well-marked dip-slope between Chunal and The Intakes. Workings [0352 9128] ¼ mile S. of Chunal show a typical section: coarse feldspathic sandstone with fine-grained flaggy bands 13 ft. The beds separating the grit from the overlying *Reticuloceras gracile* Marine Band are represented by a thickness of 2 ft of unexposed beds in the stream north-west of the Grouse Inn (see p. 226).

Between The Intakes and Brookhouses the outcrop of the Upper Kinderscout Grit is much broken by faulting, but farther south it resumes its regular character with well-marked dip-slopes. Like the Lower Kinderscout Grit it reaches its maximum development in the Hayfield area, where it is from 100 to 150 ft thick. The best sections are in several quarries south of Chinley Head; a typical section [0503 8429] near the Lamb Inn showed 28 ft of yellow massive coarse felds-

pathic sandstone, with bands of small pebbles up to ¼-in diameter throughout, scattered shale pellets and a shale lens (2 ft by 8 ft) near base, on 0 to 14 ft of massive sandstone as above, but with some finer bands and full of plant debris, on 18 ft of massive sandstone, as above. Southwards the thickness of the leaf is maintained, being about 120 ft in and around the railway-cutting at New Smithy [west end: 0474 8267].

The borehole at Whitehall Works [0355 8202] proved the base of the *Reticuloceras gracile* Marine Band at 467½ ft. Below were 40½ ft of silty micaceous mudstones with thin siltstone bands, proved to 508 ft. No Kinderscout Grit was found and it is considered that the silty beds are its lateral equivalent rather than an expansion of the few feet of beds which normally lie between the top of the grit and the *R. gracile* Band.

Between Chapel Milton and Lower Crossings the Upper Kinderscout Grit gives rise to a fairly well-defined escarpment. South of Lower Crossings the outcrop passes beneath drift.

*Martinside to Buxton.* In this area only one main bed of Kinderscout Grit is present, though this is subject to minor variation. As noted above in the Whitehall Works Borehole the upper part of the Kinderscout Grit locally passes laterally into more argillaceous and silty beds, especially around Martinside. Here the exposure of the *R. gracile* Marine Band north-east of Ridge Hall (see p. 229) lies about 100 ft above the Kinderscout Grit, the intervening beds being mainly siltstones and silty mudstones. The Kinderscout Grit at Martinside (the Upper Kinderscout Grit of the country to the north) splits southwards into three principal leaves and it is the middle one of these which persists, the other two dying out rapidly. South-east of Cowlow the grit is somewhat irregular, being split into a number of variable components. The best section here is that in a quarry [0693 7871] showing 32 ft of yellow-brown massive coarse-grained sandstone in beds up to 6 ft thick, with flaggy partings. At Cowlow (see also p. 229) there is room for some 90 ft of mainly argillaceous beds between the base of the *R. gracile* Band and the top of the Kinderscout Grit.

Between Dove Holes and the southern boundary of the district the Kinderscout

Grit consists of a more regular upper portion from 60 to 80 ft in thickness, underlain by one or more thin lower leaves; the overall thickness not exceeding 120 ft. Sections are scarce, though a quarry [0698 7702] south-west of Dove Holes showed about 25 ft of sandstone, mostly rather coarse-grained. At Blackedge Reservoir there is room for about 100 ft of argillaceous beds between the top of the Kinderscout Grit [0680 7650] and the *R. bilingue* Band; this suggests a southerly thinning from Cowlow of the beds between the *R. gracile* Band and Kinderscout Grit to perhaps 70 ft. This is further borne out by the section in Hogshaw Brook, in the Buxton (111) district, where the *R. bilingue* early form Marine Band [0598 7422] lies about 40 ft above the Kinderscout Grit suggesting that the *R. gracile* Band (which is not exposed) is at or near to its normal position close above the Kinderscout Grit. Beyond the southern edge of the district, the Kinderscout Grit thins south-west of Hogshaw Brook and dies out [0575 7385] near the Palace Hotel, Buxton.

*Todd Brook Anticline.* Towards the northern end of this structure the Kinderscout Grit, some 180 ft thick and free from shale partings, passes round the axis, forming a sharp feature. Exposures are good only near Lumbhole Mill; a typical one [9872 8039] showed brown massive sandstone with micaceous partings 12 ft, siltstone and silty mudstone with sandstone bands 7 ft, on flaggy sandstone 1 ft 2 in. The outcrop of the Kinderscout Grit is repeated by faulting ¼ mile S.W. of Bailey's Farm.

Between Kettleshulme and Hollowcowhey there is a fairly regular arrangement of the Kinderscout Grit. The calculated section near Dunge Farm is typical: upper leaf 320 ft, shale 55 ft, lower leaf 70 ft. South of Hollowcowhey the grit has irregular shale partings and there is some tendency for the lower leaves to die out. The grit gives rise to a sharp feature at Andrews Edge running south from a point [9857 7543] near Redmoor Farm. A small exposure east of Andrews Edge [9853 7514] shows flaggy sandstones with shale and siltstone partings. The outcrop passes over the axis of the Todd Brook Anticline near the southern boundary of the district and continues northwards to Nab End; from thence to Kettleshulme it is faulted out.                      I.P.S.

*East of the Derwent.* The lower leaf of the Lower Kinderscout Grit is about 150 ft thick north of Howden Dean, where it forms the main escarpment of Howden Edge above Howden Moors. There are few exposures, but loose blocks of coarse pebbly sandstone are widespread. East of this edge a series of ill-drained slacks suggests the presence of minor shale bands. Around the head of Abbey Brook the lower leaf consists of a lower bed of medium- to fine-grained well bedded and flaggy sandstone with subsidiary argillaceous bands, and an upper bed of mainly massive coarse sandstone, which rests, with apparent slight discordance in places, on up to 50 ft of siltstones and silty shales. The complete succession of the lower leaf in Bent's Clough [2000 9244], is:

|  | ft |
|---|---|
| Sandstone, massive coarse-grained .. | 55 |
| Shale, silty with siltstone bands .. | 15 |
| Sandstone, massive medium- to fine-grained .. .. .. .. | 12 |
| Shale, silty .. .. .. .. | 15 |
| Sandstone, flaggy medium- to fine-grained .. .. .. .. | 30 |
| Shale, silty, and siltstone .. .. | 10 |
| *Not exposed* .. .. .. .. | 6 |
| Sandstone, flaggy fine-grained .. | 12 |

The lower bed is not more than 30 ft thick on the moors south of Howden Dean, where loose fragments and small outcrops in old quarries are mainly of fine- to medium-grained sandstone, and no further trace of this bed is present south of Dovestone Clough. The upper bed forms a prominent feature on Lost Lad Hillend [191 914], where there are abundant loose blocks of massive medium- to coarse-grained sandstone. This feature can be traced southwards, decreasing in magnitude across the Gusset and Dovestone Clough, where fine-grained flaggy sandstone with thin siltstone bands is exposed, until it finally disappears under Whinstone Lee Tor.

The strata between the lower and upper leaves of the Lower Kinderscout Grit are about 120 ft thick north of Low Tor. Numerous small exposures of siltstone and silty shale with thin sandstone bands occur in the highest reaches of Abbey Brook 2¼ miles E.N.E. of Marebottom Farm. No exposures are present on the slopes west of Low Tor, Howshaw Tor and Back Tor, but in the upper reaches of Dovestone Clough, where the total sequence is approximately 80 ft, there are small exposures of silty shale with thin siltstone and sandstone bands. No outcrops occur under Derwent Edge farther south, where with the disappearance of the lower leaf, these strata overlie and become indistinguishable from the beds between the Shale Grit and Kinderscout Grit.

The upper leaf of the Lower Kinderscout Grit is not much more than 80 ft thick N.N.E. of Low Tor, where it forms a small but well-defined feature with scattered blocks of massive coarse-grained pebbly sandstone. Between Howshaw and Whinstone Lee Tors, on Derwent Edge, it has increased to 120 to 180 ft in thickness and forms the main escarpment overlooking the Derwent valley. The numerous crags on Derwent Edge consist of predominantly massive coarse-grained pebbly sandstone. Other features present are small- and large-scale false-bedding, generally dipping to the south-west; less commonly load and flute marks on the under surfaces of sandstones in contact with thin shaly bands or partings; shale conglomerates and large soft sub-spherical brown concretions which weather into hollows. The most common topographical features are large cliff-like faces of considerable lateral extent forming the highest scarp slopes such as those under Back Tor and Dovestone Tor and residual blocks and towers situated on the tops of the scarp, such as the Salt Cellar and the Wheel Stones. The existence of very large-scale false-bedding planes dipping south-westwards is evident along Derwent Edge, and the resulting features give a misleading impression of successive beds dipping southwards along the scarp, each being separated from the next by a well-marked 'slack'. These 'slacks', although suggestive of the presence of shales, are considered more likely to be the outcrops of soft, usually coarse-grained beds at the base of the false-bedding units. Approximately 80 ft of mainly massive coarse-grained pebbly sandstone is exposed under Dovestone Tor.

On Ladybower Tor sandstones, mainly massive coarse-grained and pebbly, but with subsidiary thin-bedded medium-grained bands, about 75 ft thick, are exposed in the landslip-scar. Continuous exposures of similar lithology occur along the roadside for 700 yd S.W. of Cutthroat Bridge. Well developed south-westerly dipping false-bedding is visible in a large quarry [2083 8684] ¼ mile N.E. of Ladybower House.

On Bamford Edge the Lower Kinderscout Grit is approximately 140 ft thick, with craggy outcrops up to 40 ft high of massive coarse-grained pebbly sandstone, in which south-westerly dipping false-bedding is fairly common. At the southern end of Bamford Edge, ¾ mile N.N.E. of Bamford church, large-scale false-bedding gives rise to 'slacks' as described on Derwent Edge. In a quarry [2153 8432] ¾ mile N.E. of Bamford church, 40 ft of sandstone, mainly massive and coarse-grained, but with some thin-bedded bands, show south-westerly dipping false-bedding.

Between Bole Hill and Scraperlow, ¾ mile E. of Hathersage, the Lower and Upper Kinderscout grits appear to be united, the total thickness of both grits decreasing southwards from 140 to 70 ft at Mitchell Field. Massive medium- to coarse-grained sandstone, locally pebbly, is well exposed in the stream east of Sheepwash Bank and on Carhead Rocks. South of Scraperlow the feature of the Lower Kinderscout Grit can be distinguished 200 yd E. of Kettle House, suggesting a thickness of about 45 ft. Sandstone, fine- to medium-grained, in part fissile, flaggy and micaceous, with siltstone partings, crops out in quarries [2400 8121] ¼ mile W.S.W. of Scraperlow. Farther south medium- to coarse-grained sandstone is visible in a roadside exposure [2404 8067] 200 yd S. of Hathersage Booths.

The **beds between the Lower and Upper Kinderscout grits** are nearly 70 ft thick north-west of Holling Dale, where they form an ill-drained 'slack' with no exposures. In the stream [2145 9215] ¾ mile N. of Foulstone Delph, the following beds rest on the ganister-like top of the Lower Kinderscout Grit, which contains rare *Stigmaria* casts:

|  | ft | in |
|---|---|---|
| Siltstone, with thin flaggy sandstone bands .. .. .. .. | 7 | 0 |
| Siltstone and silty shale .. .. | 11 | 0 |
| Shale, dark grey near base, becoming silty near top .. .. | 12 | 0 |
| Shale, dark grey; *Palaeoneilo sp.*, palaeoniscid scales and *Reticuloceras sp. nov.* .. .. .. | 11 |  |
| Shale, grey .. .. .. .. | 5 | 0 |

The Butterly Marine Band is here less fully developed than in the north-western exposures, showing only the goniatite-phase at the base.

The feature formed by a 7-ft impersistent sandstone occurs immediately above this exposure, and in a number of outcrops 220 to 480 yd farther downstream, the overlying silty shales and siltstones, with thin sandstone bands, are visible. In a stream bank [2136 9102], just north-west of Foulstone Delph, 30 ft of silty shale and siltstone are seen underlying the Upper Kinderscout Grit, and the same strata are exposed higher up this stream, where the total sequence is estimated to be 40 to 45 ft thick. On the moors farther south, only the vaguest trace of a 'slack' is present, but in the middle reaches of Highshaw Clough, these beds appear to have maintained a thickness of about 45 ft, and numerous exposures of silty shales and siltstones with thin sandstone bands are present. Silty shales, highly contorted owing to the proximity of the Moscar Fault, are seen in outcrops in the stream ¼ to ⅓ mile E.N.E. of Cutthroat Bridge.

The beds between the Lower and Upper Kinderscout Grit form a wide bench under Hordron Edge, where they are estimated to be nearly 75 ft thick. There are no exposures, but two low features suggest the development of thin sandstones. On Bamford Moor the position of these beds is uncertain. They are not exposed, but three narrow slacks suggest the existence of shaly bands separated by two substantial sandstones which, as the slacks merge southwards, evidently die out. It would appear that the two grits join at Bole Hill, as no evidence for the existence of the intervening beds can be found between Bole Hill and Scraperlow. Features south of Scraperlow suggest the separation of the two grits by intervening shaly beds, but these, estimated to be 20 to 30 ft thick, are not exposed.

The **Upper Kinderscout Grit** is 60 to 90 ft thick between Holling Dale and Bamford Moor. It forms wide gentle dip-slopes on Holling Dale and Foulstone Moor, but is poorly exposed both on the moors and on the distinct but shallow west-facing scarps. The lower reaches of streams draining Brogging Moss show sandstone varying from flaggy fine-grained to massive coarse-grained and locally pebbly, and with fairly common false-bedding. The highest beds are exposed in many outcrops in Strines Dike and in the stream between Holling Dale Plantation and Bole Edge Plantation, here mainly medium-to fine-grained well-bedded, locally ganister-like, and containing rare casts of *Stigmaria*. Farther south an exposure in Rising Clough [2189 8849] shows 10 ft of sandstone, massive coarse-grained and slightly pebbly, but with some flaggy beds and southerly dipping false-bedding, these beds being 10 to 20 ft below the top of the Upper Kinderscout Grit. A large slab of ripple-marked sandstone is present in Rising Clough 15 to 20 yd downstream from the above outcrop. In a borehole [2349 9179] north-north-east of Hallfield, the Upper Kinderscout Grit was recorded as 'grey rock' between 150 ft and the base of the borehole at 191 ft 6 in.

The Upper Kinderscout Grit is mainly massive or well-bedded medium- to coarse-grained sandstone on Hordron Edge, where abundant residual blocks and a number of small outcrops occur. On Bamford Edge, only a very subdued feature is present, a typical exposure [2130 8534] being coarse-grained pebbly sandstone 8 ft.

The **beds between the Kinderscout Grit and** *R. gracile* **Marine Band** consist of up to 13 ft of strata. A thin coal and seatearth are sometimes present at the base and these are overlain by mudstone and micaceous siltstone with thin sandstone. In a stream bank [2330 9289] nearly a mile north of Hallfield, shaly coal, with carbonaceous shale, 0 to 5 in thick, overlies grey micaceous mudstone 12 ft. The succession in another stream bank [2259 9200] west-north-west of Hallfield and for 100 yd downstream is:

|  | ft | in |
|---|---|---|
| Shale, dark grey, with plant fragments    .. .. .. .. | 1 | 0 |
| Sandstone   .. .. .. .. | | 2 |
| Mudstone, dark grey carbonaceous | | 1 |
| Mudstone, dark grey, with rootlets near top   .. .. .. .. | 4 | 0 |

Lateral variation of these beds is illustrated in an exposure [2214 9100] ½ mile E. of Foulstone Delph:

|  | ft | in |
|---|---|---|
| Coal, passing laterally into black carbonaceous shale   .. .. | | 2 |

|                                          | ft | in |
|------------------------------------------|----|----|
| Mudstone, dark grey .. 1 ft to           | 2  | 0  |
| Sandstone, with abundant included        |    |    |
| coaly fragments .. .. 4 in to            |    | 6  |
| Mudstone, dark grey .. 2 ft to           | 3  | 0  |

Another section is seen in the upper reaches of Strines Dike [2104 8974]:

|                                          | ft | in |
|------------------------------------------|----|----|
| Shale, silty dark grey .: ..             | 1  | 0  |
| Shale, black carbonaceous .. 0 to        |    | 2  |
| Coal .. .. .. .. 0 to                    |    | 0½ |
| Mudstone, grey .. .. ..                  | 2  | 0  |
| Siltstone and silty shale with thin      |    |    |
| flaggy sandstones .. ..                  | 4  | 0  |

In Rising Clough [2155 8897] these beds are absent, the marine band resting directly on the Upper Kinderscout Grit. The most complete exposures of these strata on Moscar Moor are in stream banks [2223 8736] ½ mile S. of Moscar House:

|                                          | ft | in |
|------------------------------------------|----|----|
| Shale, silty, passing into fine-grained  |    |    |
| sandstone .. .. 1½ in to                 |    | 6  |
| Coal .. .. .. .. 0 to                    |    | 2  |
| Seatearth .. .. .. 6 in to               | 1  | 0  |
| Siltstone (base not seen) .. ..          | 3  | 0  |

At an exposure [2487 8193] north-north-east of Mitchell Field only sandy seatearth 1 ft, lies between the Kinderscout Grit and the *R. gracile* Marine Band.

*Bamford and Hathersage.* Between Bole Hill and Scraperlow, the Lower and Upper Kinderscout grits cannot be separated (see above). Sandstone, medium- to coarse-grained with thin siltstone bands and partings 19 ft thick, is exposed in an old quarry [2302 8426]. South of Scraperlow the feature of the Upper Kinderscout Grit is discernible running due south, and crags and blocks of medium- to coarse-grained sandstone form a small scarp 100 to 150 yd E. of Hathersage Booths.

*Crook Hill and Win Hill.* Approximately 70 ft of the Lower Kinderscout Grit form the two craggy summits on Crook Hill. Outcrops of massive coarse-grained pebbly sandstone occur in 'edges', up to 30 ft high in the north-westerly outlier, and 40 ft in the south-easterly outlier.

The summit of Win Hill is formed by approximately 70 ft of Lower Kinderscout

Grit. Residual crags and blocks of mainly medium- to coarse-grained fairly massive sandstone are abundant.

*Offerton, Abney and Curbar.* The sandstone forming the top of Offerton Moor is considered to be Kinderscout Grit on lithological grounds and from a general consideration of the thickness relations of the underlying Shale Grit. The maximum thickness of this highest sandstone on Offerton Moor is estimated to be nearly 100 ft. Abundant blocks of medium- and coarse-grained sandstone, mainly massive, are present on and under Shatton Edge. In a quarry [2070 8106] east of Shatton Edge, 25 ft of well-jointed medium-grained sandstone are exposed. The Reform Stone [2121 8077] is part of a 400-yd long craggy edge of medium- to coarse-grained fairly massive sandstone. A number of outcrops of similar rocks are present in the upper reaches of Dunge Brook [2123 8040] and for 220 yd upstream. In a quarry [2016 8021] 19 ft of mainly massive medium-grained sandstone are exposed. A small outlier of the highest sandstone of Offerton Moor is present on High Low [222 802], and contains a number of small outcrops of medium-grained sandstone, well-bedded and even flaggy in places.

G.D.G.

*South of Shatton Edge.* The Kinderscout Grit forms a tract of moorland sloping gently to the east. A small outlier is also present at Abney Low. On Eyam Moor the grit gives rise to a long dip-slope and this, though broken by faulting near Mag Clough, extends south-eastwards to Knouchley. The best section in this area is at Stokehall Quarry [2365 7693] where grey siltstone with plants 3 ft, overlies buff-coloured coarse sandstone, massive but with micaceous partings, a few small pebbles and some soft ferruginous patches 60 ft. About 15 ft of red-stained coarse sandstone on 5 ft of shale were reported to have been proved below the exposed section.

In the south-east corner of the district the Kinderscout Grit forms an outlier in the small hill east of Calver Sough. The grit thins to some 30 ft on the northern side of Curbar, but thickens again before passing outside the district a quarter of a mile south of this village.

I.P.S.

## MIDDLE GRIT GROUP

### SHALES BELOW CHATSWORTH GRIT

*Glossop to Hayfield.* Around Chunal these beds are some 550 ft in thickness, but they thicken southwards to about 720 ft at Hayfield. The most conspicuous features are a lower series of shales with the *Reticuloceras gracile, Reticuloceras bilingue* early form and *Reticuloceras bilingue* late form bands and a sandstone up to about 40 ft thick in the middle of the group, the latter traceable south from the latitude of Chunal through Hollingworth Head to Carr Meadow where it thins out.

The lowest beds[1] are well seen in an exposure [0302 9073] in Long Clough:

|  | ft |
|---|---|
| Mudstone, lower part micaceous .. | 7 |
| Mudstone, dark, with decalcified bands; *Caneyella rugata, Hudsonoceras* cf. *ornatum, Reticuloceras bilingue* early form, mollusc spat .. | 7 |
| Mudstone, grey, with some harder platy bands .. .. .. .. | 8 |
| Mudstone, grey, pyritous .. .. | 2 |
| Mudstone, grey, with some harder platy bands .. .. | 4 |
| Mudstone, slightly silty; *C. rugata, Dunbarella speciosa*, orthocone nautiloid indet., *Anthracoceras sp., Reticuloceras gracile*, ostracods, fish debris .. .. .. about | 4½ |
| *Not exposed* .. .. .. .. | 2 |
| Upper Kinderscout Grit .. .. | — |

Farther downstream a small inlier of Upper Kinderscout Grit and *R. gracile* Marine Band is present: the section [0297 9096] is similar to that already described and extends upwards to include the *R. bilingue* early form horizon. Higher beds are present in a tributary clough [0284 9104] where the *R. bilingue* late form Marine Band is present, some 60 ft above the *R. bilingue* early form Band: the band, 3 ft thick, yielded *D.* cf. *speciosa* in addition to *R. bilingue* early form. The *Reticuloceras superbilingue* Marine Band, about 95 ft higher in the sequence, is exposed 140 yd farther upstream [0271 9098]; it is 2 ft 4 in thick and yielded *Myalina ?, Caneyella* cf. *multirugata, Dunbarella sp., Donetzoceras*

cf. *sigma, Homoceratoides* cf. *divaricatus, Gastrioceras sp., R. superbilingue.* The *R. superbilingue* Band lies about 45 ft below the sandstone of Hollingworth Head (see above) which forms the head of the clough.

Farther south, the *R. bilingue* late form Marine Band is exposed in a stream [0332 8951] 140 yd W.N.W. of Carr Meadow: 2 ft of soft shale overlie 6 in of shale with *Caneyella* cf. *rugata, Anthracoceras* or *Dimorphoceras sp., Hudsonoceras ornatum* and *Reticuloceras metabilingue* Wright. Here some 160 ft of shale separate the overlying sandstone from the Upper Kinderscout Grit and the marine band lies about 90 ft above the base. As the *R. superbilingue* Band should lie between 20 and 45 ft below the sandstone this section suggests a considerable local thinning in the beds between the two marine bands (to some 40 ft) compared with Long Clough where the thickness is 95 ft. The *R. bilingue* late form Band is better exposed in the stream [0309 8880] west of Marl House[2]:

|  | ft |
|---|---|
| Shale, grey, brown-weathering, seen | 4½ |
| Shale, blue-grey, iron-stained, with thin ironstone bands in lower half .. | 15 |
| *Not exposed* .. .. .. .. | 2 |
| Shale, dark blue-grey, iron-stained .. | 1 |
| Shale, soft dark blue-grey, 'sulphurous'; *Dunbarella* cf. *speciosa, Anthracoceras sp., Reticuloceras metabilingue, R.* cf. *bilingue* late form; the first-named is common at the base only .. .. .. .. .. | 5 |

Between Little Hayfield and Hayfield the lower part of the shale group is concealed by boulder clay. Higher beds are seen on the west side of the stream, and an exposure [0311 8766] near the mill-pond north of Bankvale Mill showed:

|  | ft | in |
|---|---|---|
| Sandstone, hard flaggy .. .. | 25 | 0 |
| Shale, poorly exposed .. about | 18 | 0 |

---

[1]Measured by Mr. C. G. Godwin.

[2]Section measured by Mr. M. J. Reynolds.

|  | ft | in |
|---|---|---|
| Shale, dark, with *C*. cf. *rugata*, cf. *Posidonia insignis* (Jackson), orthocone nautiloid, *Donetzoceras sigma*, *Gastrioceras sp.*, *Reticuloceras superbilingue*, mollusc spat | | 4 |
| Shale, grey, darker above .. .. | 1 | 6 |
| *Not exposed* .. .. .. .. | 4 | 0 |
| Shale, soft grey .. .. .. | 4 | 0 |

The marine band is the *Donetzoceras sigma* horizon, probably lying a few feet higher than that of Long Clough (see p. 226). About 20 yd south of the above section the uppermost beds are cut out by a fault which throws down 25 ft of well-bedded sandstone with siltstone partings. To the north of the fault the sandstone, which lies at or near the horizon of the sandstone of Hollingworth Head, gives rise to a steep bank above the river for a distance of 350 yd. It is not developed, however, in a small clough west of Little Hayfield where a nearly continuous section of the beds at this horizon shows them to consist mainly of silty mudstones with thin siltstone and sandstone bands.

*Hayfield to Chinley.* Traced southwards from Hayfield, there is a marked expansion in these beds to about 950 ft at Chinley.

The lowest beds are exposed in a stream [0439 8554] near The Heys:

|  | ft |
|---|---|
| Shale, grey; *C. rugata*, *Dunbarella speciosa*, *Anthracoceras* or *Dimorphoceras sp.*, *Reticuloceras gracile* .. | 4½ |
| Mudstone, sandy micaceous .. .. | 1 |
| *Not exposed* (sandy micaceous mudstone exposed in this position 80 yd upstream) .. .. .. .. | 3 |
| Upper Kinderscout Grit .. .. | — |

Higher beds are seldom well exposed, though the section in Foxholes Clough [base at 0377 8595] shows exposures of silty mudstones with thin sandstones totalling some 200 ft and overlain by the Chatsworth Grit. Farther south the col west of Hills House shows a rather exceptional development of the uppermost two-thirds of the group, three sandstones, each some 20 to 30 ft thick, being developed. These sandstones, however, die out when traced both north and south from the col.

The Otter Brook valley, north of Chinley, shows few exposures, the lower beds being obscured by boulder clay and the higher forming featureless ground on the slopes below the escarpment of the Chatsworth Grit and Rough Rock. An old quarry [0439 8357] north of The Naze shows 8½ ft of silty mudstone with thin flaggy sandstones.

*Whitehough to Chapel en le Frith.* South of Chinley these beds maintain their thickness. The most important feature is the incoming of the Roaches Grit a little north of Whitehough. The higher beds are exposed on the slopes of Eccles Pike and a good section of the lower is provided by the borehole[1] at Whitehall Works [0355 8202]:

|  | Thickness | | Depth | |
|---|---|---|---|---|
|  | ft | in | ft | in |
| Superficial deposits .. | 14 | 0 | 14 | 0 |
| Roaches Grit: sandstone with two shale bands, 13 ft at 58 ft and 10 ft at 94 ft .. .. | 110 | 0 | 124 | 0 |
| Mudstone, silty, often finely micaceous with a few thin sandstone and siltstone bands; ironstone nodules in places; *Curvirimula?* at 246 ft and from 250 ft 3 in to 251 ft 6 in; plant debris .. .. .. | 191 | 2 | 315 | 2 |
| Mudstone, some ironstone; scattered plant debris; fish scales in places; thin pyritous bands near base; *Curvirimula?* at 331 ft 9 in *Reticuloceras bilingue* Marine Band | 73 | 10 | 389 | 0 |
| Mudstone, hard dark platy; *Caneyella rugata*, *Anthracoceras* or *Dimorphoceras sp.*, *R. bilingue* .. | 1 | 4 | 390 | 4 |
| Mudstone, grey, with some ironstone; a 2-in band with *Lingula mytilloides* at 390 ft 8 in .. | 5 | 8 | 396 | 0 |

---

[1]Drilled by rock-bit to 224 ft; record from borer's log and examination of cuttings above this depth and from examination of cores below.

| | Thickness ft in | Depth ft in |
|---|---|---|
| Mudstone, hard dark; *Anthracoceras* or *Dimorphoceras sp.*, *Reticuloceras sp. nov.*, mollusc spat .. .. | 7 | 396 7 |
| Mudstone, grey blocky; rare ill-preserved goniatites; fish debris at 401 ft 8 in .. .. .. | 5 11 | 402 6 |
| Mudstone, hard grey, with a 5-in limestone bullion at 403 ft 6 in; *Dunbarella sp.* and mollusc spat .. .. | 1 2 | 403 8 |
| Mudstone, grey, with some hard bands, pyritous; fish debris in places .. .. .. | 21 10 | 425 6 |
| *Reticuloceras bilingue* early form Marine Band Mudstone, hard dark platy; *Anthracoceras* or *Dimorphoceras sp.*, mollusc spat .. | 1 0 | 426 6 |
| Mudstone .. .. | 1 6 | 428 0 |
| *No core* .. .. | 2 6 | 430 6 |
| Mudstone, grey, with hard platy bands; *Caneyella sp.*, *Dunbarella sp.*, *Anthracoceras* or *Dimorphoceras sp.*, cf. *Gastrioceras sp.*, *R. bilingue* early form, mollusc spat, palaeoniscid scale .. .. | 9 9 | 440 3 |
| Mudstone, grey, partly pyritous .. .. | 8 5 | 448 8 |
| *Reticuloceras gracile* Marine Band Mudstone, hard; *Anthracoceras* or *Dimorphoceras sp.*, *Reticuloceras sp.* .. | 4 | 449 0 |
| Mudstone; fish debris | 3 0 | 452 0 |
| Mudstone, hard, with two thin lenses of limestone; a 10-in bullion at 459 ft 6 in; *C. rugata*, *Dunbarella* *sp.*, *Anthracoceras* or *Dimorphoceras sp.,R. gracile* [evolute form] | 14 10 | 466 10 |
| Mudstone, medium grey to dark; fish debris, *Caneyella sp.*, *Anthracoceras* or *Dimorphoceras sp.*, mollusc spat .. .. | 8 | 467 6 |

The Forge Works No. 3 Borehole [0417 8219], at Chinley, proved drift to 50 ft and then "mainly shaly beds" (*fide* F. W. Cope). Specimens representing the *R. bilingue* early form Marine Band were found: *C. rugata*, *Anthracoceras* or *Dimorphoceras sp.* and mollusc spat at 109 ft and *C. rugata*, *R. bilingue* early form and mollusc spat between 113 ft and 117 ft. The *R. gracile* Marine Band was recorded as showing *C. rugata*, *Dunbarella sp.*, *Anthracoceras* or *Dimorphoceras sp.*, *R. gracile* and mollusc spat, all from 147 ft. The record of the borehole cannot, however, be regarded as complete, nor the thicknesses of the two marine bands as certain.

The Roaches Grit, 100 ft thick and faulted, forms subdued glaciated features between Bradshaw Hall and Whitehough. It is poorly exposed, though 3 ft of micaceous flaggy sandstone at the base are seen in a road-cutting [0460 8140] near Hallhill Farm. At Whitehough the lower beds were proved in a trench, the base being at [0408 8186], and flaggy sandstones with thin mudstone partings were seen west of this point for a distance of 40 yd. Some 280 ft higher a persistent sandstone about 30 ft thick is present on both north-east and south-east flanks of Eccles Pike; it is exposed behind Eccles House [0354 8156] as sandstone with siltstone partings, 6 ft thick. About 170 ft of shale separate this sandstone from the Chatsworth Grit.

*Combs and Dove Holes.* In the Combs valley the lower beds are obscured by drift. The Roaches Grit can be traced, though faulted, from Spire Hollins southwards into the axial area of the Longhill Anticline (see p. 326) where it is some 30 to 70 ft. The underlying Corbar Grit, separated from the Roaches Grit by about 150 ft of shale, is developed in the vicinity of Heylee. To the west of the Longhill Anticline exposures of the Roaches Grit are poor though sandstone near its top is exposed in a stream [0342 7886] at Spire Hollins. A stream section north-west of Heylee shows higher beds, silty mudstones and siltstones with some beds of dark platy shale, extending nearly up to the base of the Chatsworth Grit, though the uppermost beds are faulted out. From a point ["032 783"] in this section, Jackson (1958, p. 80) records *Calamites sp.*, *Lepidophloios laricinus* Sternberg, *Dunbarella elegans*, *Posidonia insignis*, *Orthoceras sp.*, nautiloid, *Gastrioceras spp.*, *Homoceratoides divaricatus* and *Reticuloceras superbilingue*. This horizon appears to lie some 200 ft above the Roaches Grit.

On the eastern limb of the Longhill Anticline the Roaches Grit is well exposed along Ridge Lane. A quarry [0380 7730] here shows 30 ft of coarse sandstone with pebbly bands. The overlying beds, mostly shales with thin sandstones, are seen in the lower part of Pyegreave Brook. Higher up the same brook 6 in of hard flaggy calcareous mudstone is exposed [0488 7806] yielding *Anthracoceras* or *Dimorphoceras sp.* and *Reticuloceras superbilingue*.

At Castle Naze a small exposure [0482 7857] showed dark-weathered shale with *Reticuloceras superbilingue*. This is at or near the Bull Hills locality where Jackson (1953, pp. 191–2) obtained *D. elegans*, *P.* cf. *insignis*, *Gastrioceras spp.*, *Homoceratoides fortelirifer* and *R. superbilingue*. This horizon is only some 30 ft above the Roaches Grit indicating rapid variations in the thickness of the beds in this part of the succession.

The Roaches Grit forms conspicuous features between Ridge Hall and Lady Low. It is developed as one main bed though there is a thin lower leaf which unites locally with it. A quarry [0556 7932] showed 22 ft of medium-grained sandstone, mostly well bedded though more massive in places and with some thin partings of silty mudstone; the undersides of some of the sandstones showed load-casts. Farther east a quarry [0650 7807] near Lady Low shows 40 ft of coarse massive sandstone. South of here the upper leaf of the grit thins, but additional thin leaves come in at a lower level, the whole ranging over some 250 ft of strata. This development continues south on to Brown Edge.

The *Reticuloceras gracile* Band is exposed in soft shales, some 4 ft thick, in a small clough [0595 7925], ¼ mile N.N.E. of Ridge Hall. It yielded *Dunbarella sp.* and *R.* cf. *gracile*. The higher part of the clough shows a nearly continuous succession of silty mudstones, siltstones and thin sandstones up to the base of the Roaches Grit. The same band was found in an exposure [0662 7878], 300 yd N.W. of Cowlow farm: 5 ft of dark shale with *C. rugata*, *D. speciosa* and *R. gracile* [evolute form] resting on 2½ ft of brown-weathered shale. Local thinning is apparent here as only 170 ft of shale separate the *R. gracile* Band from the Roaches Grit.

A stream section by Blackedge Reservoir shows beds between the Roaches Grit and Kinderscout Grit. The *R. gracile* Band at the base is not visible but the *R. bilingue* Band was found [0673 7655] with the following section: shale 2½ ft, on dark shale, decalcified in lowest 7 in, with *R. bilingue* 1 ft 7 in, on shale 4 ft. This band lies about 100 ft above the Kinderscout Grit and some 30 to 50 ft above the horizon of the *R. gracile* Band (see also p. 222). About 30 yd upstream and 30 ft higher in the succession a 3-ft band of dark slightly silty shale yielded *R. metabilingue*, indicative of the *R. bilingue* late form horizon. About 300 ft of shale (seen in the upper half as silty mudstone with thin flaggy sandstones) separate this from the lowest beds of the Roaches Grit.

*Brown Edge and Longhill.* The most complete section in this area is seen in Hogshaw Brook. The *R. gracile* Band is not exposed, but the *R. bilingue* early form Marine Band [0599 7423] (on One-inch Sheet 111) consists of 2¾ ft of shale, grey above, blocky pyritous below, with *Dunbarella sp.* and *R. bilingue* early form. This band lies some 40 ft above the Kinderscout Grit. About 60 ft higher in the sequence and 150 yd upstream [0596 7434] the *R. bilingue* Band is exposed (also on One-inch Sheet 111): shale 4 ft on dark decalcified shale with *R. bilingue* 4 ft, soft dark grey shale 2 ft. Between the latter band and the base of the Corbar Grit (see p. 230) are about 350 ft of shale, the lower half of which is seen in a few exposures near the base, where it is free of silty bands. The upper part is, as elsewhere, composed of silty shales with thin sandstones and siltstones.

Q

The Roaches and Corbar grits form a complex group of sandstones with shale partings, about 600 ft thick in Hogshaw Brook. The group thins northwards with the disappearance of the Corbar Grit; south-westwards the sandstones at the base of the Roaches Grit fail, while those in the upper part coalesce. The sandstones in the Corbar Grit are flaggy, with shale partings in the vicinity of Hogshaw Brook; the Roaches Grit is much more regular in development and it is continuous between Corbar Hill (One-inch Sheet 111) and Brown Edge. A quarry in the Roaches Grit below Light-wood Reservoir [0539 7495] shows well-bedded coarse sandstone 40 ft thick.

Above Lightwood Reservoir the *R. super-bilingue* Marine Band was found in the brook course [0547 7529] with the following section: grey mudstone 3 ft, hard flaggy calcareous mudstone with *Promytilus* cf. *foynesianus*, *Reticuloceras sp.* and palaeo-niscid scales 3 in, on grey blocky mudstone 3 in. This section lies within a foot or two of the top of the Roaches Grit and about 280 ft of shale separate it from the Chatsworth Grit.

On Corbar Hill the Corbar Grit is about 220 ft thick and is separated by 450 ft of shale from the overlying Roaches Grit, which is 150 ft thick. This succession holds in a general way for the Longhill Anticline where, however, some splitting and dying out of individual sandstone beds makes the development more complex. Underlying strata are exposed near the axis of the anti-cline in a stream [0379 7483], ¼ mile E.S.E. of Longhill Farm:

|  | ft | in |
|---|---|---|
| Shale, grey .. .. .. .. | 2 | 6 |
| Mudstone, soft grey; *Reticuloceras bilingue* .. .. .. .. | 1 | 6 |
| Mudstone, hard dark platy; *R. bilingue* .. .. .. .. | 5 | |
| Mudstone, dark to medium grey, platy; *R. bilingue* .. .. .. | 3 | 6 |
| Mudstone, hard dark platy calcar-eous and slightly micaceous; *C. rugata*, *D. speciosa* and *R. bilingue* .. .. .. .. | 4 | |
| Mudstone, grey platy slightly mica-ceous; *R. bilingue*; *D. speciosa* in a 3-in band 6 in from base .. | 2 | 0 |
| Shale, soft grey .. .. .. | 1 | 0 |

On the western side of the anticline the shales between the Roaches Grit and Chatsworth Grit expand greatly, reaching a thickness of some 600 ft. About 380 ft from the base they contain a flaggy sandstone up to 90 ft thick but of limited lateral extent.

*Todd Brook Anticline.* In this area the group reaches its maximum development, some 1500 ft in the area south of Kettles-hulme, though to the north of that place it is about 1100 ft thick. The beds are not usually very well exposed and the sandstone features are smoothed by glacial action.

The lowest beds, a few feet above the Kinderscout Grit, are seen in a stream [9890 7769] near Dunge Farm:

|  | ft |
|---|---|
| Shale, grey .. .. .. .. | 10 |
| Shale, decalcified fossiliferous .. | 0¼ |
| Shale, 'sulphurous' .. .. .. | 2 |
| Shale, dark decalcified, with *C. rugata*, *D. speciosa*, *Anthracoceras sp.* and *R. gracile* .. .. .. .. | 2½ |
| Shale, grey finely micaceous .. .. | — |

About 40 yd upstream [9893 7768] the *R. bilingue* early form Marine Band occurs:

|  | ft |
|---|---|
| Shale, grey 'sulphurous' .. .. | 2 |
| Shale, dark decalcified; *Caneyella sp.* and *R. bilingue* early form .. .. | 2½ |
| Shale, soft pale .. .. .. .. | — |

This band lies about 30 ft above the *R. gracile* Band and 180 ft below the Corbar Grit.

On the western limb of the anticline much of the shale between the Corbar and Kinder-scout grits was formerly exposed in two streams south-east of Lamaload; the sections now lie beneath Lamaload Reservoir. One [9729 7479] showed 3 ft of dark grey shale on black shale with *C. rugata*, *Anthracoceras* or *Dimorphoceras sp.* and *R. bilingue*. Farther downstream [9727 7492] a disturbed section in dark shales yielded *Homoceratoides sp.* and *R. bilingue* late form (Taylor and others 1963, pp. 6–7). A section in a tributary clough [9731 7513], now also flooded, showed black shale on soft black shale with goniatites 1 ft 6 in, on dark grey shale 8 ft; the horizon is that of *R. bilingue*.

In the northern part of the anticline the beds between the Roaches–Corbar group of grits and the Kinderscout Grit are poorly exposed. An exposure [9818 8119] in Brow-side Clough shows:

|  | ft |
|---|---|
| Shale, dark .. .. .. .. | 29 |
| *Not exposed* .. .. .. .. | 4 |
| Shale, dark .. .. .. .. | 3 |
| *Not exposed* .. .. .. .. | 1 |
| Shale, dark decalcified; *Caneyella?*, | |
| *Posidonia sp.*[juv.] and *R. bilingue* .. | 2 |
| Shale, dark, pyritous in places .. | 18 |

The marine band lies some 380 ft below the Roaches Grit, the Corbar Grit being absent (see below).

The Corbar Grit and Roaches Grit show a variable development around the Todd Brook Anticline. Both grits are readily recognizable in most places though each is much split. The Corbar Grit consists of a variable group of thin sandstone bands (a typical section in Kirby Clough [990 780] shows three such sandstones) some 300 ft thick in the central and southern part of the anticline, thinning to 90 ft at Kettleshulme and absent in the north in the nose of the anticline around Bailey's Farm. The sandstones are more flaggy than those in the Roaches Grit though on the western limb more massive and less variable beds occur; the Roaches Grit, separated from the Corbar Grit by some 250 ft of shales, mostly silty, is developed as a coarse massive sandstone, 250 ft thick at Kettleshulme. It splits both to north and south of this locality (two beds being present at Kirby Clough). On the eastern limb of the Anticline the Roaches Grit dies out southwards near Green Stack, but on the western limb it shows a more consistent arrangement, though locally split into as many as three separate leaves.

The Corbar Grit was proved in a series of boreholes for the cut-off trench for Lamaload Reservoir. The deepest of these[1] [9694 7519] showed sandstone to 70 ft, mudstone to 74 ft 6 in, sandstone to 79 ft 6 in, mudstone to 92 ft, sandstone to 125 ft, siltstone to 172 ft and mudstone to the bottom of the hole at 208 ft.

The shales lying between the Roaches and Chatsworth grits call for little comment. The thickness varies from about 400 ft at Kettleshulme to 500 ft near Dunge Farm. With the dying out of the Roaches Grit in the southeast part of the anticline the thickness of the shale group between the Corbar Grit and

Chatsworth Grit increases correspondingly to about 700 ft. These beds are well exposed in stream sections near Thursbitch [9925 7513] and consist of silty mudstones with thin flaggy sandstones and siltstones.     I.P.S.

*Bradfield and Moscar.* In this area the presence of the Heyden Rock allows the beds between the Kinderscout Grit and Chatsworth Grit to be described under three headings as below.

**Shales with *Reticuloceras gracile* Marine Band** at base. Between Bradfield Moors and Moscar Moor these beds thin southwards from 120 ft to approximately 60 ft and comprise shales and silty shales succeeded by siltstones with very thin sandstone bands. The *R. gracile* Marine Band varies from 5 to 10 ft in thickness, and consists of dark grey, rarely black, shale and mudstone, with abundant goniatites.

In the south bank of the stream [2330 9289] nearly a mile N. by W. of Hallfield, the marine band is 10 ft thick and contains *Caneyella sp., D. speciosa, Anthracoceras* or *Dimorphoceras sp.* and *R. gracile.* Good exposures of fossiliferous shale also occur in the east bank of the stream west of Bole Edge Plantation, where, for example [2259 9200] 6 ft of dark grey shale contain *L. mytilloides, C. rugata, Dunbarella sp., Anthracoceras* or *Dimorphoceras sp., R.* cf. *gracile* and *Reticuloceras gracile* late form. The marine band also crops out ¾ mile to the south-south-west at the junction of two streams [2214 9100]. One of the original localities from which this marine band was described by Bisat (1924, p. 117) is on the northern side of Raddlepit Rushes, in Strines Dike [2104 8974] where the band is 5 ft thick and yields *C. rugata, D. speciosa, Anthracoceras* or *Dimorphoceras sp.* and *R. gracile* late form. Similar faunas have been obtained from exposures in Rising Clough [2168 8886; 2171 8882; 2206 8814]. On Moscar Moor the lowest 2 ft of the marine band are exposed in a stream [2223 8736] where *Dunbarella sp.* and *R. gracile* late form have been found.

The strata overlying the *R. gracile* Marine Band are well exposed in the stream west of Bole Edge Plantation. On the east bank [2228 9166], siltstone 8 ft, overlies silty shale

---

[1]Cores examined by Dr. A. A. Wilson.

8 ft, on shale 22 ft. The east bank of Strines Dike contains frequent outcrops of these beds. At [2176 9058] 600 yd S.E. of Foulstone Delph, 55 ft of shale and silty shale are exposed, and 300 yd to the S.W. [2155 9040] the succession is: siltstone and silty shale 3 ft, fine-grained flaggy sandstone 6 ft, silty shale 8 ft, siltstone 1 ft 6 in, on shale, silty near the top, 40 ft. In the borehole [2349 9173] at Lane Head Farm near Hallfield (pp. 372–3) 62½ ft of shale were recorded between the Heyden Rock and Kinderscout Grit. Small ironstone nodules ½ to 2 inches in diameter are present in the shales about 10 ft above the top of the *R. gracile* Marine Band, some of them with pyrite centres, and at one locality [2114 8989] nodules from a band lying 7 ft above the stream bed contain fish debris.

**Heyden Rock.** Between Bradfield Moors and Strines Moor the Heyden Rock is 80 to 100 ft thick, but it thins considerably farther south and cannot be traced beyond Moscar Fields. Residual blocks of medium- to coarse-grained sandstone, locally pebbly, are scattered over the northern part of Bradfield Moors, but in small quarries in the northwestern part of Bole Edge Plantation, nearly ¾ mile N.N.W. of Hallfield, the sandstone varies between coarse-grained massive, and fine-grained thin-bedded and flaggy, with some false-bedding. The borehole near Hallfield, referred to above, commenced near the top of the Heyden Rock and penetrated 82½ ft of sandstone before passing down into shales. In the deep gorge of Strines Dike [224 907], outcrops indicate that the lower part of the Heyden Rock is medium- to coarse-grained massive or thick-bedded, but towards the top becomes medium- to fine-grained, well-bedded and, in places, even flaggy. A temporary excavation on the northern flank of the Strines Dike gorge [2256 9071] revealed the following section[1] near the top of the Heyden Rock:

|  | ft | in |
|---|---|---|
| Sandstone, fine-grained .. .. | 2 | 0 |
| Silty mudstone and siltstone about | 11 | 0 |
| Sandstone, fine-grained flaggy .. | 4 | |
| Sandstone, fine- to medium-grained, with abundant casts and moulds of *Carbonicola spp.* .. .. | 3 | |

|  | ft | in |
|---|---|---|
| Sandstone, medium- to coarse-grained .. .. .. .. | 4 | 0 |

Blocks of sandstone containing *Carbonicola spp.* are also present in walls in the area around Strines Inn [2227 9062]. Small craggy outcrops and abundant residual blocks of medium to coarse sandstone, mainly well-bedded or gently false-bedded, occur along the western edge of Strines Moor. Farther south medium-grained sandstone 11 ft is exposed in a quarry [2227 8848]. In a stream [2251 8756], 690 yd S.33°E. of Moscar House, and for 40 yd upstream small exposures indicate that the Heyden Rock consists of 15 to 20 ft of sandstone, fine-grained flaggy and micaceous, with thin bands and partings of siltstone and silty shale. The feature formed by the Heyden Rock is indistinct in this locality, and it disappears some 400 yd south of Moscar Fields.                           G.D.G.

**Beds between Heyden Rock and Chatsworth Grit.** Between Woodseats and Moscar Moor these beds are 200 to 275 ft thick and are mainly silty shales with some siltstones and two thin sandstones. The lower of these, some 20 ft thick and 60 ft above the Heyden Rock, forms the conspicuous dip-slope on which Woodseats is situated, but it dies out rapidly to the south-west. About 60 ft higher lies the upper sandstone of similar thickness; the intervening shales contain, in their middle, the *Reticuloceras superbilingue* and *Donetzoceras sigma* horizons. The latter was seen in the river bank between Strines and Dale Dike reservoirs [2336 9066][2]:

|  | ft | in |
|---|---|---|
| Shale, dark grey, rusty-weathered .. | 10 | 0 |
| Shale, black 'sulphurous'; *Caneyella rugata*, *Donetzoceras sigma* and mollusc spat .. .. .. | | 9 |
| Shale, dark grey, rusty-weathered .. | 5 | 0 |
| Shale; *Lingula sp.* and *C. rugata* .. | | 1 |
| Shale; *L. mytilloides* at base .. | | 6 |

I.P.S.

South of Sugworth Hall silty mudstones with thin siltstone and sandstone bands succeed the Heyden Rock and have been seen in a number of temporary excavations. Unidentifiable fragments of goniatites and

---

[1]Measured by Mr. C. G. Godwin.
[2]Section measured by Mr. J. Pattison.

posidoniids were noted in a temporary section [2244 8872] estimated to lie 10 to 20 ft above the Heyden Rock, at the horizon of the *Reticuloceras bilingue* late form Marine Band.

A thin flaggy sandstone is locally developed on the slopes west of Bents Farm, probably about 60 ft above the Heyden Rock and is also exposed in Pears House Clough [2247 8989] near where it is faulted against higher shales containing *R. superbilingue*. The succeeding 100 ft or more of shales and mudstones contain the *R. superbilingue* Band 10 to 40 ft above, and the *D. sigma* Band 70 to 90 ft above, their base. Masses of slipped shale containing *R. superbilingue* are present on the northern bank of Strines Reservoir [2287 9061].

From the upper 2 ft of a 13-ft shale outcrop in Pears House Clough [2252 8989], the following fossils were collected: *Caneyella sp.*, *Posidonia sp.*, *Gastrioceras sp. nov.*, *Homoceratoides fortelirifer*, *Ht.* aff. *divaricatus* and *R. superbilingue*. This exposure is at, or very near, the locality mentioned by Davies (1941, pp. 241–4) and by Pulfrey (1934, pp. 254–64). The same band also crops out 190 yd lower down the clough and 9 ft up on the south bank [2271 8992], and a similar fauna was also found near Bradfield (Eden and others 1957, p. 19). In the banks of the stream east of Bents Farm the *R. superbilingue* fauna has been found at several localities [2277 8924; 2278 8908; 2282 8900]. The most northerly of these outcrops is believed to be downfaulted with respect to the other two, but assuming that no slipping has occurred the fauna appears to be spread through 30 ft of strata (see p. 180). Approximately 40 to 50 ft above, in the same stream [2283 8896], *Caneyella sp.* and *D. sigma* occur in an 8-ft bed overlying 8 ft of unfossiliferous shale, supporting the impression that in this area the interval between the *R. superbilingue* and *D. sigma* bands has expanded to 50 ft or more (Pulfrey 1934, p. 259). The higher beds in this stream are silty shales and siltstones with a few thin hard sandstone bands, succeeded by a thin flaggy sandstone estimated to be about 20 ft thick. The *R. superbilingue–Ht. fortelirifer* fauna was collected from a temporary section farther south [2267 8832], and from extensions of this section to the south-east. The beds above the

*R. superbilingue* Band contain a few feet of fine-grained flaggy sandstone exposed in quarries [2279 8836] ¼ mile N.W. of Moscar Lodge, where it forms a distinct feature. The succeeding strata up to the Chatsworth Grit consist of silty mudstones, siltstones and thin hard sandstones. G.D.G.

*Hathersage to Curbar.* West of Stanage Edge, about 400 ft of strata separate the Kinderscout Grit from the Chatsworth Grit; these beds are obscured by head. Farther south these beds thin to 250 ft and consist mainly of shale, though the Ashover Grit makes its appearance near the south-eastern corner of the sheet at Curbar.

To the east of Hathersage the lowest of these beds are exposed in a stream [2487 8192], 210 yd N.N.E. of Mitchell Field:

| | ft |
|---|---:|
| Shale, black; *Caneyella sp.*, *R. gracile* | 3 |
| Seatearth, sandy .. .. .. | 1 |
| Sandstone, hard; top of Kinderscout Grit .. .. .. .. .. | — |

Some 250 yd upstream dark grey mudstone with ironstone nodules overlies dark unfossiliferous shale of marine aspect, possibly at or near the *R. bilingue* horizon. The uppermost beds are exposed in a landslip-scar at Callow Bank [2519 8229]:

| | ft |
|---|---:|
| Sandstone, massive medium-grained (Chatsworth Grit, main bed) .. | 15 |
| Shale, dark, poorly exposed .. .. | 6 |
| Sandstone .. .. .. .. | 1 |
| Mudstone, grey silty, poorly exposed | 12 |
| Sandstone, massive, lower leaf of Chatsworth Grit .. .. .. | 10 |
| Mudstone, grey silty .. .. .. | 4 |
| Sandstone, soft fine-grained .. .. | 2 |
| Shale, dark, fossiliferous in top 4 in and lowest 3 in; *Zygopleura sp.* and *Donetzoceras sigma* .. .. .. | 5½ |
| Shale, dark .. .. .. .. | 8 |

Farther south no sections of these beds are available until Burbage Brook is reached. Here, a section [2514 7910] in Yarncliff Wood, showed:

| | ft |
|---|---:|
| Sandstone (lower leaf of Chatsworth Grit, see p. 236) .. .. .. | — |
| *Not exposed* .. .. .. .. | 30 |

ft

Shale, dark pyritous; *Lingula sp.* and
  *D. sigma* in 3-in band at 6 in from
  top; rare *Lingula sp.* 3 ft from base    9
*Not exposed*    ..    ..    .. about   15
Shale, grey    ..    ..    ..    ..    12½
Shale, soft dark grey 'sulphurous';
  *C. multirugata, Dunbarella sp.* [juv.],
  *R. superbilingue, D. sigma* ..    ..    1½

The upper band, with *Lingula sp.* and *D. sigma*, evidently represents the *D. sigma* horizon while the lower, with *D. sigma* and *R. superbilingue*, is the *R. superbilingue* horizon.

The *Donetzoceras sigma* Band is also exposed farther downstream, near Grindleford station [2503 7869]:

ft in

Shale, dark ..    ..    ..    ..      6
Shale, soft grey    ..    ..    ..    5   6
Shale, black with a 1-in band con-
  taining *C. multirugata* and *D. sigma*    ..    ..    ..    3
Mudstone, grey    ..    ..    ..      6

At Curbar, just within the district, the Ashover Grit makes its appearance, in the middle of the shale between the Kinderscout Grit and the Chatsworth Grit. It is poorly exposed, but thickens to the south and south-east to reach its normal development in the Chesterfield district (Smith and others 1967, pp. 67–72).

## CHATSWORTH GRIT

*East of the River Goyt.* Near the northern edge of the district, the Chatsworth Grit is a yellow-brown medium-grained sandstone about 80 ft thick near Plainsteads, though thinning locally, as at Simmondley. At the latter locality a lower leaf appears and this has been included by J. V. Stephens within the outcrop of the grit postulated beneath the drift of the Glossop valley. An exposure [0241 9149] near the middle of the outcrop of the grit, shows 2 ft of ganister with rootlets and *Stigmaria* on 1¾ ft of creamy saccharoidal siliceous sandstone (see also p. 181). The grit forms a broad outcrop on Matley Moor, where it is of moderately coarse grain. A quarry [0261 9040] near Knorrs shows coarse sandstone with some red staining, which was also found [0248 8968] near Matleymoor Farm. To the east of the latter locality a lower leaf of the grit is present for three-quarters of a mile along the outcrop.

To the west of Cown Edge, the record of the working of the Simmondley Coal at Coombes Pit (see p. 236) allows the presence of an outcrop of Chatsworth Grit to be inferred, though it is entirely under drift.

At Birch Vale the Garrison Bleach Works No. 2 Borehole [0151 8682] proved the Chatsworth Grit to be 55¾ ft in thickness with its base at 200 ft depth. It thickens eastwards and is split into two leaves between Birch Vale and Hayfield. At Birch Quarry

[0302 8690] the lower leaf, some 60 ft of yellow-brown rather massive coarse sandstone with some cross-bedding, is worked.

To the west of Hills House, impersistent shale partings are present towards the base of the Chatsworth Grit. Followed southwards the feature of the grit becomes inconspicuous but it stands out again at The Naze [0414 8348], owing to a striking change in the lithology to a massive coarse cross-bedded sandstone with scattered quartz pebbles, 50 ft thick. The grit has been much quarried at Buxworth and a section [0279 8195] shows 40 ft of coarse-grained reddened sandstone. Farther to the south-east the Chatsworth Grit forms the prominent feature of Eccles Pike, the top of which shows some 18 ft of coarse pebbly sandstone with dark red staining in places.

Between Tunstead and the southern edge of the district the Chatsworth Grit gives a well-marked craggy scarp. A typical section [0306 7856] near Thorny Lee shows 30 ft of coarse red-stained sandstone, with bands of pebbles up to ¾ inch in diameter. The grit is also well exposed in an old road-cutting at Wainstones [0280 7715] and at Rake End [0257 7586]. Near the latter locality a lower leaf of sandstone appears beneath the Chatsworth Grit and joins the main bed to the south.

To the east of the main outcrop the Chatsworth Grit forms the conspicuous outlier of Combs Moss, where it is about 130 ft thick on the western side, though it probably thins eastwards. The lower part of the grit varies to some extent owing to the local development of thick flaggy beds beneath the upper coarse massive part; a typical section [0390 7642] of this type lies north of Combs Edge:

|                                                        | ft |
|--------------------------------------------------------|----|
| Sandstone, massive coarse  ..    ..                    | 32 |
| Sandstone, flaggy micaceous with partings of micaceous siltstone and shale; partings showing deep red staining   ..    ..    ..    .. | 75 |

A section [0527 7840] at Castle Naze shows 55 ft of pink-stained massive coarse sandstone, pebbly in places and with erosional base, on pink and yellow variegated well-bedded coarse- to medium-grained sandstone 9½ ft.

In the central part of the Goyt Trough the Fernilee No. 1 Borehole [0124 7823] proved the Chatsworth Grit from 349½ ft to the bottom of the hole at 476 ft. Except for the top 4¾ ft and lowest 14¾ ft, which were finer-grained and more flaggy, it was coarse and massive, frequently red-stained and with darker hematitic patches. Pebbly bands were present between 435 ft and 461¼ ft.

*West of the River Goyt.* To the west of Mellor the Chatsworth Grit forms a small outcrop in the stream at Cataract Bridge (see also p. 238). Faulted [at 9712 8878] against Woodhead Hill Rock on the west, it is present upstream to the east for a distance of 150 yd as a yellow-brown medium-grained sandstone. A larger outcrop, much obscured by drift, is present near the River Goyt between Bottom's Hall and Hague Bar. It is exposed in several streams north-east of Strines, for example south of Windybottom Farm [9733 8702] and for 190 yd upstream, the exposures covering nearly the full thickness: here it is hard and siliceous in its lower part and more flaggy above. The borehole at Grove Mill [9942 8506], New Mills, proved a thin coal on 43½ ft of sandstone, with base at 240 ft, at this horizon.

The Chatsworth Grit was quarried [9762 8402] in the faulted axial area of the Todd Brook Anticline near Stoneridge, where 15 ft of pink-brown medium-grained feldspathic sandstone are seen. The total thickness of the grit in this area is some 140 to 200 ft, but it forms subdued features often much modified by glaciation. On the western limb of the anti-cline the grit passes outside the district south-west of Charles Head. On the eastern limb it is flaggy and fine-grained between Cliff and Todd Brook. Farther south, between Claytonfold and Fivelane Ends, splitting and faulting tend to obscure the outcrop. To the south-east of Fivelane Ends it becomes coarse-grained, giving rise to the sharp feature of Windgather Rocks, south of which a roadside quarry [9952 7809] shows 35 ft of yellow massive coarse feldspathic sandstone with bands of small pebbles, some cross-bedding and patchy reddening in places. Still farther to the south it forms a marked escarpment and dip-slope. A section [9953 7633] near Oldgate Nick shows 52 ft of pink coarse feldspathic sandstone with bands of pebbles of up to ¾ in diameter and strong cross-bedding.

*Eastern area.* In the north-eastern corner of the district, the Chatsworth Grit (see also p. 181) is variable in character. To the north-east of Walker House it is well developed, though split into two leaves, the whole totalling some 60 ft. A section [2519 9171] near the base, south-east of Walker House, shows 23 ft of yellow-brown flaggy sandstone. The bed thins rapidly south-westwards and in much of Bradfield Dale it is only 25 ft thick and forms very subdued features. It thickens again south-west of Holes Clough [236 905], though from here to Moscar Lodge it maintains its fine-grained flaggy character. To the south of Moscar Lodge it again becomes coarse-grained and massive, as in the Rivelin valley (Eden and others 1957, p. 19), and forms strong features. To the east of the escarpment the grit forms a broad outcrop on Hallam Moors.                I.P.S.

Between Stanage End and High Neb the Chatsworth Grit produces a bold escarpment showing up to 60 ft of coarse massive sandstone with bands of quartz pebbles up to (rarely) 1 in diameter. The total thickness hereabouts is around 80 ft, but the lowest beds are obscured. The sandstones are finer-grained at the top. False-bedding is fairly common. A lower leaf, up to 40 ft thick, of flaggy sandstone, makes its appearance near Overstones Farm and forms a broad outcrop north of Higger Tor [256 820], where the coarser massive upper leaf forms an outlier.

I.P.S., G.D.G.

To the south-east of Hathersage the Chatsworth Grit gives rise to the strong scarp of Millstone Edge. A section in Grey Millstone Quarries [248 805] shows 60 ft of coarse grey sandstone with sporadic conglomeratic bands. Farther south the grit forms the strong escarpments of Froggatt Edge, Curbar Edge and Baslow Edge. The section in Burbage Brook given on p. 233 is continued upwards as follows:

|  |  |  |  |  | ft |
|---|---|---|---|---|---|
| Chatsworth Grit, main leaf | .. | .. |  |  | — |
| *Not exposed* | .. | .. | .. | .. | 60 |
| Sandstone, massive | .. | .. | .. | .. | 30 |
| Mudstone, grey, part exposed | about | 25 |
| Sandstone, massive, base in stream [at 2519 7916] | .. | .. | .. | 15 |

The two lowest sandstones are shown on the map as the lower leaf of the Chatsworth Grit. It is present in the valley of Burbage Brook and is probably identical with the lower leaf of Higger Tor, though the outcrops are not continuous.

## BEDS BETWEEN CHATSWORTH GRIT AND *Gastrioceras cancellatum* MARINE BAND

These beds, lying near the base of the shale slack between the Rough Rock and Chatsworth Grit, are often poorly exposed and in many places the position of the *Gastrioceras cancellatum* Band can only be inferred. Moreover, the presence in places of a sandstone close above the Chatsworth Grit leads to confusion where exposures are not good.

*Glossop to Hayfield.* At Simmondley, the Simmondley Coal, 24 to 30 in thick, lies about 40 ft above the Chatsworth Grit. Many shafts were sunk to the coal between points west and south of Simmondley village. The *G. cancellatum* Band probably lies a short distance above the coal as material from it was noted by J. V. Stephens (*in* Bromehead and others 1933, p. 70) on the tip from a shaft to the coal at Gamesley, just north of the district boundary. Farther south a sandstone lying between the Simmondley and Ringinglow coals and up to 25 ft thick is developed from a point ¼ mile N. of Plainsteads to Cloughhead [0160 9043]. A section [0203 9110] shows the Ringinglow Coal horizon: sandstone 6 in, coal 2 in, seatearth (close above Chatsworth Grit) 2 ft. The Simmondley Coal was worked as late as 1931 by adit at Cown Edge Colliery [0188 9104] under the name 'Two Sheds'; it was here 24 in thick. Farther west, at Coombes Colliery [0158 9212] it was 30 in thick. To the south-east of Lantern Pike, a coal, probably the Ringinglow, was worked from a shaft [0264 8764] near Higher Cliff. A stream section [0256 8730] near Lower Cliff shows a small thickness of flaggy sandstone on 3 ft of ganister, the whole a short distance beneath the Ringinglow Coal. The presence of the sandstone indicates that here the coal and ganister do not belong to the same cycle, an additional small cycle having made its appearance. This accords with the information from the Rod Moor No. 3 Borehole (Eden and others 1957, p. 213) where 7½ ft of dark sandy shale separate the Ringinglow Coal from the top of the Rivelin (Chatsworth) Grit. The Garrison Bleach Works No. 2 Borehole (see p. 366) showed 30 in of 'black shaly coal' at 136 ft, lying 8¼ ft above the top of the Chatsworth Grit, no other coal being recorded.

*Chinley.* North of Chinley a thin Simmondley Coal was seen in the following section [0394 8377] by the tram-track from Cracken Edge Quarry:

|  |  | ft | in |
|---|---|---|---|
| Mudstone, brown silty | .. about | 3 | 0 |
| Mudstone, black, with fusain fragments | .. .. .. about | | 1½ |
| **Coal** .. | .. .. .. .. | | 1 |
| Seatearth, with rootlets and other plant debris | .. .. .. | 2 | 6 |
| *Not exposed* .. | .. .. .. | 2 | 0 |
| Sandstone .. | .. .. .. | 8 | 0 |

The sandstone, up to about 20 ft thick and lying 15 to 20 ft above the Chatsworth Grit, is present for some 700 yd south of this locality.

On the western slopes of Eccles Pike a sandstone, up to 40 ft thick, is present between the Simmondley Coal and the calculated position of the *G. cancellatum* Marine Band. This may be a western continuation of the Redmires Flags.

*South of Whaley Bridge*. In this part of the eastern limb of the Goyt Trough exposures are usually lacking. The Ringinglow Coal was formerly worked near Rake End [0241 7575]. Near the southern boundary of the district a section [0196 7453] in a tributary of Wildmoorstone Brook showed:

|  | ft | in |
|---|---|---|
| Sandstone, ? *in situ*    ..    .. | — | — |
| Mudstone, grey silty, with thin sandstone bands ..    ..    .. | 6 | 0 |
| Mudstone, soft grey    ..    .. | 2 | 1 |
| **Ringinglow Coal**, 23 in    ..    .. | **1** | **11** |
| Seatearth, grey, darker at top    .. |  | 6 |
| Seatearth, buff rusty-weathering, becoming silty downwards    .. | 2 | 4 |
| Sandstone, buff fine-grained (top of Chatsworth Grit)    ..    .. | 2 | 6 |

The coal was formerly worked by bell-pits at this locality.

Fernilee No. 1 Borehole [0124 7823], near the axis of the Goyt Trough, proved the most complete section of these beds in the area:

|  | Thickness ft in | Depth ft in |
|---|---|---|
| *Gastrioceras cancellatum* Marine Band (see p. 240) ..    ..    .. | — — | 288  9 |
| Mudstone, pyritous in places; fish debris in lowest 7 ft 7 in    .. | 22  1 | 310  10 |
| Mudstone, medium to dark grey often platy, pyritous in places; *Planolites sp.*, *Lingula mytilloides*, productoid fragment, *Anthracoceras sp.* and mollusc spat ..    ..    .. | 7  7 | 318  5 |
| Mudstone, dark carbonaceous; *Sanguinolites sp.* ..    ..    .. | 1 | 318  6 |
| Mudstone, dark carbonaceous; some plant debris; *Curvirimula belgica* at 319 ft 5 in to 320 ft 6 in    ..    .. | 2  6 | 321  0 |

|  | Thickness ft in | Depth ft in |
|---|---|---|
| SIMMONDLEY |  |  |
| Coal        2 in |  |  |
| Cannel      7 in |  |  |
| Coal        6 in |  |  |
| Mudstone   8 in |  |  |
| Coal, dirty   1 in    .. | 2  0 | 323  0 |
| Siltstone, pale hard banded    ..    .. |  8 | 323  8 |
| Sandstone, hard flaggy siliceous    ..    .. | 2  0 | 325  8 |
| Siltstone and silty mudstone, banded..    .. | 5  7 | 331  3 |
| Sandstone, banded flaggy | 6  9 | 338  0 |
| Mudstone, grey silty; *Naiadites sp.* [juv.] at 338 ft 6 in    ..    .. | 1  0 | 339  0 |
| Siltstone, dark flaggy micaceous    ..    .. | 1  2 | 340  2 |
| Mudstone, banded silty, with thin siltstone bands; plant debris .. | 7  0 | 347  2 |
| Mudstone, grey blocky, pyritous at base    .. | 1  10 | 349  0 |
| RINGINGLOW |  |  |
| Coal (wash only) about 3 in    ..    ..    .. |  3 | 349  3 |
| Chatsworth Grit (see p. 235) ..    ..    .. | — — | — — |

Near Errwood Hall much of this group of strata is well exposed. A section [0012 7562], 1180 yd N.39°W. of the site of the hall, showed the Ringinglow Coal in the following section: soft grey shale 1 ft 4 in, coal 19 in, soft grey clay 6 in. Between here and the southern edge of the district the beds between the Ringinglow and Simmondley coals, some 25 ft thick, maintain a constant character with a few feet of shale at the base and the remainder sandstone. The Simmondley Coal is variable, a section [0039 7533], ¼ mile N.W. of Errwood Hall, showing coal 8 in on dirty coal 7 in. Farther south [0048 7501] the lower leaf passes into carbonaceous mudstone:

|  | ft | in |
|---|---|---|
| *Gastrioceras cancellatum* Band    .. | — | — |
| Shale, soft grey    ..    ..    .. | 4 | 0 |
| Shale, grey platy 'sulphurous', with some ironstone nodules ..    .. |  | 6 |
| Shale, soft grey finely micaceous    .. | 3 | 0 |
| Shale, hard dark finely micaceous platy    ..    ..    ..    .. |  | 11 |

SIMMONDLEY

| | ft | in |
|---|---|---|
| Coal, dirty 5 in .. .. .. | 5 | |
| Mudstone, dark carbonaceous .. | 4 | |
| Mudstone, dark finely micaceous .. | 4 | |
| Shale, grey-brown, silty .. .. | 1 | 0 |
| Sandstone, flaggy .. .. .. | 3 | 10 |
| Mudstone, silty with siltstone bands | 6 | 0 |
| Sandstone, flaggy .. .. .. | — | — |

To the south-west of Errwood Hall the Simmondley Coal was worked until 1933 at Castedge Colliery, where it was 15 in thick, under the name 'Little Mine'.

*Combs Moss.* The Ringinglow Coal and a small thickness of the overlying beds are present as an outlier on the eastern part of Combs Moss. A section in "one of the streams" was given by Hull and Green (1866, p. 61): black shale on coal 10 in, on "light grey underclay with lumps of raddle" 2 ft, resting on Third [Chatsworth] Grit. The coal was worked by bell-pits at the northern and southern ends of the outlier. Overlying the coal are some 10 ft of shale, above which are flaggy sandstones, also 10 ft thick, forming the highest beds in the outlier.

*Marple and New Mills.* At Cataract Bridge (see also p. 235) an exposure [9712 8877] in the stream shows the Ringinglow Coal, 12 in, on sandy seatearth overlying the Chatsworth Grit. At Strines an exposure [9759 8688] in a stream north of Greenclough Farm shows no coal at this horizon but a thin ganister-like sandstone a short distance above the Chatsworth Grit.

Near New Mills, the Grove Mill Borehole (see also p. 366) showed the following section of these beds, below the probable horizon of the Simmondley Coal:

| | Thick- ness | | Depth | |
|---|---|---|---|---|
| | ft | in | ft | in |
| Shale, etc. .. .. | 160 | 0 | 160 | 0 |
| Sandstone .. .. | 32 | 0 | 192 | 0 |
| Shale .. .. .. | 2 | 0 | 194 | 0 |
| **Ringinglow Coal** 12 in | 1 | 0 | 195 | 0 |
| Seatearth .. .. | 1 | 6 | 196 | 6 |
| Top of Chatsworth Grit | — | — | — | — |

*West of the Goyt Trough.* Near the axis of the Todd Brook Anticline an exposure [9773 8168], west of Handleybarn, showed the following section:

| | ft |
|---|---|
| Shale, grey weathered .. .. .. | 1½ |
| **Ringinglow Coal** smut 12 in .. | 1 |
| Seatearth, grey and purple mottled .. | 2 |
| Top of Chatsworth Grit .. .. | — |

The shale at the top of the section is thin and is succeeded by about 30 ft of sandstone which dies out rapidly southwards and also around the nose of the anticline. In Mather Clough 20 ft of shale, part exposed, lie between the top of the sandstone and the *Gastrioceras cancellatum* Band (see p. 241).

Just south of Todd Brook, a stream [9982 8045], 270 yd N.E. of Gap House, showed the Ringinglow Coal in the following section: sandy mudstone with thin sandstones 8 ft, coal 16 in, hard ganister with rootlets and carbonaceous streaks 8 in. A short distance above this section lies a sandstone extending upwards to the horizon of the Simmondley Coal (which is not exposed). This sandstone is a constant feature over much of the western side of the Goyt Trough. It is split between Green Head Farm and Needham Farm.

A coal smut in the position of the Ringinglow was seen in the side of the track [9955 7888] leading to Blackhillgate. To the south, sandstone occupies nearly the whole thickness of the beds between the Ringinglow and Simmondley coals, and, together with the Chatsworth Grit, forms a well-developed composite dip-slope. The Simmondley Coal was formerly worked in an adit in Mill Clough [0016 7802]. The section at this locality (see also pp. 241–2) is:

| | ft | in |
|---|---|---|
| *Gastrioceras cancellatum* Band .. | — | — |
| Shale, dark, with ironstone bands and lenses .. .. .. | 17 | 6 |
| *Obscured* (including position of Simmondley Coal) .. .. .. | 3 | 0 |
| Seatearth, siliceous with rootlets .. | 1 | 5 |
| Sandstone, hard siliceous .. seen | | 6 |

On the western limb of the Todd Brook Anticline the Ringinglow Coal was formerly worked by adit [972 793] near Charles Head. Here, as on the eastern side of the anticline, the measures between the Ringinglow and Simmondley coals contain a sandstone which appears half a mile north of Charles Head and continues south-westwards beyond the district boundary. The Simmondley Coal was worked here [9721 7937] about 1952 with the following section: grey sandstone 4 ft, brown sandy shale 6 ft, bright coal 17 in, soft seatearth.

*North-eastern area.* On the north-eastern margin of the district a great expansion of these beds takes place, some 120 ft being present in Bradfield Dale. The Ringinglow Coal, as elsewhere, lies at the base and coal debris at this horizon was seen in Holes Clough [2366 9053]. Some 90 to 100 ft of shale separate the coal from the Redmires Flags, 20 to 30 ft thick, which underlie the *Gastrioceras cancellatum* Band. Farther south the coal was stated by Green and Russell (1878, p. 41) to have been 12 in thick in an adit [2337 8932] near Moor Lodge.

At Lee Bank [231 901] the Redmires Flags are faulted out and to the south, where beds at this horizon reappear between Moor House and Hollow Meadows, they are absent. They reappear, though in an attenuated form, a short distance east of the district (Eden and others 1957, p.21). At Hollow Meadows, a thin coal, in the position of the Simmondley, is exposed in a small stream [2419 8788]: coal smut 2 in, black carbonaceous shale 3 in, on grey seatearth with rootlets 9 in.

The maximum development of the strata, about 140 ft, is reached between Brown Edge and White Path Moss. The lowest beds, with the Ringinglow Coal, were seen at several places in and around Oaking Clough. A section [2468 8610] in the stream 1300 yd N. 19°E. of Stanedge Lodge, showed: shale, with *Lingula mytilloides* at 9 in from base, 1 ft, on coal a few inches. This is the locality described by Pulfrey (1934, p. 258). A better section of the Ringinglow Coal is exposed a quarter of a mile to the north-west [2425 8632]: grey shale 3 ft, dirty coal at least 12 in, grey seatearth about 3 ft, on top of Chatsworth Grit. Farther south-west [2338 8588] the coal thickens to 24 in.

The Redmires Flags form a well-marked scarp on Brown Edge (see also pp. 182–3) and a scarp and dip-slope on High Lad Ridge. Farther south much of the outcrop of these and the underlying beds is obscured by peat. The only exposure of note [2434 8513], ¾ mile S.S.E. of High Lad Ridge, shows 6 ft of flaggy sandstone.

A stream [2566 8321] near Cowper Stone, shows abundant loose blocks of ganister representing the seat of the Ringinglow Coal.

## ROUGH ROCK GROUP

### BEDS BETWEEN *Gastrioceras cancellatum* MARINE BAND AND ROUGH ROCK

*Glossop to Long Hill.* To the south of Simmondley this group of shales, about 180 ft thick, is poorly exposed. It is also present, though much concealed by landslip, in the low ground west of Cown Edge.

To the north of Rowarth an exposure [0161 8984] in the stream shows the *Gastrioceras cumbriense* Band:

|  | ft | in |
|---|---|---|
| Shale, grey soft .. .. .. | 1 | 6 |
| Shale, hard dark platy .. .. | 1 | 0 |
| Shale, pale grey soft 'sulphurous' .. | 1 | 4 |
| *Not exposed* .. .. .. .. | 1 | 6 |
| Shale, pale grey .. .. .. | | 6 |
| Shale, soft, decalcified in top 6 in; hard below; *Caneyella multirugata, Dunbarella sp., Anthracoceras sp., Gastrioceras crenulatum, G. cumbriense* .. .. | 1 | 6 |

This band lies about 80 ft above the *G. cancellatum* horizon. Farther downstream higher beds come in; these become more arenaceous [0158 8951] as the base of the Rough Rock is approached.

Between Rowarth and Birch Vale, the thickness varies from 230 to 260 ft, with the local appearance of sandstones either joining with, or just below the Rough Rock.

From Birch Vale to Chinley the beds are about 170 ft thick, but there are no exposures of note. On the western slopes of Eccles Pike they thin to about 130 ft and a thin sandstone is locally present at the top. Farther to the south on the eastern flank of the Goyt Trough good exposures of these beds are also lacking. Near Rake End, however, Cope (1946, p. 171) records a 6-ft contorted band of shale overlying the *G. cancellatum* Band;

a contorted band 2 or 3 in thick, was also present just beneath the marine band.

The Fernilee No. 1 Borehole [0124 7823] shows the following (abridged) section of the lower part of these beds:

| | Thickness | | Depth | |
|---|---|---|---|---|
| | ft | in | ft | in |
| Strata uncertain (not cored to 140 ft and cores badly smashed and ?caved below) .. | — | — | 167 | 0 |
| Mudstone, pale silty, with thin siltstone bands .. | 21 | 0 | 188 | 0 |
| Mudstone, with ironstone bands .. .. .. | 68 | 0 | 256 | 0 |
| Mudstone, medium to dark grey, with thin ferruginous bands .. | 1 | 1 | 257 | 1 |
| *Gastrioceras cumbriense* Marine Band | | | | |
| Mudstone, medium to dark grey, with thin ferruginous bands in upper part; lowest 1 ft hard and platy; *Lingula mytilloides, Palaeoneilo?* .. | 11 | | 258 | 0 |
| Mudstone, dark grey: cf. *Angyomphalus sp., Caneyella sp.* [juv.], *Anthracoceras sp., Gastrioceras crenulatum, G. cumbriense* .. .. | 2 | 0 | 260 | 0 |
| Mudstone, medium to dark grey pyritous; *Planolites ophthalmoides* 3 in from base .. | 6 | 9 | 266 | 9 |
| Mudstone, grey, with some ironstone bands; *Cochlichnus kochi* at 274 ft, 'fucoids' below 283 ft, fish debris from 283 ft 2 in to 284 ft 3 in | 18 | 9 | 285 | 6 |
| *Gastrioceras cancellatum* Marine Band | | | | |
| Mudstone, blocky to 287 ft 2 in, hard dark platy calcareous below; *Caneyella sp., Dunbarella sp., Agastrioceras carinatum, Anthracoceras sp.,* | | | | |

| | Thickness | | Depth | |
|---|---|---|---|---|
| | ft | in | ft | in |
| *Gastrioceras crencellatum,* mollusc spat .. .. .. | 1 | 10 | 287 | 4 |
| Mudstone, hard dark platy calcareous; *Anthracoceras sp.* and mollusc spat at 288 ft to 288 ft 2 in; *Gastrioceras* cf. *crencellatum* and mollusc spat at 288 ft 6 in; *Gastrioceras cancellatum* and *Reticuloceras superbilingue* at 288 ft 9 in .. .. | 1 | 5 | 288 | 9 |
| (see p. 237 for underlying beds). | | | | |

The Fernilee No. 2 Borehole [0119 7863] shows the following (abridged) section of the beds underlying the Rough Rock:

| | Thickness | | Depth | |
|---|---|---|---|---|
| | ft | in | ft | in |
| Rough Rock (see p. 245) .. | — | — | 303 | 9 |
| Mudstone, 5-in band with plants at 306 ft 2 in; micaceous below .. | 4 | 8 | 308 | 5 |
| Sandstone, medium- to fine-grained turbulent-bedded .. .. | 2 | 0 | 310 | 5 |
| Mudstone, with micaceous and silty bands; some plant debris .. | 6 | 5 | 316 | 10 |
| Mudstone, paler above, darker and with some micaceous streaks below; a little pyrite and some plant debris including *Mariopteris sp.* at 320 ft; cf. *Cochlichnus sp.* at 323 ft 8 in, 324 ft 9 in and 331 ft 8 in; *Naiadites sp.* [juv.] at 325 ft 9 in to 331 ft 8 in | 15 | 8 | 332 | 6 |
| Siltstone, banded argillaceous .. .. | 1 | 2 | 333 | 8 |
| Sandstone, banded flaggy | 3 | 8 | 337 | 4 |
| Mudstone, banded silty | 1 | 1 | 338 | 5 |
| Mudstone, silty, with interbedded siltstone and sandstone .. .. | 8 | 11 | 347 | 4 |

|  | Thick-ness | | Depth | |
|---|---|---|---|---|
|  | ft | in | ft | in |
| Mudstone, dark above, silty micaceous below; plant with attached *Spirorbis*, cf. *Cochlichnus sp.*, *Naiadites sp.* [juv.] 347 ft 9 in to 349 ft 1 in .. .. .. | 6 | 6 | 353 | 10 |
| Mudstone, silty, with siltstone bands .. .. | 4 | 8 | 358 | 6 |
| Mudstone, with dark bands and silty in places, with thin siltstones; some plant debris; thin ironstone bands in places below 367 ft 9 in; cf. *Cochlichnus sp.* at 371 ft 1 in to 373 ft 8 in; *Naiadites sp.* [juv.] at 373 ft 7 in | 35 | 6 | 394 | 0 |
| Siltstone, banded micaceous, with silty mudstone partings .. | 6 | 0 | 400 | 0 |
| Mudstone, banded silty | 13 | 2 | 413 | 2 |
| Mudstone, with ironstone bands and lenses .. | 12 | 4 | 425 | 6 |
| Mudstone, banded silty, with thin siltstone and sandstone bands; cf. *Cochlichnus sp.* at 436 ft 6 in and 439 ft 2 in .. | 17 | 3 | 442 | 9 |
| Sandstone, fine-grained flaggy .. .. .. | 1 | 11 | 444 | 8 |
| Mudstone, banded silty | 17 | 6 | 462 | 2 |
| Siltstone, banded .. | 1 | 10 | 464 | 0 |
| Mudstone, thin silty and ferruginous bands; cf. *Cochlichnus sp.* at 467 ft 7 in and 469 ft .. | 6 | 6 | 470 | 6 |
| Mudstone, banded, silty and micaceous .. | 1 | 9 | 472 | 3 |
| Mudstone, dark to pale with ironstone bands and lenses (bottom of borehole) .. .. | 52 | 3 | 524 | 6 |

It seems likely that the lowest beds lie close above the horizon of the *G. cumbriense* Band. If this is the case, the thickness of the shales between the *G. cancellatum* Band and the Rough Rock is of the order of 260 ft.

*Marple and New Mills.* Beds from the *Gastrioceras cancellatum* Band to the Rough Rock are present, though much obscured by drift, along the Goyt valley between Marple Bridge and New Mills. Their thickness is inferred to be about 280 ft. The only good exposures are in the brook between Brook Bottom and Strines. Here the uppermost beds are seen for a distance of 170 yd downstream from the base of the Rough Rock [9851 8637] at Brook Bottom as mudstones with frequent silty and sandy bands. The Shaw Farm Borehole [9905 8661] proved alternating sandstones and mudstones from the base of the Rough Rock at 381 ft 6 in to the bottom of the borehole at 453 ft 9 in. Lower beds are seen [9800 8650] east of Strines station as dark shale overlying silty mudstones with thin siltstones.

*West of the River Goyt.* At the northern end of the Todd Brook Anticline these beds are about 220 ft thick. Exposures in Mather Clough show much of the group. Here the *Gastrioceras cancellatum* Band is exposed [9770 8213] as 1½ ft of soft dark shale with *Caneyella sp.* [juv.], *Dunbarella sp.*, *Agastrioceras carinatum*, *Gastrioceras* cf. *crencellatum* resting on 1¾ ft of grey shale. A thin contorted band, ½ in to 2 in thick and 2 to 3 ft below the *G. cancellatum* Band is exposed nearby. About 120 yd downstream and 20 ft higher in the succession, the *Gastrioceras cumbriense* Band is exposed [9774 8225] in the following section: grey shale with a few ironstone lenses 13 ft, decalcified shale with *Dunbarella sp.*, *Edmondia?*, *Gastrioceras crenulatum*, *G. cumbriense* 6 in, on grey shale 8 ft.

Farther south, the *Gastrioceras cumbriense* Band, about 40 ft above the *G. cancellatum* horizon, was found by R. H. Price in a stream [9740 8054], nearly ½ mile W. by S. of Handley Fold; it consisted of 7 in of black shale with *Caneyella sp.*, *Gastrioceras crenulatum* and *G. cumbriense*. The same band was also found [9713 7945] north-west of Charles Head.

On the western limb of the Goyt Trough the beds are well seen in the section in Mill Clough (this continues the section on p. 238 downstream to grid reference 0024 7807):

|  | ft | in |
|---|---|---|
| Shale, hard platy .. .. .. | — | 8 |
| Shale, soft .. .. .. .. | — | 9 |

|  | ft | in |
|---|---|---|
| *Gastrioceras cumbriense* Marine Band | | |
| Shale, hard dark decalcified with *Aviculopecten* aff. *losseni, Anthracoceras sp.* and *G. cumbriense* .. .. .. .. | 1 | 0 |
| Shale, hard dark platy, with thin contorted band at base .. .. | 5 | 0 |
| Mudstone, grey .. .. .. | 4 | 5 |
| Ironstone .. .. .. 1 in to | | 1½ |
| Shale, grey platy, with thin ironstone lenses .. .. .. | 13 | 0 |
| Shale, grey platy .. .. .. | 2 | 0 |
| Shale, grey platy, with contorted bands up to 4 in thick .. .. | 1 | 6 |
| *Gastrioceras cancellatum* Marine Band | | |
| Mudstone, medium grey, lowest 1 ft dark decalcified, with *Agastrioceras carinatum* and *G. crencellatum* .. .. .. | 3 | 1 |

The *G. cancellatum* Band can be traced southwards from this locality continuously for nearly a mile on the eastern bank of the stream.

The *G. cumbriense* Band was also found in a small exposure [0097 7776], 170 yd S.S.E. of Oldfield: dark decalcified shale with *Dunbarella sp., Posidonia?* and *G.* cf. *cumbriense* 1 ft, on soft grey shale 6 ft.

The section in Shooter's Clough, near the southern margin of the district [0057 7467], and from thence 25 yd downstream, shows both *G. cancellatum* and *G. cumbriense* bands:

|  | ft | in |
|---|---|---|
| Mudstone, dark grey shaly; *Parallelodon sp.* [*P. cancellatus* (Martin) group] at base .. .. | 2 | 0 |
| Mudstone, black shaly; *Anthracoceras* or *Dimorphoceras sp., G.* cf. *crenulatum, G. cumbriense, Homoceratoides* aff. *divaricatus* .. | | 5 |
| Mudstone, grey .. .. .. | | 7 |
| Mudstone, hard; *Lingula sp.* .. | | 2 |
| Shale, grey .. .. .. .. | 7 | 0 |
| *Not exposed* .. .. .. .. | 9 | 0 |
| Shale, dark grey platy .. .. | 3 | 0 |

|  | ft | in |
|---|---|---|
| Shale, soft 'sulphurous' .. .. | 2 | 0 |
| Mudstone, dark, part decalcified; *Caneyella multirugata, Dunbarella elegans, G. cancellatum* and *R. superbilingue* .. .. .. | 2 | 0 |
| Shale, hard grey .. .. about | 4 | 0 |

*North-eastern area.* On the eastern slopes of Bradfield Dale about 130 ft of shale lie between the base of the *G. cancellatum* Band and the Rough Rock.

Close above the Redmires Flags, dark fossiliferous shale at the horizon of the *G. cancellatum* Band was formerly exposed in the upper part of Holes Clough [2383 9038]. In the upper reaches of the clough the *G. cumbriense* Band, about 70 ft higher in the succession, is exposed [2392 9031] in the following section[1]:

|  | ft |
|---|---|
| Shale, rusty-weathered .. .. | 6 |
| Shale, dark decalcified; 1½-in ironstone at 6 in from top; *Caneyella sp., D. elegans, G. cumbriense, G. crenulatum, Ht.* aff. *divaricatus* .. | 1¼ |
| Shale, with rare *L.* cf. *mytilloides* .. | 2 |

The latter band, seen to 10 in, is also exposed in the hillside a little farther west [2370 9032], where it yielded *Caneyella sp., D. elegans, G. cumbriense* and *G. crenulatum.*

At Hollow Meadows an exposure in a stream [2422 8798] showed 1½ ft of black shale on 2 ft of soft black shale with goniatites and *Dunbarella sp.* From its relations to the thin coal noted on p. 239, and in the absence of the Redmires Flags, this band is considered to be the *Gastrioceras cancellatum* horizon. Farther east, an exposure [2535 8779] by the main road, showed the same band: grey shale 1 ft 6 in, on black shale with *C.* cf. *multirugata, Dunbarella sp.* and *G.* cf. *crencellatum* 6 in.

In the area from Brown Edge to White Path Moss, farther south, about 100 ft of beds lie between the *Gastrioceras cancellatum* Marine Band and the Rough Rock; they are much obscured by peat.

---

[1]Measured by Mr. J. Pattison.

## ROUGH ROCK

*Charlesworth and Hayfield.* Between Charlesworth and Rowarth, the Rough Rock, 50 ft thick, forms a long tongue-like outcrop showing strong features, particularly in the landslip-scar of Coombes Rocks. An old quarry [0206 9188], west of Sitch, shows 40 ft of coarse sandstone, massive in the lower part but more thinly bedded above.

Around Rowarth a lower leaf of flaggy fine-grained sandstone, up to about 20 ft thick, is present beneath the upper massive part of the Rough Rock. As the outcrop is traced northwards from the village the lower bed joins the upper. An old quarry [0169 8950], in the lower bed north of Lower Harthill, shows 11 ft of fine-grained flaggy sandstone. This bed dies out rapidly southwards but reappears over a distance of half a mile north and east of Lantern Pike.

*East of the Goyt Trough.* Between Birch Vale and Foxholes Clough [034 856] the Rough Rock Flags, up to 20 ft thick, are developed beneath the Rough Rock. The latter maintains its usual features, a typical exposure in the upper part of Foxholes Clough [0348 8557] showing 40 ft of massive coarse sandstone.

As the Rough Rock is traced southwards from Foxholes Clough, the Rough Rock Flags appear again beneath it [0385 8495], near Hills Farm, and continue south as far as Cotebank, reaching a maximum development in the area north-east and east of Chinley Churn. In this area the Rough Rock Flags were formerly much worked for flagstones and in several places the workings were continued underground for a short distance. The best section is at the northern end of the quarries [0374 8432]:

|  | ft |
|---|---|
| Rough Rock | |
| Sandstone, coarse yellow-brown cross-bedded .. .. .. | 26 |
| *Not exposed* .. .. .. .. | 35 |
| Sandstone, as above .. .. | 7 |
| Rough Rock Flags | |
| Sandstone, yellow-brown flaggy with coarse bands at base .. | 30 |

Between Buxworth and Tunstead Milton the Rough Rock is poorly exposed, being mostly faulted out north of Hilltop, while south of it the features are modified by glaciation. At Cadster [0214 7982] a section, part quarried, at the head of the clough, shows 49 ft of flaggy, cross-bedded sandstone. To the west of Ladder Hill the Rough Rock forms a marked dip-slope; south of here, however, the outcrop is not conspicuous until the vicinity of Rake End, where a quarry [0232 7613] shows 15 ft of massive cross-bedded sandstone. A lower leaf develops here and continues southwards to merge with the main bed in a long dip-slope east of Goyt's Bridge.

Near the southern boundary of the district, Wildmoorstone Brook, in its lower reaches, shows almost continuous sections of the Rough Rock. The easternmost of these [0166 7467] shows 22 ft of massive coarse sandstone on 7 ft of flaggy micaceous sandstone and siltstone. Massive sandstones are exposed farther downstream and an 18-ft bed with plant stems is present [0149 7476] towards the middle of the Rough Rock.

*North-western area.* Here the Rough Rock crops out on the eastern limb of the Romiley Anticline, on and around Idle Hill. An exposure [9731 9326] near the base shows 30 ft of brown-speckled massive medium-grained sandstone, with partings of brown sandy mudstone.

At Broadbottom, on the northern edge of the area, the Rough Rock is some 85 ft thick. It is well exposed in the gorge of the River Etherow at Besthill Bridge where a quarry [9971 9375] shows 30 ft of massive coarse sandstone. To the south, in the Etherow valley, the outcrops are largely concealed by drift.

*Marple and New Mills.* At Mill Brow a stream has cut deeply into the Rough Rock, which is here well exposed. The lowest beds, 2½ ft of flaggy sandstone resting on silty mudstones and flaggy sandstones, are seen [9782 8947] near Primrose Mill. A typical section of higher beds occurs [9788 8953] a little to the east:

|  | ft |
|---|---|
| Boulder clay .. .. .. .. | 6 |
| Sandstone, flaggy ? *in situ* .. .. | 0½ |
| Mudstone, brown silty .. .. | 3 |

ft

Sandstone, yellow-brown fine-grained,
with several partings of silty mudstone ..    ..    ..    ..    ..    14
Sandstone, massive wedge-bedded
medium-grained feldspathic    ..    14

Farther east a quarry [9907 8969] near Lee
shows 40 ft of massive medium-grained
sandstone, feldspathic and with some cross-
bedding. Near Mellor a borehole [9895
8833], put down about 1947 by the then
Directorate of Opencast Coal Production,
showed the following section of the Rough
Rock:

|  | Thick-ness | | Depth | |
|---|---|---|---|---|
|  | ft | in | ft | in |
| 'Shale' beneath Six-Inch | | | | |
| Mine Coal    ..    .. | — | — | 36 | 6 |
| Sandstone    ..    .. | 4 | 0 | 40 | 6 |
| Shale    ..    ..    .. | 2 | 0 | 42 | 6 |
| Sandstone    ..    .. | 57 | 6 | 100 | 0 |
| Sandstone with shale | | | | |
| partings    ..    .. | 15 | 0 | 115 | 0 |
| Sandstone    ..    .. | 85 | 0 | 200 | 0 |

This borehole, which appears not to have
proved the base of the rock, thus shows it to
be at least 163 ft 6 in thick; this is the greatest
thickness known in the district. The beds
between 100 and 115 ft are probably those
which develop into a shale parting farther
west, at Tarden. This parting continues
southwards to disappear near Capstone.
Near the latter locality another borehole
[9846 8700] by the Directorate of Opencast
Coal Production in 1949 showed:

|  | Thick-ness | | Depth | |
|---|---|---|---|---|
|  | ft | in | ft | in |
| Clay (see p. 246).. | — | — | 49 | 0 |
| Sandstone    ..    .. | 23 | 0 | 72 | 0 |
| Clay    ..    ..    .. | | 6 | 72 | 6 |
| Sandstone    ..    .. | 19 | 0 | 91 | 6 |
| Clay    ..    ..    .. | | 6 | 92 | 0 |
| Sandstone    ..    .. | 48 | 6 | 140 | 6 |
| Shale ('muddy fireclay' | | | | |
| fide F. W. Cope)    .. | 5 | 0 | 145 | 6 |
| Sandstone    ..    .. | 7 | 6 | 153 | 0 |

The seatearth may represent the horizon of
the Sand Rock Mine Coal, not otherwise
developed in the area.

The Shaw Farm Borehole [9905 8661]
proved the Rough Rock from 266 ft 3 in
to 381 ft 6 in. Much of the sandstone was
coarse and massive and there was a noticeable
amount of red staining; a 27-in pebbly band
occurred at 286 ft 10 in. Below the Rough
Rock and to the bottom of the borehole at
453 ft 9 in there occurred a group of sandstones, siltstones and silty mudstones (see p.
241) comparable with the Rough Rock Flags.
Farther south-east the Rough Rock forms
subdued glaciated features. It is partially
exposed in the railway-cutting ½ mile W. of
New Mills, but is cut off by faulting farther
to the east.

On the south side of the Goyt valley the
railway-cutting by Newtown station [9951
8473] shows 30 ft of yellow-brown massive
coarse sandstone on silty mudstone with thin
sandstones. A borehole at Disley Paper
Mills [9805 8528] proved the Rough Rock to
be 75½ ft; the record was 'fakes' (probably
flaggy sandstone or siltstone) from 324 ft to
332 ft 6 in; sandstone (top 34 ft 6 in with
some red staining) to 399 ft 6 in, and shale to
403 ft.

*Todd Brook Anticline.* On the western flank
of the Todd Brook Anticline the Rough Rock
forms a marked feature from near Bolder
Hall to Brink Farm where faulting carries it
outside the present area. The best exposure
is in a quarry [9736 8206], south-east of
Lantern Wood, which shows 15 ft of coarse-
grained pale sandstone with a little mica and
some purple banding. The lower part of the
Rough Rock, exposed to the north-east, is
flaggy and micaceous.

The eastern flank of the Todd Brook Anticline shows a more continuous development
of the Rough Rock. At Lane Ends a split
appears, as at Capstone (see above). A borehole [9823 8385] at Spencer Hall shows: sandstone to 11 ft, 'fireclay' to 24 ft and sandstone to 54 ft. Between here and Whaley
Moor the Rough Rock, some 100 ft thick,
gives a well-marked escarpment which
includes Black Hill. A section [9875 8300]
¼ mile N. of Black Hill, shows:

ft

Sandstone, massive cross-bedded
coarse feldspathic; rare feldspar
crystals and quartz pebbles up to
½ in diameter    ..    ..    ..    12

ft

*Not exposed* .. .. .. .. 6

Sandstone, yellow-brown massive fine-
to medium-grained feldspathic .. 22

To the south of Todd Brook the Rough
Rock thins to about 20 ft and may even in
places be absent (though the outcrop is shown
on the map as continuous). The outcrop is
traceable with difficulty as far south as
Blackhillgate [997 790] where it is faulted out.
To the south-east a rapid thickening takes
place, the Rough Rock being well developed
on Hoo Moor, where it gives rise to a long
dip-slope.

The Fernilee No. 2 Borehole [0119 7863]
proved the Rough Rock to be 110 ft 3 in
thick, with the top 1 ft hard and ganister-like
with rootlets.

Near Goyt's Bridge a series of shallow
boreholes along the line of the dam for the
Errwood Reservoir proved the Rough Rock
and the results may be summarized:

ft

Sandstone, mainly massive .. .. 93

Shale .. .. .. .. 7½ ft to 12

Sandstone, mainly flaggy with partings of siltstone and silty mudstone,
more micaceous below .. .. 40

The lower leaf appears at outcrop ¼ mile N.W.
of Errwood Hall. Followed southwards it
joins with the main leaf of the Rough Rock
near the hall.

*North-eastern area.* The Rough Rock,
some 50 to 60 ft thick, forms a well-marked
escarpment overlooking Bradfield Dale.
Good sections are rare, but the lowest beds
are seen in a quarry [2450 9106] ¼ mile S.W.
of Tor: coarse sandstone 15 ft, on medium-
grained sandstone 4 ft. Farther south the
Rough Rock forms a broad dip-slope on
Bradfield Moors. Near Moscar Heights a
quarry [2407 8830] shows coarse and yellow-
brown massive sandstone 25 ft; nearby
exposures are cross-bedded.

Between Brown Edge and Stanedge Pole
the Rough Rock is split into two leaves by a
shale band up to about 25 ft thick. The shale
thins rapidly eastwards and is absent east of
Rud Hill. A section of the upper leaf [2508
8528] shows 2½ ft of yellow-brown flaggy
sandstone on 4 ft of hard yellow-brown
massive sandstone with prominent joints.
The lower leaf is more flaggy and micaceous.

## BEDS BETWEEN ROUGH ROCK AND *Gastrioceras subcrenatum*

### MARINE BAND

*Western area.* These beds, only a few feet in
thickness, are in general poorly exposed. A
coal smut in the position of the Six-Inch
Mine was seen close above the Rough Rock
[0059 9146] ¼ mile N. of Robin Hood's
Picking Rods on Ludworth Moor. Farther
north a stream exposure [0047 9196], ¼ mile
S.W. of Holehouse, shows 6 in of sandy
seatearth on the top of the Rough Rock; the
coal, if present, is not exposed. To the west of
Mellor a stream section in Knowle Wood
[9816 8866] shows:

|  | ft | in |
|---|---|---|
| *Gastrioceras subcrenatum* Marine Band (see p. 276) .. .. .. .. | — | — |
| Mudstone, hard dark grey 'blocky' and slightly silty .. .. .. | 4 | |
| Shale, dark grey finely silty and micaceous; plant debris .. .. | 0½ | |
| SIX-INCH MINE | | |
| **Coal**   5½ in   .. .. .. | **5½** | |

|  | ft | in |
|---|---|---|
| Mudstone, hard black carbon- aceous; pyrite lenses up to ½ in .. | | 1 |
| Seatearth, dark grey and grey-brown carbonaceous; rootlets and *Stig- maria* .. .. .. .. | 1 | 0 |
| Sandstone, brown siliceous mica- ceous; rootlets .. .. .. | | 4 |
| Seatearth, buff and grey sandy mica- ceous; rootlets .. .. about | 1 | 9 |

The section is cut off downstream by a fault
but the lowest beds in it are within a foot or
two of the Rough Rock.

At Mill Brow a section [9821 8976], in the
stream, shows a thin coal smut in the position
of the Six-Inch Mine, close above the Rough
Rock. Nearby the coal was formerly worked
at Ludworth Fireclay Mine [9815 8979] where
the section was: good coal 4 in, dirt and
brass (pyrite) lumps 5 in, bastard coal 6 in,

R

fireclay 2 ft 6 in, on inferior fireclay 9 in. Half a mile to the east, Hollyhead Colliery [9870 8996] had the following section: shale roof, coal 5 in, dirt 3 in, coal 5 in, on fireclay. The coal was worked under the name 'Two Sheds'; this should not be confused with the 'Two Sheds' of Cown Edge Colliery (see p. 236).

The Six-Inch Mine continues southwards from Mellor. The borehole near Capstone, referred to on p. 244, proved the following section:

| | | Thickness | | Depth | |
|---|---|---|---|---|---|
| | | ft | in | ft | in |
| Shale .. .. .. | | — | — | 42 | 4 |
| **Coal** | 3 in | | 3 | **42** | **7** |
| Shale and fireclay | 16 in | 1 | 4 | 43 | 11 |
| **Coal 'mixture'** | 8 in | | 8 | **44** | **7** |
| Fireclay .. .. .. | | 1 | 5 | 46 | 0 |
| Clay .. .. .. | | 3 | 0 | 49 | 0 |
| Rough Rock .. .. | | — | — | — | — |

The Shaw Farm Borehole [9905 8661], near Brook Bottom, proved a complete section of these beds:

| | Thickness | | Depth | |
|---|---|---|---|---|
| | ft | in | ft | in |
| *Gastrioceras subcrenatum* Marine Band (see p. 276) .. .. .. | — | — | 249 | 10 |
| SIX-INCH MINE | | | | |
| **Coal**, bright .. 7 in | | 7 | **250** | **5** |
| **Coal**, dirty, 1-in pyrite lens at 250 ft 6½ in, 2 in | | 2 | **250** | **7** |
| **Coal**, bright .. 2½ in | | 2½ | **250** | **9½** |
| Mudstone, dark carbonaceous, with coal streaks | | 7 | 251 | 4½ |
| **Coal**, bright, some pyrite on joints .. 5½ in | | 5½ | **251** | **10** |
| Mudstone, dark carbonaceous with some plant debris, coal streaks and a little pyrite .. about | | 6 | 252 | 4 |
| *No core* .. .. .. | 1 | 5 | 253 | 9 |
| Seatearth, dark carbonaceous; some pyrite .. | | 6 | 254 | 3 |
| Seatearth, pale slightly silty; rootlets .. .. | | 9 | 255 | 0 |
| Mudstone, some plant debris; silty to 260 ft 5 in; poorly bedded to 257 ft 6 in .. .. | 9 | 0 | 264 | 0 |

| | Thickness | | Depth | |
|---|---|---|---|---|
| | ft | in | ft | in |
| Mudstone; silty to 264 ft 8 in; silty and interbedded with sandstone and siltstone to base .. | 2 | 3 | 266 | 3 |
| Rough Rock, (see p. 244) | — | — | — | — |

The Disley Paper Mills Borehole [9805 8528] (see p. 364) proved fireclay and strains of coal' from 319 to 324 ft, resting on the Rough Rock. At Fernilee, a stream section [0191 7851] (see also p. 275) showed:

| | ft | in |
|---|---|---|
| *Gastrioceras subcrenatum* Marine Band .. .. .. .. | — | — |
| *Not exposed* .. .. .. .. | 2 | 6 |
| Seatearth, pale brown sandy, with thin sandstone lenses up to 2 in thick .. .. .. .. | 2 | 2 |
| Rough Rock, hard siliceous finegrained sandstone .. .. | — | — |

The Fernilee No. 2 Borehole [0119 7863] (see p. 366) proved 1 ft 9 in of slightly silty buff and grey seatearth, more silty towards base, resting on the Rough Rock at 193 ft 6 in.

*North-eastern area.* This forms a westward extension of the Sheffield Pot Clay area (see Eden and others 1957). The Pot Clay has, in recent years, been worked extensively on Bradfield Moors. A section at the southern end of the workings in 1963 [2400 8950] showed:

| | ft | in |
|---|---|---|
| *Gastrioceras subcrenatum* Marine Band (see p. 291) .. .. .. | — | — |
| Shale, dark grey; black pyritous, with carbonaceous partings in lowest 2 in; '*Anthraconaia*' *lenisulcata, Geisina arcuata* and palaeoniscid scale .. .. | | 10 |
| POT CLAY COAL | | |
| **Coal**, pyritous at top 1¼ in to | | **2** |
| POT CLAY | | |
| Seatearth, grey, a few small ironstone nodules near base; rootlets | 2 | 9 |
| Seatearth, pale grey-brown; rootlets and small ironstone nodules 1 ft to | 2 | 0 |
| Sandstone, ganister-like; rootlets .. 1 ft to | 2 | 3 |
| Seatearth, pale grey silty in places; rootlets .. .. .. .. | 2 | 3 |

Farther east, the Pot Clay has been mined extensively from Wheatshire Pit [main adit at 2508 9004]. Here the section is:

|                          |         | ft | in |
|--------------------------|---------|----|----|
| *Gastrioceras subcrenatum* Marine Band | .. .. .. .. | — | — |
| Shale, black .. | .. .. .. | | 9 |
| POT CLAY COAL | | | |
| Coal .. | .. .. ..1 in to | | 2 |
| Seatearth ('black clay') | .. 3 in to | 2 | 6 |

|                          |       | ft | in |
|--------------------------|-------|----|----|
| Seatearth ('pot clay', restricted use) | 0 to | 4 | 0 |
| Seatearth, silty ('brick clay') | 0 to | 4 | 0 |
| Mudstone, flaggy micaceous ('shoddy') .. | .. .. .. | — | — |

The thickness of the Pot Clay (i.e. the three components of the seatearth listed above) varies from $3\frac{1}{2}$ to 6 ft.                    I.P.S.

# PETROGRAPHY OF SANDSTONES IN THE MILLSTONE GRIT SERIES

Though arenaceous rocks form only part of the Millstone Grit Series in the district, being exceeded in thickness by shales and mudstones, they are better exposed and of more economic interest. Petrographical studies, therefore, have been focused on the sandstones. Some 72 specimens have been studied in thin section and heavy mineral separations, supported by X-ray diffraction where necessary. Density and effective porosity determinations (see Table 6) were made by Mr. K. S. Siddiqui. Densities of whole rock substance (d)[1] were determined by a direct method using accurately cubed specimens, and confirmed by a standard immersion method for porous rocks (Holmes 1921, pp. 43–5). Effective porosity was determined by an adaptation of a standard technique (Holmes ibid., pp. 47–9). Mineral compositions have been determined by point-counter, though the results have been expressed as ranges and averages in view of the sampling and other errors incurred in the method.

The general inapplicability of the term 'grit' to the sandstones in the Millstone Grit Series has been discussed by Dunham (*in* Stephens and others 1953, p. 111) though sporadic minor secondary siliceous overgrowths impart a rough texture to certain of the present sandstones.

The arenaceous members of the Millstone Grit Series exhibit two main lithofacies within the district.

*Quartzitic lithofacies.* Quartzitic sandstones ('crowstones') in the Edale Shales are stable and highly mature, being composed of well-graded, tightly packed, interlocking and secondarily silicified resistates, with the absence of feldspars and their decomposition products. Porosities are accordingly low. Metamorphic and igneous quartz with a range of conspicuous lithic particles (quartzites pure and argillaceous; chert; mylonite; volcanic material) are predominant, with traces of heavy minerals, especially zircon and tourmaline. Less stable minerals (garnet, apatite) are very rare.

*Subarkosic lithofacies.* Immature feldspathic sandstones, ranging from $R_1$ to G in age and of 'fluvial grit' lithofacies, are the principal arenaceous members of the Millstone Grit Series. They are poorly sorted, of medium sand grade (British Standards Institution 1957, p. 104) with sporadic rounded gravel, strewn with medium to coarse sand grades, poorly or false-bedded, and composed of patchily interlocking quartz of igneous origin, orthoclase, microcline-

---

[1](d) = the lower density of a rock in bulk, including air-spaces.

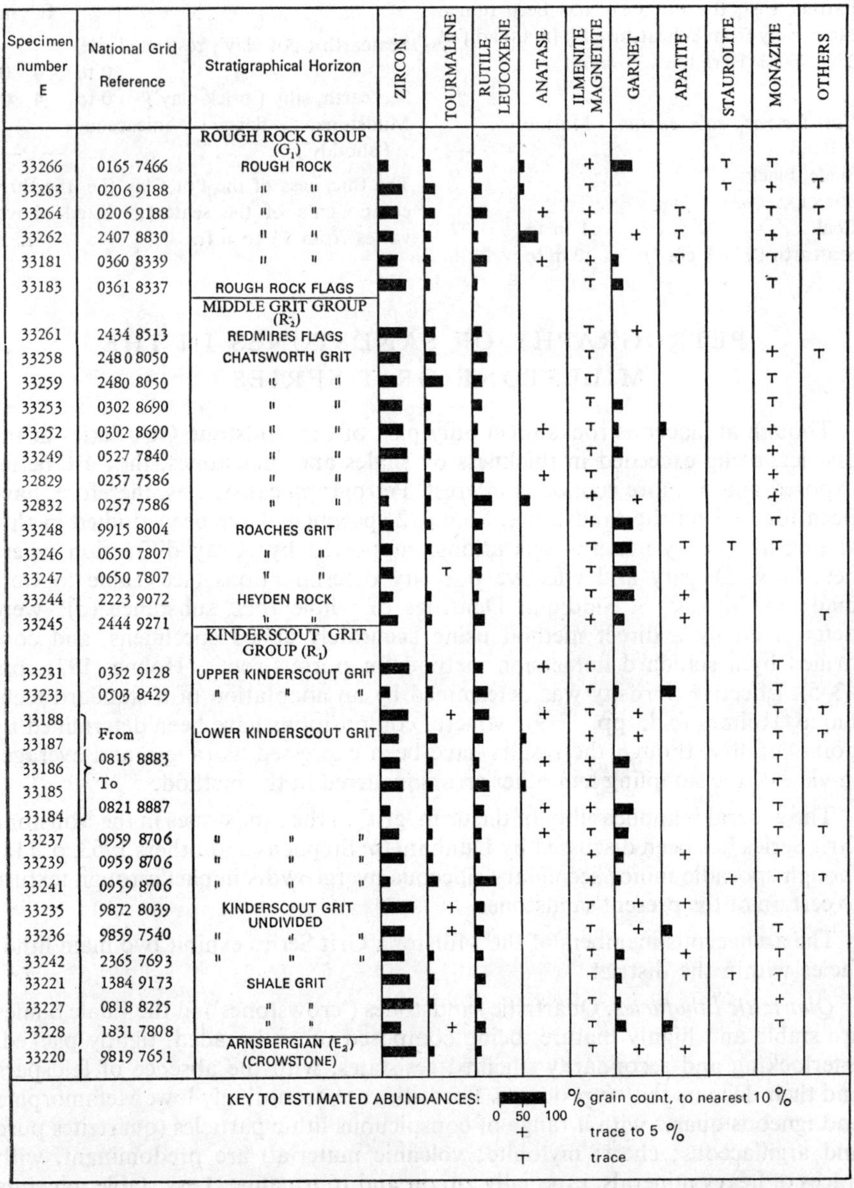

| Specimen number E | National Grid Reference | Stratigraphical Horizon | ZIRCON | TOURMALINE | RUTILE LEUCOXENE | ANATASE | ILMENITE MAGNETITE | GARNET | APATITE | STAUROLITE | MONAZITE | OTHERS |
|---|---|---|---|---|---|---|---|---|---|---|---|---|
| | | **ROUGH ROCK GROUP (G₁)** | | | | | | | | | | |
| 33266 | 0165 7466 | ROUGH ROCK | ■ | ■ | ■ | | + | ■ | | T | T | |
| 33263 | 0206 9188 | "  " | ■ | ■ | ■ | | T | | | T | + | T |
| 33264 | 0206 9188 | "  " | ■ | ■ | ■ | + | + | | T | T | T | T |
| 33262 | 2407 8830 | "  " | ■ | ■ | ■ | T | T | + | T | | + | T |
| 33181 | 0360 8339 | "  " | ■ | ■ | ■ | + | + | ■ | T | | T | |
| 33183 | 0361 8337 | ROUGH ROCK FLAGS | ■ | ■ | ■ | | T | ■ | | | T | |
| | | **MIDDLE GRIT GROUP (R₂)** | | | | | | | | | | |
| 33261 | 2434 8513 | REDMIRES FLAGS | ■ | ■ | ■ | | T | + | | | | |
| 33258 | 2480 8050 | CHATSWORTH GRIT | ■ | ■ | ■ | | T | | | | + | T |
| 33259 | 2480 8050 | "  " | ■ | ■ | ■ | | T | | | | T | |
| 33253 | 0302 8690 | "  " | ■ | ■ | ■ | | T | ■ | | | T | |
| 33252 | 0302 8690 | "  " | ■ | ■ | ■ | + | T | ■ | ■ | | + | |
| 33249 | 0527 7840 | "  " | ■ | ■ | ■ | | T | ■ | | | + | |
| 32829 | 0257 7586 | "  " | ■ | ■ | ■ | + | T | | | | + | |
| 32832 | 0257 7586 | "  " | ■ | ■ | ■ | | + | + | | | + | |
| 33248 | 9915 8004 | ROACHES GRIT | ■ | ■ | ■ | | + | | | | T | |
| 33246 | 0650 7807 | "  " | ■ | ■ | ■ | | T | ■ | T | T | T | |
| 33247 | 0650 7807 | "  " | ■ | ■ | ■ | T | T | ■ | | | | |
| 33244 | 2223 9072 | HEYDEN ROCK | ■ | ■ | ■ | | T | ■ | + | | | |
| 33245 | 2444 9271 | "  " | ■ | ■ | ■ | | + | ■ | + | | | T |
| | | **KINDERSCOUT GRIT GROUP (R₁)** | | | | | | | | | | |
| 33231 | 0352 9128 | UPPER KINDERSCOUT GRIT | ■ | ■ | ■ | + | + | | | | T | |
| 33233 | 0503 8429 | "   "   " | ■ | ■ | ■ | | T | | ■ | | T | |
| 33188 | | | ■ | ■ | + | | + | T | + | | T | T |
| 33187 | From | LOWER KINDERSCOUT GRIT | ■ | ■ | ■ | | + | T | + | | T | T |
| 33186 | 0815 8883 | | ■ | ■ | ■ | | + | + | ■ | | T | |
| 33185 | To | | ■ | ■ | ■ | | T | | | | T | |
| 33184 | 0821 8887 | | ■ | ■ | ■ | + | + | ■ | | | T | T |
| 33237 | 0959 8706 | "   "   " | ■ | ■ | ■ | + | T | ■ | | | T | |
| 33239 | 0959 8706 | "   "   " | ■ | ■ | ■ | | + | ■ | + | | T | |
| 33241 | 0959 8706 | "   "   " | ■ | ■ | ■ | | + | ■ | | + | T | T |
| 33235 | 9872 8039 | KINDERSCOUT GRIT UNDIVIDED | ■ | + | ■ | | T | + | + | | T | |
| 33236 | 9859 7540 | "   "   " | ■ | + | ■ | | T | + | ■ | | T | |
| 33242 | 2365 7693 | "   "   " | ■ | ■ | ■ | | T | | + | | T | |
| 33221 | 1384 9173 | SHALE GRIT | ■ | ■ | ■ | | T | | T | T | T | |
| 33227 | 0818 8225 | "   " | ■ | ■ | ■ | | T | | | | + | |
| 33228 | 1831 7808 | "   " | ■ | ■ | ■ | + | T | | | | T | |
| 33220 | 9819 7651 | **ARNSBERGIAN (E₂) (CROWSTONE)** | ■ | ■ | ■ | | T | + | + | | | |

KEY TO ESTIMATED ABUNDANCES:   ■   % grain count, to nearest 10 %  
0  50  100  
\+   trace to 5 %  
T   trace

FIG. 17. *Heavy mineral abundances in sandstones of the Millstone Grit Series*

microperthite, very minor acid plagioclase, primary micas (muscovite, hydrobiotite-vermiculite) and secondary matrix, especially kaolinite. Secondary cement is usually restricted to sporadic quartz overgrowths. Heavy mineral suites, though sharing certain common factors with those of the 'crowstones', contain more conspicuous proportions of garnet and apatite and are of granitic affinity. The heavy minerals themselves are more common and generally coarser than in the quartzitic sandstones.

With the exception of certain sandstones in the beds below the Kinderscout Grit, the predominant lithofacies is subarkosic, with conspicuous amounts of feldspars comparing generally with Namurian sandstones throughout the Pennine region (Gilligan 1920; Eden and others 1957, p. 15). Gilligan (ibid., p. 259) described the sandstones as arkosic, though this term is not warranted in the light of modern classifications of feldspathic sandstones (Pettijohn 1957, p. 322). The mean feldspar (plus derived kaolinite) content of the present specimens is approximately 14 per cent by volume (Table 7). Greensmith (1957, p. 408) noted that the feldspar content of arenaceous rocks in the Millstone Grit Series in the Smeekley (No. 3) Borehole [2967 7653] about 3 miles to the southeast of the present district, is not more than 7 per cent.

Within the subarkosic units, finer grained, flaggy to laminated, evenly bedded, commonly graded, argillaceous and micaceous sandstones have been considered (p. 162) as parts of turbidite sequences, though petrological properties in themselves are not positive criteria (cf. Bouma 1964, p. 247). These sandstones, however, bear close analogy to those thinner arenaceous beds in the Shale Grit considered by Walker (1966) to be a distinctive facies of turbidite units.

Turbiditic sandstones apart, however, the major lithological change occurs at the top of the Edale Shales. A similar change was recorded at a somewhat higher horizon in the Macclesfield (110) and Stoke-on-Trent (123) districts (Harrison in Evans and others 1968, p. 91), where mature quartzitic sandstones range through the E and H sequences around Macclesfield, and up to the middle of $R_2$ near Stoke-on-Trent, being succeeded by beds of subarkosic fluvial-grit type. Post-$R_1$ sandstones of the Macclesfield and present districts contain broadly consanguineous major and accessory components. The present specimens from the Kinderscout Grit, Chatsworth Grit and Rough Rock in particular, are, however, apparently richer in labile constituents, namely feldspars, kaolinite, biotite and chlorite. The Roaches Grit is more closely comparable in the two districts. Around Chesterfield (One-inch Sheet 112), slight petrographical differences were observed (Harrison in Smith and others 1967, p. 269) between the Ashover Grit and the Chatsworth Grit. Other variations have been noted between the Roaches and Chatsworth grits of the present district, though these variations are slender in supporting differences of provenance. The Chatsworth Grit is closely comparable in both areas confirming the constancy of lithology noted above.

In conclusion, the present study of the subarkosic lithofacies confirms previous petrographical studies of the Namurian deltaic deposits in the Central Province (Gilligan 1920; Greensmith 1957; Allen 1960; Shackleton 1962; Reading 1964; Walker 1966).

This lithofacies is of granitic or granitoid-gneissic affinity and on sedimentological and petrological evidence, a northern or north-eastern provenance was deduced by Gilligan (1920, p. 281), and accepted by subsequent workers. Evidence has, however, been presented above (p. 180) for a southerly derivation of the Roaches Grit, the 'Wales-Brabant Massif' being the most likely quarter. A similar provenance has been proposed for the quartzitic lithofacies of the 'crowstones' following earlier work in adjacent districts (Evans and others 1968; Holdsworth 1963). It seems unlikely that a complete and relatively sudden cessation of sediment supply to the Central Province occurred at the close of

TABLE 6. *Physical Properties of Sandstones in the Millstone Grit Series*

| E No. | Stratigraphical horizon | Density gm/cc (d) | Ratio of absorption (A) | Porosity = A x d |
|---|---|---|---|---|
| | *Rough Rock Group (G)* | | | |
| 33265 | Rough Rock .. .. .. | 2·32 | 5·27 | 12·2 |
| 33181 | Rough Rock .. .. .. | 2·43 | 3·72 | 9·0 |
| 33178 | Rough Rock .. .. .. | 2·47 | 8·02 | 19·8 |
| 33183 | Rough Rock Flags .. | 2·44 | 3·14 | 7·7 |
| | Means: | 2·42 | 5·04 | 12·2 |
| | *Middle Grit Group (R₂)* | | | |
| 33261 | Redmires Flags .. .. | 2·26 | 6·53 | 14·8 |
| 33253 | Chatsworth Grit .. .. | 2·36 | 4·17 | 9·8 |
| 32830 | Chatsworth Grit .. .. | 2·50 | 1·95 | 4·9 |
| 33256 | Chatsworth Grit .. .. | 2·18 | 7·26 | 15·8 |
| 33260 | Chatsworth Grit .. .. | 2·31 | 5·06 | 11·7 |
| 33969 | Roaches Grit .. .. | 2·44 | 1·81 | 4·4 |
| 33246 | Roaches Grit .. .. | 2·46 | 2·33 | 5·8 |
| 33989 | Roaches Grit .. .. | 2·43 | 3·71 | 9·0 |
| 33244 | Heyden Rock .. .. | 2·32 | 4·69 | 10·9 |
| | Means: .. .. | 2·36 | 4·17 | 9·7 |
| | *Kinderscout Grit Group (R₁)* | | | |
| 33233 | U. Kinderscout Grit .. | 2·40 | 3·01 | 7·2 |
| 33186 | L. Kinderscout Grit .. .. | 2·51 | 4·33 | 10·9 |
| 33239 | L. Kinderscout Grit .. .. | 2·43 | 3·39 | 8·2 |
| 33242 | Kinderscout Grit (undivided) | 2·32 | 5·87 | 13·6 |
| | Means: .. .. | 2·42 | 4·15 | 9·9 |
| 33222 | Shale Grit .. .. .. | 2·46 | 1·44 | 3·5 |
| 33226 | Shale Grit .. .. .. | 2·42 | 3·28 | 7·9 |
| 33229 | Shale Grit .. .. .. | 2·33 | 5·11 | 11·9 |
| | Means: .. .. | 2·40 | 3·28 | 7·8 |
| | *Beds below the Kinderscout Grit Group (E–H)* | | | |
| 33219 | 'Crowstone' .. .. .. | 2·46 | 1·10 | 2·7 |

E₂ (later, in R₁ times, nearer the landmass) in favour of the northerly-derived fluvial sediment furnished to the Pennine delta. Continuing, though waning, southerly-derived detritus was probably fed into the Central Province in post-R₁ times, perhaps accounting largely for the Roaches Grit, the 'Wales-Brabant Massif' forming a nearby, extensive landmass at least until Lower Westphalian times (Wills 1951, p. 33). The sandstones within the Edale Shales, like the 'crowstones' of Staffordshire, are evidence that the drainage area of the Massif consisted, at least in part, of igneous and quartzitic rocks, providing the stable, mature, feldspar-free detritus. Prior to, and following this episode, however, feldspathic rocks must have been exposed and subjected to denudation in the 'Wales-Brabant Massif', for the Astbury Sandstone of the Macclesfield district and the Roaches Grit are both of this character.

## DETAILS

### BEDS BELOW KINDERSCOUT GRIT
### GROUP (E–H)

Two specimens (E 33219, 33220) of sandstones were examined from disused quarries, north-east of Redmoor Farm [9815 7590] and on Fox Hill [9819 7651] respectively. These are well-bedded, very hard, indurated, quartzitic and of fine to medium grain, composed of closely packed, mainly interlocking subrounded to subangular clastic grains exhibiting sutured contacts and secondary quartz overgrowths in places (Plate XX, fig. 6). The porosity (Table 6) is accordingly low, particularly in comparison with the higher sandstones. The grains show a generally high degree of sphericity, ignoring diagenetic and secondary processes, and represent a highly mature, stable sediment. Sorting varies from a high degree in the first specimen (E 33219) with average grain diameter 0·2 mm to moderate in the second, with cross-bedded units of fine to coarse sand. Metamorphic quartz predominates, showing strain and general turbidity, being charged with submicroscopic gaseous, liquid and other inclusions; igneous quartz is subordinate. Rock particles include quart-

zite (coarse, strained, interlocking, evenly graded quartz aggregates); argillaceous quartzite (with interstitial orientated illite); inequigranular quartz-mosaics of igneous derivation; cherty silica, highly charged with leucoxenic and ferric oxide dust; sheared quartzite and very turbid felsitic particles. The clastic grains show derivation from a complex metamorphic-igneous terrain. Modal analyses are given in Table 7. Heavy detrital minerals indicate an acid igneous provenance. They include rounded zircon, and tourmaline grains up to 0·2 mm across, with pleochroism: $\omega$ = olive-green, dark brown to black; $\varepsilon$ = pale brown to colourless. Rutile, garnet, leucoxene and apatite are sparse. The matrix is generally quite minor and includes illite (patchily iron-stained), goethite, fine cherty silica and a trace of kaolinite. The specimens compare closely with highly mature quartzitic lithic sandstones ('crowstones') from the Macclesfield (One-inch Sheet 110) and Stoke-on-Trent (One-inch Sheet 123) districts (Harrison in Evans and others 1968, pp. 83, 89).

### KINDERSCOUT GRIT GROUP (R₁)

#### SHALE GRIT

In the northern part of the district, the sandstones include well-bedded, medium to fine subarkoses and curly-bedded, laminated fine subarkoses with intercalations and pellets of carbonaceous silty mudstone of turbidite lithofacies (Walker 1966). These macroscopic interlaminations and pellets grade into mere films only 0·5 mm thick. Clay minerals, especially brown hydrobiotite and its alteration products, illite and chlorite, are conspicuous as primary matrix. Coarser, gravelly, detritus occurs in subarkosic rocks from Kinder Bank and Breck Edge com-

parable with the more massive, higher fluvial-grits. In the south-eastern part of the district (Hucklow Edge; Eyam Edge), the sandstones are well-bedded, fine to coarse grade and subarkosic. Quartz predominates with subordinate microcline[1], orthoclase and sodic plagioclase and their alteration products, differing markedly from the 'crowstones' described above. Modal analyses (Table 7) give mean values within the ranges obtained by Walker (1966) on similar sandstones in the Shale Grit. Rock particles, though lower in proportion than in the

[1]Throughout this account the term 'microcline' includes microperthite.

EXPLANATION OF PLATE XX

PHOTOMICROGRAPHS OF SANDSTONES IN THE MILLSTONE GRIT SERIES
(all taken under partly uncrossed polars)

1. Arkose, Rough Rock ($G_1$). Quarry [0360 8339] on east side of Chinley Churn. E 33178. An interlocking mosaic of evenly sorted, medium sand, subrounded quartz, orthoclase, microcline-microperthite, rock particles, with intergranular brown mica (?hydrobiotite: centre field) and muscovite.                × 45

2. Subarkosic medium sandstone, in Roaches Grit ($R_2$). Quarry [0556 7932] N.W. of Ridge Hall. E 33969. Evenly sorted, tightly packed, quartz, potash feldspars and rock particles; quartz grains are rimmed with secondary silica, and orthoclase by authigenic feldspar.                × 45

3. Arkose in Upper Kinderscout Grit ($R_1$). Quarry [0352 9128] at Chunal. E 33231. Poorly sorted, subrounded coarse sandy quartz, polygranular quartz, orthoclase, microcline-microperthite, are scattered in a finer matrix of micas and kaolinite.                × 48

4. Arkose in Lower Kinderscout Grit ($R_1$). Kinder Downfall [0820 8885]. E 33186. Fine gravelly sandstone composed of closely packed quartz, potash feldspars and rock particles, with interlocking contacts and sparse matrix.                × 45

5. Subarkosic micaceous sandstone in Shale Grit ($R_1$). Quarry [0818 8225] on Breck Edge. E 33227. Fine sandy, subrounded to angular quartz, rock particles, microcline-microperthite, orthoclase and micas are closely packed with little matrix.                × 45

6. Quartzitic sandstone ('crowstone') in Edale Shales ($E_2$). Quarry [9815 7590] north-east of Redmoor Farm. E 33219. Fairly well-sorted, closely packed inter-locking clastics are composed of quartz and quartzose rock particles cemented in places with silica; feldspars are absent.                × 44

PLATE XX

1

2

3

4

5

6

PHOTOMICROGRAPHS OF SANDSTONES IN THE MILLSTONE GRIT SERIES

'crowstones' of the Edale Shales, range from chert to quartzite. Heavy minerals differ also, with conspicuous garnet and apatite, though like the 'crowstones', zircon is predominant, with subordinate tourmaline.

*Alport Castles landslip-scar* [1384 9173]. A specimen (E 33221) from the topmost sandstone (22 ft) (p. 212) is grey, hard, well-compacted and slightly unevenly bedded with conspicuous fine micas on bedding planes. Clastic grains (0·03 to 0·7 mm) are tightly packed in places with sutured contacts and elsewhere separated by pellicles or strained sheaves of clay minerals. Quartz is mainly of igneous origin, though on account of tight packing, some granulation and strain has taken place. Other polygranular quartz includes rock particles, quartzite and igneous quartz-aggregate. Feldspars including highly sericitized and kaolinized orthoclase, microcline and acid plagioclase are conspicuous. The intergranular matrix includes muscovite, kaolinite, turbid brown clay minerals, probably mainly hydrobiotite[1] and sparse chlorite. Heavy minerals include clear, colourless euhedral, and sparser rounded, purple, grains of zircon; prismatic tourmaline ($\omega$ = dark green, $\varepsilon$ = pale pink; $\omega$ = dark brown, $\varepsilon$ = colourless); yellow to red-brown rutile; colourless, step-etched garnet; and traces of staurolite and apatite. A specimen (E 33222), 30 ft from the top of the section, of a massive sandstone (15 ft) is closely similar to E 33221. A third sample (E 33223) from the underlying sandstone (39 ft) and 70 ft from the top of the section, however, is a fine subarkose intercalated with carbonaceous illite-mudstone. The sandstone units contain subangular quartz (mainly igneous), rock particles, subordinate potash feldspars with abundant interstitial colourless illite, brown hydrobiotite, green chlorite, with transgressive muscovite, fine intergranular kaolinite and a little leucoxene. The muscovite shows replacement of quartz through pressure-solution. Mudstone intercalations (1 to 2 mm thick) are of predominant illite fibres orientated parallel to the bedding, carbonaceous dust, fine kaolinite and fine silty quartz. Modal analyses of the first two specimens (E 33221–2) gave, in volume per cent: quartz 70, 74; rock particles

5, 9; orthoclase 5, 4; microcline 2, 3; plagioclase 1, 1; kaolinite 4, 1; muscovite, illite 2, 1; hydrobiotite, biotite 10, 5; secondary $SiO_2$ 1, 1; $Fe_2O_3$ tr, 1. The third specimen was too heterogeneous for meaningful modal analysis.

*Quarry on Kinder Bank* [0506 8802]. Two specimens (E 33224–5) are gravelly, coarse–medium sandstones, hard and dense. Igneous and polygranular quartz with subordinate kaolinized orthoclase, plagioclase and fresh microcline, are subangular to subrounded, closely packed and interlocking in part, with finer sporadic interstitial kaolinite, muscovite and silty quartz. Modal analyses of the two specimens gave quartz 71, 68 per cent; rock particles 11, 14; orthoclase 2, 6; microcline 12, 5; plagioclase tr, 1; kaolinite 1, 2; muscovite, illite 2, 2; hydrobiotite, biotite 1, tr; secondary $SiO_2$ tr, 1; $Fe_2O_3$ tr, tr.

*Roadside quarry, Breck Edge* [0818 8225]. A specimen (E 33226) of massive, brownish grey, subarkosic sandstone is hard, dense and gravelly. The poorly sorted (gravel to fine sand), closely packed interlocking clastics include igneous and polygranular quartz (quartzite, argillaceous quartzite, chert) with subordinate orthoclase, albite-oligoclase and microcline. The interstitial matrix consists of kaolinite, muscovite, illite and hydrobiotite. A second specimen (E 33227) of a more thinly bedded sandstone is also hard, massive and dense, with conspicuous micas on bedding planes. It is finer grained (0·1 to 0·2 mm average granularity), the clastic grains forming interlocking clusters with some secondary silicification, the cores ranging from rounded to subangular. The matrix contains abundant strained sheaves of brown-green hydrobiotite, and muscovite, with disseminated ferric oxide and leucoxene (Plate XX, fig. 5). Heavy minerals comprise colourless, little-abraded zircon; rutile; tourmaline ($\omega$ = dark green to dark brown, $\varepsilon$ = colourless; $\omega$ = black, $\varepsilon$ = pale pinkish brown); clear, colourless garnet; leucoxene. Modal analyses of the two specimens gave quartz 66, 66 per cent; rock particles 13, 10; orthoclase 4, 4; microcline 8, 3; plagioclase 1, 0·5; kaolinite 2, 1; muscovite, illite 1, 2; hydrobiotite, biotite 3, 10; secondary $SiO_2$ 1, 2; $Fe_2O_3$ 1, 1; heavy minerals tr, 0·5.

---

[1]'Hydrobiotite' denotes the variably altered products of detrital biotite and may include vermiculite (see p. 262).

*Roadside quarry, Hucklow Edge* [1831 7808]. A specimen (E 33228) is a hard, compact, dense sandstone of closely packed, interlocking fine–medium (0·05–0·2 mm) sandy clastics, showing some secondary silicification. Igneous quartz predominates, with subordinate orthoclase, microcline and rock particles. Plentiful brownish hydrobiotite, granular kaolinite and transgressive muscovite occur interstitially. Heavy minerals include euhedral, colourless zircon; red-brown angular rutile exhibiting geniculate twins; colourless garnet; fresh, rounded to angular apatite; and tourmaline ($\omega$ = dark brown to dark green, $\varepsilon$ = colourless). A modal analysis gave quartz 63 per cent; rock particles 8; orthoclase 4; microcline 6; plagio-clase tr; kaolinite 2; muscovite, illite 2; hydrobiotite 13; secondary silica 1; $Fe_2O_3$ 1.

*Eyam Edge* [2251 7651]. One specimen (E 33229) is a hard, dense, pale buff, subarkosic sandstone, composed of tightly packed, poorly sorted interlocking clastics (coarse to fine sand-grade) with microstylolitic contacts and a little secondary silicification. Igneous quartz predominates with subordinate rock particles (chert, quartzite), potash feldspars, intergranular brown hydrobiotite and muscovite. A modal analysis gave quartz 77 per cent; rock particles 8; orthoclase 4; microcline 5; plagioclase tr; kaolinite 1; muscovite, illite 1; hydrobiotite 3; secondary $SiO_2$ tr, 1; $Fe_2O_3$ tr.

## KINDERSCOUT GRIT

The Lower Kinderscout Grit, in its thick development in the type area, is an immature, coarse, gravelly and markedly feldspathic sandstone of fluviatile lithofacies, comparing closely with higher sandstones in the Millstone Grit Series. Certain leaves in the south of the district, however, are less coarse-grained and more argillaceous, like the sandstones in the Shale Grit. The lithofacies appears, therefore, to be somewhat diachronous. Porosities (Table 6) are, on average, higher than those of the Shale Grit specimens measured, and considerably higher than the quartzitic 'crowstones'. Ten specimens collected serially at Kinder Downfall and Crowden Bank are typically hard, gravelly, subarkosic, coarse to fine sandstones. The rounded gravel is composed mainly of quartz individuals or composite grains, and particles of quartz-feldspar, granite, vein-quartz and sparser metasediments. Sand-grade components include predominant quartz and composite quartz, subordinate potash feldspars with conspicuous microcline, and minor amounts of muscovite, kaolinite, hydrobiotite and a little secondary silica (Table 7). Heavy minerals comprise predominant colourless (less common purple) zircon, conspicuous garnet,

subordinate rutile-leucoxene, and minor tourmaline, apatite, ilmenite. Monazite, though sparse, is ubiquitous.

Specimens from the Upper Kinderscout Grit range from feldspathic to subarkosic (Table 7) and are mainly finer grained than those from the Lower Kinderscout Grit, except in the northern part of the district. In places shale pellets, with conspicuous mica flakes and carbonaceous material on bedding planes, occur in thinly bedded sandstones resembling the Shale Grit specimens described. Sandy constituents, however, generally resemble those of Lower Kinderscout Grit with dominant quartz individuals and polygranular aggregates, conspicuous potash feldspars and variable amounts of clay minerals, with kaolinite stemming chiefly from altered feldspars. The heavy mineral suites are similar to those of the lower division, and though variable, appear to contain rather lower proportions of garnet.

Specimens of undifferentiated Kinderscout Grit from the south-east part of the district are similar to those from the upper division, except that garnet is quite conspicuous among the heavy minerals.

### LOWER KINDERSCOUT GRIT

*Kinder Downfall* (from [0815 8883] to [0821 8887]). Five specimens (E 33184–8) are medium to fine gravelly subarkosic sand-stone, hard and well-compacted (Plate XX, fig. 4). Gravel is rounded and scattered amongst rounded to angular, closely packed,

Volume per cent (pores and holes ignored)

| | No. of specimens analysed | Quartz individual and composite grains Range | Mean | Rock particles Range | Mean | Orthoclase Range | Mean | Microcline Range | Mean | Plagioclase Range | Mean | Muscovite Illite Range | Mean | Biotite Hydrobiotite (Vermiculite) Range | Mean | Kaolinite Range | Mean | Chlorite Range | Mean | Goethite Range | Mean | Secondary SiO₂ Range | Mean |
|---|---|---|---|---|---|---|---|---|---|---|---|---|---|---|---|---|---|---|---|---|---|---|---|
| Rough Rock (G₁) | 10 | 67–84 | 76 | tr–2 | 1 | 1–4 | 2 | 3–12 | 9 | 0–1 | ½ | ½–2 | 1 | 1–10 | 3 | 3–9 | 5 | — | — | tr–1 | ½ | tr–3 | 1 |
| Rough Rock Flags (G₁) | 2 | 65–72 | 68 | tr–½ | — | 3–4 | 4 | 5–6 | 6 | tr–½ | — | 2–3 | 3 | 10–18 | 14 | — | 3 | — | — | tr–1 | ½ | tr–1 | ½ |
| Redmires Flags (R₂) | 1 | — | 92* | 5 | ½ | — | 1 | — | 2 | — | ½ | — | 3 | — | 1 | — | 1 | — | tr | — | ½ | — | 1 |
| Chatsworth Grit (R₂) | 15 | 65–87 | 77 | tr–1 | 1 | 0–3 | 1 | 2–22 | 10 | tr–½ | ½ | 0–4 | 1 | 0–21 | 3 | ½–8 | 4 | — | tr | 0–1 | ½ | 1–4 | 2 |
| Roaches Grit (R₂) | 8 | 75–88 | 82 | ½–3 | 1 | 1–4 | 2 | 1–13 | 5 | tr–2 | 1 | tr–3 | 2 | tr–7 | 3 | tr–3 | 1 | — | tr | tr–2 | ½ | 1–4 | 2 |
| Heyden Rock (R₂) | 2 | 69–76 | 73 | — | 1 | 2–5 | 4 | 7–12 | 10 | 1–3 | 2 | — | 1 | 1–5 | 3 | 1–11 | 6 | — | tr | — | tr | — | ½ |
| Kinderscout Grit (R₁) Upper | 4 | 64–85 | 71 | 1–4 | 3 | 2–4 | 3 | 3–14 | 8 | — | tr | 2–3 | 2 | 2–19 | 8 | 1–4 | 2 | — | tr | tr–2 | 1 | 1–2 | 1 |
| Kinderscout Grit (R₁) Lower | 10 | 65–84 | 76 | 0–4 | 2 | tr–4 | 2 | 9–19 | 14 | tr–2 | tr | tr–1 | 2 | tr–4 | 2 | 0–4 | 1 | — | tr | 0–2 | tr | 0–4 | 2 |
| Kinderscout Grit (undiffer.) | 1 | — | 84* | — | tr | — | ½ | — | 10 | — | 0 | — | 1 | — | 3 | — | ½ | — | tr | — | 0 | — | 1 |
| Shale Grit (R₁c) | 8 | 62–77 | 69 | 5–14 | 10 | 2–6 | 4 | 2–12 | 5 | 0–1 | ½ | 1–2 | 2 | tr–13 | 6 | tr–4 | 2 | — | tr | tr–1 | ½ | 0–2 | 1 |
| Beds below Kinderscout Grit Group (E–H) | 2 | 71–82 | 77 | 12–17 | 15 | — | 0 | — | 0 | — | 0 | — | 1 | — | 0 | — | tr | — | tr | 0–3 | 1 | 4–7 | 6 |

*Figures in this row are the single analysis.

coarse sand with intergranular angular sand and silt. The quartz grains show interlocking contacts and some secondary siliceous overgrowths. Graded- and cross-bedding is developed in places. Rock particles, especially in the gravel grades, commonly include polygranular quartz and quartz-feldspar grains of acid igneous derivation; graphic granite; vein-quartz; sparser sheared quartzite and argillaceous siltstone. Quartz, particularly pellucid, relatively unstrained and of igneous origin, predominates in the sand grades. Sparse irregular inclusions include brown biotite. Feldspars comprise conspicuous microcline, minor orthoclase and a trace of sodic plagioclase, all showing varying degrees of saussuritization and kaolinization. Soft, creamy-white or brown-stained kaolinite is common in the rock matrix, which also includes flakes of muscovite, illite, hydrobiotite, sparse hydrocarbons, and goethite. Heavy minerals include conspicuous euhedral to subrounded, colourless (and more rarely rounded, purple) zircon; pale pink garnet, with refractive index ranging from $1 \cdot 797$ to $1 \cdot 802$, though commonly about $1 \cdot 799$; dark- red to yellow-brown rutile and plentiful authigenic leucoxene; yellow tabular anatase; very sparse ilmenite; tourmaline, commonly with pleochroism; $\omega =$ dark brown, $\varepsilon =$ colourless to pale yellow; $\omega =$ dark green, $\varepsilon =$ colourless; $\omega =$ dark yellowish brown, $\varepsilon =$ colourless; $\omega =$ indigo blue,

$\varepsilon =$ colourless. Modal analyses of all five specimens gave, in volume per cent: quartz 69–78; rock particles 2–3; orthoclase 1–3; microcline 12–17; plagioclase tr –1; kaolinite 1–4; muscovite, illite tr–2; hydrobiotite 1–4; secondary silica 1–4; $Fe_2O_3$ tr–2. In light of the coarse granularity and patchy kaolinization, however, these ranges are to be regarded as only approximate.

*Crowden Brook and Crowden Tower* [0959 8706]. Five serial samples (E 33237–41) collected from the main lithologies are greybuff (speckled cream), medium to coarse subarkosic sandstones with sporadic rounded to subangular gravel up to 5 mm diameter. Clastic materials are generally closely packed with interlocking contacts in places, minor secondary silicification, and intergranular kaolinite. Monogranular and polygranular igneous quartz predominates in all grades with subordinate fresh microcline and generally kaolinized orthoclase. Interstitial matrix includes muscovite, conspicuous hydrobiotite and kaolinite. The specimens closely resemble those from Kinder Downfall and this extends to the heavy mineral suites. Modal analysis of the five specimens gave the following very approximate ranges; quartz 73–84 per cent; rock particles tr–2; orthoclase tr–4; microcline 9–19; plagioclase tr; kaolinite 1; muscovite tr–2; hydrobiotite tr–6; secondary silica 0–3; $Fe_2O_3$ tr.

## UPPER KINDERSCOUT GRIT

*Roadside quarry, ¼ mile S. of Chunal* [0352 9128]. Specimens (E 33230–1) of flaggy, and in places, false-bedded subarkose display conspicuous mica flakes on bedding planes. Medium sand (0·3 mm average diameter) predominates, consisting of sporadically closely packed secondarily silicified igneous quartz, lesser metamorphic quartz, and conspicuous feldspars — microcline, kaolinized orthoclase and albite–oligoclase (Plate XX, fig. 3). Elsewhere intergranular clay minerals are conspicuous with flakes of hydrobiotite and muscovite orientated parallel to the bedding planes, accounting for the flaggy lithology, with abundant granular secondary kaolinite. Heavy minerals include dominant colourless, angular (rarely polycyclic, rounded, purple) zircon; angular tourmaline ($\omega =$ dark brown, dark green,

black); dark red-brown rutile; traces of monazite and ilmenite. A modal analysis (E 33230) gave quartz 65 per cent; rock particles 4; orthoclase 4; microcline 14; plagioclase tr; kaolinite 4; muscovite 2; hydrobiotite 4; secondary $SiO_2$ 1; $Fe_2O_3$ 2.

*Roadside quarry, by the Lamb Inn, Chinley* [0503 8429]. Specimens (E 33232–3) are massive, well-bedded, fine to coarse subarkosic sandstones, though seams of fine gravel also occur. The coarser specimen (E 33232) is composed of moderately packed, poorly sorted clastic grains of mainly igneous quartz, saussuritized and kaolinized orthoclase; microcline; plagioclase; micro-pegmatite intergrowths, with intergranular kaolinite, hydrobiotite, muscovite and a trace of probable glauconite. A specimen (E 33233) of thinly bedded, pale grey sandstone is of finer grade (0·1 to 0·2 mm), with closely packed, interlocking quartz, microcline, orthoclase, plagioclase and intergranular muscovite, kaolinite and hydrobiotite. Muscovite, associated with carbonaceous staining, occurs on bedding planes. Heavy minerals comprise dominant colourless (rare purplish), subhedral zircon; step-etched, colourless garnet with n = 1·795; angular tourmaline

($\omega$ = dark blue-green; dark green, $\varepsilon$ = pink); rutile, apatite, leucoxene and a trace of monazite. An X-ray powder photograph (NEX 365) of separated leucoxene gave a rutile pattern. A modal analysis of the second specimen gave quartz 85 per cent; rock particles 1; orthoclase 2; microcline 3; plagioclase 1; kaolinite 2; muscovite 2; hydrobiotite 2; secondary silica 2; $Fe_2O_3$ tr.

## KINDERSCOUT GRIT (UNDIVIDED)

*Lumbhole Mill, Kettleshulme* [9872 8039]. Specimens (E 33234–5) of grey-buff hard, compacted gravelly subarkosic sandstones consist of fine (2 to 4 mm diameter) rounded quartz and feldspar gravel, and rounded to subangular, poorly sorted, sandy clastics— igneous quartz individuals and aggregates, metamorphic quartz, microcline, kaolinized orthoclase, albite-oligoclase. These are closely packed in groups with interlocking contacts and patchy secondary silicification. Biotite forms platy inclusions in igneous quartz. The conspicuous fine matrix in one specimen (E 33235) consists of angular quartz, feldspar, muscovite, hydrobiotite and kaolinite with ferric oxide and carbonaceous dust. Heavy minerals include predominant colourless, euhedral, or abraded zircon, with sparse well-rounded, purple varieties; angular, step-etched garnet; rutile; rounded prisms of apatite; tourmaline prisms, with $\omega$ = dark green, $\varepsilon$ = colourless; $\omega$ = dark yellow-brown, $\varepsilon$ = colourless; traces of ilmenite and monazite. A modal analysis (E 33234) gave quartz 71; rock particles 4; orthoclase 2; microcline 9; plagioclase tr; kaolinite 1; muscovite 3; hydrobiotite 8, secondary $SiO_2$ 1·5; $Fe_2O_3$ 0·5.

*Old quarry 500 yd S.S.E. of Redmoor Farm, Andrews Edge* [9859 7540]. Flaggy, wedge-bedded, argillaceous subarkose (E 33236) shows ripple marks (wave-length 1·5 cm) and fine micas on bedding planes. Graded, wedge-bedded units consist of fine to coarse sandy quartz (mainly igneous), minor potash

and sodic feldspars closely packed in places with sutured contacts, and separated elsewhere by abundant primary clay-mineral matrix of turbid brown-green hydrobiotite, muscovite with sparse chlorite. Mica-rich layers separate the wedges of graded sediment, and consist of finely interlaced muscovite and brown-green hydrobiotite. Heavy minerals (up to 0·2 mm) comprise colourless, euhedral zircon (rare rounded, purple zircon); red-brown, rounded rutile; rounded apatite; etched garnet; tourmaline ($\omega$ = dark brown and dark green); traces of monazite, and ilmenite. These minerals are concentrated in the micaceous partings. A modal analysis gave quartz 64 per cent; rock particles (siltstone, chert, quartzite) 3; orthoclase 2; microcline 3; plagioclase 2; kaolinite 1; muscovite 3; hydrobiotite 19; secondary $SiO_2$ 1; $Fe_2O_3$ 2.

*Stoke Hall Quarry* [2365 7693]. Specimens (E 33242–3) are fairly compact, hard, subarkosic with finely scattered muscovite and secondary kaolinite. Igneous quartz predominates, with conspicuous microcline, minor rock particles, kaolinized orthoclase and a trace of plagioclase. The matrix includes undulose hydrobiotite and muscovite with granular kaolinite. Heavy minerals are similar to those described above from the Upper Kinderscout Grit, except that the conspicuous garnet content compares more closely with specimens of the Lower Kinderscout Grit from Kinder Downfall and Crowden Brook.

## MIDDLE GRIT GROUP (R₂)

### HEYDEN ROCK

Two specimens (E 33244, 33245) were examined from Strines Moor [2223 9072] and Emlin Dyke [2444 9271] respectively.

Both samples are subarkosic, hard and well-compacted, with interlocking medium sand, and some wedge-bedded gravel in the second specimen. Igneous quartz predominates in both, with minor metamorphic quartz and secondary overgrowths, coarse, fresh microcline, kaolinized orthoclase, sodic plagioclase, muscovite and finely granular interstitial secondary kaolinite. Sparse rock particles of perhaps igneous origin occur in E 33244, and one grain appears to have been completely leucoxenized. Heavy minerals include conspicuous garnet and zircon, with minor rutile, leucoxene, a little tourmaline, monazite, minor apatite, and a trace of ilmenite. The two suites more closely resemble those of the Roaches Grit than others of the Middle Grit Group, within the overall granitic aspect. Modal analyses of the two specimens (Table 7) show an overall similarity, allowing for the variations, to the Middle Grit Group as a whole.

## ROACHES GRIT

Sampling presents great difficulty owing to the variability of the Roaches Grit (see p. 179) but the few specimens described below indicate certain lithological variations within the group. The Corbar Grit, underlying the Roaches Grit, has not been examined in detail on account of its lesser stratigraphical significance. In general, the sandstones are hard, well-compacted with interlocking clastics and sporadic secondary silica and feldspathic cement.

As noted above (p. 249) the coarser beds, which are gravelly in places, are closely similar to the subarkosic 'fluvial-grit' facies of the higher sandstones of the Middle Grit Group. The present specimens from Lady Low, Ridge Lane and Longhill in particular are typical fluvial grits, being gravelly, coarse to medium subarkoses with predominant igneous quartz (polygranular and individuals) and conspicuous potash feldspars, particularly microcline, ranging from gravel to fine sand. The coarser constituents are rounded, the finer subrounded to subangular; all are fairly closely packed with interlocking contacts in places, and with generally conspicuous intergranular, secondary, microcrystalline kaolinite. Rock particles, other than igneous, are sparse. Other minor intergranular minerals include muscovite and hydrobiotite.

The finer facies, illustrated by specimens from Kettleshulme and Ridge Hall, are feldspathic fine to medium, quartzitic sandstones differing (granularity apart) from the gravelly subarkosic facies by a generally minor feldspar content (plagioclase being conspicuous in relation to microcline-orthoclase), close packing of clastic grains, with sutured microstylolitic contacts and conspicuous secondary overgrowths of silica and feldspar around quartz and feldspar cores, respectively. Rock particles are quite minor and include a range of sedimentary and metamorphic lithologies. Intergranular micas and clay minerals are generally conspicuous, particularly hydrobiotite.

Heavy mineral suites of both gravelly subarkose and feldspathic sandstones appear to be similar, with predominant garnet; subordinate zircon (mainly little-abraded colourless subhedra, and polycyclic purplish grains); minor tourmaline ($\omega$ = dark green, brown, black), rutile-leucoxene, and traces of ilmenite and monazite.

*Quarry near Longhill* [0409 7515]. A specimen (E 33968) is a gravelly subarkose, with rounded quartz and feldspar gravel 2 to 4 mm diameter. Clastic grains are closely packed in places with microstylolitic contacts, and quite minor intergranular clay minerals. Igneous quartz predominates in gravel and sand, with subordinate fresh microcline, kaolinized orthoclase and a little sodic plagioclase. A modal analysis gave quartz 75 per cent; non-igneous rock particles 2; orthoclase 2; microcline 13; plagioclase 1; kaolinite 1; muscovite 3; hydrobiotite 1; secondary $SiO_2$ 1; $Fe_2O_3$ 1.

*Quarry opposite Cold Springs Farm, Buxton* [0447 7462]. A typical specimen [E 33989] is a hard compacted, fine gravelly subarkose closely similar to the above, except that the feldspars have been more kaolinized. A modal analysis gave quartz 83 per cent; rock particles 1; orthoclase 1; microcline 5; plagioclase 1; kaolinite 3; muscovite 2; hydrobiotite 3; secondary $SiO_2$ 0·5; $Fe_2O_3$ 0·5.

*Crag south of Lady Low* [0650 7807]. The coarse massive sandstone (40 ft) described on p. 229 consists (E 33246) of interlocking medium to coarse igneous quartz and feldspars (microcline being conspicuous, with altered orthoclase and albite), with sporadic secondary silica overgrowths on quartz, and intergranular brown hydrobiotite, kaolinite aggregates and muscovite. A modal analysis gave quartz 78 per cent; rock particles 1; orthoclase 3; microcline 10; plagioclase 3; kaolinite 1; muscovite 1; hydrobiotite 2; secondary $SiO_2$ 1. Soft sandy concretionary masses (cf. 'mare's balls') occur up to 6 ft in diameter and a specimen (E 33247) taken from a hard concretion core consists of a clastic assemblage similar to the host rock including the heavy mineral suite, except the matrix and the clastics (especially feldspars) have been partially replaced by dolomite. Petrographical descriptions of 'mare's balls' occurring in Crawshaw Sandstone were given in the Chesterfield Memoir (Harrison *in* Smith and others 1967, p. 278). These concretions were considered to have formed through the selective replacement of calcite by goethite, in calcite-enriched masses. Dolomite enrichment may explain the formation of the present concretions though the source of the migratory carbonate ions is unknown and the process may be attributed to diagenesis.

*Disused quarry, Ridge Lane* [0380 7730]. A specimen (E 33967) of the coarse sandstone exposed (30 ft) with pebbly bands (p. 229) is hard, well-compacted grey-buff subarkosic, with wedges of fine gravel and coarse sand. The clastic constituents and texture are closely similar to the other gravelly subarkoses described above. A modal analysis gave quartz 83 per cent; rock particles 3; orthoclase 2; microcline 5; plagioclase 2; kaolinite 0·5; muscovite 1; hydrobiotite 2; secondary $SiO_2$ 1·5.

*Quarry, 550 yd N.30°W. of Ridge Hall* [0556 7932]. Specimens (E 33969–71) of the medium-grained, well-bedded sandstone (with thin partings of silty mudstone), are hard, quartzitic and subarkosic, with fine mica sprinkled on bedding planes. Sand of medium grade (0·2 to 0·6 mm) predominates with intercalations of coarse sandy units showing rough graded bedding. Clastic grains are tightly interlocking in clusters, with microstylolitic contacts and elsewhere prominent secondary quartz overgrowths. Feldspars in one section (E 33969) show authigenic feldspar overgrowths. Igneous and subordinate metamorphic quartz predominates, with subordinate potash feldspars (microcline, orthoclase) and rock particles (including cherty silica); muscovite flakes marginally replace quartz grains due to pressure-solution (Plate XX, fig. 2). Matrix is conspicuous in another section (E 33971) with ferric oxides, anatase, leucoxene, turbid and pale brown hydrobiotite, and much finely granular kaolinite and illite. Modal analyses gave quartz 80–88 per cent; rock particles 1–2; orthoclase 2–4; microcline 1–5; plagioclase 0·5–2; kaolinite tr–1·5; muscovite tr–2; hydrobiotite tr–6; secondary $SiO_2$ 1–4·5; $Fe_2O_3$ tr–2.

*Quarry at Kettleshulme* [9915 8004]. A feldspathic, hard, fine to medium sandstone (E 33248). This specimen consists of closely packed, well-sorted (0·2 mm average granularity) clastics. The clastic assemblage is similar to that of the specimen from Ridge Hall, and intergranular minerals include muscovite, kaolinite, goethite, hydrobiotite, and leucoxene. A modal analysis gave quartz 81 per cent; rock particles 1; orthoclase 2; microcline 5; plagioclase tr; kaolinite 1; muscovite, illite 2; hydrobiotite 7; secondary $SiO_2$ 1. A heavy mineral suite (Fig. 17) is comparable with those from the coarser gravelly sandstones of the Roaches Grit.

## CHATSWORTH GRIT

The Chatsworth Grit exhibits marked lithological variations. In its coarser development, it is a compacted, hard, gravelly, medium to coarse sandstone with subrounded to rounded particles of moderate sphericity, sporadically interlocking, with minor secondary silicification, and with variable intergranular matrix. Clear, pellucid quartz individuals and aggregates of undoubted igneous origin predominate with subordinate clear, conspicuous microcline, minor turbid kaolinized and sericitized orthoclase, and traces of plagioclase. Granular kaolinite, generally intermixed with illite, muscovite and hydrobiotite, forms the matrix. The lithology is subarkosic with an average feldspar and kaolinite content of 15 per cent (Table 7). Finer lithologies such as that (E 33251) from Hayfield show less-rounded clastics, lower proportions of feldspars, and concomitant increases in proportions of clay minerals, especially the degraded micas—illite and hydrobiotite—which impart a flaggy texture. Heavy mineral suites follow the general trends of the older sandstones of the Millstone Grit Series, though proportions of garnet are apparently somewhat lower compared with the Roaches Grit and Heyden Rock. The red facies is differentiated from the grey only by secondary impregnation of hematite, in pores and along bedding planes. The impregnations are clearly post-sedimentation and post-diagenesis for they lie outside secondary quartz overgrowths. Colour apart, the facies do not apparently differ.

*Birch Quarry, Hayfield* [0302 8690]. Sandstones from the lower leaf of the Chatsworth Grit exhibit lithologies ranging from coarse, massive, yellow-brown and subarkosic to fine-grained and argillaceous. Three specimens (E 33252 to 33254) of the predominantly coarse sandstone (average granularity 0·5 mm) are composed of clusters of poorly sorted interlocking clastics—igneous quartz, microcline and variably kaolinized orthoclase, with matrix composed of fine kaolinite, muscovite, leucoxene and illite. The clay minerals marginally replace etched resistates. Secondary quartz overgrowths occur in places. Flaggy beds (E 33254) contain scattered mica flakes, clay minerals and ferric oxides concentrated along bedding planes imparting fissility. The average granularity of this specimen (0·3 mm) is a little lower than that of the more massive beds. The fine sandstone (E 33251) differs markedly in its low average granularity (0·06 mm), the angular, dispersed resistates and abundance of matricial hydrolyzates—hydrobiotite, illite, chlorite and muscovite. The clastic assemblage, however, is similar to the other three samples from this locality, with predominant igneous quartz and subordinate potash feldspars. Heavy minerals from two samples (E 33252–3) include: zircon (colourless, unabraded crystals to well-rounded grains; also purple grains); tourmaline with pleochroism $\omega$ = black, $\varepsilon$ = pale brown; $\omega$ = dark brown, $\varepsilon$ = colourless); rutile in subangular yellow to red-brown prisms; apatite; garnet (in E 33252) as colourless, step-etched particles; monazite, which though sparse, forms conspicuous yellowish grains. Modal analyses gave, in volume per cent: quartz 65–84; rock particles tr–1; microcline 2–13; orthoclase 2–3; plagioclase tr–2; kaolinite 2–8; muscovite tr–1; hydrobiotite, chlorite 1–21; secondary $SiO_2$ 1–4; $Fe_2O_3$ tr–2.

*Roadside exposure at Rake End* [0257 7586]. Four specimens (E 32829–32) taken at intervals of 3 to 4 ft through the exposure of the red facies of Chatsworth Grit are lithologically similar. Only the two higher samples (E 32829–30) are reddish in colour, owing to hematite disseminated in pores and along bedding planes. The lower samples are both pale grey with fairly sparse hematite. Gravel is generally rounded to subrounded, consisting of quartz and white to pink feldspars. The sandy detritus is closely packed in places with interlocking contacts; quartz grains show sporadic secondary silicification. Clear igneous quartz predominates with sparse polygranular particles, conspicuous potash feldspars—microcline and altered orthoclase. Micas are generally minor and much of the matrix consists of secondary kaolinite and illite. Graphite was found in one sample (E 32832) as soft, black, metallic flakes (X-ray powder photograph NEX 463). Heavy detrital minerals from two samples (E 32829, 32832) include zircon (colourless subhedral, and purple, rounded grains); tourmaline ($\omega$ = deep olive green, $\varepsilon$ = colourless; $\omega$ = dark yellowish brown); rutile (deep red-brown and opaque leucoxenized pseudomorphs); anatase (yellow tablets); garnet (colourless, step-etched) and conspicuous monazite in rounded yellowish grains.

Modal analyses of three of the specimens (E 32830–2) gave quartz 79–85 per cent; rock particles 1; orthoclase 1–2; microcline 8–12; plagioclase tr; kaolinite 0·5–3; muscovite, illite tr–1; secondary $SiO_2$ 1–3; $Fe_2O_3$ tr–0·5.

*Roadside quarry Matley Moor* [0190 9020]. A specimen (E 33255) is well-bedded, pale grey, with alternating units of fine gravel, fine to medium sandstone. Sorting is poor. Gravel layers include interlocking or discrete grains of quartz and feldspars (mainly microcline), with coarse muscovite and much brown kaolinite-illite matrix. Sandy constituents are closely similar and the rock is a gravelly subarkose.

A modal analysis gave quartz 71 per cent; rocks 1; orthoclase 1; microcline 14; plagioclase 0·5; kaolinite 6; illite, muscovite 3; hydrobiotite 3; secondary $SiO_2$ 0·5.

*Exposure at Castle Naze* [0527 7840]. A specimen (E 33249) of the coarse-grained pink-stained facies (p. 235), is gravelly and subarkosic; it closely resembles the specimens described from Rake End. Heavy minerals include colourless (rarely purple) zircon; tourmaline ($\omega$ = dark brown; dark blue); garnet; rutile, leucoxene and a trace of monazite. A modal analysis gave quartz 87 per cent; rocks 1; orthoclase 1; microcline 9; plagioclase trace; kaolinite 0·5; secondary $SiO_2$ 1; $Fe_2O_3$ 0·5.

*Section at The Naze, Chinley* [0414 8348]. A thinly bedded specimen (E 33250) from the coarse, gravelly massive cross-bedded sandstone (50 ft) exposed (p. 234), is moderately hard and compacted, with hematitic laminae and dispersed kaolinite. Clastic grains are mainly subrounded, moderately sorted with local secondary overgrowths and interstitial clay minerals. The lithology is subarkosic, a modal analysis giving quartz 78 per cent; rocks 1; orthoclase 1; microcline 13; kaolinite 4; hydrobiotite 1; secondary $SiO_2$ 2.

Specimens were also examined from the grey Chatsworth Grit in the eastern part of the district.

*Grey Millstone Quarries, Hathersage* [2480 8050]. Three specimens (E 33258–60) were examined of the uppermost pebbly flags and underlying coarse, massive-bedded sandstone. All are subarkosic, with variable amounts of gravel and well-bedded or flaggy structures. Rounded to subrounded gravel includes quartz and potash feldspars. Sandgrade clastics (mainly subrounded of moderate sphericity) range from 0·3 to 0·6 mm, are closely packed in clusters, with interlocking contacts and some grains show secondary quartz overgrowths. Igneous quartz predominates with subordinate though conspicuous feldspars (clear, fresh microcline; turbid orthoclase). Kaolinite and micas are common in the matrices, with scattered leucoxene. Heavy minerals from one sample (E 33259) include colourless zircon crystals; angular tourmaline ($\omega$ = black, $\varepsilon$ = pink; $\omega$ = olive green, $\varepsilon$ = colourless); rutile, monazite, leucoxene and a trace of likely topaz. Modal analyses gave quartz 72–76 per cent; rocks 0·5–1; orthoclase tr–1; microcline 7–22; plagioclase tr; kaolinite 3–9; illite, muscovite 1–4; hydrobiotite, chlorite 0·5–1; secondary $SiO_2$ 1–2.

*Exposure on Stanage Edge* [2380 8459]. Two specimens (E 33256–7) are subarkosic and of predominantly medium sand grade (0·6 mm). The subrounded clastics are moderately sorted and tightly packed in clusters, with intergranular iron-stained kaolinite-illite. Quartz, potash feldspars (predominantly microcline), micas and secondary kaolinite with some ferric oxide are the main components. Modal analyses gave quartz 79, 76 per cent; rocks 1, tr; orthoclase tr; microcline 8, 15; kaolinite 5, 2; muscovite, illite tr; hydrobiotite, chlorite 4, 3; secondary $SiO_2$ 2, 3; $Fe_2O_3$ 1, 0·5.

## REDMIRES FLAGS

These are predominantly fine to medium-grained, well-bedded, fissile sandstones. A specimen (E 33261) from an exposure [2434 8513] near Stanedge Lodge is fine grained, feldspathic, grey and fairly compact, with ferric oxide streaks and coarser sandy laminae. Rather angular interlocking clastics (predominantly igneous quartz; very minor orthoclase, microcline and albite-oligoclase) average 0·1 mm; quartz is secondarily silicified in places. The potash feldspars are variably kaolinized. Matrix includes detrital muscovite and hydrobiotite. Heavy minerals include: predominant zircon (colourless crystals to rounded grains); tourmaline ($\omega$ = black, $\varepsilon$ = pale brown; $\omega$ = dark green, $\varepsilon$ = colourless); garnet, rutile, leucoxene and a trace of monazite. A modal analysis (Table 7) indicates a low feldspar content, which is, however, comparable with those of the finer grained sandstones of the Middle Grit Group.

s

# ROUGH ROCK GROUP ($G_1$)

## ROUGH ROCK FLAGS

The fine-grained (average granularity 0·1 mm) evenly and thinly bedded arenaceous rocks (E 33182–3) underlying the Rough Rock at Chinley Churn [0361 8337] are partly laminated, with abundant muscovite and hydrobiotite flakes on bedding planes. Resistate grains are well-sorted and mainly subangular with igneous quartz (showing minor secondary overgrowths), minor rock particles (chiefly chert), microcline, sericitized orthoclase, sodic plagioclase, and contorted micas aligned along the bedding. The last-named, which are commonly finely inter-layered, comprise muscovite, illite and degraded hydrobiotite. Secondary matrix includes kaolinite and leucoxene. Heavy minerals include zircon, garnet, tourmaline, apatite, rutile and traces of monazite and ilmenite. Modal analyses gave quartz 65, 72 per cent; rocks 1, 1; microcline 5, 6; orthoclase 3, 4; plagioclase 0·5, 0·5; kaolinite 3, 3; illite 3, 2; hydrobiotite 18, 10; secondary $SiO_2$ 1, 1.

## ROUGH ROCK

The Rough Rock is of relatively uniform lithology in the district, though local differences are shown in the presence of gravel, overall grading and rarely in the presence of a dolomite-hematite secondary cement. In the northerly exposures, pale grey to buff, well-consolidated, fairly indurated medium to coarse, subarkose predominates, with sporadic rounded gravel. Certain specimens exhibit cross-bedding while others are well bedded, with graded units 1 to 2 cm thick in places. Coarser clastics are rounded and tightly packed in clusters with sutured contacts and local secondary quartz overgrowths occur. Pockets and granules of cream kaolinite from the alteration of potash feldspars are macroscopically conspicuous. Subarkosic in character, the mean content of feldspars plus kaolinite is about 16 per cent. Clear, igneous quartz in individuals or composite grains is predominant. Primary matrix includes muscovite, illite and hydrobiotite. The secondary matrix, especially leucoxene and anatase marginally replaces clastic and secondary quartz. Dolomite in the two borehole specimens described below, extensively replaces clastic grains. Heavy minerals include: colourless (rarely purple) zircon; rounded brown and green tourmaline grains, up to 0·2 mm diameter; step-etched garnet (n = 1·799).

Brown micas are common in the primary matrix of the more flaggy beds. These micas vary from clear, strongly pleochroic (Z = deep red-brown; X = colourless) with refractive index (β) = 1·630 and 2V (—) 0 to 12°, to dark brown, very turbid and with inde-terminable optics. They are commonly finely interlayered with each other and muscovite, and are intercalated with resistate grains. Flakes concentrated from one specimen (E 33178) were examined by X-ray diffractometry untreated, glycerol saturated and heated in air at 110° and 700°C. The results, by Mr. K. S. Siddiqui, identify the less altered brown micas as hydrobiotite and the highly turbid products as vermiculite.

*Chinley Churn* [0360 8339]. Specimens (E 33177–81) sampled at 3 to 4 ft intervals through the 30 ft section of Rough Rock exposed, are subarkosic to arkosic (average granularity 0·2 mm) with fairly well-sorted, subrounded to subangular clastics tightly packed in places with microstylolitic contacts and secondary quartz overgrowths (Plate XX, fig. 1). Igneous quartz predominates with subordinate feldspars (fresh microcline, turbid orthoclase, sericitized albite-oligoclase) and interstitial contorted micas, (muscovite, illite, hydrobiotite), leucoxene and anatase. Modal analyses gave quartz 67–80 per cent; rock particles 1–2; orthoclase 1–4; microcline 3–12; plagioclase 0·5–1; kaolinite 2–8; illite 1–2; hydrobiotite, chlorite 2–10; secondary $SiO_2$ 1–3; $Fe_2O_3$ tr–1. Heavy minerals include zircon, rutile, garnet, tourmaline, monazite, leucoxene, anatase and apatite.

*Old quarry, Cown Edge* [0206 9188]. Three specimens (E 33263–5) are pale grey-buff, fairly hard and compact, ranging from fine gravelly to coarse and medium sandstone. The coarsest specimen (E 33265) is also the most poorly-sorted, while a bedded specimen (E 33263) exhibits graded units 1 to 2 cm thick. Within these units the clastic grains range from 0·1 to 0·4 mm diameter and show interlocking contacts and some secondary silicification, with intergranular kaolinite. The third specimen (E 33264) is even-grained with more secondary silica. The major and accessory mineral suites are closely comparable with predominant igneous (lesser metamorphic) quartz; conspicuous feldspars —microcline, orthoclase, perthite; sparse polygranular quartz and other rock particles, except in the graded arkose (E 33263) which contains more conspicuous particles of leucoxenic siltstone, quartzite and acid lava. Matricial constituents include muscovite, finely granular kaolinite and dark brown, pleochroic hydrobiotite. Heavy mineral suites for two specimens (E 33263, 33264) differ from the remainder in having little or no garnet. Modal analyses gave quartz 70–84 per cent; rock particles tr–1; orthoclase 1–2; microcline 6–12; kaolinite 3–9; illite 1–2; hydrobiotite 1–4; secondary $SiO_2$ 0·5–2; $Fe_2O_3$ 0·5–1.

*Quarry near Moscar Heights* [2407 8830]. Fine, rounded gravel (3 mm) is scattered in predominant fine to coarse sand in a fairly compact subarkose (E 33262), which closely resembles those described above. Intergranular kaolinite and illite are abundant. The only difference noted is the presence of conspicuous tablets of yellow anatase. Garnet is sparse (cf. two specimens, E 33263–4, from Cown Edge). However, although the present specimen was collected from the eastern margin of the district, some 13 to 15 miles from the other localities, which all lie towards the western margin, no significant differences were noted, the anatase no doubt representing a local redistribution of authigenic titania. A modal analysis gave quartz 79 per cent; rock particles 1; orthoclase 1; microcline 12; plagioclase 0·5; kaolinite 5; illite 0·5; hydrobiotite 0·5; secondary $SiO_2$ 0·5.

*Exposure in stream, S.E. of Goyt Bridge* [0165 7466]. Fairly hard, compacted coarse subarkose (E 33266) consists of sparse fine, rounded gravel, and predominant poorly-sorted, subangular to subrounded fine to coarse sand, exhibiting sutured contacts in places, with some secondary silicification. Predominant igneous quartz (with conspicuous biotite and microcline inclusions), subordinate partly kaolinized potash feldspars, rock particles (chert) and fine intergranular micas (muscovite, hydrobiotite, kaolinite) comprise the rock. Garnet is particularly conspicuous among the heavy minerals (Fig. 17). A modal analysis gave quartz 83 per cent; rocks 1; orthoclase 1; microcline 10; kaolinite 2; illite 0·5; hydrobiotite 2; secondary $SiO_2$ 0·5.

*Shaw Farm Borehole* [9905 8661]. Specimens (E 38257–8) of Rough Rock from 288½ and 360 ft respectively are dark purple, dense, well-compacted and hard, differing considerably from the specimens described above. Igneous quartz predominates with subordinate microcline, orthoclase, and interstitial micas. Dolomite associated with hematite forms the secondary cement and extensively replaces clastic grains, especially orthoclase. In view of this replacement, modal analyses were not made.        R.K.H.

# REFERENCES

ALLEN, J. R. L. 1960. The Mam Tor Sandstones: a "turbidite" facies of the Namurian deltas of Derbyshire, England. *J. sedim. Petrol.*, **30**, 193–208.

BARNES, J. and HOLROYD, W. F. 1897. On the occurrence of a sea-beach at Castleton, Derbyshire, of Carboniferous Limestone age. *Trans. Manchr geol. Soc.*, **25**, 119–32.

BISAT, W. S. 1924. The Carboniferous goniatites of the North of England and their zones. *Proc. Yorks. geol. Soc.*, **20**, 40–124.

———— 1932. On some Lower Sabdenian goniatites. *Trans. Leeds geol. Ass.*, **5**, 27–36.

———— 1950. The junction faunas of the Viséan and Namurian. *Trans. Leeds geol. Ass.*, **6**, 10–26.

BISAT, W. S. and HUDSON, R. G. S. 1943. The Lower *Reticuloceras* (R$_1$) goniatite succession in the Namurian of the North of England. *Proc. Yorks. geol. Soc.*, **24**, 383–440.

BOUMA, A. H. 1964. *Turbidites. Development in Sedimentology*, **3**, 247–54.

BRITISH STANDARDS INSTITUTION 1957. *Site Investigations*. British Standard Code of Practice CP 2001.

BROMEHEAD, C. E. N., EDWARDS, W., WRAY, D. A. and STEPHENS, J. V. 1933. The geology of the country around Holmfirth and Glossop. *Mem. geol. Surv. Gt Br.*

CHALLINOR, J. 1921. Notes on the geology of the Roches District. *Trans. North Staffs. Nat. Field Club*, **55**, 76–87.

COLLINSON, J. D. and WALKER, R. G. 1967. *in Geological Excursions in the Sheffield region and the Peak District National Park*. Ed. R. Neves and C. Downie, University of Sheffield.

COPE, F. W. 1946. Intraformational contorted rocks in the Upper Carboniferous of the southern Pennines. *Q. Jl geol. Soc. Lond.*, **101**, 139–76.

COSGROVE, M. E. *in* Ramsbottom, Rhys and Smith. 1962. q.v.

DAVIES, W. 1941. On a boring in the Millstone Grit Series at Hallam Head, Sheffield. *Proc. Yorks. geol. Soc.*, **24**, 241–4.

DIETZ, R. S., EMERY, K. O. and SHEPARD, F. P. 1942. Phosphorite deposits on the sea floor off southern California. *Bull. geol. Soc. Am.*, **53**, 815–47.

DUNHAM, K. C. *in* Stephens and others. 1953. q.v.

EAGAR, R. M. C. 1953. Variation with respect to petrological differences in a thin band of Upper Carboniferous non-marine lamellibranchs. *Lpool Manchr geol. J.*, **1**, 161–90.

EARP, J. R., MAGRAW, D., POOLE, E. G., LAND, D. H. and WHITEMAN, A. J. 1961. Geology of the country around Clitheroe and Nelson. *Mem. geol. Surv. Gt Br.*

EDEN, R. A., STEVENSON, I. P. and EDWARDS, W. 1957. Geology of the country around Sheffield. *Mem. geol. Surv. Gt Br.*

EDWARDS, W. 1932. The country around the head of the Derwent. *Proc. Geol. Ass.*, **43**, 179–82.

EVANS, W. B., WILSON, A. A., TAYLOR, B. J. and PRICE, D. 1968. Geology of the country between Macclesfield and Crewe. *Mem. geol. Surv. Gt Br.*

FAREY, J. 1811. *A general view of the agriculture and minerals of Derbyshire*. Vol. 1. London.

FEARNSIDES, W. G. and TEMPLEMAN, A. 1932. A boring through Edale Shales to Carboniferous Limestone and pillow lavas, at Hope Cement Works, near Castleton, Derbyshire. *Proc. Yorks. geol. Soc.*, **22**, 100–21.

FORD, T. D. 1952. New evidence on the correlation of the Lower Carboniferous reefs at Cronkston, North Derbyshire. *Geol. Mag.*, **89**, 346–56.

GAUNT, G. D. 1960. *in Summ. Prog. geol. Surv. Gt Br.* for 1959, 32–3.

GIBSON, W. and WEDD, C. B. 1913. The geology of the northern part of the Derbyshire Coalfield and bordering tracts. *Mem. geol. Surv. Gt Br.*

GILLIGAN, A. 1920. The petrography of the Millstone Grit Series of Yorkshire. *Q. Jl geol. Soc. Lond.*, **75**, 251–94.

GREEN, A. H., FOSTER, C. LE NEVE and DAKYNS, J. R. 1887. The Carboniferous Limestone, Yoredale rocks, and Millstone Grit of North Derbyshire. 2nd edit. with additions by A. H. Green and A. Strahan. *Mem. geol. Surv. Gt Br.*

————— and RUSSELL, R. 1878. The geology of the Yorkshire Coalfield. *Mem. geol. Surv. Gt Br.*

GREENSMITH, J. T. 1957. Lithology, with particular reference to cementation of Upper Carboniferous sandstones in northern Derbyshire. *J. sedim. Petrol.*, **27**, 405–16.

HARTLEY, J. 1957. Jarosite from Carboniferous shales in Yorkshire. *Trans. Leeds geol. Ass.*, **7**, 19–23.

HODSON, F. 1957. Marker horizons in the Namurian of Britain, Ireland, Belgium and Western Germany. *Publs Ass. Étude Paléont. Stratigr. houill.*, **24**, 1–26.

HOLDSWORTH, B. K. 1963. Pre-fluvial, autogeosynclinal sedimentation in the Namurian of the southern Central Province. *Nature, Lond.*, **199**, 133–5.

———— 1964. The 'crowstones' of Staffordshire, Derbyshire and Cheshire. *North Staffs. J. of Field Studies*, **4**, 89–102.

———— 1965. The Namurian goniatite *Nuculoceras stellarum* (Bisat). *Palaeontology*, **8**, 226–30.

HOLMES, A. 1921. *Petrographic methods and calculations*. London.

HUDSON, R. G. S. 1944. The faunal succession in the *Ct. nitidus* zone in the mid-Pennines. *Proc. Leeds Phil. Lit. Soc.*, **4**, 233–42.

———— and COTTON, G. 1943. The Namurian of Alport Dale, Derbyshire. *Proc. Yorks. geol. Soc.*, **25**, 142–73.

———— ———— 1945. The Carboniferous rocks of the Edale Anticline, Derbyshire. *Q. Jl geol. Soc. Lond.*, **101**, 1–36.

———— and MITCHELL, G. H. 1937. The Carboniferous geology of the Skipton Anticline. *Summ. Prog. geol. Surv. Gt Br.* for 1935, (2), 1–45.

HULL, E. and GREEN, A. H. 1864. On the Millstone-Grit of North Staffordshire and the adjoining parts of Derbyshire, Cheshire and Lancashire. *Q. Jl geol. Soc. Lond.*, **20**, 242–67.

———— ———— 1866. The geology of the country around Stockport, Macclesfield, Congleton and Leek. *Mem. geol. Surv. Gt Br.*

JACKSON, J. W. 1923. On the correlation of Yoredales and Pendlesides. *Naturalist, Hull*, No. 801, 337–8.

———— 1925. The relation of the Edale Shales to the Carboniferous Limestone in North Derbyshire. *Geol. Mag.*, **62**, 267–74.

———— 1926. The goniatite zones below the Kinder Scout Grit in North Derbyshire. *Naturalist, Hull*, No. 834, 205–7.

———— 1927. The succession below the Kinder Scout Grit in North Derbyshire. *J. Manchr geol. Ass.*, **1**, 15–32.

———— 1953. *Reticuloceras reticulatum* mut. *superbilingue* Bisat at Combs, near Chapel-en-le-Frith, Derbyshire. *Lpool Manchr geol. J.*, **1**, 191–3.

———— 1958. Further records of *Reticuloceras reticulatum* mut. *superbilingue* in the Combs valley, near Chapel-en-le-Frith, Derbyshire. *Lpool Manchr geol. J.*, **2**, 80.

MORRIS, T. O. 1929. The Carboniferous Limestone and Millstone Grit Series of Stoney Middleton and Eyam, Derbyshire. *Proc. Sorby scient. Soc.*, **1**, 37–67.

PEACOCK, J. D. and TAYLOR, K. 1966. Uraniferous collophane in the Carboniferous Limestone of Derbyshire and Yorkshire. *Bull. geol. Surv Gt Br.*, No. 25, 19–32.

PETTIJOHN, F. J. 1957. *Sedimentary Rocks*. 2nd edit. New York.

PONSFORD, D. R. A. 1955. Radioactivity studies of some British Sedimentary rocks. *Bull. geol. Surv. Gt Br.*, No. 10, 22–44.

PULFREY, W. 1934. A boring in the Millstone Grits, Rod Moor, Sheffield. *Proc. Yorks. geol. Soc.*, **22**, 254–64.

RAMSBOTTOM, W. H. C. 1958. A new goniatite *Homoceratoides fortelirifer* of Millstone Grit Upper *Reticuloceras* age. *Bull. geol. Surv. Gt Br.*, No. 15, 29–31.

RAMSBOTTOM, W. H. C., RHYS, G. H. and SMITH, E. G. 1962. Boreholes in the Carboniferous rocks of the Ashover district, Derbyshire. *Bull. geol. Surv. Gt Br.*, No. 19, 75–168.

READING, H. G. 1964. A review of the factors affecting the sedimentation of the Millstone Grit (Namurian) in the central Pennines. *Developments in Sedimentology*, 1 (*Deltaic and Shallow Marine Deposits*), Amsterdam, 340–6.

SHACKLETON, J. S. 1962. Cross-strata of the Rough Rock (Millstone Grit Series) in the Pennines. *Lpool Manchr geol. J.*, 3, 109–18.

SHIRLEY, J. and HORSFIELD, E. L. 1945. The structure and ore deposits of the Carboniferous Limestone of the Eyam district, Derbyshire. *Q. Jl geol. Soc. Lond.*, 100, 289–308.

SIMPSON, I. M. and BROADHURST, F. M. 1969. A boulder bed at Treak Cliff, north Derbyshire. *Proc. Yorks. geol. Soc.*, 37, 141–51.

SMITH, E. G. 1955. in *Summ. Prog. geol. Surv. Gt Br.* for 1954, 36.

————— RHYS, G. H. and EDEN, R. A. 1967. Geology of the country around Chesterfield, Matlock and Mansfield. *Mem. geol. Surv. Gt Br.*

SORBY, H. C. 1859. On the structure and origin of Millstone Grit in South Yorkshire. *Proc. Yorks. geol. polytech. Soc.*, 3, 669–75.

STEPHENS, E. A. 1952. On the 'Rough Rock' and Lower Coal Measures near Crich, Derbyshire. *Proc. Yorks. geol. Soc.*, 28, 221–7.

STEPHENS, J. V. MITCHELL, G. H. and EDWARDS, W. 1953. Geology of the country between Bradford and Skipton. *Mem. geol. Surv. Gt Br.*

STOBBS, J. T. 1905. in Life-zones in the British Carboniferous rocks: Report of the committee (drawn up by Wheelton Hind). *Rep. Br. Ass. Advmt Sci.* for 1904 (Cambridge), 231–4.

TAYLOR, B. J., PRICE, R. H. and TROTTER, F. M. 1963. Geology of the country around Stockport and Knutsford. *Mem. geol. Surv. Gt Br.*

TREWIN, N. H. 1968. Potassium bentonites in the Namurian of Staffordshire and Derbyshire. *Proc. Yorks. geol. Soc.*, 37, 73–91.

TROTTER, F. M. 1952. Sedimentation facies in the Namurian of north-western England and adjoining areas. *Lpool Manchr geol. J.*, 1, 77–112.

TRUEMAN, A. E. 1947. Stratigraphical problems in the coalfields of Great Britain. *Q. Jl geol. Soc. Lond.*, 103, lxv-civ.

WALKER, R. G. 1966. Shale Grit and Grindslow Shales: Transition from turbidite to shallow water sediments in the Upper Carboniferous of northern England. *J. sedim. Petrol.*, 36, 90–114.

WHITEHURST, J. 1778. *An enquiry into the original state and formation of the earth.* London.

WILLS, L. J. 1951. *A Palaeogeographical Atlas.* London.

WRAY, D. A., STEPHENS, J. V., EDWARDS, W. and BROMEHEAD, C. E. N. 1930. The geology of the country around Huddersfield and Halifax. *Mem. geol. Surv. Gt Br.*

YATES, P. J. 1962. The palaeontology of the Namurian rocks of Slieve Anierin, Co. Leitrim, Eire. *Palaeontology*, 5, 355–443.

*Chapter V*

# COAL MEASURES

## INTRODUCTION

COAL MEASURES occupy much of the axial region of the Goyt Trough in the western part of the district. North of Disley the flattening of the syncline gives rise to a broader spread of Coal Measures which is continuous with that described in the Stockport area (Taylor and others 1963) and hence with the Lancashire Coalfield. In the north-eastern corner of the district small areas of Coal Measures form westward extensions of the outcrops described in the Sheffield district (Eden and others 1957).

The maximum thickness of Coal Measures present is about 800 ft near New Mills (see Fig. 18). These beds fall entirely within the Lower Coal Measures (see Stubblefield and Trotter 1957); nearly all lie within the *Anthraconaia lenisulcata* Zone. The lowering of the base of the *Carbonicola communis* Zone to the horizon of the Pasture Mine Marine Band, proposed by Eagar (1956, pp. 355–6), results in some 140 ft of shale above the Milnrow Sandstone at Gowhole now falling within this zone. The succession shows only slight lateral variation and corresponds closely with that proved in Lancashire (e.g. Jones and others 1938) and on the eastern side of the Pennines (Eden 1954). The lithology of these beds, and particularly their cyclic character, have been described by numerous workers, for example Wray (1929), Eagar (1952) and Eden (1954). In the present district eleven cyclothems are present, though not all are complete. The past working of three coals, in the lower half of the succession, has provided records of shafts and seam thicknesses which have been drawn on in this account. At present, working of coal is carried out on a small scale at Ludworth Moor only.

## GENERAL STRATIGRAPHY
### THE GOYT TROUGH

The **beds between the base of the** *G. subcrenatum* **Marine Band and the Woodhead Hill Rock** vary in thickness from 74 ft at Mellor to 150 ft at Birch Vale; they are thickest where the overlying Woodhead Hill Rock is thin, the interval between marine band and Yard Coal being fairly constant. The marine band at the base of the group is everywhere well developed, and is underlain by a small thickness of shale referred to the Millstone Grit (see pp. 185–6), resting on a coal or seatearth overlying the Rough Rock. The marine band contains pectinoids and conodonts in addition to *G. subcrenatum*. A contorted band of the type described by Cope (1946) has been found to overlie the marine band at several localities. The upper part of the group is composed mainly of shales which call for little comment; two thin sandstones are present in these beds in the Mellor area.

The **Woodhead Hill Rock** varies from 130 ft at Fernilee to as little as 50 ft at Hayfield, the thicker developments being for the most part rather coarse and

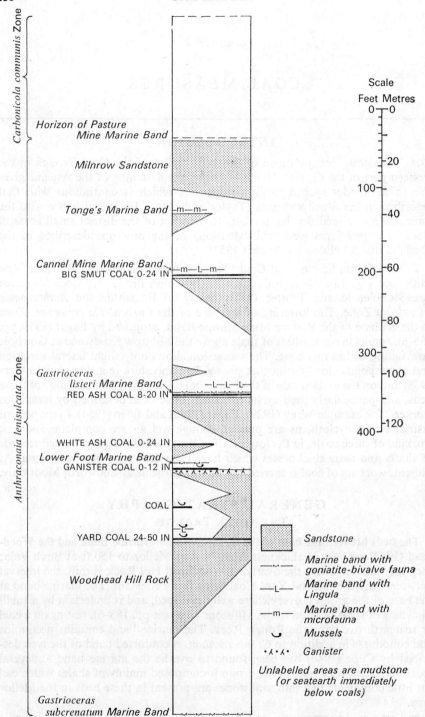

FIG. 18. *Generalized section of the Coal Measures of the area between Whaley Bridge and Chisworth*

massive, while the thinner are finer grained and flaggy, recalling the Crawshaw Sandstone of the eastern part of the district (see below). Around Fernilee red staining has been noted at a number of places. The sandstone splits northwards on Chinley Churn, and the lower leaf dies out. With the absence at New Mills of the Yard Coal, the Woodhead Hill Rock unites with the overlying Yard 'roof-rock'.

Up to 4 ft of seatearth usually lie between the Woodhead Hill Rock and the Yard Coal. This is often rather sandy, especially in the lower part.

The **Yard Coal** is also known as the Bassy, Big Mine, Mountain Mine of Mellor, Compstall Two Feet and Kiln of Whaley Bridge. It varies between 24 and 54 inches in thickness, averaging about 36 in (Fig. 19). Although in places of inferior quality, this is the most widely worked seam in the area. In its thickest development, around Whaley Bridge, it contains much dirt in thin partings. Between Gowhole and Birch Vale, where a sandstone roof to the seam is present, washouts are frequent, though of limited extent; they are usually confined to the upper part of the seam but at New Mills the seam appears to be completely missing. The Yard Coal has been much worked in the Goyt Trough, especially at Whaley Bridge, between Gowhole and Thornset and at Mellor.

The **beds between the Yard and Ganister coals** are usually between 80 and 90 ft thick, but are rather thicker in places where much sandstone is present. They are composed of three cyclothems.

In the southern part of the area, the basal beds are grey to dark grey shales, with some ironstone bands, and an abundant mussel fauna has been found a few feet above the Yard Coal at several places, as elsewhere in the Pennine region (see Eagar 1952, pp. 23–56). The most conspicuous forms are allied to *Carbonicola fallax* Wright and *C. protea* Wright. Overlying these beds and about 12 ft above the coal is a thin *Lingula* band recorded by Eagar (1952, pp. 27–8) at Fernilee and, south of the present district, at Goyt's Moss (Cope 1949, p. 475) (One-inch Sheet 111). This is the Lower Bassy Marine Band (Magraw 1957, p. 38). Near Botney Works, Whaley Bridge, a band referred to this horizon lies some 25 ft above the Yard Coal. In the north-western corner of the district an 18-in coal occurs about 40 ft above the Yard and probably just below the horizon of the Upper Bassy Marine Band, present at Goldsitch Moss (Cope 1949, p. 477).

The uppermost part of this group of strata consists usually of sandstone, up to about 15 ft in thickness, underlying the Ganister Coal. At Taxal, a second sandstone appears in the underlying shales and coalesces northwards with that underlying the Ganister Coal, sandstone forming the whole of the group in the area between Newtown and Birch Vale; in the detailed description of these beds this has been referred to as the Yard 'roof-rock'. Around Birch Vale it is pebbly in places, an unusual feature for sandstones in this part of the succession.

The **Ganister Coal** (Lower Foot) is a thin seam varying from about 7 to 12 inches in thickness, which is present in the south and west parts of the Goyt Coalfield and in the north near Ludworth, though it has nowhere been worked. It is underlain by a persistent ganister, 6 to 21 in thick, which has proved most useful for mapping purposes.

Around New Mills and Birch Vale the Ganister and White Ash (see below) both fail and it has been observed that their absence coincides closely with the development of the thick sandstone, above the Yard Coal, in the cyclothems beneath the Ganister Coal horizon. This is well brought out by the sections of Burn'd Edge New Pit, High Lee and Cave Abdullam collieries (see Fig. 20). It is evident that the presence of a thick mass of sandstone in the cyclothems underlying the Ganister Coal or its horizon influenced subsequent deposition so far as the Ganister and White Ash coals are concerned, probably through the effect of differential rates of compaction. This is supported by the fact that both the overlying cyclothems are affected.

Beds between the Ganister and White Ash coals. Some 30 ft of beds separate the Ganister and White Ash coals, the lower part being shale and the upper a sandstone immediately underlying the White Ash Coal. At Sandy Lane, in the north-west part of the district, the roof of the Ganister Coal yields a non-marine fauna and the Lower Foot Marine Band has been found as an 8-in band with a fauna including *Lingula*, goniatites, bivalves and conodonts, about 8 ft above the coal. This is the only record of the band in the district and it appears to be absent in most places, though exposures are generally infrequent at this horizon. West of the district, at Pott Shrigley in the Stockport (98) district, the beds immediately above the Ribbon Mine (or Ganister) show a similar sequence of first non-marine forms and then the Lower Foot Marine Band, though the latter is there only 1 ft 9 in above the coal (Eagar and Pressley 1954). Fragmentary mussels have been found close above the Ganister Coal at Fernilee, near Whaley Bridge.

The **White Ash Coal** (Sixteen Inch of Whaley Bridge, Smithy of Whaley Bridge) is somewhat variable in thickness, being from 12 to 24 inches in the Whaley Bridge area. It was worked until recently, together with the 5½ ft of fireclay beneath it, at Furness Vale, and was formerly got around Whaley Bridge. Like the Ganister Coal, the White Ash is absent between Newtown and Birch Vale (see p. 269), though it is present on the western margin of the district. Around Chisworth, in the northern part of the area, where the Ganister Coal is again present, the White Ash fails to re-appear, and its absence here corresponds with a local thickening of the sandstone beneath the Red Ash. It has been suggested by Cope (1949, p. 471) that the seam has been washed out by the overlying sandstone and study of the stream section at Sandy Lane confirms this.

The **beds between the White Ash and the Red Ash coals**, some 60 ft in thickness, consist of shale in the lower part and a somewhat variable sandstone above. The latter is reported to have contained galena, in a vein up to 24 in wide in Whaley Bridge Colliery (Binney 1851).

The **Red Ash Coal** (Little of New Mills, Eighteen Inch of Whaley Bridge, First Mine of Hockerley) is the second most widely worked coal in the area, despite its frequently sulphurous character. In much of the central area of the Goyt Trough the seam is of a very constant thickness of 18 in though it is known to vary somewhat locally (from 14 in at Low Leighton to 22 in at Disley). The coal is a much more persistent seam than either the White Ash or Ganister coals, and has not been found to fail anywhere in the district.

The **beds between the Red Ash and Big Smut coals** are about 120 ft in thickness. The *Gastrioceras listeri* Marine Band, about 18 in thick, seems to be present everywhere in the roof of the Red Ash Coal. The band is well exposed only at Chisworth, though it is seen in part in two localities farther south;

however, debris of marine shale is present wherever the Red Ash Coal has been worked. A thin band with *Lingula* and/or fish remains, lying some 12 ft above the main *G. listeri* Marine Band, is regarded as a late phase of the band rather than evidence of a separate cyclothem at this horizon. A similar development is known in nearby areas, for example in the Chesterfield area (Smith and others 1967, p. 111), where in places the *Lingula* band is separated from the main *G. listeri* Marine Band by a non-marine phase with *Geisina*. The marine band is overlain by a uniform series of shales with thin ironstone bands and these by variable sandstones, up to about 35 ft in thickness, which lie at the top of the group.

In the present area, the beds from the *G. listeri* Marine Band to the Big Smut Coal constitute one cyclothem. In neighbouring areas, however, two cyclothems are present; for example, in the Sheffield district (Eden and others 1957, p. 44) and in Lancashire, where a thin coal, the Inch Mine, is often present (Magraw 1957, p. 23) at the top of the lower cyclothem.

The **Big Smut Coal** is much more variable in thickness than the seams below, ranging from 0 to 13 in. It has been worked at Ludworth only. At most localities within the district there is a single coal at this horizon, contrasting with the Lancashire development of two seams, the Upper Mountain Mine below and the Cannel Mine above. In the adjacent Stockport area, Taylor (1963, p. 18) suggested that the Big Smut Coal was the equivalent of the Upper Mountain Mine and that the Cannel Mine had disappeared. This seems unlikely as the Cannel Mine Marine Band occurs close above the single coal at Woodend (see p. 289); this seam is considered to be the Upper Mountain Mine and Cannel Mine united. In places the Big Smut splits again, the section at Whaley Bridge showing two seams tentatively equated with the Cannel Mine and Upper Mountain, separated by 24 ft of measures. Two seams, 11 ft 10 in apart, were also recorded in the Chadkirk Borehole (Magraw 1957, p. 23). Splitting at an equivalent horizon (that of the Forty Yard Coal) in Yorkshire and Derbyshire has been noted by Eden (1954, p. 95). The Big Smut is also present as a single seam in the Pott Shrigley area, west of the present district where it is about 18 in thick (Cope 1949, p. 473). Boreholes in North Staffordshire (Magraw 1957, pp. 29–31) have, on the other hand, shown the presence of a split seam at this horizon.

The **beds between the Big Smut Coal and Milnrow Sandstone** are about 80 to 100 ft in thickness in all but the northern outcrops where they are only 30 ft. The local thinning may possibly be due to the development of a washout base to the Milnrow Sandstone. The beds are most fully developed in the central area of the Goyt Trough at and east of Gowhole. Here they are made up of two cyclothems; the lower is some 80 ft thick and the upper extends to the top of the Milnrow Sandstone (which is described separately below). Outside the Gowhole area the lower shaly part of the upper cyclothem is missing and probably washed out, the Milnrow Sandstone resting directly on the lower.

At the base of these beds the Cannel Mine Marine Band occurs in places, a short distance above the Big Smut Coal. The band is variable in character. Locally it consists of up to 2 ft 6 in of shale with foraminifera but a thin *Lingula* band is present in places at the base and a fish-phase sometimes occurs beneath this. The Cannel Mine Marine Band fails at some localities, being represented by barren pyritous shale. A thin 'contorted band' has been found above the marine band. The succeeding beds are shales, capped by a thin sandstone where the lower cyclothem is complete. The sandstone crops out over a limited area east of Gowhole, where it is overlain by a seatearth over 2 ft in thickness in Beard Hall cutting.

The upper cyclothem carries at its base Tonge's Marine Band, known only in the present area from the Beard Hall cutting. The band consists of 7 in of shale with foraminifera. The remainder of the cyclothem up to the Milnrow Sandstone is made up of shale, locally pyritous. A microfaunal band representing Tonge's Marine Band has also been recorded by Magraw from the Chadkirk Borehole (1957, p. 21) just west of the present district.

The **Milnrow Sandstone** falls within the cyclothem with Tonge's Marine Band at the base. About 70 ft in thickness, it is present in isolated outcrops in the axial area of the Goyt Trough and small outcrops also lie within the district on the western margin. The sandstone is usually more massive and coarser grained than is typical of the sandstones of the Coal Measures of the area, and it also shows a characteristic jointing causing it to weather out into large blocks.

The **beds above the Milnrow Sandstone,** consisting of some 150 ft of shales, crop out only in a small area near Gowhole. They are entirely concealed by drift and are only known from the record of a drainage level. By comparison with the sections in Lancashire (Magraw 1957), the horizon of the Pasture Mine Marine Band should lie at the base of the shales.

## THE EASTERN AREA

The small outcrop of Coal Measures on the eastern margin of the district is continuous with the outcrops of the Sheffield area described by Eden and others (1957). The chief features of the stratigraphy are summarized below.

The **beds between the base of** *G. subcrenatum* **Marine Band and the Crawshaw Sandstone** range from 110 to 120 ft in thickness. The lowest beds are well seen in open workings for the Pot Clay near Ughill. The *G. subcrenatum* Marine Band, about 18 in thick, is present at the base. The lower part of the succeeding shales is also seen in these workings; the exposed beds are grey shales with ironstone bands. The upper part of the shale group is not exposed.

The **Crawshaw Sandstone** is a flaggy micaceous sandstone some 30 ft in thickness. Its type locality lies just within the present district.

The **Soft Bed or Coking Coal.** Although present within the sheet boundary there are no details of this seam. A short distance east of the present district the coal is 6 to 15 inches in thickness and underlain by a workable fireclay (Eden and others 1957, p. 38).

**Beds above the Coking Coal** are present in an outlier at Gibraltar Rocks, Ughill, where they consist of some 20 ft of shale, overlain by 24 ft of coarse massive sandstone. The sandstone is probably that underlying the Clay Coal (Middle Band) horizon of Sheffield.                                                        I.P.S.

## PALAEONTOLOGY

In this district only the lower part of the Westphalian is present, the measures comprising the *A. lenisulcata* Zone and the basal part of the *C. communis* Zone. Most of the important faunal marker-bands recognized in these beds in the adjoining areas are represented and are described below in upward sequence.

The *Gastrioceras subcrenatum* **Marine Band** as developed in the Goyt Trough is less than 2 ft in thickness and contains a goniatite-pectinoid fauna typical of the central part of the Pennine basin. The collections made from stream sections in Knowle Wood and near Fernilee, and also from nearby boreholes, provide the following composite faunal list: *Caneyella multirugata* (Jackson), *Dunbarella papyracea* (J. Sowerby), *Posidonia* cf. *gibsoni* Salter, *P. sp. nov.* [right valve with prominent anterior ear], orthocone nautiloids, *Anthracoceras arcuatilobum* (Ludwig), *Gastrioceras subcrenatum* (Frech), *G. sp. nov.* [widely-spaced lirae], *G. sp.* [crenulate lirae], *Homoceratoides sp.*, mollusc spat, *Hindeodella sp.*, plat-formed conodonts and palaeoniscid scales including *Elonichthys sp.* There is no obvious faunal differentiation within the band, although the basal portion is the most fossiliferous. A thin layer with *Lingula* occurs immediately below the goniatite-pectinoid fauna in the Fernilee No. 2 and Shaw Farm boreholes (pp. 275–6).

A similar assemblage is known in the east of the district from an opencast clay working on Bradfield Moors. The underlying musselband was also found (see p. 246).

In the Fernilee No. 2 Borehole, *Curvirimula sp. nov.* [cf. Calver *in* Price and others 1963, p. 67, pl. iv, figs. 3–5] and fish remains were found 50 ft and 60 ft respectively above the marine band.

In the south Pennine region the measures **above the Yard Coal (or Bassy Mine) horizon** are distinguished by the presence of well-developed musselbands. These comprise three distinct cyclothems which have been referred to in Lancashire as the Lower, Middle and Upper Division of the Bassy Mine sequence, and are separated by two thin *Lingula*-bearing horizons known as the Lower and Upper Bassy marine bands (Eagar 1947; 1952, pp. 48–52; Magraw 1957, pp. 27, 38; see also Fig. 18). Although the full sequence is thought to be present in the district the beds are not continuously exposed and the following faunal summary is based on information from scattered localities (p. 269).

The fauna of the 'Lower Division' is represented by *Carbonicola artifex* Eagar, *C. declinata?* Eagar, *C.* cf. *discus* Eagar, *C. fallax, C.* cf. *haberghamensis* Wright, *C.* aff. *protea, Curvirimula sp.* and the ostracod *Geisina arcuata* (Bean). The presence of forms recalling the *discus/haberghamensis* group in the collections from the banks of the Goyt, north-west of Fernilee (p. 280), appears to be anomalous as elsewhere these species are characteristic of the 'Middle Division' fauna (see also Eagar 1956, pp. 334–5).

The **Lower Bassy Marine Band** was recorded at the above locality by Eagar (1952, pp. 27–8) and is also known from the banks of the Etherow (p. 282) and in Mather Clough (p. 280). *Lingula* and palaeoniscid scales are the only fossils collected; in addition to *L. mytilloides* J. Sowerby there are several examples which are broader than the typical form and *Lingula* spat is common. Similar features are shown by the *Lingula* band occurring in the right bank of the Goyt at Botney Bleach Works and this is tentatively identified as the Lower Bassy Marine Band.

Collections of the 'Middle Division' fauna have been made from the beds up to 8 ft above the Lower Bassy Marine Band in the banks of the Etherow (p. 282); the assemblage includes *C. artifex, C.* cf. *discus, C. fallax, C.* aff. *protea, C.* cf. *pilleolum* Eagar, *C. sp.* cf. *acuta* (J. Sowerby) and *Curvirimula sp.*, asso-ciated with scales of *Elonichthys sp.* and *Rhadinichthys sp.*

The Upper Bassy Marine Band has not been recognized in the district.

The musselbands occurring some 80 ft above the Yard Coal in the stream at Sandy Lane (p. 283) are believed to be the equivalent of the 'Upper Division' of the Bassy sequence. The *Carbonicola* species recognized include *C. artifex*, *C.* aff. *fallax*, *C.* cf. *pilleolum*, *C.* aff. *protea* and *C. rectilinearis* Trueman and Weir. *Curvirimula sp.*, *Naiadites sp.* cf. *obliquus* Dix and Trueman and *Geisina arcuata* are also present.

The **Lower Foot Marine Band** and underlying musselband are present some 40 ft higher in the sequence at Sandy Lane. The musselband occurs in silty mudstones 6 ft 6 in above the Lower Foot Coal; preservation of the mussels is poor, but the following fauna has been identified: *Spirorbis sp.*, *Carbonicola* cf. *artifex*, *C.* cf. *limax* Wright, *C.* aff. *obliqua* Wright, *C. sp.* (*protea* group) and *Geisina arcuata*. This assemblage is similar to that recorded from Goyt's Moss (Cope 1949, p. 475; Eagar 1952, p. 29) and at Pott Shrigley (Eagar and Pressley 1954, p. 67). The overlying marine band is separated from this *Carbonicola*-band by 12 in of mudstone containing *Curvirimula sp.* and *Geisina arcuata*. Both these fossils characteristically occur in strata adjacent to marine horizons.

The fauna of the Lower Foot Marine Band is varied and composed of *Planolites ophthalmoides* Jessen, *Lingula mytilloides*, *Caneyella multirugata*, *Posidonia gibsoni*, *Anthracoceras sp.*, *Gastrioceras sp.*, *Hindeodella sp.*, platformed conodonts and fish remains including *Elonichthys sp.* It is distinguished from the assemblage above the Bullion Mine (see below) by the absence of *Dunbarella* and foraminifera. The goniatites are also well represented, but show an unusual preservation in which the siphuncle is present in relief. Conodonts are common.

The *Gastrioceras listeri* **Marine Band** has been found at several localities (pp. 286-8) and contains the typical goniatite-pectinoid fauna of the Westphalian basin facies (Calver 1968, pp. 15, 27), namely *Caneyella multirugata*, *Dunbarella papyracea*, *Posidonia gibsoni*, *P. sp. nov.* [right valve with prominent anterior ear], *Anthracoceras arcuatilobum*, *Gastrioceras listeri* (J. Sowerby), *Homoceratoides* aff. *divaricatus* (Hind) and mollusc spat. This rich basal fauna ranges through up to 2 ft of mudstone but thin bands, containing *Lingula sp.*, *Caneyella sp.* [juv.], *Posidonia sp. nov.*, mollusc spat, conodonts and rare orthocone nautiloids, occur at intervals up to 8 ft above the base of the marine band. Locally these bands are interspersed with layers containing fish remains, plant debris and other non-marine fossils; the latter include *Curvirimula*, *Naiadites* and *Geisina*, which are also associated with the regression stage of this marine incursion elsewhere in the Pennine province (Magraw 1957, p. 30; Calver *in* Smith and others 1967, p. 111).

The **Cannel Mine Marine Band** is well known in the district (pp. 289-90); the section at Woodend provides a good example of the fauna, which is mainly a *Lingula*/microfaunal assemblage. The collections show that foraminifera, including large *Ammodiscus*, are characteristic of the band which also contains *Lingula mytilloides*, rare *Cypridina*, abundant *Hindeodella* and platformed conodonts. Fish remains including *Rhabdoderma sp.* and *Rhadinichthys sp.* also occur. The presence of the large ostracod *Cypridina* is of interest as the genus is not common in the marine bands of the Lower Coal Measures.

**Tonge's Marine Band** is represented by a thin band of silty mudstone, containing small *Ammodiscus*, palaeoniscid scales and pyrite-filled burrows, which was found in the raillway-cutting near Beard Hall (p. 290). The horizon is placed some 20 ft below the local base of the Milnrow Sandstone and a comparison can be made with the sequence in the Chadkirk Borehole (p. 272) in which abundant foraminifera and conodonts were recorded from Tonge's Marine Band (Magraw 1957, p. 21). M.A.C.

# DETAILS

## THE GOYT TROUGH

*Gastrioceras subcrenatum* **Marine Band and overlying shales.** In the south these beds are present as a narrow strip along the axis of the Goyt Trough for a mile south of the Fernilee Reservoir dam. A small outlier also occurs at Bunsal Cob half a mile north-east of Goyt's Bridge. Boreholes along the line of the puddle trench for the Fernilee Reservoir dam (details communicated by Stockport Corporation Waterworks) show some 90 ft of shale between the Rough Rock and Woodhead Hill Rock. Half a mile downstream, Fernilee No. 2 Borehole [0119 7863] passed through the whole group:

| | ft | in | ft | in |
|---|---|---|---|---|
| Base of Woodhead Hill Rock .. .. .. | — | — | 115 | 1 |
| Mudstone, silty with thin siltstone bands .. | 10 | 2 | 125 | 3 |
| Mudstone, grey, some ironstone bands; small mussels at 130 ft 2 in, 134 ft and 153 ft 10 in; *Cochlichnus?* 152 ft 10 in and 153 ft 10 in .. | 63 | 3 | 188 | 6 |
| Mudstone, grey platy, with pyrite specks; fish debris 188 ft 8 in; pyritized worm-tubes 188 ft 10 in; *Planolites ophthalmoides* Jessen 188 ft 11 in .. .. | 1 | 6 | 190 | 0 |
| *G. subcrenatum* Marine Band | | | | |
| Mudstone, dark hard platy, with a little pyrite; hard, dark, and slightly silty in lowest 3 in; *Dunbarella papyracea, Posidonia* cf. *gibsoni, P. sp. nov., Anthracoceras arcuatilobum, Gastrioceras subcrenatum,* mollusc spat | | | | |

| | ft | in | ft | in |
|---|---|---|---|---|
| and palaeoniscid scales including *Elonichthys sp.* .. .. | 1 | 4 | 191 | 4 |
| Mudstone, grey slightly silty and with pyrite blebs in lowest 1 in; *Lingula mytilloides* | | 5 | 191 | 9 |
| Millstone Grit Series (see pp. 240–1, 245–6) | — | — | — | — |

Nearly the whole thickness of the beds is exposed, though intermittently, in a stream east of Fernilee. The following section of the lowest part was measured [0191 7851] 300 yd E. of the main road:

| | ft | in |
|---|---|---|
| Shale, grey .. .. .. .. | | 11 |
| Shale, dark grey contorted .. .. | | 6½ |
| Shale, black (top 2 in hard; softer and decalcified below); *Caneyella multirugata, Dunbarella sp., Posidonia sp., Anthracoceras sp., Gastrioceras subcrenatum, Hindeodella sp.* .. .. .. .. | 1 | 1 |
| Millstone Grit Series (see p. 246) | — | — |

The contorted bed is of the type described by Cope (1946).

Farther north, along the eastern limb of the Goyt Trough, exposures in the group are poor. A thickness of 90 to 100 ft is maintained in most places, for example on Chinley Churn. In the outlier north of Birch Vale these beds are about 150 ft thick, and the overlying Woodhead Hill Rock is thin; the *Gastrioceras subcrenatum* Marine Band is nowhere clearly exposed, though shale with *Dunbarella* was seen, close above the Rough Rock, at one point [0195 8858].

On the western limb of the Goyt Trough, near Kettleshulme, beds of this group, though not exposed, can be traced for about half a mile southwards from Wright's Farm, but are then cut off by a fault. Between Kettleshulme and Disley the shales are about 100 ft thick; exposures are lacking or inconspicuous. At Higher Disley [9855 8391] debris of marine shale was seen close above the Rough Rock.

South of Mellor, where a lower leaf of the Woodhead Hill Rock is developed, these beds are as little as 50 to 100 ft in thickness. The Shaw Farm Borehole [9905 8661] ½ mile N.W. of Eaves Knoll, showed the following abridged section:

| | Thick- ness | | Depth | |
|---|---|---|---|---|
| | ft | in | ft | in |
| Base of Woodhead Hill Rock (see p. 277) .. | — | — | 165 | 0 |
| Mudstone, silty and micaceous, with frequent red and lilac staining | 23 | 0 | 188 | 0 |
| Sandstone (no core recovered) .. .. | 20 | 4 | 208 | 4 |
| Mudstone, grey .. .. | 33 | 8 | 242 | 0 |
| Mudstone, grey soft, with some ironstone .. | 3 | 0 | 245 | 0 |
| Mudstone, dark platy, with pyrite specks in lowest 1 in .. .. | 1 | 7 | 246 | 7 |
| Mudstone, grey with fucoid markings and occasional pyrite specks | 1 | | 246 | 8 |
| Mudstone, grey .. .. | | 5 | 247 | 1 |
| Mudstone, platy; thin pyrite lens 247 ft 6 in .. | | 7 | 247 | 8 |
| Mudstone, grey contorted | | 6 | 248 | 2 |

*G. subcrenatum* Marine Band

Mudstone, hard calcareous, with *Dunbarella sp.*, *Posidonia* cf. *gibsoni*, *P. sp. nov.*, mollusc spat and fish debris, including *Elonichthys sp.* in top 3 in; below, with *Pseudozygopleura?*, *Caneyella multirugata*, *Dunbarella papyracea*, *P.* cf. *gibsoni*, *P. sp. nov.*, *Anthracoceras sp.*, *Gastrioceras subcrenatum, G. sp. nov.* [widely-spaced lirae], *G. sp.*

| | Thick- ness | | Depth | |
|---|---|---|---|---|
| | ft | in | ft | in |
| [crenulate lirae], *Homoceratoides sp.*, mollusc spat and fish remains .. .. | 1 | 0 | 249 | 2 |
| Mudstone, grey with some pyrite .. | | 5 | 249 | 7 |
| Mudstone, grey platy with thin pyrite lens; *Lingula sp.* and fish debris including *Elonichthys sp.* .. | | 3 | 249 | 10 |
| Millstone Grit Series (see p. 377) .. .. | — | — | — | — |

Farther north, beds overlying the Six-Inch Mine are well seen in a stream [9816 8866] in Knowle Wood:

| | ft | in |
|---|---|---|
| Shale, evenly bedded .. .. | 1 | 6 |
| Shale, contorted .. ¾ in to | | 2 |
| Shale, evenly bedded; *Dunbarella papyracea, Posidonia sp., Anthracoceras sp., Gastrioceras sp.*, mollusc spat, *Elonichthys sp.* [scale] .. | | 4 |
| Shale, black soft decalcified 1¼ in to | | 1½ |
| Shale, black platy small 'bullions' up to 1 inch in top 3 in; *Posidonia* cf. *gibsoni, P. sp. nov.*, orthocone nautiloid, *Anthracoceras sp., Gastrioceras subcrenatum, G. sp. nov.*, mollusc spat, *Elonichthys sp.* [scale] .. .. .. .. | | 10 |
| Shale, grey 'sulphurous'; *Caneyella multirugata, Dunbarella papyracea, D. sp., Posidonia* cf. *gibsoni, P. sp. nov., Anthracoceras sp., Gastrioceras subcrenatum*, mollusc spat and palaeoniscid scales | | 6½ |
| Mudstone (see p. 245) .. .. | — | — |

Near Hambleton Fold about 130 ft of shale separate the Rough Rock and Woodhead Hill Rock. Here, features suggest the presence of two thin beds of sandstone within the shale, but these do not appear to persist laterally.

**Woodhead Hill Rock.** To the east of Fernilee Reservoir the Woodhead Hill Rock is the highest horizon exposed in this part of the Goyt Trough. The rock also caps the small outlier of Bunsal Cob, half a mile north-east of Goyt's Bridge. Boreholes and excavations for the puddle-trench of the dam at Fernilee [01237770 to 01487773] proved the following general sequence:

|  | ft |
|---|---|
| Sandstone, with a shale parting, 0 to 30 ft, about 30 ft down .. .. | 130 |
| Sandstone, described as 'Passage Beds'; flags with shale partings, resting on shale .. .. .. | 30 |

The top of these beds is mapped as lying just below the horizon of the Yard Coal and they must represent virtually the full thickness of the Woodhead Hill Rock.

From the dam northwards to within a quarter of a mile from Horwich End the Woodhead Hill Rock is conspicuous along the banks of the Goyt. Fernilee No. 2 Borehole [0119 7863] (see also p. 275) proved the Woodhead Hill Rock below 28 ft of drift to a depth of 115 ft 1 in, the hole commencing a few feet below the top of the sandstone which was, for the most part, coarse and massive with current-bedding in places. The river closely follows the line of the faulted anticline described on p. 324, but the throw of the structure is small enough for it to lie within the thickness of the sandstone. A typical section [0114 7855] on the left bank of the river is:

|  | ft |
|---|---|
| Sandstone, yellow massive, medium-grained .. .. .. .. | 15 |
| Sandstone, yellow; numerous plant impressions, some carbonized .. | 6 |
| Sandstone and shale, in beds up to 6 in, on shale .. .. .. about | 14 |

To the north of Fernilee the Woodhead Hill Rock forms prominent features on both limbs of the Goyt Trough, dip-slopes being well marked at Taxal Wood on the western limb and Black Edge Plantation on the east. At the northern end of Taxal Wood the rock has been much quarried, a typical section [9996 8016] being:

|  | ft |
|---|---|
| Sandstone, coarse and massive about | 25 |
| Sandstone, well-bedded, with irregular siltstone and silty mudstone partings up to 9 in thick, some red staining; a 6-ft lens with plant stems .. .. | 12 |
| *Not exposed* .. .. .. about | 27 |
| Sandstone, flaggy .. .. .. | 5 |

The section starts close above the base of the Woodhead Hill Rock.

Between Chinley and Hayfield the Woodhead Hill Rock forms a well-marked scarp, at the top of the composite feature of Chinley Churn (see Plate XVIII B). At Throstle Bank its thickness is about 80 ft, though it thins to some 50 ft at Hayfield. At the northern end of Chinley Churn a shale parting is present near the base, and the sandstone beneath disappears when traced northwards. At Birch Vale a good section of the uppermost part of the Woodhead Hill Rock is exposed beneath the Yard Coal and its seatearth (see p. 279) in the quarry [0230 8660] 300 yd S.S.E. of the station: flaggy sandstone 2 ft, on coarse well-bedded sandstone with a few hard spherical ferruginous masses near the base, 45 ft. The outlier north of Birch Vale shows no special features so far as the Woodhead Hill Rock is concerned; it is here about 50 ft thick.

Farther west another outcrop of Woodhead Hill Rock, somewhat faulted, forms the high ground of Mellor Moor. Here it is split into two leaves. North of Mellor an outlier of the rock caps a small hill.

Between Smithylane and Ernocroft Wood the Woodhead Hill Rock, here about 80 ft thick, forms a well-marked scarp. The best section is in Loads Quarry [9765 9028] which shows, beneath the boulder clay cover, 16 ft of cross-bedded sandstone with irregular ferruginous patches. In the north-western corner of the district extensive outcrops of the rock are present beneath drift, between Ernocroft Wood and Hodgefold. The sandstone is seen in small exposures only, in particular in a stream at and above Botham's Hall.

At New Mills the Woodhead Hill Rock forms the sides of the 80-ft gorge cut by the Goyt. It is here coarse and massive and, with the local absence of the Yard, is united with the sandstone overlying that coal. The Shaw Farm Borehole [9905 8661] see also pp. 377–8) proved the Woodhead Hill Rock between 74 ft 5 in and 165 ft: sandstone (not recovered) 9 ft 10 in, shale (not recovered) 6 ft 9 in, on pale brown, coarse, feldspathic sandstone, with some red staining 74 ft. South of New Mills these beds form a long dip-slope at Longside Plantation. A typical section in a quarry [9901 8343] by Cart House, showed 12 ft of yellow massive medium-grained sandstone.

T

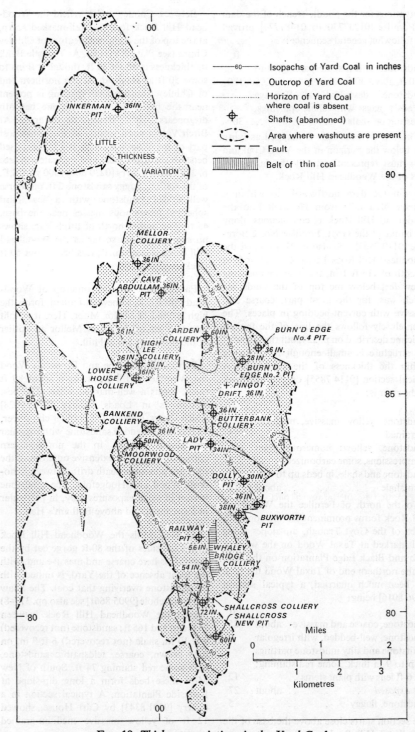

FIG. 19. *Thickness variations in the Yard Coal*

A borehole at Disley Paper Mills [9805 8528] (see p. 364) proved 3 ft of coal, certainly the Yard, overlying 'shale' 48 ft on sandstone 54 ft. The record is anomalous in that the Yard Coal appears everywhere to lie only a few feet above the Woodhead Hill Rock and it seems that the beds described as 'shale' are more likely to be soft flaggy sandstone for the greater part.

On the western side of the Todd Brook Anticline the escarpment of the Woodhead Hill Rock runs southwards from Lyme Park to Sponds Hill before passing out of the district.

**Yard Coal.** The Yard Coal (Fig. 19) is present to the north of Fernilee, where the top few inches are exposed in a stream [0145 7864], near Fernilee Hall. It is also recorded by Cope (1949, p. 471) as having been temporarily exposed in the bed of the Goyt [1005 7898] $\frac{1}{2}$ mile N.W. of Fernilee Hall, where the topmost 18 in were stated to be "layers of bright coal intercalated with fissile carbonaceous shale". The seam was 46 in thick in a shaft [0123 7893] at Fernilee, but thickens northwards to 72 inches in Shallcross Colliery Winding Shaft [0112 8024].

At Whaley Bridge Colliery (Winding Pit) [0122 8117], the Yard Mine, known also as the 'Kiln' in the vicinity from its use in lime burning, is recorded as 54 in thick; in workings to the north it thins to only 16 to 36 in.

To the west of the Goyt, the Yard Coal, about 42 in thick, was worked on a small scale until 1948 in the upper part of Diglee Clough at Diglee. Farther north, Moorwood Colliery [9965 8396] worked 'Kiln Coal' 9 in on coal 41 in. Only a quarter of a mile south-west, however, workings from outcrop in Higher Disley Colliery proved cannel 12 in, coal 20 in, dirt 1 in, coal 12 in. The records at Bankend Colliery [0002 8432] and of the Disley Paper Mills Borehole [9805 8528] (see above) both showed thicknesses of 36 in.

At Dolly Pit [0216 8305] north of Buxworth, the Yard was 30 in thick and the tips show much carbonaceous shale from this horizon. At crop the seam was formerly worked by adits [0281 8319; 0292 8330] near Throstle Bank and was here 45 in thick. Farther north an exposure [0322 8467] west of Whiterakes shows the following section: coal with dirt partings 36 in, beds not exposed a few inches, buff sandy seatearth with thin bands of siliceous sandstone 2 ft 6 in, on flaggy sandstone. At Gowhole the seam is 34 in at Lady Pit [0141 8395]. North of Gowhole, with the development of the sandstone roof (see p. 269), the Yard Coal is frequently affected by washouts. This, however, has not prevented a large area of it being worked between Gowhole and Birch Vale. At Pingot Colliery [adit, 0154 8527], $\frac{1}{4}$ mile E.S.E. of Low Leighton, the seam was 36 in thick at the point [0174 8523] where the coal was reached in the cross-measures drift leading to the workings.

At Birch Vale Quarry [0230 8660] (see also p. 277) the following section was measured:

|  | ft |
|---|---|
| Sandstone, coarse massive .. .. | 10 |
| YARD **Coal 18 in** .. .. .. | 1$\frac{1}{2}$ |
| Seatearth, pale grey, on brown sandy seatearth .. .. .. .. | 4 |
| Mudstone, grey-brown, silty .. .. | 2 |
| Woodhead Hill Rock .. .. .. | — |

The upper part of the coal is probably washed out at this locality. The section at Arden Colliery [0125 8642], Birch Vale was unusual: coal 36 in, clay 17 in[1], coal 7 in, fireclay. In view of the rather constant thickness of the seam hereabouts, except where washouts are present, it appears likely that the lower leaf of the coal is a local development rather than a split off the main seam.

Between Birch Vale and Rowarth the Yard Coal is present in an outlier, now largely worked out. The following section was recorded in a shaft [0213 8760] north-east of Wethercotes: 'rock' 48 ft, shale 3 ft, coal 36 in. Between here and Thornset Fields a few feet of shale separate the coal from the overlying sandstone and no washouts are recorded. At Thornset Fields the coal was found to be washed out near outcrop.

North of New Mills, Cave Abdullam Pit [0021 8736], showed 30 to 36 in of coal with a 'rock' roof; in the workings the seam was found to be washed out in many places. In

---

[1]Shown as 7 in on graphic section, Abandonment Plan 6886. The written section on this plan is coal 3 ft, clay 17 in, coal 7 in, fireclay, and it seems likely that '7 in' should be read as '17 in'.

workings at and north of New Mills the seam was known as the 'Mountain'. On the north-western side of New Mills, High Lee Colliery [9969 8580] proved the coal to be 36 inches in thickness. To the north-east, however, old workings are unknown in the town and the nearly continuous sections in sandstone along the River Sett fail to show any sign of the seam. It is therefore likely that the coal is here washed out, the thick roof-rock being united with the Woodhead Hill Rock.

Workings in Chisworth Colliery, between Chisworth and Gun Farm, showed the following section (Abandonment Plan 3433): 'bastard cannel' 12 in, top coal (House Coal) 12 in, bottom coal (Smithy Coal) 12 in. The cannel is stated to be of irregular occurrence.

The Yard Coal is poorly exposed[1] north-east of Compstall beneath the section [9731 9116] in the bank of the Etherow (see also p. 282):

| | ft | in |
|---|---|---|
| Shale .. .. .. .. .. | — | — |
| **Coal,** alternating with dark car- | | |
| bonaceous seatearth .. seen | 6 | |
| *Not exposed* .. .. .. about | 3 | |
| Seatearth, darker at top, with iron- | | |
| stone nodules and *Stigmaria* .. | 1 | 0 |
| *Not exposed* .. .. .. about | 2 | 0 |
| Mudstone, sandy; some rootlets .. | 7 | 0 |
| Woodhead Hill Rock .. .. | — | — |

Farther north the Yard was 36 in thick in a stream [9829 9236], north-west of Stirrup. It was formerly worked to some extent north and north-east of Botham's Hall, but details are lacking.

**Beds between Yard and Ganister coals.** South of Whaley Bridge these beds were passed through in Shallcross Colliery New Pit [0146 7947]: 'rock' 11 ft 3 in, shale 18 ft, 'rock' 19 ft, shale 30 ft. A similar, though slightly thicker, section was proved at Whaley Bridge Colliery (see Fig. 20).

The lowest beds are dark shales, with abundant mussels a few feet above the Yard Coal. They are exposed in a small cliff [1005 7898] on the right bank of the Goyt ½ mile N.W. of Fernilee (see Cope 1949, p. 471). Here 10 ft of mudstone, with the base 3 ft above the Yard Coal, yielded abundant

*Carbonicola* cf. *artifex, C. discus, C. fallax ?, C. haberghamensis* and *Curvirimula sp.* At this locality a band with *Lingula* has been found by Eagar, 12 ft above the Yard Coal (1952, pp. 27–8). Dark shales 25 ft thick overlying the Yard Coal are also exposed in a stream [0145 7864] at Fernilee (see p. 279) yielding in the lowest 6 ft: *Carbonicola* cf. *discus, C.* aff. *fallax, C. haberghamensis* and *Geisina arcuata.*

Beds referred to a slightly higher horizon are exposed a mile farther downstream in the right bank of the Goyt at Botney Bleach Works [0102 8039]:

| | ft | in |
|---|---|---|
| Shale, grey .. .. .. .. | 2 | 0 |
| Sandstone, yellow soft flaggy .. | 1 | 0 |
| Sandstone, yellow-brown hard well- | | |
| bedded .. .. .. .. | 13 | 0 |
| Mudstone, silty with large concre- | | |
| tions and ironstone nodules up to | | |
| 1 ft by 4 in .. .. .. | 4 | 0 |
| Shale, grey with ironstone bands | | |
| and nodules up to 3 in .. .. | 18 | 0 |
| Shale, dark platy 'sulphurous'; *Ling-* | | |
| *ula sp.* and fish debris in lowest | | |
| 1 in .. .. .. .. | | 3 |
| Shale, soft and broken .. ¾ in to | | 0½ |
| Shale, black 'sulphurous' .. .. | | 5½ |
| Mudstone, dark poorly bedded | | |
| blocky and silty .. .. about | 1 | 2 |
| Mudstone, dark slightly silty, with a | | |
| few ironstone nodules .. about | 8 | 0 |

The sandstone in this section is that present in Shallcross Colliery New Pit (see Fig. 20) and by comparison the *Lingula* band lies some 10 to 12 ft above the Yard Coal and is the Lower Bassy Marine Band. The mussels normally present between *Lingula* band and coal were not found.

The sandstone noted above is conspicuous at outcrop in an outlier at Taxal and farther south in two small outliers around Overton and Normanswood. Though thin, it can be traced almost continuously from Whaley Bridge to Cart House, one mile south-east of Disley, where it is faulted off.

Near the western margin of the district the beds overlying the Yard Coal are exposed in Mather Clough [9793 8286]:

---

[1]Section measured by Mr. C. G. Godwin.

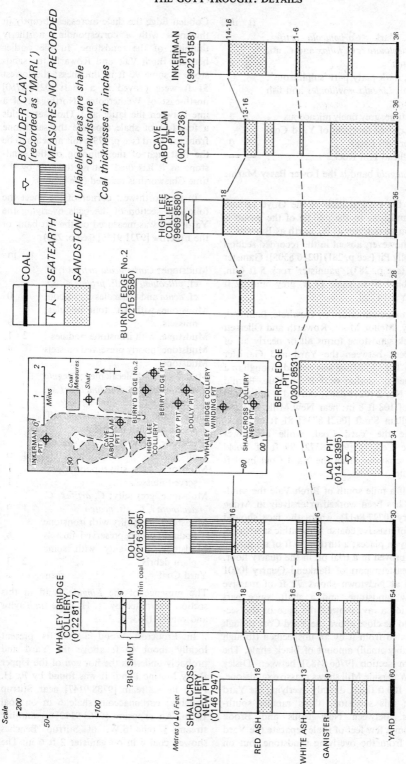

FIG. 20. Sections of shafts in the Coal Measures of the Goyt Trough

|  | ft | in |
|---|---|---|
| Shale, dark; *Carbonicola artifex*, *C. declinata?, C. fallax* and *C.* aff. *protea* .. .. .. .. | 4 | 0 |
| Shale, dark platy part 'sulphurous'; small *Lingula mytilloides* and fish debris .. .. .. .. | | 3 |
| Mudstone, grey finely micaceous .. | 2 | 3 |
| *Not exposed* (horizon of Yard Coal at or near base) .. .. .. | 9 | 0 |
| Woodhead Hill Rock .. .. | — | — |

The *Lingula* band is the Lower Bassy Marine Band.

On the eastern limb of the Goyt Trough the sandstone in the middle of the group is present at outcrop as far north as Buxworth. It is, however, absent in the recorded section of Dolly Pit (see p. 281) [0216 8305]: Ganister Coal (see p. 283), 'gannister' rock 5 ft 6 in, grey bind 20 ft, shale 17 ft, grey bind 52 ft 6 in, Yard Coal.

In an area bounded by Gowhole, Bankend, Disley, Mellor Moor, Rowarth and Ollersett a thick sandstone forms all or nearly all of the beds between the Yard and Ganister coals. The sandstone was passed through in a number of shafts: near Birch Vale, Burn'd Edge No. 2 [0215 8580] showed a full thickness of 108 ft 8 in; near New Mills, in Cave Abdullam Shaft [0021 8736] 81 ft of 'rock' overlie the Yard Coal, while at Lower House Colliery [9945 8572] 99 ft of 'rock' were separated from the Yard Coal by 9 ft of shale.

Half a mile south of Birch Vale the sandstone has been worked extensively in Arden Quarry [0227 8611], where the face showed 65 ft of massive coarse feldspathic sandstone, pebbly in places; a further 35 ft of sandstone were proved in a trial pit in the quarry floor. The eastern part of Bankend Quarry [0001 8431] at Newtown shows 72 ft of massive yellow sandstone, and a shaft, since part quarried away, showed the base of the section to lie close above the Yard Coal though separated from it by an unspecified (though probably small) amount of black shale. The stream section [9766 8485] between Disley and Waterside Mill shows massive sandstone, some 100 ft thick, directly overlying the Yard Coal; this sandstone thins rapidly southwards. Between New Mills and Brook Bottom a few feet of shale separate the Yard Coal from the overlying sandstone, but on

Cobden Edge the shale increases abruptly in thickness with a corresponding northerly thinning of the sandstone. In the outlier between Birch Vale and Rowarth the sandstone is some 90 ft in thickness; the lowest 51 ft were proved in a shaft [0213 8760] north-east of Wethercotes. North-east of a line between the latter and Thornset Fields a few feet of shale separate the sandstone from the Yard (see p. 279); this appears to be the first sign of the thinning of the sandstone as it has died out northwards by the time Chisworth is reached (see below).

In the north-west corner of the sheet the following section of the beds overlying the Yard Coal was measured on the left bank of the Etherow [9731 9116] (see p. 280):

|  | ft | in |
|---|---|---|
| Mudstone; *Carbonicola artifex?, C.* cf. *pilleolum, C.* aff. *protea, C. sp.* cf. *acuta* and fish scales .. about | 4 | 0 |
| Mudstone, with ironstone nodules; mussels .. .. .. .. | | 9 |
| Mudstone, with ironstone nodules | 2 | 1 |
| Mudstone, poorly preserved mussels | | 4 |
| Mudstone .. .. .. .. | | 6 |
| Mudstone, dark decalcified; *Lingula mytilloides* and *Rhadinichthys sp.* [scales] .. .. .. .. | | 2 |
| Shale, dark platy; *Carbonicola* aff. *protea* and fish fragments .. | | 3 |
| Mudstone, grey silty; *Carbonicola* aff. *fallax* .. .. .. | 1 | 0 |
| Mudstone, grey silty; poorly preserved mussels .. .. .. | | 1 |
| Mudstone, grey silty; *C. artifex, C. declinata?, C.* aff. *protea* .. | 1 | 9 |
| Mudstone, dark silty with ironstone nodules; poorly preserved mussels | | 6 |
| Mudstone, grey silty with some plant debris .. .. .. | 2 | 3 |
| Yard Coal .. .. .. seen | | 1 |

The presence of the *Lingula* band in this section was noted by R. H. Price (*in* Taylor and others 1963, p. 32).

In Ernocroft Wood a coal is present locally about 40 ft above the Yard and probably underlies the horizon of the Upper Bassy Marine Band. It was found by R. H. Price in a stream [9788 9147] near Stirrup Benches: carbonaceous shale 6 in on coal 18 in. An exposure [9763 9112] in another stream ⅓ mile S.W. of Stirrup Benches showed coal 6 in on ganister 2 ft 6 in. The beds between this coal and the Yard are shale, containing at the base [9757 9116] the characteristic mussel band.

The highest beds in this group of strata are exposed in a stream at Sandy Lane[1] [9876 9197]:

Ganister Coal (see p. 284).

|  | ft | in |
|---|---|---|
| Ganister with rootlets .. .. | 2 | 0 |
| *Not exposed* .. .. .. .. | 2 | 3 |
| Mudstone, grey silty; occasional rootlets .. .. .. about | 3 | 0 |
| *Not exposed* .. .. .. about | 6 | 0 |
| Mudstone, silty micaceous, with several ironstone bands up to 3 in thick; plant debris .. .. | 6 | 0 |
| Mudstone, grey-brown silty micaceous, with a band of ironstone nodules 3 ft from base .. .. | 12 | 0 |
| Mudstone, medium to dark grey, somewhat silty, with ironstone nodules in lowest 6 in; some fish debris .. .. .. .. | 3 | 0 |
| Mudstone, medium to dark grey; fish and plant debris .. .. | 2 | 0 |
| Mudstone, silty; some fish debris; plant debris in upper 2 ft, rather micaceous below; a few mussels at base .. .. .. .. | 4 | 8 |
| Mudstone; *Spirorbis sp., Carbonicola* cf. *artifex, C.* aff. *fallax, C.* cf. *pilleolum, C.* cf. *protea, C. rectilinearis, Curvirimula sp., Naiadites sp.* cf. *obliquus* and *Geisina arcuata* .. .. .. .. | 2 | 9 |

The beds below the Ganister Coal are also seen in Chew Wood in a stream section [9927 9247][1]:

|  | ft | in |
|---|---|---|
| Ganister Coal .. .. .. | — | — |
| Seatearth, siliceous, with rootlets .. |  | 3 |
| Seatearth, micaceous, with rootlets |  | 6 |
| Seatearth, siliceous in lower 1 ft to 1 ft 6 in .. .. .. .. | 4 | 0 |
| Siltstone, grey argillaceous .. | 2 | 9 |
| Mudstone, silty, lowest 3 ft showing some reddening .. .. .. | 4 | 9 |
| Clay, brown .. .. .. .. |  | 2 |
| Mudstone, silty, with pink variegation .. .. .. .. | 4 | 6 |

|  | ft | in |
|---|---|---|
| Mudstone, rather silty at top, with mussels; fish debris in a thin band near base .. .. .. .. | 2 | 7 |
| Mudstone, grey silty .. .. |  | 1 |
| Mudstone, with mussels; plant debris near base .. .. .. | 2 | 3 |
| Mudstone, silty, with ironstone nodules .. .. .. .. | 2 | 1 |

Here the highest beds with mussels occur only 16 ft 11 in below the Ganister Coal, as compared with about 40 ft at Sandy Lane. This probably indicates a local thinning at Chew Wood, although the possibility of variation in the development of the mussel bands cannot be completely discounted.

**Ganister Coal.** This seam and its floor form an easily recognized horizon south of Chinley. In Shallcross Colliery New Pit [0146 7947] the coal was 9 in thick and was here known as the 'Cannel Mine'. A road-cutting [0089 7985] east of Taxal, showed:

|  | ft | in |
|---|---|---|
| Shale .. .. .. .. about | 8 | 0 |
| Shale, black .. .. .. .. |  | 2 |
| Coal .. .. .. .. .. |  | 7 |
| Ganister .. .. .. .. |  | 9 |
| Seatearth, buff .. .. .. | 1 | 3 |
| Seatearth, sandy .. .. .. | 1 | 6 |
| Seatearth, buff .. .. about | 6 | 0 |

In Whaley Bridge Colliery Winding Pit [0122 8117] 9 in of coal were recorded at this horizon. At Railway Pit [0109 8180] the coal was 12 in thick, its greatest thickness in the area. In Whaley Lane [0043 8151] the coal and its seat were exposed: coal 6 in, hard grey-white ganister with rootlets 6 in, soft flaggy yellow ganister-like sandstone with rootlets 9 in. The ganister was also seen [9988 8223], north-west of Whaley Bridge, where it was 10 in thick and [9982 8281] near Diglee where it reached a thickness of 1 ft 9 in. In Mather Clough [9791 8302] the coal is absent, but its horizon is represented by 1 ft of ganister with rootlets. At Dolly Pit [0216 8305] the section was coal 6 in on 'gannister rock' 5 ft 6 in; the latter probably includes some sandstone since true ganister is everywhere much thinner at this horizon. The ganister, grey-white and with rootlets, is

---

[1]Measured by Mr. C. G. Godwin. The position given is that of the base of the section.

exposed in a track [0267 8326] north of Hill Farm.

In an area, including New Mills and Birch Vale, coinciding closely with that in which the underlying Yard 'roof-rock' is developed (see p. 282), the Ganister Coal fails. However, it is evident from shaft sections that the horizon of the Ganister Coal lies at the top of the rock (see Fig. 20). Sections at Burn'd Edge No. 2 Pit, Cave Abdullam, High Lee and Mellor collieries fail to record the Ganister Coal. The significance of the absence of the coal (and also the White Ash Coal) where the 'roof-rock' of the Yard is thick, has already been discussed (pp. 269–70).

The Ganister Coal or its seat reappear in the north-western part of the district and are traceable north of Ludworth Intakes. The identity of the coal is proved by the occurrence of the Lower Foot Marine Band above it in the stream section at Sandy Lane (see also pp. 283 and 285). Here the coal is 8½ in thick and rests on at least 2 ft of ganister with rootlets. In Chew Wood (see p. 283) it is only 1½ in thick and the ganister is absent. At Glossop Vale Ganister Mine [9971 9215] Chisworth, the coal was absent, the section being: shale, ganister 2 to 16 in, fireclay 30 in. In the Inkerman Shaft at Chisworth Colliery [9922 9158] Hull and Green (1866, p. 24) record the section as coal 1 to 6 in on 2 ft of ganister.

**Beds between Ganister and White Ash coals (south of Chinley and New Mills).** Owing to the absence of the White Ash Coal north of Chinley and New Mills all the beds between the Ganister Coal or its horizon and the Red Ash are described later under a single heading.

South of Chinley and New Mills the beds between the Ganister and White Ash are from 26 to 45 ft thick, averaging 30 ft (see p. 270). Typical sections are in Shallcross Colliery New Pit [0146 7947] where they consist of 11 ft 11 in of 'rock' on shale 17 ft, and Whaley Bridge Winding Pit [0122 8117], 'rock' 18 ft on shale 12 ft. This group of measures expands at Railway Pit, Whaley Bridge [0109 8180]; 'rock' 33 ft, shale 12 ft. A section through these beds, beneath the seatearth of the White Ash Coal in Higgin's Clough [9783 8515], north-east of Disley,

showed banded sandy mudstone 1 ft 3 in on flaggy sandstone with silty mudstone partings 22 ft. A quarter of a mile to the north-east a section [9803 8548] on the right bank of the Goyt shows flaggy sandstone 18 ft on silty mudstone 15 ft beneath a seatearth at the horizon of the White Ash Coal.

**White Ash Coal and its seatearth.** This is present only in the area south of Chinley and New Mills (see also p. 270).

At Shallcross Colliery New Pit the White Ash was 13 in and stated[1] to rest directly on 'rock'. It was formerly worked at Shallcross Colliery where the thickness varied from 12 to 18 in. In Whaley Bridge Colliery Winding Shaft it was 16 in thick and was here known as the 'Smithy'; in the workings it was 17 to 24 in, but thinned northwards, being only 6 to 9½ inches in the vicinity of Whaley Print Works. The coal was exposed [0094 7963] west-south-west of Shallcross Manor:

|  | ft | in |
|---|---|---|
| Shale, grey-brown, weathered .. | 1 | 0 |
| Coal .. .. .. .. about | | 5 |
| Seatearth, sandy micaceous, with rootlets and thin sandstone bands | 5 | 0 |
| Siltstone, banded argillaceous .. | | 8 |
| *Not exposed* .. .. .. .. | 2 | 0 |
| Sandstone, flaggy .. .. .. | | 10 |

At Ringstones Colliery (Bottom Pit) [0058 8248] west of Bridgemont the White Ash was worked but details of the thickness are lacking. At Furness Vale Colliery the seam was got until about 1963 in conjunction with its seatearth. The workings were reached by a cross-measures drift [entrance 0049 8338] and the White Ash was intersected at 235 yd in [0027 8339] where it showed the following section: shale, coal 18 in, fireclay 5 ft 6 in, shale and rock bind. The White Ash is not recorded in Dolly Pit, Buxworth [0216 8305]; at its inferred position, however, the walls of the shaft are bricked, probably due to the seeping of water from the sandstone underlying the coal.

Near Disley the section in Higgin's Clough (see above) [9783 8515] shows the White Ash Coal and its seatearth: coal 14½ in, dark seatearth 2½ in, sandy seatearth with bands and lenses of ganister 1 ft 3 in.

---

[1]Graphic section on Abandonment Plan 2105.

Beds between White Ash and Red Ash coals (south of Chinley and New Mills). Typical sections are in Shallcross Colliery New Pit [0146 7947] where they consist of 15 ft 9 in of 'rock' on 39 ft of shale and 'bind', and Whaley Bridge Winding Pit [0122 8117] where they were 'stone' 24 ft on shale 36 ft. The presence of a vein of galena in the sandstone under the Red Ash Coal at this locality is discussed on p. 317. At "Mr. Srigley's pit" [probably 0106 8185], about ½ mile N. of Whaley Bridge, the shales above the White Ash were stated by Hull and Green (1866, p. 25) to contain fish remains, ostracods and plants.

Northwards these beds thicken, being 74 ft at Ringstones Colliery Bottom Pit [0058 8248]. At Dolly Pit, Buxworth [0216 8305] the precise thickness is uncertain as the White Ash Coal was not found, but the sandstone underlying the Red Ash Coal was 35 ft thick. Half a mile north-east of Dolly Pit the sandstone appears to thicken further, to perhaps as much as 45 ft, and it occupies a substantial outcrop west of Chinley Churn.

Beds between Ganister and Red Ash coals (north of Chinley and New Mills). The thickness varies between about 80 and 100 ft. To the south of Birch Vale, the section in Burn'd Edge No. 2 Pit [0215 8580] showed 15 ft 6 in of 'rock' on 65 ft 6 in of 'bind'. At outcrop the thickness of the sandstone appears to vary abruptly in places; near Shedyard it is as much as 50 ft but it thins rapidly westwards to as little as 10 ft north of Beard Hall. From the latter locality northwards to Low Leighton the shales beneath the sandstone underlying the Red Ash Coal contain a lower sandstone up to about 15 ft in thickness. The top of this sandstone probably marks the horizon of the White Ash Coal.

At New Mills this group of strata thickens, the section at High Lee Colliery [9969 8580] being 'rock' 48 ft on shale 48 ft. Farther north the thickness is even greater, as at Cave Abdullam Colliery [0021 8736] where the section was 'rock' 12 ft 8 in, shale 10 ft, 'rock' 10 ft, 'bind' 30 ft, 'rock' 4 ft, 'bind' 8 ft, black shale 26 ft. The uppermost two sandstones are normally united hereabouts and the 4-ft sandstone is taken to mark the horizon of the White Ash Coal.

At Chisworth Colliery the Inkerman Shaft [9922 9158] (Hull and Green 1866, p. 24) proved 44 ft of coarse flaggy sandstone on 40 ft of shale between the Ganister and Red Ash coals. North of Chisworth a quarry [9966 9198] shows 26 ft of the sandstone, medium- to coarse-grained, cross-bedded and with ferruginous patches. The sandstone thickens westwards, being as much as 70 ft near Stirrup. In this vicinity, as suggested by Cope (1949, p. 471), it seems probable that the White Ash Coal has been washed out by the sandstone.

The important stream section at Sandy Lane [9876 9197] shows the best exposures in the district of the lower beds and the only exposure of the Lower Foot Marine Band:

|  | ft | in |
|---|---|---|
| Sandstone, with some mudstone partings .. .. .. about | 15 | 0 |
| *Not exposed* .. .. .. about | 30 | 0 |
| Mudstone, dark .. .. .. | 9 | 0 |
| Shale, soft with listric surfaces .. | | 1 |
| Mudstone, dark; fish debris .. | 2 | 3 |
| Mudstone, dark with a thin carbonaceous band; *Planolites ophthalmoides, Lingula mytilloides, Caneyella multirugata, Posidonia gibsoni, Anthracoceras sp., Gastrioceras sp.* [juv.], *Hindeodella sp.*, platformed conodonts and fish remains including *Elonichthys sp.* | | 8 |
| Mudstone, dark soft .. .. | | 1 |
| Mudstone, silty, unbedded; *Curvirimula sp., Naiadites?, Geisina arcuata* and fish remains .. .. | 1 | 0 |
| Mudstone, silty unbedded .. .. | | 3 |
| Sandstone, with thin mudstone parting .. .. .. .. .. | | 10 |
| Mudstone, silty, dark at base, with ironstone bands; uppermost 10 in with *Spirorbis sp., Carbonicola* cf. *limax, C.* cf. *artifex, C.* aff. *obliqua, C. sp.* (? *protea* group), *Geisina arcuata* .. .. .. | 5 | 4 |
| Mudstone, dark with plant debris .. | | 3 |
| Ganister Coal (see pp. 383–4) .. — | | — |

Red Ash Coal. At Shallcross Colliery New Pit [0146 7947] the coal was 18 in thick and is recorded as resting directly on 'rock'. At Whaley Bridge Colliery Winding Pit [0122 8117], where it is also 18 in, it was worked extensively as far east as Horwich House. Other records at Whaley Bridge are Railway Pit [0109 8180] 16 in and Wharf Pit [0107 8121] 20 in.

In Furness Vale Colliery the Red Ash Coal was 18 inches in No. 1 Shaft [0049 8339] but only 15 inches in a cross-measures drift in from the main adit (which lies within a few feet of the shaft). Near Disley the seam was worked by adit and from a shaft in Roach Hey Wood [9786 8554]; here it was 22 in thick and rested directly on 'rock'. North of Buxworth the Red Ash has been worked in several places under the name 'Little Mine'. At Dolly Pit [0216 8305] it was 16 in thick and underlain by 4 ft 8 in of 'clay' resting on 'rock'. In an adit at Marsh Lane [0063 8460] near Beard Hall a similar thickness was recorded. A small amount of the coal was worked by adit at Pingot Clough Colliery [for example 0157 8524] near Low Leighton where it was 14 in thick and was stated to rest on 'ganister'. Nearby a stream section [0164 8532] showed: shale roof (see below), coal 9 in, obscured 4 in, coal 1 in, sandy seatearth 4 in, ganister 1 in, sandy micaceous seatearth 6 in.

The coal was 16 in thick in Burn'd Edge No. 2 Pit [0215 8580]. It was absent, however, in an exposure [0154 8613] near Over Lee which showed: silty shale 1 ft 6 in, flaggy siltstone 4 in, hard ganister 6 in, passing laterally into seatearth.

Near New Mills the Red Ash was worked at High Lee Colliery [9969 8580] where it was 18 inches in thickness, and Cave Abdullam Colliery [0021 8736] where it was 13 to 16 in. The downcast shaft of Mellor Colliery [9970 8808], proved the Red Ash to be 16 inches.

Near the northern edge of the district, the Red Ash Coal is exposed at several places around Chisworth. A section [9835 9187] near Stirrup, showed 16 in of coal on a thin seatearth. In the Inkerman Shaft [9922 9158] Hull and Green (1866, p. 24) state that the coal, at 37 ft 4 in, was 14 to 16 inches in thickness and rested directly on sandstone.

**Beds between Red Ash and Big Smut coals.** The southernmost outcrops are in the outlier of these beds at Shallcross where they consist of about 30 ft of sandstone on 95 ft of shale. At Whaley Bridge Colliery Winding Pit [0122 8117] the section was 'stone' 6 ft, 'bind' 12 ft, 'stone' 18 ft, shale 90 ft. At Railway Pit [0109 8180] 12 ft 6 in of 'rock' overlay 106 ft 2 in of shale. Shale with marine fossils was turned out from adits to the Red

Ash Coal [0190 8060] near Over Leigh and [0091 8181] south of Hockerley. The lowest beds are exposed in the upper part of Ringstone Clough [0008 8221]:

| | ft | in |
|---|---|---|
| Shale, with thin ironstone bands and lenses; a thin band with fish debris at about 12 ft above base .. | 13 | 0 |
| Not exposed .. .. .. about | 1 | 0 |
| Siltstone, hard calcareous; Dunbarella sp., Anthracoceras sp. and Gastrioceras listeri .. .. | | 3 |
| Shale, silty; Dunbarella sp. .. | | 8 |

The fish band near the top of this section is probably the equivalent of the thin marine or near-marine band at a similar height above the G. listeri Marine Band in the Disley area (see p. 287).

Between Ringstone Clough and Diglee the sandstone in the upper part of these beds expands greatly and only some 30 ft or so of shale separate it from the Red Ash Coal. North of Diglee, however, normal conditions prevail and a shaft [0028 8309], starting just below the base of the sandstone, met the White Ash Coal at a depth of 150 ft; the Red Ash should here lie at a depth of about 73 ft.

In Furness Vale Colliery the roof of the Red Ash Coal was examined in a cross-measures drift near the shaft [0038 8339] (see where it yielded Dunbarella sp., Gastrioceras listeri and Homoceratoides aff. divaricatus. Higher beds, grey shales with some ironstone and rare fish remains including Elonichthys sp., are present in Furness Clough [0038 8335], a thickness of 35 ft being exposed. The overlying strata are well seen in the yard of Messrs. R. E. Knowles' brickworks [top at 0060 8346]:

| | ft | in |
|---|---|---|
| Sandstone, flaggy .. .. .. | 5 | 0 |
| Mudstone, sandy, with thin sandstones up to 2 in thick .. .. | 2 | 6 |
| Sandstone, yellow hard fine-grained | 1 | 5 |
| Siltstone .. .. .. .. | 1 | 4 |
| Mudstone, grey silty .. .. | | 9 |
| Obscured .. .. .. about | 16 | 0 |
| Sandstone, well-bedded, medium-grained .. .. .. .. | 15 | 0 |
| Mudstone, grey silty, with thin sandstone bands .. .. .. | 9 | 0 |
| Mudstone, grey silty .. .. | 7 | 0 |

Shale, grey, with ironstone bands (continuous with Furness Clough section, see above) .. .. — —

ft in

A small thickness of the sandstone extends westwards on the slopes of Broadhey Hill.

To the south of Disley the beds overlying the Red Ash Coal are exposed in Bollinhurst Brook and a tributary near the reservoir. Beds immediately above the G. listeri Marine Band are exposed in the roof of an old adit [9786 8333]: dark grey pyritous shale on dark decalcified shale with Posidonia sp. nov., conodonts including Hindeodella sp. and fish remains. Nearby, R. H. Price (in Taylor and others, 1963, p. 40) recorded the following section [9783 8338]: "36 yd south (upstream) from Bollinhurst reservoir, Disley, where blue shale 4 in overlies black shale containing Lingula mytilloides, an orthocone nautiloid, Geisina arcuata, fish remains including Rhabdoderma and Palaeoniscid scales and an Acanthodian spine 1½ in, blue shale 3 ft, blue shale with Caneyella sp. 2 in, and blue shale". The upper band also contains Caneyella sp. [juv.] and Elonichthys sp. The higher marine horizon is near or at the position of the thin marine band 12 ft above the Red Ash Coal in Higgin's Clough (see below). Higher beds in this group are exposed in the tributary north-east of Bollinhurst Brook. These are shales at the base but become silty upwards and include flaggy sandstones of somewhat uncertain thickness at the top. There is room for some 130 to 150 ft of beds between the Red Ash Coal and the horizon of the Big Smut at this locality.

North of Disley a tributary to Higgin's Clough provides a good section of the shales overlying the Red Ash Coal. The lowest beds are not now visible but were seen by R. H. Price in 1949 [9786 8530]: shale with Caneyella multirugata, Dunbarella papyracea, Posidonia sp. nov., Anthracoceras sp., Gastrioceras listeri and mollusc spat 3 in, carbonaceous shale 2 in, Red Ash Coal. Price (see Taylor and others 1963, p. 40) found the following section of the overlying beds, the base about 6 ft above the coal: "blue shale 4 ft, over contorted shale (crozzle) 6 in, blue shale 6 ft, black shale with fish scales and Caneyella sp. 1 in and blue pyritous shale"

6 ft[1]. The band with Caneyella sp. contains mollusc spat and in addition one specimen of Naiadites sp.

Farther up the tributary clough some 80 ft of shale with ironstone in places are exposed. Higher beds are seen underlying the Big Smut Coal and its seatearth in the left bank of the Goyt [9782 8570] near Woodend House:

| | ft | in |
|---|---|---|
| Seatearth (see p. 288) passing down | — | — |
| Sandstone, fine-grained flaggy .. | 3 | 7 |
| Siltstone, banded with thin sandstone bands (3-in ferruginous band 2 ft 3 in from base) .. .. | 7 | 6 |
| Mudstone, silty .. .. .. | 1 | 0 |
| Obscured .. .. .. about | 6 | 0 |
| Sandstone, massive .. .. .. | 6 | 0 |

In the axial region of the Goyt Trough, the sandstone at the top of these beds can be traced, though faulted, from Green Head to Cold Harbour, half a mile south of Birch Vale. It is conspicuously developed and at least 70 to 80 ft thick on Downs Bank, ¼ mile N.E. of Beard Hall. Here a lower sandstone up to 15 ft thick is present between it and the Red Ash Coal, and the greater part of the beds between the Red Ash and Big Smut coals is sandstone. The lower sandstone is also present above the Red Ash Coal on the south-west side of Beard Hall. The top of this sandstone, about 20 ft above the Red Ash, is probably just below the position of the minor marine phase already noted above the Red Ash Coal.

Farther north, the lowest beds, with the Gastrioceras listeri Marine Band, are exposed in Pingot Clough [0164 8532]: dark grey shale 1 ft, on hard decalcified shale with Caneyella multirugata, Dunbarella papyracea, Posidonia gibsoni, P. sp. nov., Anthracoceras sp., Gastrioceras listeri 1 ft 6 in, on the Red Ash Coal. Near Birch Vale an exposure [0156 8604] near Over Lee, shows the top of the marine band: soft black shale 8 in on shale with marine fossils 3 in. Burn'd Edge No. 2 Shaft [0215 8580] passed through 127 ft 8 in of shale before reaching the Red Ash Coal.

In the area north of New Mills the sandstone underlying the Big Smut horizon crops out over a wide area, forming a long dip-

---

[1]Thickness omitted from work quoted.

slope east of Broadhurst Edge. Smaller outcrops are present on Mellor Moor and at
Eaves Knoll. In this area, tips from shafts to
the Red Ash Coal show marine shale, in
many places accompanied by carbonate
bullions up to about 9 in diameter, a feature
seldom present in the *G. listeri* Marine Band
in the district. The Cave Abdullam Shaft
[0021 8736], starting at the base of the sandstone, penetrated 72 ft of black shale before
reaching the Red Ash Coal; from this a
thickness of some 75 ft for the sandstone
may be inferred.

To the west of Chisworth, beds overlying
the Red Ash Coal form an outlier. A section
[9835 9187] south of Stirrup (see p. 286)
showed shale with *Caneyella multirugata*,
*Dunbarella sp., Posidonia* cf. *gibsoni, P. sp. nov.*,
*Gastrioceras listeri* and *Elonichthys sp.* [scale],
overlying the Red Ash Coal. In the Inkerman
Shaft [9922 9158] the 36 ft of beds from the
top to the Red Ash Coal are described by
Hull and Green (1866, p. 24) as sandstone
but, as pointed out by Cope (1949, p. 471),
this is certainly a mistake. Near Chisworth
a long strike-section in a stream shows the
beds above the Red Ash Coal in several
places. A typical section [9940 9203] near
Chew, showed, close above the position of
the Red Ash Coal, 2 ft of shale with *Caneyella multirugata, Dunbarella papyracea,
Posidonia gibsoni, P. sp. nov., Anthracoceras
arcuatilobum* and *Gastrioceras listeri*.

**Big Smut Coal and its seatearth.** The
southernmost section in the area is that in
Whaley Bridge Colliery Winding Pit [0122
8117] where the seam is apparently split, the
section being coal 9 in, fireclay 6 ft, grey
bind 18 ft, coal a few inches[1]. The upper coal
is tentatively correlated with the Cannel
Mine and the lower with the Upper Mountain Mine of the Lancashire succession. The
section at Railway Pit [0109 8180] shows a
single coal, 18 inches in thickness, overlain
by drift and lying 120 ft above the Red Ash.
This is considered to represent the split seam
of the Whaley Bridge Winding Pit, here
re-united.

Part of the Big Smut Coal is exposed in the
upper part of Ringstone Clough [0024 8219]
where a few inches of coal smut rest on at
least 6 ft of seatearth with rootlets.

To the south of Disley a section [9796 8331]
in a confluent of Bollinhurst Brook shows the
seat of the Big Smut Coal: grey seatearth 1 ft,
ganister with some rootlets, soft and argillaceous in places, 2 ft, on hard flaggy sandstone with rootlets. A section 100 yd upstream from this locality (see p. 289) shows
that the coal is absent here. Three-quarters of
a mile farther west in Lyme Park the Big
Smut has been worked by bell-pits in Coalpit
Clough. North of Disley the section [9782
8570] on the left bank of the Goyt at Woodend (see also p. 287), shows shale, coal 13 in,
buff 'sulphurous' seatearth with rootlets
1 ft 6 in, grey silty seatearth with rootlets
4 ft 2 in, on sandstone.

In the central area of the Goyt Trough the
Big Smut Coal is irregularly developed. Near
Green Head an exposure in the side of a
track [0218 8339] showed about 12 in of coal
smut resting on seatearth 1 ft 6 in. Farther
north, a thin coal was present at this horizon
in the stream section [0198 8543] west of
Moor Lodge: sandstone 1 ft 6 in, coal smut
about 2 in, sandy seatearth 3 in, seatearth,
slightly sandy, with rootlets 3 ft, flaggy siltstone. Nearby, the upper surface of the underlying sandstone was hard and ganister-like,
with rootlets. Near Cheetham Hill a section
[9950 8808] north-west of Whitehouse Farm,
showed shale (see below), beds not exposed
1 ft, silty seatearth 1 ft 6 in; there was no
evidence of the coal being present here.

At Ludworth the Big Smut and its seatearth have been (and are) worked at Ludworth Moor Colliery. The thickness of the
coal in the workings is usually 8 to 10 in,
though a section [9964 9082], measured
underground, showed it to be thinner locally:
pale slightly micaceous mudstone 10 in, coal
6 in, soft seatearth with rootlets 2 ft, silty
micaceous seatearth 3 ft, hard silty unbedded
mudstone 2 ft 3 in. The Big Smut is, in this
area, of very variable thickness. Opencast
workings [9980 9081] only 170 yd E. of the
last-mentioned section, showed it to be
absent. However, a nearby exposure [9983
9073] in a stream showed 2 ft of coal smut
resting on seatearth.

**Beds between the Big Smut Coal and Miln-
row Sandstone.** At Whaley Bridge Colliery
Winding Pit [0122 8117] 41 ft of 'grey bind

---

[1]Shown as thin coal on Abandonment Plan 1721.

and shale' overlie the split Big Smut Coal. North-west of Whaley Bridge the following section, in beds with the Cannel Mine Marine Band, was measured in the upper part of Ringstone Clough [0028 8222]:

| | ft | in |
|---|---|---|
| Shale, grey .. .. .. .. | 15 | 0 |
| Shale, contorted .. .. 4 in to | | 5 |
| Shale, grey .. .. .. .. | | 2 |
| Shale, black decalcified, with fish debris .. .. .. .. | | 3½ |
| Shale, grey soft, with foraminifera.. | 2 | 9 |
| Shale, dark grey hard .. .. | | 9 |

The base of the section lies close above the horizon of the Big Smut Coal.

To the south of Disley, the section [9804 8328], in a tributary of Bollinhurst Brook, shows the lowest beds:

| | ft | in |
|---|---|---|
| Shale, rather dark finely micaceous, with oxidized pyrite blebs .. | | 10 |
| Shale, grey platy 'sulphurous' .. | 1 | 1 |
| Shale, pale grey soft .. ¼ in to | | 0¾ |
| Mudstone, grey .. .. .. | | 1½ |
| Shale, pale grey soft .. ¼ in to | | 0½ |
| Mudstone, grey .. .. .. | | 9 |
| Shale, pale grey soft .. ¾ in to | | 1¾ |
| Mudstone, grey .. .. .. | | 10 |
| Shale, pale grey soft .. | | 0½ |
| Mudstone, grey very finely micaceous, with occasional pyrite blebs; fish debris at 9 in from base about | 3 | 6 |
| Ganister, pyritous .. .. .. | | 5 |

No evidence of the Cannel Mine Marine Band was found in this section, though the presence of fish debris at the base, close above the horizon of the Big Smut Coal, provides a link with the section at Woodend (see below). Another section of the lowest beds was measured south of Disley by R. H. Price (in Taylor and others 1963, p. 40) in the bank of the Stockport Corporation reservoir [9692 8341] west of Cockhead: black shale 1 ft, black shale with Caneyella sp. [juv.] 1 in, black shale with ironstone 4 ft. The section lies about 6 ft above the Big Smut Coal. To the north of Disley the lowest beds, with the Cannel Mine Marine Band, are exposed in the section [9782 8570] at Woodend:

| | ft | in |
|---|---|---|
| Shale, dark, with ironstone bands up to 1½ in thick, occasional 'fucoids' near base .. .. .. .. | — | — |
| Shale, contorted .. .. 9 in to | 1 | 1 |
| Shale, grey pyritous and 'sulphurous' with pyrite blebs and 'fucoids' .. .. .. .. | 5 | 0 |
| Shale, black hard: foraminifera including Ammodiscus sp., Lingula mytilloides, Cypridina sp., abundant conodonts including Hindeodella sp. and platformed types, and fish scales including Elonichthys sp. and Rhadinichthys sp. .. | | 3 |
| Mudstone, with fish debris .. .. | | 2 |
| Big Smut Coal (see p. 288) .. | — | — |

The beds between the Big Smut Coal and Milnrow Sandstone show their thickest and most complete development, about 100 ft, around Gowhole. The lowest beds are exposed in a stream section in Hutfall Plantation [0195 8387] which showed the following:

| | ft | in |
|---|---|---|
| Shale, grey, reddened along joints and bedding in lowest 2 ft .. | 6 | 0 |
| Shale, pyritous, hard below and softer above; foraminifera, including Ammodiscus sp., and Rhabdoderma sp. [scale] .. .. | 1 | 8 |
| Shale, grey contorted, slickensided and broken .. .. 5 in to | | 7 |
| Shale, hard, blocky; Lingula sp., conodonts including Hindeodella sp., platformed types and fish debris .. .. .. .. | | 3½ |
| Mudstone, grey slightly silty, well-jointed .. .. .. .. | 1 | 3 |

North-east of the last exposure some 40 ft of shale intervene between the base of the group and the sandstone, here perhaps 15 ft thick, which lies below the horizon of Tonge's Marine Band. This sandstone is developed locally in the central part of the Goyt Trough between The Haugh and the neighbourhood of Shedyard and also in an outlier at Whaley Bridge and in the cutting near Beard Hall. The latter exposure, starting [0087 8429] due south of Beard Hall, shows the following section:

ft in

Milnrow Sandstone (see below) .. — —
Shale, grey; some pyrite blebs in
   lowest 5 in    ..    .. about   20   0
Tonge's Marine Band
   Shale, dark grey slightly silty, with
     pyrite-filled burrows; foramini-
     fera including *Ammodiscus*, and
     palaeoniscid scales    ..    ..        7
Shale, grey ..    ..    ..    ..     1   3
Shale, black hard; somewhat car-
   bonaceous, slightly silty ..    ..       9
Seatearth, grey-brown    ..    ..     2   3
Sandstone, fine-grained hard sili-
   ceous    ..    .. seen about     4   0

This is the only locality in the district where
Tonge's Marine Band has been found. It was,
however, noted in the Chadkirk Print Works
Borehole (Magraw 1957, p. 21) a little west
of the district boundary. The seatearth
represents the horizon of the Norton Coal
of the Sheffield area.

Near Moor Lodge, south of Birch Vale,
these beds consist of about 70 ft of shale.

Near Mellor, the Cannel Mine Marine
Band appears to fail in the section at Cheetham Hill [9950 8808] (see also p. 288) where a
seatearth at the horizon of the Big Smut
Coal is overlain by 4 ft of shale with fish
debris including scales of *Elonichthys sp.* and
*Rhadinichthys sp.*

On Ludworth Moor, opencast workings
(now filled in) for coal and fireclay [9980
9081] showed the following section of the
beds above the horizon of the Big Smut
Coal (here absent):

ft in

Shale, grey ..    ..    .. about   9   0
Shale, platy, with foraminifera including *Ammodiscus sp.*, *Lingula*
   *mytilloides* and fish remains including palaeoniscid scales ..    ..       4
Shale, soft weathered 'sulphurous',
   with small decomposed pyrite
   blebs    ..    ..    ..    ..       5
Mudstone, grey    ..    ..    ..     3   1
Mudstone, dark micaceous blocky     6
Seatearth, pale, with sporadic rootlets (horizon of Big Smut Coal) ..    1   3

**Milnrow Sandstone.** The southernmost
outcrop of this sandstone is in a faulted
outlier on the western limb of the Goyt
Trough at Hockerley, north-west of Whaley

Bridge, where it forms a long dip-slope.
Weathered blocks of the sandstone are
present in the fields here, though exposures
are few. A section [0010 8198] near Stone-
heads showed 5 ft of coarse sandstone.

A large outcrop of Milnrow Sandstone,
known in part from exposures and in part
calculated, as much of it is drift covered, lies
just west of the axis of the Goyt Trough
between Bridgemont and Gowhole. The
lowest beds are exposed in a quarry [0055
8250] at Ringstone as 12 ft of medium-
grained micaceous sandstone with coarse
bands and a little current-bedding. The
sandstone has been much worked in Furness
Vale Quarry; a section [0093 8306] at the
south end showed 30 ft of medium-grained
sandstone with some cross-bedding.

To the north of the last-mentioned area the
Milnrow Sandstone is well exposed in the
railway-cutting south of Beard Hall [top of
section, 0100 8416]:

ft

Sandstone, flaggy    ..    ..    ..    6
Sandstone, very massive, hard, felds-
   pathic, somewhat flaggy at base    ..   25
Shale, mainly overgrown; shale visible
   at top and base    ..    ..    ..    6
Sandstone, medium-grained, hard
   feldspathic; rather massive    ..   19
Shale, (see above)    ..    ..    ..    —

Just east of the axis of the Goyt Trough,
several faulted outliers of Milnrow Sandstone are present near Shedyard Piece [021
842], south-west of Piece Farm, near Moor
Lodge and east of Cold Harbour. In all of
these there is a tendency for the sandstone to
weather out into roughly rectangular blocks
several feet in width. The maximum thickness
of sandstone present in these outliers is about
20 ft. A typical section [0231 8451] in the
southernmost outlier shows 6 ft of coarse
massive sandstone.

On the western edge of the district a small
outcrop of Milnrow Sandstone, near Disley,
passes south-westwards into the Stockport
(98) district. A small faulted outcrop is also
present half a mile farther north.

On Ludworth Moor the Milnrow Sandstone is exposed in a quarry [9967 9076];
coarse, well-bedded feldspathic sandstone 14
ft. Two smaller, faulted outcrops are present
between Brown Low and Ludworth Moor,
but these are ill-exposed.

Beds above Milnrow Sandstone. These beds are confined to a narrow drift-covered outcrop east of Gowhole and Beard Hall. The only information regarding them is provided by a section (on Duchy of Lancaster Plan No. B17) of the cross-measures drift (Jowhole or Gowhole Tunnel) running from the River Goyt [0103 8385] to Lady Pit. This commenced in 'rock', the Milnrow Sandstone, but met shale above this at about 140 yd in from the entrance. The shale was faulted out at about 360 yd in. This information, together with the dips given and general structural considerations, suggests a thickness of 150 ft for the shale.

## THE EASTERN AREA

*Gastrioceras subcrenatum* **Marine Band and overlying shales.** These beds occupy a small area on the eastern margin of the district between Gibraltar Rocks and Moscar. The thickness of the group is shown by levels in the Pot Clay (see p. 186) to be 110 ft, ½ mile E.N.E. of Moor Lodge, and 120 ft ½ mile N. of Crawshaw. The group appears to be thinner in the vicinity of Moscar though the precise thickness here is difficult to estimate

Workings in the Pot Clay have enabled the overlying beds to be seen at two places. A large open working [2400 8950] south of Sugworth Road gave the following section:

|  | ft | in |
|---|---|---|
| Shale, silty, with a band of ironstone nodules up to 3 in at base .. | 17 | 0 |
| Shale, pale grey and silty at top, darker below; abundant ironstone nodules    ..    ..    .. | 12 | 0 |
| Siltstone, hard ferruginous    .. |  | 2 |
| Shale, with hard ferruginous silty bands    ..    ..    ..    .. | 4 | 0 |
| Shale, hard ferruginous, silty 3 in to |  | 5 |
| Shale, grey, dark and hard below    .. | 9 | 0 |
| Shale, dark calcareous; *Caneyella multirugata, Dunbarella papyracea, Posidonia gibsoni, P. sp. nov.*, orthocone nautiloid, *Anthracoceras sp., Gastrioceras subcrenatum, G. sp.* [prominent, elongate nodes], *G. sp. nov.*, mollusc spat, *Hindeodella sp.*, platformed conodont, *Elonichthys sp.* [scales]    .. | 1 | 8 |
| Shale (see p. 246)    ..    ..    | — | — |

At Wheatshire Mine [main adit 2508 9004], east of Turner Walls, the section is dark shale with marine fossils 1 ft 6 in, dark shale 9 in, coal 1 to 2 in, seatearth 3 ft 6 in to 6 ft, micaceous flags. A small area of shale, at the base of this group, is present at White Path Moss. It is, however, totally obscured by peat.

**Crawshaw Sandstone.** This sandstone gives rise to an elevated tract of moorland, between Moscar and Gibraltar Rocks, which forms part of a faulted outlier extending eastwards into the Sheffield district (One-inch Sheet 100). The full thickness, about 50 ft, is seen only at Gibraltar Rocks. An old quarry [2408 8947] east of Moor Lodge, shows a typical section: fine-grained micaceous flaggy sandstone with very micaceous partings 8 ft. Near Turner Walls a shale parting is present towards the base, but does not persist laterally.

**Coking or Soft Bed Coal.** This coal is present as an outlier at and east of Gibraltar Rocks and a faulted outlier north and northwest of Ughill. No sections of the coal are present in either outcrop, though it is stated by Green and Russell (1878, p. 98) to be between 15 and 16 inches in thickness "north of Ughill". Its seatearth rests directly on the Crawshaw Sandstone.

**Beds above the Coking Coal.** In the small outcrop within the present district some 20 ft of shale overlie the Coking Coal. Debris from an old tip [2525 9086] at Gibraltar Rocks shows dark shale with fish remains, probably from the roof of the coal, and similar debris is to be seen from an old working 400 yd to the east. The shale is overlain at Gibraltar Rocks by up to 24 ft of massive coarse sandstone. This is probably the sandstone underlying the Clay Coal horizon in the Sheffield district (Eden and others 1957, p. 39).    I.P.S.

## REFERENCES

BINNEY, E. W. 1851. Remarks on a vein of lead found in the Carboniferous strata in Derbyshire, near Whaley Bridge. *Trans. Manchr lit. phil. Soc.*, ser. 2, **9**, 125–9.

CALVER, M. A. 1968. Distribution of Westphalian marine faunas in northern England and adjoining areas. *Proc. Yorks. geol. Soc.*, **37**, 1–72.

COPE, F. W. 1946. Intraformational contorted rocks in the Upper Carboniferous of the southern Pennines. *Q. Jl geol. Soc. Lond.*, **101**, 139–76.

———— 1949. Correlation of the Coal Measures of Macclesfield and the Goyt Trough. *Trans. Instn Min. Engrs*, **108**, 466–83.

EAGAR, R. M. C. 1947. A study of a non-marine lamellibranch succession in the *Anthraconaia lenisulcata* Zone of the Yorkshire Coal Measures. *Phil. Trans. R. Soc.*, Ser. B, **233**, 1–54.

———— 1952. The succession above the Soft Bed and Bassy Mine in the Pennine region. *Lpool Manchr geol. J.*, **1**, 23–56.

———— 1956. Additions to the non-marine fauna of the Lower Coal Measures of the north-midlands coalfields. *Lpool Manchr geol. J.*, **1**, 328–69.

———— and PRESSLEY, E. R. 1954. A section in the Lower Coal Measures near Pott Shrigley, Cheshire. *Mem. Proc. Manchr lit. phil. Soc.*, **95**, 66–8.

EDEN, R. A. 1954. The Coal Measures of the *Anthraconaia lenisulcata* Zone in the East Midlands Coalfield. *Bull. geol. Surv. Gt Br.*, No. 5, 81–106.

———— STEVENSON, I. P. and EDWARDS, W. 1957. Geology of the country around Sheffield. *Mem. geol. Surv. Gt Br.*

GREEN, A. H. and RUSSELL, R. 1878. The geology of the Yorkshire Coalfield. *Mem. geol. Surv. Gt Br.*

HULL, E. and GREEN, A. H. 1866. The geology of the country around Stockport, Macclesfield, Congleton and Leek. *Mem. geol. Surv. Gt Br.*

JONES, R. C. B., TONKS, L. H. and WRIGHT, W. B. 1938. Wigan District. *Mem. geol. Surv. Gt Br.*

MAGRAW, D. 1957. New boreholes into the Lower Coal Measures below the Arley Mine of Lancashire and adjacent areas. *Bull. geol. Surv. Gt Br.*, No. 13, 14–38.

PRICE, D., WRIGHT, W. B., JONES, R. C. B., TONKS, L. H. and WHITEHEAD, T. H. 1963. Geology of the country around Preston. *Mem. geol. Surv. Gt Br.*

SMITH, E. G., RHYS, G. H. and EDEN, R. A. 1967. Geology of the country around Chesterfield, Matlock and Mansfield. *Mem. geol. Surv. Gt Br.*

STUBBLEFIELD, C. J. and TROTTER, F. M. 1957. Divisions of the Coal Measures on Geological Survey maps of England and Wales. *Bull. geol. Surv. Gt Br.*, No. 13, 1–5.

TAYLOR, B. J., PRICE, R. H. and TROTTER, F. M. 1963. Geology of the country around Stockport and Knutsford. *Mem. geol. Surv. Gt Br.*

WRAY, D. A. 1929. The Carboniferous succession in the central Pennine area. *Proc. Yorks. geol. Soc.*, **21**, 228–87.

# Chapter VI

# INTRUSIVE IGNEOUS ROCKS

INTRUSIVE IGNEOUS rocks are present within the Carboniferous Limestone of the Chapel en le Frith district as sills, dykes and vents. The intrusions were first described by Geikie (1897) who noted the presence of the Speedwell Vent at Castleton and certain features of the Peak Forest and Tideswell Dale sills. The more detailed descriptions of Arnold-Bemrose (1894, 1907) are widely quoted below.

## SILLS

Five sills of olivine-dolerite are present in the district. None is completely concordant and this is particularly true of the lower contacts. The horizons at which the sills are intruded vary, but in two cases they are closely associated with the Lower Lava, suggesting stratigraphical control by the latter. The Waterswallows Sill (80 ft thick) is the thickest sill proved with certainty, the 600 ft of dolerite penetrated in the Black Hillock Shaft in Potluck Sill being ascribed to a feeder. Strong columnar jointing was noted in the Waterswallows Sill, near and parallel to the margin, and conspicuous spheroidal weathering is a common feature.

Alteration of the roof rocks of the Peak Forest Sill is well developed at Damside Farm where at least 15½ ft of marmorized limestones occur together with the development of nodule-like masses of secondary silica; higher beds show some dolomitization. Elsewhere, for example around the Potluck Sill, the lack of contact alteration contrasts strongly with that at Peak Forest.

The field relations show the sills to be in general of earlier age than the faulting (except in so far as the postulated penecontemporaneous faulting is concerned) and also that they have preceded mineralization. Whole-rock K-Ar age determination of the Waterswallows Sill (see also p. 296), by Dr. N. J. Snelling, gave an average age of 311 ± 6 m.y. This places (see Francis and Woodland 1964, p. 227) the age of the intrusion at about the Namurian–Westphalian boundary.

*Waterswallows Sill.* The presence of a sill at Waterswallows was first recognized by Arnold-Bemrose (1907, p. 273). The rock was not then well exposed though it has since been extensively worked to a maximum vertical thickness of 80 ft in Waterswallows Quarry (Stevenson and others 1970).

The sill is rather more than half a mile across. The upper surface lies within the outcrop of the Lower Lava and is only gently discordant; the lower surface is irregular in the quarry floor and very strongly discordant on the east side of the quarry [086 752]. The floor of the quarry shows the irregular lower surface of the intrusion, small bosses of limestone being visible in the lowest levels. These were best seen in a cutting used as a sump [0860 7504] (see below). An exposure of limestone [0869 7500] in the east wall of the quarry contains *D. septosa* and is either the Upper or Lower *D. septosa* Band; in either case a fault of general northerly trend is inferred as separating this from the main area of the quarry floor at about the horizon of the base of the Lower Lava. The fault predates the intrusion which showed horizontal columnar structure against it in the north-eastern part of the quarry [085 752] (now obscured). In the lower part of the quarry vertical columnar joints are present above the base of the dolerite where the floor of the intrusion is more nearly horizontal (Plate XXI B).

The highly irregular lower contact of the dolerite in the sump cutting (see above) has been figured by Moseley (1966, p. 284) and the following description is partly based on this author's work. The intrusion rests on the following sequence of rocks (in downward succession): (*d*) Calcitized amygdaloidal lava ('lumb') with calcite- and chlorite-filled vesicles, with much pyrite near contact in euhedral crystals up to 2 mm, 0 to 6 ft; (*c*) Limestone, pale apparently little altered, 0 to 10 ft; (*b*) Limestone breccia, 0 to 2½ ft; (*a*) Tuff, pale green calcareous, 0 to 10 ft. The 'lumb' is manifestly not part of the dolerite intrusion and would appear to be part of the Lower Lava. The tuff probably represents the 'toadstone-clay' usually present beneath lava flows in the Derbyshire area. Marmorization of the wall of the intrusion is very slight.                     I.P.S.

In hand-specimen the Waterswallows dolerite is a medium-grained, dark crystalline rock and shows little variation though rare chlorite-filled vesicles occur in places.

Throughout its exposed thickness at Waterswallows Quarry the sill exhibits an overall petrographical uniformity with only local textural or mineral variations. Arnold-Bemrose (1907, p. 273) noted a variation from coarse-grained olivine-dolerite, in which feldspar predominates with augite, olivine forming small phenocrysts, through an intermediate type, with smaller feldspars and augite grains and coarser olivine, to a fine-grained olivine-basalt, with minute feldspars, small augite grains and relatively coarse olivine crystals. Though all these varieties have been found in the present samples (E 32810; 32821–6; 32842–3)[1], medium-grained, dark grey, olivine-dolerite with evenly distributed phenocrysts (1 to 3 mm length) predominates throughout most of the intrusion (Plate XXII, fig. 6). The specific gravities of nine medium-grained specimens range from 2·71 to 2·92 (mean: 2·84; standard deviation 0·19). Phenocrysts, which form on average about 10 per cent of the predominant dolerite, consist mainly of pseudomorphs after olivine crystals, with optically continuous remnants of olivine set in a chlorite-carbonate-hematite base. Feldspar also in places forms phenocrysts. The groundmass consists principally of a mesh of pale green to pale brown anhedral augite (averaging 0·05 mm) and plagioclase laths (averaging 0·2 by 0·02 mm). A subophitic texture is developed in places, though not apparently detected by Arnold-Bemrose (ibid., p. 273). In places, and particularly in the basaltic variety, the plagioclase needles are fluxioned around olivine crystals. The groundmass plagioclase is predominantly labradorite near $Ab_{40}An_{60}$, though in places it has been chloritized and carbonated. The main accessory mineral is skeletal opaque iron oxide, probably mainly titaniferous magnetite. Small vesicles (averaging 0·4 mm diameter) are sporadically developed, and are filled with zoned chlorites. The vesicles do not, apparently, become more pronounced or relatively abundant towards the top of the sill.

---

[1]Sliced Rock Collection of the Institute of Geological Sciences.

(L 485)

A. Thermally metamorphosed limestones overlying dolerite sill, near Damside Farm, Peak Forest

PLATE XXI

B. Quarry in dolerite sill, Water Swallows

(L 239)

Modal analyses of eight of the freshest samples of medium-grained, olivine-dolerite, taken transversely through the sill gave the following mean abundances, in volume per cent (standard deviations in parentheses): plagioclase 46·5 (2·47); clinopyroxene 30·3 (2·81); olivine, unreplaced 0·8 (0·30); magnetite-ilmenite 6·5 (1·62); chlorite 15·5 (1·91) and carbonate 0·4 (1·36). The relatively high chlorite content indicates the partly altered state of the dolerite.

TABLE 8. *Chemical analyses of intrusive igneous rocks*

|  | (1) | A | (2) | B |
|---|---|---|---|---|
| $SiO_2$ | 48·98 | 48·23 | 45·44 | 49·04 |
| $Al_2O_3$ | 14·06 | 14·56 | 14·74 | 11·86 |
| $Fe_2O_3$ | 3·36 | 2·68 | 1·57 | 4·31 |
| FeO | 7·27 | 8·28 | 5·64 | 7·81 |
| MgO | 8·26 | 8·75 | 4·59 | 5·60 |
| CaO | 8·62 | 9·36 | 11·70 | 11·37 |
| $Na_2O$ | 2·62 | 2·74 | 2·28 | 2·19 |
| $K_2O$ | 0·44 | 0·58 | 0·18 | 0·40 |
| $H_2O > 105°C$ | 1·76 | 2·44 | 3·40 | } 2·24 |
| $H_2O < 105°C$ | 1·75 | 0·69 | 1·39 | |
| $TiO_2$ | 1·74 | 1·19 | 1·54 | 3·75 |
| $P_2O_5$ | 0·19 | 0·16 | 0·19 | 0·26 |
| MnO | 0·15 | 0·16 | 0·06 | 0·24 |
| $CO_2$ | 0·36 | 0·07 | 6·41 | 1·16 |
| Total S calculated as $FeS_2$ | 0·15 | 0·07 | 0·49 | — |
| Allow. for minor constituents | 0·19 | 0·17 | 0·21 | — |
| TOTALS | 99·90 | 100·13 | 99·83 | 100·23 |

1. Waterswallows Sill; olivine-dolerite (E 32810; 32821–6; 32842–3). Waterswallows Quarry [086 750], 1½ miles N.E. of Buxton. Analyst: W. H. Evans. Lab. No. 1981.
2. Buxton Bridge south dyke; chloritized, carbonated tholeiite (E 32811–5); below Tunstead Quarry [1055 7506], ½ mile S.E. of Buxton Bridge. Analysts: P. R. Kiff and G. A. Sergeant. Lab. No. 1977.
A. Olivine-dolerite sill; (S 50480A). Greigston Water Hole, Fifeshire. Analysts: J. M. Nunan, W. H. Evans and G. A. Sergeant (*in* Ann. Rep. Inst. geol. Sci. for 1966, p. 96). Lab. No. 1995.
B. Buxton Bridge north dyke; tholeiite. Analyst: W. A. Deer (*in* Cope 1933, p. 419).

A chemical analysis (by W. H. Evans, Table 8, Analysis 1) was made of a combined serial sample of chips taken at intervals of 2 to 3 ft transversely across the sill, but excluding contacts and altered rock. The analysis compares generally with those of other olivine-dolerites and olivine-basalts, and in particular with that from Greigston Water Hole, Fifeshire (*in* Ann. Rep. Inst. geol. Sci. 1966, p. 96). The normative mineral molecules (in percentages) of the present sill are: quartz 2·20; orthoclase 2·61; albite 22·17; anorthite 25·27; diopside 11·23 (wo 5·83, en 3·87, fs 1·53); hypersthene 23·00 (en 16·62, fs 6·38); magnetite 4·87; ilmenite 0·82; apatite 0·34; calcite 0·82; pyrite 0·20. Though a small amount of residual olivine occurs in the modal analysis, none appears in the norm. The dolerite is apparently saturated with respect to silica, for 2·20 per cent quartz appears in the norm, though the $SiO_2$ content is insufficient to form modal quartz. Comparing the Waterswallows dolerite with the very large number of published analyses of basaltic rocks, the present analysis lies within the tholeiite group and this is confirmed by the plot of total alkalies against $Al_2O_3$ for the silica range 47·51 to 50·00 (Kuno 1960, p. 130).

TABLE 9. *Trace elements in intrusive igneous rocks*

| | Parts per million | | |
|---|---|---|---|
| | (1) | A | (2) |
| Ba | 80 | 90 | 110 |
| Co | 35 | 33 | < 10 |
| Cr | 310 | 210 | 230 |
| Cu | 66 | 23 | 120 |
| Ga | 18 | 16 | 38 |
| Li | 25 | 20 | 78 |
| Ni | 230 | 160 | 110 |
| Rb | — | — | 14 |
| Sr | 220 | 200 | 130 |
| V | 170 | 230 | 220 |
| Zr | 100 | 90 | 90 |
| B | 10 | 13 | 11 |
| F | 320 | 440 | 850 |

All of the above elements, except Li, B and F, were determined spectrographically.

1. Waterswallows Sill; olivine-dolerite (E 32810; 32821–6; 32842–3). (Details as in Table 8). Spectrographic analysis by C. Park.
2. Buxton Bridge south dyke; chloritized and carbonated tholeiite. (E 32811–5). (Details as in Table 8). Spectrographic analysis by C. Park.
A. Olivine-dolerite sill; (S 50480A). Greigston Water Hole, Fifeshire. Spectrographic analysis by C. Park. (*in* Ann. Rep. Inst. geol. Sci., for 1966, p. 96).

The minor elements of the sill, mainly determined by optical spectrography (analyst: C. Park), are compared in Table 9 (Analysis 1) with those determined in the olivine-dolerite sill from Greigston Water Hole, Fifeshire (Ann. Rep. Inst. geol. Sci. 1966, p. 96), and those in the Buxton Bridge dyke, discussed further below. There is overall close agreement between the Greigston and present analyses, except for Ni, Cr and Cu. These elements are enriched in the early-crystallizing magmatic minerals (Lundergårdh 1949, p. 7) and are probably localized in the olivine and ilmenite-magnetite, though the differences between the two sills are difficult to explain.

In addition, the following analyses for K-Ar age determination (by Dr. N. J. Snelling) were made on two samples (E 32842, 32843 respectively) of the sill: K = 0·491 per cent; 0·512 per cent; $^{40}Ar$ = 0·0107 p.p.m.; 0·0123 p.p.m. The potassium values are a little lower than that (Table 8, col. 1) for the bulk sample (see also p. 293). R.K.H.

*Peak Forest Sill.* This sill, first described by Geikie (1897, p. 17) is present in the axial area of the Peak Forest Anticline and is the largest intrusion in the district; its contact phenomena were noted by Barnes (1902). The lateral extent of exposures is over a mile and a quarter in an east–west direction and debris from tips [098 790] on the western side near Bee Low, shows an extension of the sill below a roof of $D_1$ limestone here. Only the uppermost part of the sill is exposed and its thickness is uncertain. The upper contact is slightly transgressive.

The sill is exposed in two main areas: an arcuate area north, east and south of Peak Forest in which the more important exposures occur; and a western area near Backlane Farm.

At Peak Forest, $3\frac{1}{2}$ ft of dolerite were proved in a temporary excavation by the main road [1123 7934]. To the north and north-east evidence for the outcrop is confined to debris at a number of points and to the presence of impermeable rock at Old Dam where a well [1155 7960] proved 'toadstone' to a depth of 8 ft. The best exposures lie to the south-east of Damside Farm, the contact with the overlying limestones being exposed [1161 7880] (see also pp. 41–2) near an old dam. Here at least $15\frac{1}{2}$ ft of the overlying limestones are marmorized to a greater or lesser extent and also show the development of nodular masses of secondary silica (Plate XXI A). The highest beds in the section, 4 ft thick and $34\frac{1}{2}$ ft above the dolerite are strongly dolomitized; the dolomitization is considered to be an alteration feature as the $S_2$ limestones in the vicinity are of the pale crinoidal (Peak Forest) facies. Dolerite farther below the upper contact is exposed south of Damside Farm [1154 7872]. Exposures of the uppermost part of the sill and the contact with the overlying limestone are also seen farther to the south-west [111 785]. Much of the southern edge of the outcrop is faulted and elsewhere the contact shows minor faulting including a displacement along the line of Oxlow Rake.

The western outcrop forms a broad hollow west of Backlane Farm. Exposures are confined to a series of sink-holes near the edge of the outcrop, one [1069 7893] north-west of Backlane Farm showing 6 ft of spheroidal-weathering dolerite. Nearby [1073 7895] the dolerite is faulted against $S_2$ limestones and the fault is mineralized with calcite and galena in a clay matrix. The western margin of the outcrop also shows mineralization with a line of old workings extending across the upper contact of the sill; it is, however, uncertain to what extent the vein was worked within the sill. I.P.S.

A specimen (E 35687) from an exposure [1110 7854] 200 yd W.S.W. of New-houses Farm, is a dark grey olivine-dolerite with specific gravity 2·86. Common pseudomorphs after olivine, retaining in places optically-continuous olivine remnants, form individual phenocrysts and glomeroporphyritic aggregates up to 4 mm length. Unaltered olivine is near forsterite with 2V near 90°. These crystals are scattered through a subophitic groundmass of clinopyroxene plates and plagioclase laths (averaging 0·9 by 0·2 mm across), the last-named consisting of labradorite near $Ab_{37}An_{63}$ (Plate XXII, fig. 4). The clinopyroxene (augite) is pale pink-brown, allotriomorphic and coarser plates poikilitically enclose labradorite laths. There are scattered opaque laths and anhedra of magnetite and ilmenite. The interstitial material is largely composed of green chlorite. A modal analysis gave (in per cent): olivine 2·0; chloritized olivine 16·5; augite 11·4; labradorite 59·4; magnetite-ilmenite 3·1; chlorite 7·6.

A second specimen (E 35686) from a temporary section [1123 7934] at Peak Forest, is by contrast very altered, though recognizably an olivine-dolerite. Olivine phenocrysts have been completely altered to chlorite aggregates. Groundmass plagioclase is albitized and charged with chlorite, while there is much interstitial chlorite. Clinopyroxene, however, is colourless, fresh and little altered. Opaque ores have been extensively leucoxenized.

Both the Peak Forest and Potluck sill specimens (see below) are closely similar petrographically though the enrichment in olivine (both replaced and unaltered) contrasts with the specimens of the Waterswallows Sill described above.

R.K.H.

*Potluck Sill.* This sill, described by Arnold-Bemrose (1907, pp. 272–3), occupies an extensive area on Tideswell Moor. Exposures are lacking for the most part and the main evidence is from abundant debris on the outcrop. There is, however, a small exposure [1364 7833] near Pittle Mere. Black Hillock Shaft [1410 7822] proved 600 ft of "toadstone" (Green and others 1887, p. 134), the base not being reached, and the shaft at this locality probably followed a feeder to the sill.

The part of the Potluck Sill by Black Hillock shows an alignment parallel to White Rake. Workings have extended some distance into the sill but the vein appears to fail a short distance west of Black Hillock Shaft.          I.P.S.

A specimen (E 35685) from the exposure near Pittle Mere is of a coarse (gabbroic) facies of ophitic olivine-dolerite, dark grey (mottled green) with macroscopically visible phenocrysts 2 to 3 mm across and groundmass feldspars (up to 1 mm length). The specific gravity is 2·85. In section, the rock is fresh, with feldspar phenocrysts charged with myrmekitic inclusions of clinopyroxene and magnetite, up to 5 mm length, and partly replaced olivine crystals up to 2 mm length (Plate XXII, fig. 3). Unreplaced olivine forming the phenocryst cores is colourless, with 2V near 90° indicating a composition near the Mg end-member, forsterite. This is bordered by chlorites and veined by opaque oxides. Feldspar laths form a groundmass mesh, average 0·6 by 0·1 mm, and are of labradorite composition, near $Ab_{37}An_{63}$. The laths are partly in ophitic texture with clinopyroxene, or held poikilitically in the coarser clinopyroxene plates, the latter attaining 1·5 mm across, and with refractive index $(\beta) = 1·694$ $\pm$ ·003, 2E (+) = $92\frac{1}{2}°$, is augite. Accessory orthopyroxene is entirely altered to chlorite. Irregular granules of opaque accessories consist of ilmenite and magnetite. There is some interstitial yellow-brown chlorite, dark and turbid in places, which may well be primary. A modal analysis gave (in per cent): olivine 1·1; replaced olivine 6·8; augite 18·3; labradorite 67·8; opaques 1·5 and chlorite (groundmass) 4·5.          R.K.H.

*Mount Pleasant Sill.* The full extent of this sill was not recognized by Arnold-Bemrose (ibid., p. 277) though the westernmost exposure was noted. The presence of the larger outcrop recorded here is based on debris ploughed up at the western end and on weathered material elsewhere [1262 7865; 1300 7875]. In addition, debris of compact dark dolerite is present in a sink-hole at the eastern end of the outcrop [1315 7883]. The sill is markedly cross-cutting and shows a striking alignment parallel to Shuttle Rake and to two east–west faults south of Hernstone Lane Head. The unusual degree of cross-cutting and the alignment with Shuttle Rake strongly suggest control by the latter. Shuttle Rake may thus be inferred to be an early (pre-Namurian) feature, subsequently mineralized.

*Tideswell Dale Sill.* A small area of this sill falls within the present district. The sill is capped by the Lower Lava and the northern margin is faulted off.

## DYKES

Dykes are known with certainty only from Great Rocks Dale where two are present. The dyke-rock is texturally distinct from the sills and has been described as a tholeiite by Cope (1933, p. 418) and compared to both Tertiary and Permo-Carboniferous tholeiites of Scotland and Northumberland. As the Upper and Lower lavas of Miller's Dale are ascribed to fissure-eruptions (p. 20) it seems most reasonable to consider that the dykes acted as feeders for these.

The amount of marmorization associated with the dykes is variable, though normally up to 6 in are affected near the walls. Near the south dyke, however, 10 ft of limestone beneath the Lower Lava are strongly marmorized in the vicinity of the dyke. This is attributed to metasomatic alteration of the limestone by the dyke beneath the impermeable lava cover.

A thin and linear outcrop of igneous rock is present on Tideswell Moor between the Mount Pleasant and Potluck sills. Debris from this has been proved by auger. The outcrop is interpreted as a dyke which acted as a feeder for either Upper or Lower lava.

*Buxton Bridge.* Here two tholeiite dykes are intruded into the Chee Tor Rock. The more northerly, trending E.41°N. and now poorly exposed, is best seen in a quarry approach road [0974 7565] north-east of Buxton Bridge. This locality was described by Cope (1933, pp. 414–22). The dyke is about 12 ft wide, is weathered near the contacts and shows spheroidal weathering in the central part. Near the edges xenoliths of fine sandstone and dolomite were noted. The wall rock shows marmorization up to 6 in from the contact.

The more southerly dyke is best seen by the railway [1005 7506] nearly $\frac{1}{2}$ mile S.E. of Buxton Bridge. The dyke trends E. 23°N. and is traceable for some 400 yd west of the railway. A small offshoot is present near the eastern end. The width of the main dyke is about 15 ft and much of the intrusion is deeply weathered. Alteration of the wall-rock is confined to the nearest few inches to the intrusion.                                                                I.P.S.

The southerly dyke is a markedly amygdaloidal, dark grey, tholeiitic dolerite. Amygdales are mainly spheroidal (2 to 3 mm average diameter) and contain black masses of chlorite, or white carbonates and chalcedony, sporadically rimmed with black chlorite. Towards the contacts, phenocrysts of white feldspar laths become abundant, attain 6 mm length, and are commonly fluxioned, giving rise to a coarse, near-gabbroic texture. Towards the centre of the dyke, the rock is finer-grained, dense and aphyric, though still containing isolated amygdales and streaks of amygdaloidal basalt. Specific gravity measurements of the freshest samples collected across the width of the dyke, range from 2·697 to 2·705 (mean 2·700; standard deviation 0·004). These low values, compared with the northerly dyke (Cope 1933, p. 415) are due to the altered state of the dyke. Thin sections (E 32811–5) show mainly fresh, fluxional or randomly arranged feldspar laths in glomeroporphyritic aggregates and individual phenocrysts exhibiting shadow zoning, both normal and reversed (Plate XXII, fig. 5). Cores of crystals are mainly of labradorite composition ($Ab_{41}An_{59}$) with refractive index ($\beta$) = 1·563. Groundmass feldspar laths are mainly fresh, randomly orientated or fluxioned locally, and average 0·1 by 0·3 mm. They are also of similar labradorite composition. Primary ferromagnesian minerals have been replaced in all the sections examined, and their original nature can only be inferred by pseudomorphs. These include granular, eight-sided clinopyroxene crystals (averaging 0·3 mm) and prismatic crystals after likely orthopyroxene, in subophitic arrangement with the labradorite laths. The pseudomorphs consist chiefly of chlorite-group minerals and carbonates. Accessory ilmenite, forming randomly arranged slender laths, is mainly restricted to the groundmass. Much of the last-named, however, consists of chlorites, carbonate and leucoxene. Around amygdales and along fractures, dark grey to brown, microcrystalline or devitrified glassy material charged with feldspar microlites, opaque ores, leucoxene and chlorite, is conspicuous. The interstitial chlorites are pleochroic principally in shades of pale olive green to bright green, and in places are markedly replaced by microglobular chalcedony, the "bacteria-like" bodies of Cope (ibid., p. 417), which in turn is replaced by carbonate (Plate XXII, fig. 5). Amygdales contain chlorites, chalcedony, quartz and carbonates, chlorite generally rimming radial-fibrous silica, or coarse, undulose quartz aggregates. Elsewhere amygdales contain zoned chlorites (green to brown) with or without quartz and carbonate. A complete chemical analysis (by P. R. Kiff and G. A. Sergeant; Table 8, Anal. 2) was made of a sample of combined rock specimens, taken transversely across the dyke, excluding the highly altered contacts. The analysis compares closely with that (by W. A. Deer) of the north dyke (Table 8, Anal. B). In particular the total alkalies are comparable, as are magnesia and lime. The main differences are the lower silica and higher carbon dioxide and alumina of the south dyke, attributed to its higher degree of chloritization and carbonation. The plots of $Al_2O_3$ against ($Na_2O + K_2O$) for the respective silica contents of both rocks, lie well within the tholeiite field as delineated by Kuno (1960, p. 130). Cope (ibid., pp. 417–9) has drawn attention to the similarity between the north dyke and tholeiites of the Brunton type, though the more altered state of the south dyke precludes direct comparisons.

Spectrographic analyses (by C. Park) of the same sample of the south dyke, are listed in Table 9 (Analysis 2). The relative abundances of the rarer metals are generally of similar order to those of the Waterswallows Sill (Anal. 1) and of the olivine-dolerite from Fifeshire (Anal. A), except for increased Cu, Ga and particularly F in the present dyke. The reason for these increases is not apparent, since no discrete minerals containing these elements were detected.

R.K.H.

## VENTS

Three vents are known in the district: all appear to lie in areas which have been subject to intra-Viséan movement, two lying on the axis of a small monocline at Monk's Dale (considered to be a penecontemporaneous feature) and the third, the Speedwell Vent, lying within the marginal province at Castleton (see p. 36). In the latter area two other vents are inferred to be present, one of these from boreholes only; as no neck deposits have been definitely proved they are described under the appropriate stratigraphical heading (see pp. 61–2).

The vent rock is a fine agglomerate or lapilli-tuff which is, in the small exposures noted, much altered. The vent activity is clearly referable to the episode of formation of the extrusive volcanic rocks and the vent intrusions are only considered here on morphological grounds.

The Monk's Dale Vent [130 753] was originally described by Arnold-Bemrose (1907, p. 251). It forms a semi-elliptical outcrop showing a maximum diameter of 240 yd. The country rock is the Chee Tor Rock but its relation with the vent-fill cannot be observed. Exposures are poor though fine agglomerate was formerly (1959) seen in the roadway [1300 7525]. The rock was olive green, calcareous and contained limestone lapilli up to 1 in and smaller cognate lapilli.

A second vent, of somewhat smaller proportions was discovered during the course of the present work some ¼ mile N.W. of the Monk's Dale Vent. This vent is exposed in several places and again shows both cognate and accidental lapilli up to ¾ in [e.g. 1263 7558].

Both vents lie along a small but sharp monoclinal fold and it is suggested that this feature was penecontemporaneous and is associated with the vulcanism. The relation of vents and folds is illustrated in Fig. 14.                    I.P.S.

Specimens (E 35688, E 35688 A) from the northern vent [1263 7558] are calcareous tuffs with angular lapilli, scattered in a fine-grained, foraminiferal and calcareous matrix (Plate XXII, fig. 1). The tuffaceous particles are poorly sorted, unstratified, angular, and comprise golden-yellow, chloritized pumice and microlitic lava. The pumice contains ellipsoidal to spheroidal vesicles (up to 0·3 mm mean diameter), filled with colourless, radial-fibrous chlorite and separated by yellow chlorite exhibiting undulose extinction and anomalous polarization colours. Leucoxene dust is particularly concentrated around tuffaceous particles, vesicles and microlites. Lava particles are vesicular and contain swarms of feldspar microlites fluxioned around phenocrysts, all replaced by calcite. There are abundant irregular shards of yellow chlorite and finely comminuted pumice. The calcareous matrix comprises organic remains (foraminifera, shell fragments, crinoid stems) with foraminiferal limestone and calcite, all set in a dusty microcrystalline calcite matrix. Much of the dust may be leucoxenic.                    R.K.H.

The Speedwell Vent, first noted by Geikie (1897, p. 16) is some 200 yd in diameter and intrusive into apron-reef and Beach Beds. The best exposure [1429 8254] is by the path on the western side. The rock consists of limestone lapilli up to 1 in diameter and smaller cognate lapilli in a dark green matrix.

Despite the acceptance by Arnold-Bemrose (1907, pp. 250–1), Shirley and Horsfield (1940, p. 294) and other workers of the intrusive nature of the Speedwell Vent there has existed some possibility that it might be in fact a tuff-mound similar to that described by Shirley and Horsfield (1940, p. 283) in Earle's Quarry. However, recent geophysical work (Wilkinson 1967, p. 49) indicates that the junctions are steeply inclined and confirms the original interpretation.
                    I.P.S.

Rocks collected from the Speedwell Vent [1432 8245], approximately 800 yd S.W. of Castleton, include black chert, silicified limestone, agglomerate and olive-green lapilli-tuff. A section of the latter (E 33468) consists of angular lapilli of chloritized pumiceous lava (carbonated fluxioned feldspar laths 0·9 by 0·2 mm; vesicles filled with radial-fibrous green to yellow-brown chlorite in an opaque base of green chlorite and pyrite); chloritized, carbonated crystal fragments and finely comminuted lava, all cemented by coarsely crystalline, interlocking calcite, showing radial extinction. Lava particles are commonly surrounded by microcrystalline dusty carbonate which might, perhaps, represent incipient marmorization.

EXPLANATION OF PLATE XXII

PHOTOMICROGRAPHS OF IGNEOUS ROCKS
(all taken under uncrossed polars)

1. Calcareous tuff. Northern vent [1263 7558], Monks Dale. E 35688. Angular particles of yellow, chloritized pumiceous lava (bottom right; middle left, of field) are scattered in a calcareous matrix.                               ×33

2. Tuffaceous calcite-siltstone. Litton Tuff. Wardlow Mires No. 1 Borehole [1850 7553] at 450¼ ft depth. E 35758. Angular pyroclasts mainly of pumice (top centre and centre of field) are scattered in a dark, turbid limestone matrix.        ×134

3. Olivine-dolerite. Potluck Sill. Exposure [1364 7833] near Pittle Mere. E 35685. Coarse plates of augite (upper left) poikilitically enclose labradorite laths. A heavily chloritized olivine crystal is seen in the bottom of the field. The ground-mass is a mesh of labradorite laths and granular magnetite.                  ×30

4. Olivine-dolerite. Peak Forest Sill. Exposure [1110 7854] 200 yd west-south-west of Newhouses Farm. E 35687. Clusters of partly chloritized, partly unreplaced, olivine crystals (upper right of field) are scattered with anhedral titanaugite (bottom right) in a mesh of labradorite laths and interstitial chlorite, ilmenite and magnetite.                                                        ×33

5. Amygdaloidal, chloritized, carbonated tholeiite. Buxton Bridge south dyke [1005 7506]. E 32814. This consists of a felt of carbonated feldspar laths and intergranular chloritized mafic minerals, with scattered amygdales (bottom left) containing carbonates rimmed with chlorite, and which are generally surrounded by a tholeiitic selvedge.                                          ×32

6. Olivine-dolerite. Waterswallows Sill. Waterswallows Quarry [086 750]. E 32823. Chloritized olivine phenocrysts (bottom right of field) are markedly veined by hematite and scattered in a fine mesh of labradorite laths, which characterize a locally fine-grained facies of the intrusion.                            ×33

PLATE XXII

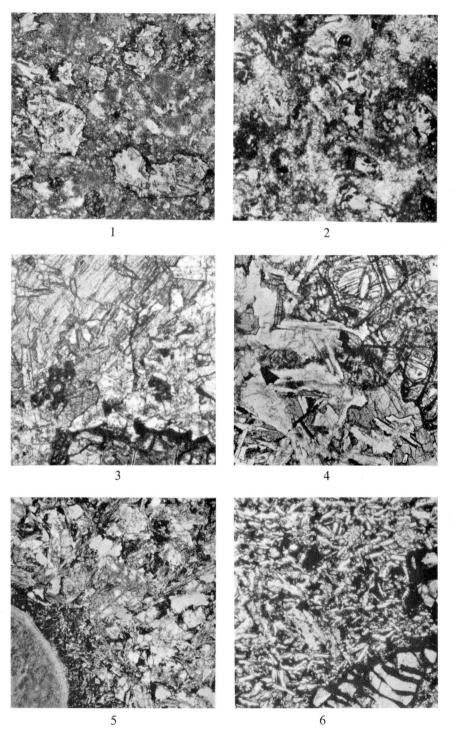

1

2

3

4

5

6

PHOTOMICROGRAPHS OF IGNEOUS ROCKS

## PETROGENESIS OF SILLS AND DYKES

On the petrographical and chemical data presented, both the Waterswallows Sill and Buxton Bridge dykes are related to the saturated tholeiite group of basic igneous rocks. In this respect the definition of 'tholeiite' has been reviewed (Yoder and Tilley 1962, p. 353) to comprise basic rocks with essential augite or subcalcic augite, plagioclase ($Ab_{50}$) and iron oxides. Olivine may be absent as a subordinate constituent (as in the case of the Waterswallows, Peak Forest and Potluck sills) and an interstitial vitreous acid residuum (intersertal texture) or a quartzo-feldspathic intergrowth may be present. Intersertal texture has previously been regarded by many petrologists as the most critical feature distinguishing tholeiites. The augite may be zoned to subcalcic augite with pigeonite, or hypersthene, or both. The analyses indicate that both the sill and dykes have relatively small alkali contents, a further feature of tholeiites in contrast to the alkali-basalts. As Yoder and Tilley point out, however, (ibid., p. 353) there is a "continuum between all basalt types". Though, therefore, there are differences in mineralogy and texture between the sills and dykes described, these appear to be insignificant in relation to their geochemistry. Mode of intrusion may well have governed certain of these differences. For example, intersertal texture in the dykes described above, is closely related to late-stage vesicle development, and may be related to the thermal conductivity of the wall-rocks during intrusion (McBirney 1963, p. 466). In conclusion, both dykes and sills may have been comagmatic with the basic extrusives, all stemming from a subcrustal reservoir of alkali-poor tholeiitic magma.                                R.K.H.

## REFERENCES

ARNOLD-BEMROSE, H. H. 1894. On the microscopical structure of the Carboniferous dolerites and tuffs of Derbyshire. *Q. Jl geol. Soc. Lond.*, **50**, 603–44.

―――― 1907. The toadstones of Derbyshire; their field-relations and petrography. *Q. Jl geol. Soc. Lond.*, **63**, 241–81.

BARNES, J. 1902. On a metamorphosed limestone at Peak Forest. *Trans. Manchr geol. Soc.*, **27**, 317–21.

COPE, F. W. 1933. A tholeiite dyke near Buxton, Derbyshire. *Geol. Mag.*, **70**, 414–22.

FITCH, F. J. and MILLER, J. A. 1964. The age of the paroxysmal Variscan orogeny in England. *in* "The Phanerozoic Time-Scale". *Q. Jl geol. Soc. Lond.*, **120 S**, 159–75.

FRANCIS, E. H. and WOODLAND, A. W. 1964. The Carboniferous period. *in* "The Phanerozoic Time-Scale". *Q. Jl geol. Soc. Lond.*, **120 S**, 221–32.

GEIKIE, A. 1897. *The ancient volcanoes of Great Britain.* Vol. 2. London.

GREEN, A. H., FOSTER, C. LE NEVE and DAKYNS, J. R. 1887. The geology of the Carboniferous Limestone, Yoredale Rocks, and Millstone Grit of North Derbyshire. 2nd edit. with additions by A. H. Green and A. Strahan. *Mem. geol. Surv. Gt Br.*

INSTITUTE OF GEOLOGICAL SCIENCES. 1967. Annual Report for 1966.

KUNO, H. 1960. High-alumina basalt. *J. Petrology*, **1**, 121–45.

LUNDEGÅRDH, P. H. 1949. Aspects to the geochemistry of chromium, cobalt, nickel and zinc. *Årsb. Sver. geol. Unders.*, **43**, Ser. C., 3–56.

MCBIRNEY, A. R. 1963. Factors governing the nature of submarine volcanism. *Bull. volcan.*, **26**, 455–69.

MOSELEY F. 1966. The volcanic vents and pocket deposits of Derbyshire. *Mercian Geol.* **1**, (3), 283–5.

SHIRLEY, J. and HORSFIELD, E. L. 1940. The Carboniferous Limestone of the Castleton–Bradwell area, North Derbyshire. *Q. Jl geol. Soc. Lond.*, **96**, 271–99.

SMITH, E. G., RHYS, G. H. and EDEN, R. A. 1967. Geology of the country around Chesterfield, Matlock and Mansfield. *Mem. geol. Surv. Gt Br.*

STEVENSON, I. P., HARRISON, R. K. and SNELLING, N. J. 1970. Potassium-argon age determination of the Waterswallows Sill, Buxton, Derbyshire. *Proc. Yorks. geol. Soc.*, **37**, 445–7.

WILKINSON, P. 1967. *in Geological Excursions in the Sheffield Region and the Peak District National Park.* Ed. R. Neves and C. Downie. University of Sheffield.

YODER, H. S. and TILLEY, C. E. 1962. Origin of basalt magmas: an experimental study of natural and synthetic rock systems. *J. Petrology.*, **3**, 342–532.

# Chapter VII

# EPIGENETIC MINERAL DEPOSITS

## GENERAL

MINERALIZATION in the Chapel en le Frith district is with few exceptions confined to that part of the Carboniferous Limestone outcrop which lies north-east of a line drawn from Sparrowpit to Litton. This area forms the northern part of a more extensive mineralized tract extending southwards to Wirksworth along the eastern side of the limestone outcrop.

No attempt is made here to summarize the extensive literature dealing with mineral deposits and mining in the area. Some of the more important sources of information, however, are as follows. Farey (1811) listed many of the mineral occurrences and mines, and distinguished rake veins, pipe-veins and flats. Green and others (1887, pp. 128–38) described many of the veins in the area and noted the principal occurrences of lead and zinc ores and gangue minerals (ibid., pp. 121–7, 161–3). Wedd and Drabble (1908) noted the zonary arrangement of gangue minerals (see below). The account by Green and others was largely drawn on by Carruthers and Strahan (1923, pp. 41–59) in describing the geology of the lead and zinc ores. The occurrence of baryte was described by Wilson and others (1922, pp. 64–6), and more recently, the fluorite deposits have been described by Dunham (1952, pp. 82–9, 92–4 and 109–10).

The present account deals only with the geology of the mineral deposits and mining details have in general been omitted. The deposits occur in veins, pipe-veins and flats and the structure and occurrence of these are discussed below and illustrated in Fig. 21.

**Ore minerals.** The main sulphide ore is galena. This is usually quite coarsely crystalline, though fine-grained or sheafy varieties occur in places and are known as 'steel-ore'. Sphalerite is absent from most of the veins but occurs in places near the margin of the limestone outcrop, for example Earl Rake, Bradwell and Odin Vein, Castleton. Pyrite is present in places in the fluorite zone (see below). Ford and Sarjeant (1964, p. 129) record pyrrhotite and marcasite from Ladywash Mine (Old Edge Vein 380 Level West). Chalcopyrite is noted by the same authors (ibid., pp. 129–30) from New Engine Mine, Eyam, and from Pin Dale; it has also been observed during the present work in tipped material from the Ladywash Mine (Hucklow Edge Vein).

Secondary ores are in most cases little developed. The Kittle End Vein, Castleton, is stated by Carruthers and Strahan (1923, p. 50) to have yielded 'calamine' (smithsonite) in commercial quantities. This mineral has also been noted by Green and others (1887, p. 132) in Shuttle Rake and by Ford and Sarjeant (1964, p. 133) from Old Edge Vein. Ford and Sarjeant (1964, p. 130) record greenockite from Bradwell Moor and hemimorphite (ibid., p. 138) from Black Hole Mine and from the Old Edge Vein at Eyam. In addition they have

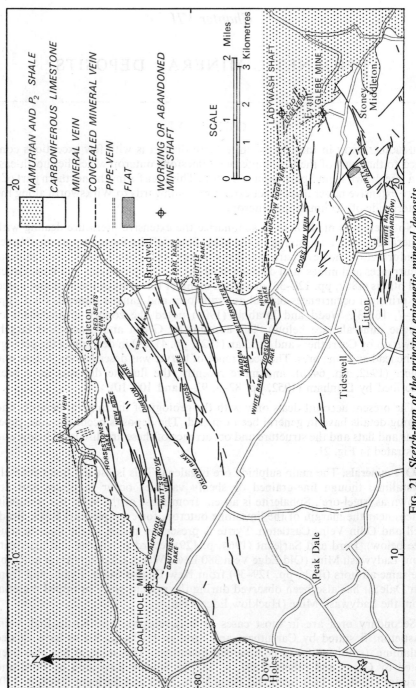

FIG. 21. Sketch-map of the principal epigenetic mineral deposits

noted the occurrence of cerussite at White Rake (Wardlow) (ibid., p. 132); this mineral has also been found *in situ* during the present work in Old Edge Vein and White Rake. Anglesite is recorded by Ford and Sarjeant (1964, p. 134) from the last vein also. Pyromorphite occurs in White Rake (see p. 314) and wulfenite has been recorded by Mawe (1802, p. 107) from an unspecified locality, probably near Castleton. Malachite and azurite are present in the Old Edge Vein but have not been observed *in situ* during the present work. Hydrozincite has been recorded by Braithwaite and Ryback (1963) from Nether Water Mine. Limonite is of moderately common occurrence as an alteration product of both pyrite and marcasite; these sulphides also alter to goethite and the polymorph of the latter, lepidocrocite, has been confirmed from Oxlow Rake.

**Gangue Minerals.** The main gangue minerals are fluorite, baryte and calcite. Three different varieties of fluorite have been recognized by Mueller (1954b, pp. 528-30) (*a*) "pyritic fluorite" with pyrite and chalcopyrite inclusions in clear fluorite (*b*) turbid fluorite, the turbidity being due to numerous gas bubbles, (*c*) purple fluorite. The "pyritic fluorite" was considered by Mueller to have been formed at the highest temperature and the purple fluorite at the lowest. These conclusions are suggested by the paragenesis and form of deposits of the different varieties, especially the association of the "pyritic fluorite" with the greatest development of sulphide minerals and the maximum development of the purple varieties outside the vein, or as thin films and veinlets in areas in the calcite zone (see below). Detailed work on trace elements in fluorites has been summarized by Dunham (1952, pp. 9–12); yttrium is the commonest of these and its presence was confirmed spectrographically in two specimens from the present district. Derbyshire fluorites are little fluorescent, and Ford (1955, p. 46) has observed that the presence of rare earths affects this phenomenon, but has little connection with the colour. The coloration of the purple varieties has been shown by Blount and Sequeira (1919, p. 705) to be due to hydrocarbons and this has been confirmed and elaborated by Mueller (1954b, pp. 530–5).

The colour-zoned radiating ornamental varieties known as 'Blue John' have been described by Ford (1955, pp. 37–49). 'Blue John' occurs in pipe-veins and cavities and not in the veins proper; it is attributed by Ford to a phase separate from the main mineralization and probably earlier than it.

Baryte occurs both in a coarsely crystalline form and in a fine-grained massive form. Schnellmann and Willson (1947, p. 10) considered that the latter tended to predominate nearest the calcite zone. Veining in Namurian sandstones of 'crowstone' type in the Todd Brook Anticline shows unusually large tabular crystals of baryte.

Calcite is the most widely developed gangue mineral and occurs both in a coarsely crystalline form and as columnar masses. The latter are especially characteristic of some of the larger and wider veins such as Dirtlow Rake and Moss Rake and probably result from the hydrothermal redistribution of wall-rock from lower levels. In places, however, veins of coarsely crystalline rhombic calcite, free from other minerals, occur.

Siderite in a finely cellular form has been noted as a minor constituent of some veins in the Bradwell area.

The presence of glauconite in small quantities in scrins in Cave Dale and near the Odin Vein has been described by Young and others (1968).

**Hydrocarbons.** Hydrocarbons occur in small quantity in many of the veins and in one instance (see below) as a larger deposit. A detailed study of these substances has been made by Mueller (1954a) in the Derbyshire Orefield. The hydrocarbons are interpreted as being derived from the action of hydrothermal solutions on organic sediments.

The only hydrocarbon occurring in any quantity in the district is that present just beneath the original position of the shale cover at the top of Windy Knoll Quarry, Castleton. The deposit is flat-like in form and occurs as an impregnation of limestone with viscous hydrocarbon in cavities. Mueller (ibid., pp. 290–2) describes the Windy Knoll flat as an olefinite, though he also recognizes five other types of hydrocarbon in pockets and veins at Windy Knoll. Mueller has also noted (1960, pp. 123–8) a relationship between the trace-element content of uranium in the hydrocarbons and the chemical composition of the latter.

**Mineral Zones.** Wedd and Drabble (1908, pp. 501–35) first noted the zonary arrangement of the gangue minerals along certain veins in the area. Dunham (1952, p. 83) refers to a "fluorite zone" at the eastern side of the mineralized area and in which the maximum development of fluorite occurs. To the west it is possible to recognize a baryte zone with conspicuous baryte and little fluorite and farther west still a calcite zone where both fluorite and baryte are, in general, lacking. Calcite occurs in all three zones.

The zones of gangue minerals have been more accurately defined by Mueller (1954b, p. 527) according to the vein content (in percentages) as follows:

(a) fluorite zone: fluorite 10–50: baryte 2–10: calcite 40–85

(b) baryte zone: fluorite 1–10: baryte 10–40: calcite 55–85

(c) calcite zone: fluorite 0–1: baryte 0–10: calcite 80–99

The zones, however, should only be taken as a general illustration of the tendencies of gangue mineral development, there being many anomalies and reversals of the normal sequence. It should be noted that recent working along White Rake and Cross Low Vein has shown that fluorite extends westwards beyond the original limit of Mueller.

The distribution of sulphides also follows the zonal pattern, though galena itself is widespread throughout the mineralized area. Sphalerite is most common in the fluorite zone and rare elsewhere. Pyrite and chalcopyrite are stated to be restricted to the fluorite zone.

The zonal arrangement reflects the temperature of formation of the minerals, the fluorite zone indicating the highest temperature and the calcite zone the lowest. A similar thermal zoning has been described by Dunham (1934, pp. 694–700) in the North Pennine Orefield.

**Form of ore-bodies.** The mineral deposits of the district (Fig. 21) may be classified as follows:

(a) Veins. (b) Pipe-veins. (c) Flats. (d) Cavity fillings and impregnations.

(a) *Veins* (or rakes) are the most common, the smaller forms being known as scrins. Two main trends of veins are apparent in the present area, the major veins and associated minor veins trending from E.–W. to about E.20°N.–W.20°S. (swinging to E.30°N.–W.30°S. in the vicinity of Ladywash Mine) while the minor cross-veins, developed mainly in the area south of Eyam and Foolow, trend N.W.–S.E. Only one of the latter group of veins, the Old Edge Vein, reaches the proportions of the east–west veins. The major veins are, in general, normal faults while the cross-veins usually fail to show any vertical displacement. All veins tend to show horizontal slickensiding on the walls suggesting that there has been at least some lateral movement; this includes the major veins with demonstrable vertical throw. Dunham (*in* discussion Schnellmann and Willson 1947, p. 22) has suggested that the extensive calcite filling of the larger veins (such as Moss Rake) may be due to a metasomatic replacement of the limestone. The phenomenon of explosive slickensides on vein walls at Odin Mine and at Highcliffe Mine was noted by Green and others (1887, pp. 137–8) and attributed to the presence of stressed and metastable baryte. In the northern part of the district, the main veins run roughly parallel to the reef-belt, suggesting a possible deep-seated structural connection between the two.

In most cases the veins are vertical, or nearly so. Old Edge Vein and Hucklow Edge Vein have variable, and sometimes reversed, hades in the limestone, but overall verticality is shown in many places where the underground and surface positions are compared.

(*b*) *Pipe-veins*. The presence of pipe-veins (or 'pipes') was first noted by Farey (1811, p. 243), who stated that they differed from flats in having a connection to an ore-channel or leader. The definition was elaborated by Green and others (1887, p. 122) as "a pipe, inclined as the beds are inclined" and probably formed by the original enlargement of the intersection of a joint and a bedding-plane by solution. Varvill (1937, p. 490) has described the shapes of such strata-controlled deposits, and Traill (1939, pp. 867–79) has given a detailed description of 'pipes' in Millclose Mine. The latter author compared the 'pipe-veins' (and also the flats) to the 'manto' or blanket deposits of the Mexican Orefield described by Prescott (1926) and Fletcher (1929).

In the present district there seems no need to postulate the presence of a continuous connection between pipe-veins and feeders, as the former may well themselves have acted as channels for ore-deposition, particularly in the Bradwell area where they are inclined. The Bradwell pipe-veins show a preferred lithology of country rock, being both in coarsely and very coarsely crinoidal limestones; this is in line with the conclusions of Traill (1939, p. 879) at Mill Close, that coarse-grained limestones were favourable for the emplacement of ore-bodies. The presence of shale in a pipe-vein at Smalldale, Bradwell (see p. 314), confirms that they formed close beneath the Namurian cover. The recognition of the importance of the pre-Namurian unconformity helps (see p. 162) to fit this fact into the stratigraphical pattern for the area. The pipe-veins cut across the main vein system and are parallel or nearly parallel to the north-westerly cross-veins, though the latter are absent in this vicinity. The pipe-veins do not appear to be related to the joint-system. In some cases there are records of pipe-veins containing large amounts of ore, for example, at Old Ladywash Mine (Dunham 1952, p. 84).

(*c*) *Flats*, like pipe-veins, are strata-controlled, but they lack the linear element of the latter. Only one major example is inferred to exist in the present district; this is at Burnt Heath where there are large numbers of unaligned shafts. Elsewhere flats may develop as replacement deposits in the walls of the larger veins.

(*d*) *Cavity fillings and impregnations*. Deposits of this type are, in general, not developed in association with the normal pattern of mineralization in the

v

present district, except in the case of pipe-veins. The 'Blue John' fluorite deposits of Castleton, however, which appear to result from a separate phase of mineralization (Ford, 1955, p. 39), show these forms. The larger and workable 'Blue John' deposits have all been in cavity fillings, the smaller and unworkable occurrences are as impregnations and veinlets.

**Alteration of wall-rocks.** The wall-rocks are in general little modified. What alteration there is, is confined to a very variable degree of silicification. The most striking example is on the southern side of Dirtlow Rake where the silicification of a preferred horizon and its subsequent partial erosion have given rise to a striking belt of silica blocks. Elsewhere silicification appears to have been much less marked and it shows as a scattered development of silica blocks along the lines of the veins.

The Dirtlow Rake occurrence overlies cherty limestones of the Monsal Dale Beds and a redistribution of the silica from these appears to have taken place. However, the presence of lesser degrees of silicification along veins in chert-free $D_1$ limestones suggests that part of the silica is of hydrothermal origin.

Another type of alteration of wall-rock which is of note is fluoritization, which results in some minor replacement of the walls of fluorite-bearing veins. Recent work by Ineson (1969) has shown variation in trace elements in limestone wall-rocks, including those at the Treak Cliff Mine and Odin Mine; the most striking changes were depletion in strontium and increase in zirconium, adjacent to the veins.

**Supergene effects.** Secondary ores of lead and zinc are little developed in most of the veins. However, concentrations of these are present in White Rake or High Rake and in White Rake (Wardlow). The first of these two occurrences extends eastwards from the point where the vein cuts the outcrops of the Lower and Upper lavas and it is suggested that the presence of a perched water-table above the lavas allowed oxygenated surface water to react with the original galena and sphalerite. A higher perched water-table above the Litton Tuff may have been responsible for the eastward extension of the secondary ores. The latter are, however, little developed where the vein (here the Hucklow Edge Vein) passes beneath the Namurian cover. In the case of White Rake the occurrence of white ore (cerussite) may again be due to the presence of a perched water-table above the Litton Tuff.

The smithsonite deposit (see p. 312) of the Kittle End Vein, Castleton, must lie close to the upper limit of the main water-table, though sufficiently above it to permit the oxidizing effect of surface water to alter the original sphalerite.

**Control of mineralization.** Green and others (1887, p. 122) noted that the richer part of the veins lies close below the shale cover. For example (ibid., p. 136) the Hucklow Edge Vein was most productive from 25 to 40 fathoms below the shale. More recent authors such as Schnellmann and Willson (1947, p. 7) and Varvill (1959, pp. 181–2) have also accepted the prime importance of the blanketing effect of the shale cover in the formation of ore-bodies. In addition, Shirley and Horsfield (1945, pp. 302–4) have maintained that many of the veins in the Eyam area are located on shallow anticlines. The present work confirms that these conclusions apply in the case of Hanging Flat Vein, Middlefield Rake and Cross Low Vein, but the relations of White Rake (Wardlow) and Dirty Rake are less clear. A good example of structural control has been noted in the Bee Low area, where a local east–west anticline bears veins on both northern and southern flanks in an area which is otherwise almost free of mineralization.

The eccentric arrangement of thermal zones in relation to the limestone outcrop indicates a concealed source to the east, from which the mineralizing fluids rose westwards in a "mainly lateral, up-dip movement" (Dunham 1952, p. 83).

**Age of mineralization.** The youngest sedimentary rocks cut by veins are of Lower Coal Measures age. The mineralization also post-dates the period of sill intrusion which has been shown (p. 293) to be of late Namurian or early West-phalian age. No isotopic age determinations have been carried out on galena from the present district. Moorbath (1962, pp. 319–20) determined a mean modal age of 180 ± 30 m.y. for galenas from the Matlock area and postulated a Mesozoic age for the mineralization. More recently Fitch and Miller (1964, p. 170) have suggested that the primary mineralization of the southern Pennines is 255 ± 14 m.y. in age and associated with the Variscan (Hercynian) orogeny.

There is evidence to suggest that some of the vein fractures were initiated long prior to the mineralization. Changes in stratigraphy across Dirtlow Rake (see p. 34) suggest that this was operative in $P_1$–$B_2$ times. There is also a striking alignment of parts of sills parallel to Shuttle Rake and White Rake (p. 298) which would suggest that these were already formed in late Namurian or early West-phalian times.

## DETAILS

**Castleton and Perryfoot.** This area lies to the north and north-west of Oxlow Rake and Dirtlow Rake. Widespread mineralization has occurred, largely as veins of predominant trend about E.20°N. The strongest vein in the area, Coalpithole Vein, is of arcuate trend, varying from the predominant north-easterly direction in the east to a W.20°N. trend in the west. On Gautries Hill, Coalpithole Rake is well seen as a large openwork, up to some 20 ft in width. The vein continues eastwards as Watt's Grove Vein or vein-plexus to join Dirtlow Rake. Tips show calcite, galena and baryte over much of this tract, with fluorite in addition in the vicinity of, and to the east of, Portway Mines [1285 8101]. Gautries Rake branches off Coalpithole Rake on the south, and Old Wham Vein lies parallel to Watt's Grove Vein on the north side of it. On and near the Old Wham Vein, near Portway Mine [128 811], Ford (1967b, pp. 57–62) described a quartz-rock, mainly lying in a solution cavity, but nearly *in situ*, and result-ing from silicification of the vein wall-rock. Debris indicates similar occurrences of silici-fied limestones between this locality and the vicinity [125 816] of Slitherstone Mine.

New Rake lies parallel to and about half a mile north of the eastern part of Watt's

Grove Vein. The eastern part is relatively simple in arrangement with only minor branching and *en échelon* veining. To the west of Slitherstone Mine [120 814], however, New Rake joins an irregular plexus of veins and scrins and a branch of this complex diverges to the south, approaching Dirtlow Rake. Exposures of the veins are few, one [118 814] north of Eldon Hill shows a 3- to 5-ft open-work up to 24 ft deep; the only veinstuff remaining here is some calcite on the walls. Baryte is abundant in much of the vein-system, fluorite being largely restricted to the eastern end. Blocks of silicified limestone, resulting from alteration of the vein walls, are common just east of Slitherstone Mine.

Horsestones Rake lies north of New Rake. Fluorite is more abundant over much of the rake than in New Rake. At the western end an open working [1143 8196] on Snels Low shows an average width of $3\frac{3}{4}$ ft with calcite-bearing baryte-galena veins up to 3 in across (one in the middle and one on each wall). Eastwards, Horsestones Rake passes into Faucet Rake. The latter is stated by Ford (1965, p. 105) to occur as a 4-ft calcite vein with a little galena in the 'Bottomless Pit' Cavern [1400 8235] in the Speedwell crosscut; the cavern is aligned parallel to the vein. A series of worked smaller veins in this vicinity includes Shackhole Scrin, Longcliff Rake, 2 to 3 ft wide in the Speedwell crosscut (ibid., p. 104) and Little Winster Vein, 3 in of yellow fluorite with some galena. These veins are either parallel or slightly convergent to the main trend of Horsestones Rake.

In the neighbourhood of Castleton a number of veins and scrins lie on the line of Horsestones Rake. These include a strong east-west vein, Red Seats Vein, the spoil from which shows calcite, sphalerite and galena. It is uncertain how far this vein continues to the west, but a thin vein on the same line is present in the quarry at the entrance to Cave Dale. At this locality a scrin parallel to the vein shows on its northern wall an occurrence of glauconite associated with hydrocarbon, galena, sphalerite, baryte and fluorite along stylolites and veinlets (Young and others 1968). Kittle End Vein, which diverges from Red Seats Vein, and is possibly a branch off it, is mentioned by Carruthers and Strahan (1923, p. 50) as having yielded commercial quantities of 'calamine' (smithsonite).

The Odin Vein is visible as a wide east–west openwork cutting the northernmost tip of the limestone outcrop [134 834]. To the west of the outcrop the trend of the vein swings to W.12°S. and its position may be traced by the presence of a series of shafts commencing in the Edale Shales. Debris from a typical example [1270 8332] shows fluorite, baryte, calcite, galena and sphalerite. The openwork fails to show a good section of the vein, but masses of purple fluorite, up to some 3 ft across, are present in places in the walls. Ford (1955, p. 39) noted the usual westward replacement of fluorite by baryte and calcite; the fluorite was either pale blue-green and often slickensided, or less often, saccharoidal, clear or pale blue and apparently recrystallized. The fluorite was associated with a little sphalerite and baryte in addition to galena. Pilkington (1789, p. 174) records native sulphur and pyrite; goethite and marcasite are also recorded (Smith 1962, p. 22) in the Odin Vein, and Hall (1868, p. 55) notes the presence of vivianite; Mawe (1802, p. 47) records selenite from the vein. These occurrences are evidently all from within the oxidized part of the deposit. The presence of wulfenite is regarded as unconfirmed by Ford and Sarjeant (1964, p. 140) though it is here

considered that it may have been derived from the Edale Shales which contain a high molybdenum concentration at one horizon. The presence of explosive slickensides was noted by Whitehurst (1778, pp. 185–9) (see also p. 309).

The deposits of purple fluorite ('Blue John') of the Treak Cliff area are distinct from the veins described above. The account by Ford (1955, pp. 35–60) has been largely drawn on here. The majority of the occurrences are in pipe-veins a short distance below the original position of the Namurian shale cover. The most important localities are enumerated below.

In Treak Cliff Cavern [1359 8314] the 'Blue John' deposits lie discontinuously in a pipe-vein of north-north-easterly trend and some 50 ft below the position of the shale cover. The fluorite occurs as pockets or vughs up to 4 ft across, showing colour banding. In one part of the deposit (ibid., p. 41) fibrous limonite, associated with a little fine pyrite, occurs in the vughs. In places, intergrowths of fluorite and baryte occur, and elsewhere Ford (ibid., p. 43) notes an intergrowth of fluorite and calcite. Open workings [1351 8298] above the cavern entrance show similar pockets of 'Blue John' up to 3 ft and many irregular veinlets of dark blue fluorite.

Old Tor Mine [1349 8275] lies near the southern end of another pipe of north-north-westerly trend and extending downwards to at least 100 ft (ibid., p. 38). A little baryte is associated with the 'Blue John' (ibid., p. 40). Open workings [133 829] behind Treak Cliff lie on the northern continuation of this pipe.

Blue John Mine [1318 8318] lies in an irregular system of pipe-veins, some, such as Waterhole Pipe (ibid., p. 38), being up to 130 ft in depth. Large pockets of 'Blue John' up to 20 ft by 10 ft have been noted here by Royse (1945).

*Bradwell, Peak Forest and Little Hucklow.* Dirtlow Rake (Plate XXIII A) forms one of the most prominent veins in the district and it is characterized over its whole course by a downthrow to the south of up to 200 ft. To the west of its junction with Watt's Grove Vein it is known as Oxlow Rake. Much of the length of Dirtlow Rake shows a complex system with up to three parallel or sub-parallel veins. Where the rake cuts the reef-

(L 486)

A. Open workings for lead: Dirtlow Rake, Castleton

PLATE XXIII

B. Workings for fluorite in a pipe-vein:
near Hazlebadge Hall, Bradwell

(L 492)

belt in Pin Dale a strong parallel vein takes over the major throw and a series of cross-courses at 45° suggests lateral movement (see also p. 309). Substantial displacement of the limestone-shale boundary is noted along this line. Within the reef-complex, fluorite pockets are developed in the vicinity of the vein and there is some secondary chert in nodule form. Wide and deep openworks west of Pin Dale show little of the vein, though large masses of columnar calcite can be seen in places. Ford (1967a, p. 59) notes a width of 8 ft of fluorite at a point [156 822] near the Siggate road. The main productive part of the vein appears to have been along the walls; one place where this may be observed [1451 8154] shows a 3-ft vein of baryte, calcite and galena on the south wall with horizontal slickensiding at the contact. Another section [1470 8169] a little farther east shows a 22 ft-wide excavation with large masses of radiating columnar calcite and veins of white fluorite and galena up to 6 in across. A nearby section [1476 8173] across the vein is as follows: north wall (horizontal slickensides; dip 70° south), thin fluorite-baryte-galena vein, shattered limestone with fluorite, baryte and galena about 5 ft, columnar calcite 2 ft, obscured 6 ft, columnar calcite 2½ ft, excavated 10 ft, south wall. A conspicuous feature of Dirtlow Rake is the associated silicification which has produced a belt of silica-rock blocks, more or less *in situ*, and extending over a width of up to 400 yd parallel to the vein on its south side. A section [1542 8199] shows 15 ft of brown compact silica-rock with traces of fluorite. Ford (1967a, p. 60) noted a section at the head of Pin Dale [155 819] showing quartz-rock with idioblastic cubes of blue fluorite and empty casts of productoids. Some of the blocks contain nodules of chert (Arnold-Bemrose, 1898, p. 176). The silicification (see also p. 87) is considered to have followed a preferential horizon in the Upper Monsal Dale Beds, and though Ford (1967b, p. 60) states that the quartz-rock is cut off by faulting along Dirtlow Rake, it may equally well be that rising silica-rich solutions were emplaced in the preferred horizon on the downthrow side only, and failed to reach it on the upthrow side.

Oxlow Rake shows a complex arrangement over its north-eastern half, with many splits and leaders. Open sections in the vein are here lacking, but the tips show calcite, baryte and galena and in addition some secondary iron oxide near the intersection of the Upper and Lower lavas. Debris from here [1315 8061] includes marcasite altering to lepidocrocite, itself altering to goethite. At the western end of the vein [1250 7996] openworks up to 15 ft wide are visible but the vein has been largely removed.

Long Rake lies midway between Dirtlow and Moss rakes and converges westwards towards the latter. Sections in the vein are lacking and information regarding it is scanty. Some cross-veining is present and at the eastern end, in the vicinity of Earle's Quarry, a number of small parallel veins accompany the main vein. Two unnamed veins, of general easterly trend, lie between Long Rake and Dirtlow Rake.

Moss Rake is a strong vein of east-north-east trend extending from Bradwell to the vicinity of Brocktor, south of which it is continued by an *en échelon* mineralized fault with a downthrow to the south. Some probably varying throw occurs along this vein, the western end showing a definite downthrow to the south while elsewhere Carruthers and Strahan (1923, p. 52) have recorded a northerly hade. Moss Rake ends at the Namurian boundary, which it does not displace. A striking feature is the large quantity of calcite present in most places. Sections in the vein are numerous. At Outlands Head [1662 8076] quarrying operations have revealed the vein as two parallel mineralized faults in a 25-ft belt, with much columnar calcite and thin veins of fluorite, baryte and galena. Near Batham Gate [1515 8027] the Rake consists of up to 14 ft of columnar calcite on the north side flanked by a fluorite-baryte-calcite-galena zone up to 2½ ft wide on the south. Farther west [1381 7976] an open working, 18 ft wide, in the vein, shows a 9-ft mass of columnar calcite associated with some fluorite; there was slight fluoritization of the south wall of the vein. To the west, blocks of silicified limestone are present along the line of the vein and were noted by Arnold-Bemrose (1898, p. 178) at "Brock Tor".

To the south of Moss Rake a series of parallel and apparently discontinuous veins is present. One of these is seen in the quarry [1714 8054] on the west side of Bradwell Dale as a 10 ft shatter zone with interstitial mineralization, mainly fluorite and calcite, with subsidiary baryte, siderite and galena.

Earl Rake is a thin but persistent vein which cuts Bradwell Dale and is known as Hill Rake on the east side of the valley. Tips in Bradwell Dale show fluorite, calcite, galena and some sphalerite. At the west end of its course [160 800] the vein probably joins a plexus of veins, including Scrin Rake, which converge southwards towards Shuttle Rake.

Shuttle Rake is a 3-mile *en échelon* vein system extending from near the upper end of Bradwell Dale to Tideswell Moor. At the western end the vein lies along a fault, throwing down to the north. The vein has been worked for fluorite on Hucklow Moor. Carruthers and Strahan (1923, p. 53) record the occurrence of "a little calamine" (smithsonite) in this vein. Maiden Rake lies parallel to, and some half mile south of, Shuttle Rake, and is of similar lateral extent; the eastern part is called Nether Water Vein. Sections in the vein are lacking. Pilkington (1789, p. 174) states that a 4-in layer of "native sulphur" was found in this vein on Tideswell Moor, and Braithwaite and Ryback (1963) have confirmed the presence of sulphur at Nether Water Mine [1708 7909].

Two pipe-veins are present in the Bradwell area (see p. 309). The more northerly runs through Smalldale Head, is about a mile in length and trends about W.20°N. Old workings suggest that it may reach a width of 200 yd at one place [161 814]. Part of the pipe-vein, in coarsely crinoidal limestone, is exposed in a working [1605 8146] south of Earle's Quarries and shows a somewhat irregular flat, up to 7 ft thick, of high-grade white fluorite with some calcite and galena; there is some suggestion of the flat splitting southwards. Some shale is present in the north-west part of the excavation about 25 ft below ground level (see p. 309) and suggests that the pipe-vein formed close below the shale cover. At Smalldale Head [1647 8132] workings show a thickness of 4 ft of white fluorite with some galena. The country rock is again coarsely crinoidal limestone. The pipe-veins may be traced by old workings for a further ¼ mile to the south-east.

On the eastern side of Bradwell Dale a strong pipe-vein is present in the Eyam Limestones flat reef immediately below the original position of the shale cover (Plate XXIII B). The flat reef is very coarsely crinoidal in places. The pipe-vein trends N.40°W. and shows a maximum width of some 30 ft. It is cut by Hill Rake but its relations with this are obscure. At one point [1737 8031] an unbroken arch of limestone has been left above the workings and a similar feature can be seen where the pipe-vein reaches the side of Bradwell Dale [1725 8041]. Here there is a more or less clearly defined floor to the pipe some 10 to 15 ft below the roof, though farther east the depth is greater. The pipe-vein is filled by brown clay with abundant loose fluorite gravel. Some of the fluorite is in cubes up to 1 in and with purple colouration; some baryte is present.

*The Hucklow Edge Vein System.* This is some 5 miles in length and trends mainly east–west. Over most, if not all, of its length, the system is a fault with downthrow to the north. The western part is exposed and is known successively from the west as White Rake, Tideslow Rake and High Rake. At Great Hucklow it passes under the Namurian and is then known as the Hucklow Edge Vein. At Milldam Mine [1766 7796] Green and others (1887, p. 136) record a northerly downthrow of about 80 ft. In Twelve Meers Mine [2067 7762] there is a northerly downthrow of 10 ft.

At the western end of the system White Rake can be traced to the vicinity of the Black Hillock Shaft [1410 7822] in Potluck Sill (see also p. 298). To the east of the sill, White Rake is here interpreted as throwing the Lower Lava on the south against the Upper Lava on the north. Contrary to previous belief the vein has been found to persist in the lava, a recent open-work [1451 7816] showing a width of 3 to 5 ft of soft saccharoidal fluorite with some baryte, galena, bright green pyromorphite and cerussite; calcite was rare or absent. The overall fluorite content was 50 per cent. The vein is stated by Green and others (1887, p. 133) to derive its name from the occurrence in it of 'white ore' or cerussite. On the north-west of Tides Low [148 781] the vein splits locally, leaving a 'horse' of limestone in the middle and with the fault on the north side. Between this locality and Windmill, the vein, here known first as Tideslow Rake, then as High Rake, shows extensive tips with much calcite, though baryte, fluorite, galena and secondary ores are also present.

A description of the workings at Milldam Mine [1766 7796], Great Hucklow, was given by Green and others (1887, pp. 135–6) who stated that the vein had a width of 150 ft at the surface but narrows downwards to 30 ft at a depth of 198 ft. The gangue was mainly calcite with some baryte, and galena occurred mainly in two thin veins near the north, and another near the south wall. The galena vein was seen by these authors to be 1 in thick on the south wall but it was stated to have reached a width of 3 ft in one place.

To the east of Great Hucklow the Hucklow Edge Vein is concealed by the Namurian. Its line may be followed by a series of shafts but little precise information is available. Green and others (ibid., p. 137) noted a north-westerly branch off the Hucklow Edge Vein near Old Grove Shaft [1895 7798]; Back of the Edge Shaft [1846 7830] in Bretton Clough lies on the line of this vein.

To the east of Bretton the Hucklow Edge Vein is recognizable in the Namurian as a fault with northerly downthrow. The underground position of the vein in the limestone is known from the alignment of shafts and near Ladywash Shaft from the workings themselves. A comparison of the surface and underground positions indicates a near-vertical hade and at Silence Mine [1880 7799] Green and others (ibid., p. 137) even note a southerly hade. The vein continues eastwards along a fault in the Namurian as far as the River Derwent but mineralization at depth is unproved. The Hucklow Edge Vein is now being worked from Ladywash Mine to the west of its intersection with the main crosscut. At the latter point [2159 7768] it is stoped out and 2 to 5 ft in width (Dunham 1952, p. 87). At Twelve Meers Mine the vein was proved 20 ft wide on top of the Cressbrook Dale Lava.

The Old Edge Vein runs close and nearly parallel to the Hucklow Edge Vein at its west end [208 775], but when traced eastwards, swings south and diverges from it. Part at least of the vein is traceable as a fault, with northerly downthrow in the Namurian, vertical at the west end, but showing a normal average hade farther east, though variations in hade occur within the limestone. The main workings from Ladywash Mine have been in this vein. The main vein, up to 15 ft wide, is mainly fluorite, but some calcite, baryte and galena also occur with small amounts of pyrite; quartz and chalcedony are known (Dunham 1952, p. 87). Robinson (in discussion Varvill 1959, p. 228) states that the vein reaches a width of 20 ft with the greatest concentration, though small, of galena, close beneath the shale cover. An interesting feature of the vein, noted by Dunham (ibid., p. 87), is the presence of two generations of fluorite, the earlier brecciated by movement along the vein and subsequently re-cemented by the later. During the present work cerussite has been confirmed by an X-ray powder photograph (X3628) of a specimen from the central part of the vein [2206 7707] and chalcopyrite has been seen in tipped material from the vein. Ford and Sarjeant (1964, p. 133) have recorded the presence of smithsonite in the Old Edge Vein. These authors (ibid., p. 129) also record marcasite and pyrrhotite from the "Ladywash Mine" (probably also from Old Edge Vein).

The Old Edge Vein is connected to Hucklow Edge Vein by Ladywash Vein. This is stated by Dunham (ibid., p. 87) to be 4 ft wide and stoped out in the vicinity of the crosscut [2175 7751].

*Foolow, Eyam and Middleton Moor.* In this area the dominant veins are Cross Low Vein and Middlefield Rake in the north, and Dirty Rake and White Rake (Wardlow)[1] in the south.

The northernmost vein in this tract, Black Hole Vein (also called Little Pasture Vein) lies beneath the Namurian cover. It trends somewhat north of east and may possibly represent a continuation of the Ladywash North Easterly Vein, displaced by the Old Edge Vein.

Cross Low Vein runs nearly east–west from Stanley Moor to Dustypit Mine [2075 7707]. Near the western end a working for fluorite [1799 7693] showed the vein to be 2 to 4 ft in width with a flat of uncertain thickness on the north wall. In Silly Dale the vein can be seen to have no throw but at its eastern end it probably has a small downthrow to the

---

[1]So called to distinguish it from the westerly continuation of the Hucklow Edge Vein.

north. Over most of its length tips on the vein show much fluorite and some baryte and galena; calcite is little developed. Between Stanley Moor and Foolow there lies on the south of Cross Low Vein, and partly joining it, a system of smaller veins in which two components may be recognized, one parallel to the larger vein and the other a series of cross-veins of north-westerly trend. A little east of Dusty Pit the vein swings to a trend about E.20°N. and is known as Dusty Pit Vein. Highcliffe Mine [2128 7723] lies at the easternmost end of this and is one of the localities where Whitehurst (1778, pp. 185–9) and Green and others (1887, pp. 137–8fn) recorded explosive slickensides, the slickenside material being "chiefly barytes".

Middlefield Rake runs nearly parallel to Cross Low Vein and from 200 to 400 yd south of it. The western end lies in Linen Dale and the vein continues east to Town End, Eyam. Sections are lacking but the tips show a similar composition to Cross Low Vein, with inconspicuous calcite. At the eastern end, the rake has been found at depth west of Glebe Shaft with good fluorspar and lead values (Dunham 1952, p. 87). There is no surface evidence of the vein in this area.

Ashton's Pipe, described by Dunham (1952, p. 84) as "a barren leader", extends from Glebe Shaft in a N.30°W. direction. Where seen in the crosscut there is no evidence of mineralization on the pipe but it appears to have been an important feeder which gave rise to flats with fluorite and galena at Old Ladywash Mine [2158 7705].

Hanging Flat Vein is a short east–west vein midway between Middlefield Rake and Dirty Rake.

Dirty Rake runs E.15°N. from Middleton Moor to Middleton Dale and is associated with a number of smaller, sub-parallel veins mainly on the south side. Sections are in general lacking, though the tips show baryte and fluorite with little calcite. An open working [2207 7571] in Middleton Dale shows the vein to be 20 ft wide at the top but narrowing down to 10 ft. An adit has been driven in on a 2-ft vein within the main part of the rake and this shows fluorite with some limonite.

White Rake (Wardlow) runs from Tansley Dale, Litton, eastwards for 2½ miles, and lies close to Dirty Rake near its eastern end. On the west side of Cressbrook Dale the vein lies along a normal fault with a southerly downthrow of about 50 ft; on the eastern side of the dale, however, the throw appears to be slight. A plexus of small parallel veins is associated with White Rake in this area. Some ¾ mile east of Wardlow [194 747] the vein cuts a knoll-reef without either vertical or horizontal displacement. Tips from the rake show abundant calcite and some fluorite and baryte. Ford and Sarjeant (1964, p. 133) record anglesite and cerussite from this vein. An open working [1916 7479] near Seedlow Mine showed a width of 9 ft.

Between Middleton Moor and Hanging Flat a number of north-westerly cross-veins, mostly short, link the major veins. At Burnt Heath [203 756] the presence of a flat, elongated parallel to the cross veins and some 400 by 150 yd in dimensions, is inferred from the occurrence in this area of a large number of shafts showing no evidence of linear arrangement. The tips from these show variable amounts of baryte, fluorite and calcite.

Watergrove Mine is stated by Green and others (1887, p. 138) to have been in an irregular pipe-vein. Forefield Shaft [1882 7577] is shown[1] as having been sunk 265 ft to the floor of the pipe which was regular, suggesting some form of stratigraphical control; the upper surface of the pipe is shown as very irregular. A comparison with Wardlow Mires No 1 Borehole (see p. 93) shows the horizon of the pipe to be some 50 ft above the Upper *Girvanella* Band.

*Area west of Tideswell and Peak Forest.* This lies outside the main area of mineralization. Veins are scattered and only the more important are described below.

Edge Rake trends in a nearly east–west direction from Wheston for over half a mile towards the Brook Bottom valley. At the western end [135 765] there is a group of smaller parallel and cross-veins. The tips show calcite, with some baryte, fluorite and galena.

In Hay Dale [1207 7696] a mineralized fault with downthrow to the north and trend

---

[1]Abandonment plan H/S 10405.

E.20°S. has been worked for calcite in a 6 to 20 ft openwork. At least 12 ft width of calcite can be seen in broken veins above the openwork. No other minerals were noted here.

Isolated occurrences of veins to the west and south-west of Peak Forest show a general alignment with Dirtlow Rake, Moss Rake and Shuttle Rake, which converge towards this area. The occurrences are important as showing that the Peak Forest Sill pre-dated the mineralization (see below and p. 297) and are all situated in the western outcrop of the sill. The most northerly occurrence is a mineralized fault [104 791] close to the line of Moss Rake and tips from here show abundant calcite. The throw of the fault is northerly in contrast to the southerly throw of Moss Rake. Near Ivy House [1073 7895] a north-westerly fault at the boundary of the dolerite shows calcite with some galena in a clay matrix. At the western side of this outcrop, an east–west line of workings along a vein is traceable into the outcrop of the sill [105 788]. From these localities a somewhat discontinuous line of veins extends westwards, running close to Lodesbarn. These veins lie close to, and a little south of, the axis of the small easterly-trending Bee Low Anticline. Some small east–west veins are also present on the northern flank of the anticline on Barmoor.

Near the southern edge of the district, workings indicate some mineralization along the line of a fault running about W.10°N. from the Heathydale Ward with northerly downthrow. The tips show traces of fluorite near Tideswell Dale but calcite and galena only at the western end near Monk's Dale.

**Occurrences of mineralization in post-Viséan strata.** At Froggatt a linear excavation in the Chatsworth Grit up to 6 ft in width and 15 ft in depth extends from a point [2555 7739] near the main road W.40°S. for a distance of about 500 yd. No ore or gangue material has been seen but the nature of the excavation confirms that it is in fact an open working along a vein. It is almost in line with Dirty Rake though nearly

two miles from the limestone outcrop. Further, the extrapolation of Hucklow Edge Rake would also intersect the excavation. It appears likely, therefore, that one or both of these structures extends eastwards at depth and allowed the mineralizing fluids to reach the Chatsworth Grit.

The occurrence of a vein of galena in the Coal Measures at Horridge End, near Whaley Bridge was noted by Binney (1851, pp. 125–9). According to him the vein was present in colliery workings in beds ranging from the horizon of the Yard Coal (or lower) to that of the Red Ash. The vein was stated to be up to 9 in wide and to contain only galena with traces of pyrite; there was no gangue. The trend of the vein was given as north-easterly. Recent examination of a plan (B.1) in the possession of the Duchy of Lancaster has fixed the occurrence more closely; the vein extended from a point [0120 8112] near the Downcast (Winding) Shaft of Whaley Bridge Colliery S.30°E. for a distance of 75 yd. Hull and Green (1866, p. 25) state that the vein reached as much as 24 inches in width and was best developed in the sandstone underlying the Red Ash Coal. These authors also note the presence of another vein at "Mr. Srigley's pit" (Railway Pit), again in the sandstone underlying the Red Ash Coal.

To the south of Whaley Bridge, traces of mineralization have been proved in two boreholes. Fernilee No. 1 [0124 7823] showed a galena-coated joint in mudstone at a depth of 306 ft, and at a horizon between the *Gastrioceras cancellatum* Marine Band and the Simmondley Coal. In Fernilee No. 2 Borehole [0119 7863] abundant veinlets with galena and kaolinite were present from $489\frac{1}{2}$ ft to the bottom of the hole at $524\frac{1}{2}$ ft in strata lying above the horizon of the *Gastrioceras cumbriense* Marine Band.

In the axial area of the Todd Brook Anticline a quarry [9815 7590] near Jenkin Chapel shows veining by coarsely tabular baryte in beds of 'crowstone' type. The veins do not exceed a few inches in thickness.          I.P.S.

## REFERENCES

ARNOLD-BEMROSE, H. H. 1898. On a quartz rock in the Carboniferous Limestone of Derbyshire. *Q. Jl geol. Soc. Lond.*, **54**, 169–83.

BINNEY, E. W. 1851. Remarks on a vein of lead found in the Carboniferous strata in Derbyshire, near Whaley Bridge. *Trans. Manchr lit. phil. Soc.*, ser. 2, **9**, 125–9.

BLOUNT, B. and SEQUEIRA, J. A. 1919. 'Blue John' and other forms of Fluorite. *J. chem. Soc.*, **115**, 705–9.

BRAITHWAITE, R. S. W. and RYBACK, G. 1963. Notice of exhibit (of Sulphur and Hydrozincite, etc.) from Derbyshire. *Mineralog. Soc. Circ.* 121.

CARRUTHERS, R. G. and STRAHAN, A. 1923. Lead and zinc ores of Durham, Yorkshire and Derbyshire. *Mem. geol. Surv. spec. Rep. Miner. Resour. Gt Br.*, **26**.

DUNHAM, K. C. 1934. The genesis of the North Pennine ore deposits. *Q. Jl geol. Soc. Lond.*, **90**, 689–720.

———— 1952. Fluorspar. 4th edit. *Mem. geol. Surv. spec. Rep. Miner. Resour. Gt Br.*, **4**.

FAREY, J. 1811. *A general view of the agriculture and minerals of Derbyshire*. London.

FITCH, J. F. and MILLER, J. A. 1964. The age of the paroxysmal Variscan orogeny in England. *in* "The Phanerozoic Time-Scale". *Q. Jl. geol. Soc. Lond.*, **120S**, 159–75.

FLETCHER, A. R. 1929. Mexico's lead-silver manto deposits and their origin. *Engng Min. J.*, **127**.

FORD, T. D. 1955. Blue John fluorspar. *Proc. Yorks. geol. Soc.*, **30**, 35–60.

———— 1965. The Speedwell Cavern, Castleton, Derbyshire. *Trans. Cave Res. Grp Gt Br.*, (2), **4**, 101–24.

———— 1967a. *in Geological Excursions in the Sheffield Region and Peak District National Park*. Ed. R. Neves and C. Downie. University of Sheffield.

———— 1967b. A quartz-rock-filled sink-hole in the Carboniferous Limestone near Castleton, Derbyshire. *Mercian Geol.*, **2**, 57–62.

———— and SARJEANT, W. A. S. 1964. The Peak District Mineral index. *Bull. Peak Distr. Mines hist. Soc.*, (3), **2**, 122–50.

GREEN, A. H., FOSTER, C. LE NEVE and DAKYNS, J. R. 1887. The geology of the Carboniferous Limestone, Yoredale Rocks and Millstone Grit of North Derbyshire. *Mem. geol. Surv. Gt Br.*

HALL, T. M. 1868. *Mineralogist's Directory*. London.

HULL, E. and GREEN, A. H. 1866. The geology of the country around Stockport, Macclesfield, Congleton and Leek. *Mem. geol. Surv. Gt Br.*

INESON, P. R. 1969. Trace-element aureoles in limestone wallrocks adjacent to lead–zinc–barite–fluorite mineralization in the northern Pennine and Derbyshire ore fields. *Trans. Instn Min. Metall.*, sect. B, **78**, B29–40.

MAWE, J. 1802. *The Mineralogy of Derbyshire*. London.

MOORBATH, S. 1962. Lead isotope abundance studies on mineral occurrences in the British Isles and their geological significance. *Phil. Trans. R. Soc.*, Series A, **254**, 295–360.

MUELLER, G. 1954a. The theory of genesis of oil through hydrothermal alteration of coal type substances within certain Lower Carboniferous strata of the British Isles, *C. r. Congr. géol. int.*, sect. 12, fasc. 12, 279–327.

———— 1954b. The distribution of coloured varieties of fluorite within the thermal zones of Derbyshire mineral deposits. *C. r. Congr géol. int.*, sect. 13, fasc. 15, 523–39.

———— 1960. The Distribution of Uranium in naturally fractionated organic phases. *Rep. 21st Int. geol. Congr.*, (15), 123–32.

PILKINGTON, J. 1789. *View of the present state of Derbyshire*. 2 vols.

PRESCOTT, B. 1926. The underlying principles of the limestone replacement deposits of the Mexican province. *Eng. and Min. J.*, **122**, 246–53.

ROYSE, J. 1945. *Ancient Castleton caves*. Castleton.

SCHNELLMANN, G. A. and WILLSON, J. D. 1947. Lead-zinc mineralization in North Derbyshire. *Trans. Instn Min. Metall.*, **485**, 1–14.

SHIRLEY, J. and HORSFIELD, E. L. 1945. The structure and ore deposits of the Carboniferous Limestone of the Eyam district, Derbyshire. *Q. Jl geol. Soc. Lond.*, **100**, 289–308.

SMITH, M. E. 1962. The Odin Mine, Castleton, Derbyshire. *Bull. Peak Distr. Mines hist. Soc.*, (6), **1**, 18–23.

TRAILL, J. G. 1939. The geology and development of Mill Close Mine, Derbyshire. *Econ. Geol.*, **34**, 851–89.

VARVILL, W. W. 1937. A study of the shapes and distribution of the lead deposits in the Pennine limestones in relation to economic mining. *Trans. Instn Min. Metall.*, **46**, 463–559.

————— 1959. The future of lead–zinc and fluorspar mining in Derbyshire. *in* Future of non-ferrous mining in Gt. Britain and Ireland. *Instn Min. Metall. Symposium.*

WEDD, C. B. and DRABBLE, G. C. 1908. The fluorspar deposits of Derbyshire. *Trans. Instn Min. Engrs*, **35**, 501–35.

WHITEHURST, J. 1778. *An inquiry into the original state and formation of the earth.* London.

WILKINSON, P. 1950. Allophane from Derbyshire. *Clay Miner. Bull.*, **1**, 122–3.

WILSON, G. V., EASTWOOD, T., POCOCK, R. W., WRAY, D. A. and ROBERTSON, T. 1922. Barytes and witherite. 3rd edit. *Mem. geol. Surv. spec. Rep. Miner. Resour. Gt Br.*, **2**.

YOUNG, B. R., HARRISON, R. K., SERGEANT, G. A. and STEVENSON, I. P. 1968. An unusual glauconite associated with hydrocarbon in reef-limestones near Castleton, Derbyshire. *Proc. Yorks. geol. Soc.*, **36**, 417–34.

*Chapter VIII*

# STRUCTURE

THE STRUCTURES in the Carboniferous rocks of the Chapel en le Frith district (see Fig. 22) are mainly of Hercynian age but they reflect in their distribution and intensity the prior existence of two different structural units which had already been affecting sedimentation in both Lower (see p. 11) and to a lesser extent in Upper Carboniferous times. The most striking of these units is the stable block of the Derbyshire Dome over which shelf sedimentation occurred in the Lower Carboniferous; this is nearly coincident with the present Lower Carboniferous outcrop and has a characteristic tectonic style. A consideration of the stratigraphy and structure makes it clear that this stable block, like the Askrigg and Alston blocks farther north, underlies the present outcrop of the Carboniferous Limestone. This has been called the Derbyshire Block by Kent (1966, p. 335), and geophysical work by White (1948, pl. 18) has shown that it is characterized by strong positive gravity anomalies. Pre-Carboniferous rocks have been proved in the Woo Dale Borehole (Cope 1949) to lie at the relatively shallow depth of 892 ft. The area of the Derbyshire Block now forms the axial part of a broad anticline which is the southern part of the Pennine uplift and George (1962, p. 44) has regarded the stable block as foreshadowing the main Hercynian structure. It is customary to refer to the anticlinal limestone outcrop as the Derbyshire Dome and its northern part falls within the Chapel en le Frith district.

The remainder of the district, now the outcrop of the Upper Carboniferous, shows two contrasting structural areas; a western area of thick sedimentation and strong folding and faulting, and a central and eastern area showing gentle folding and slight faulting. Part of the central area, the 'Edale Gulf' of Kent (1966), behaved as a subsiding basin of sedimentation in late Viséan times but the facies of the lowest known Viséan strata (see p. 37) shows affinities with that of the shelf area and suggests a possible northern continuation of the Derbyshire Block at depth.

## STRUCTURE OF THE DERBYSHIRE BLOCK

The Lower Carboniferous outcrop is characterized in general either by gentle folding or by areas with predominant low dips. The marginal areas, however, show a predominance of high dips, which are largely depositional (see p. 17) in the Castleton area, the remainder being indicative of movement of the margins of the block in late Viséan times. In the Great Hucklow and Eyam area, dips at the Namurian-Viséan junction are small and it is inferred that the margin of the block here lies below the Namurian outcrop.

**Pre-Namurian structures.** The influence of pre-Namurian movements has been most acute in the marginal areas corresponding to the edges of the Derby-

320

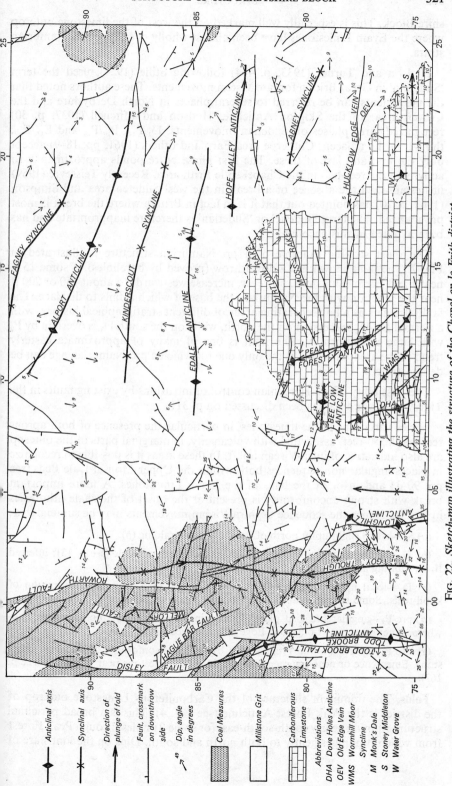

Fig. 22. Sketch-map illustrating the structure of the Chapel en le Frith district

shire Block. This is especially well marked on the west of the limestone outcrop where the Eyam Limestones have been almost wholly overlapped by Namurian strata.

Hudson and Turner (1933, p. 464) following Stille (1924), used the term 'Sudetian' in Great Britain for late Viséan movements. These authors noted that the movements can be referred to several phases in North Derbyshire and the Craven area. In the Skipton Anticline, Hudson and Mitchell (1937, p. 30) recognized three phases of 'Sudetian' movements, $D_1/P_{1a}$, $P_{1c}/P_{2a}$ and $E_{1a}/E_{1b}$, though in the adjacent Clitheroe area Earp and others (1961, pp. 18–9) accept only the first and last of these. The first phase corresponds approximately to activity in the reef-belt in the Chapel en le Frith area. Recently Teisseyre (1959) has established the absence of a break in the west Sudeten area and Simpson (1962, p. 70) has pointed out that it is in fact in Britain where the break is most pronounced. The term 'Sudetic' or 'Sudetian' is therefore inappropriate and has been avoided in this account.

The most striking example of a pre-Namurian structure is illustrated in Fig. 7. The fault shown has a downthrow (proved by boreholes) of some 15 ft near Field Farm. This must, however, increase westwards to about 150 or 200 ft near the junction with the Namurian, the base of which it fails to displace. The fault also acts as a line separating areas of different stratigraphical history, with a normal $D_1$–$D_2$ succession on the north, while on the south $D_1$ is overlain by $P_2$ with large unconformity. The fault is one of many of approximate easterly trend in the area, though it is the only one for which a pre-Namurian age can be demonstrated.

The possibility of pre-Hercynian control of intrusions by existing faults in the Tideswell Moor area has been discussed on p. 311.

Facies variations in the limestones, in particular the presence of both apron-reef and knoll-reef, associated with vulcanicity, in marginal parts of the outcrop are also indicative of intra-Viséan uplift. In these areas it is possible to recognize in places angular non-sequences both within the Upper Monsal Dale Beds (see pp. 90–1) and in or at the base of the Eyam Limestones. A more important break with strong unconformity is present at the onset of the shale lithofacies in late $P_2$. Hence the sequence of intra-Viséan movements may be summarized:

(a) $S_2$—broad slight uplift north of Peak Forest, heralding (b).

(b) $D_1$—strong progressive local uplift in Castleton reef-belt (see p. 13); inferred $B_2$–$P_1$ faulting at Pin Dale.

(c) $D_2$—local uplift in marginal areas in a broad zone, leading to formation of knoll-reefs. Some strong local movements giving non-sequences.

(d) pre-$P_2$—angular break at places in marginal areas, possible non-sequence elsewhere. Emergence near Longstone Edge anticline (see p. 17).

(e) late $P_2$—(pre-Namurian) represented by strong unconformity of shales on limestone. Emergence or near-emergence of the salient of limestone near Mam Tor (see p. 162).

Folds. The dominant structure of the Carboniferous Limestone outcrop of the district is the Peak Forest Anticline (see Fig. 4). This is a broad periclinal structure running nearly south-south-east for some $2\frac{1}{2}$ miles through Peak Forest from which it plunges gently to both north and south. Dips on the limbs are of the order of 5° to 15°. To the south and south-west of this structure and slightly oblique to it lies a gently folded area with first the shallow Wormhill Moor Syncline extending in a south-easterly direction from Peak Dale through Hargatewall. The Dove Holes Anticline lies parallel to and west of the syncline extending from Dove Holes south-south-eastwards to Orient Lodge. Dips on the limbs of these structures are of the order of 5° to 10°.

The Bee Low Anticline is a local east–west structure which contrasts sharply in direction with the structures already noted. The anticline runs for a distance of 1½ miles, disappearing at the eastern end on the western limb of the Peak Forest Anticline while to the west the plunging end of the structure produces a small limestone salient in the Millstone Grit outcrop, where the fold dies out rapidly. Dips on the flanks of the Bee Low Anticline are up to 15° away from the limestone boundary and as much as 40° at this boundary though it should be recalled that original depositional dips of up to 28° occur in places at the edge of the block. The absence of limestones of $D_2$ age on the southern limb of the anticline indicates that the folding was initiated in pre-Namurian times though the main fold is certainly post-Namurian. The Bee Low Anticline is the only example of an east–west fold in the limestone area in the present district, though strong folding on this trend is present in the asymmetrical anticline of Longstone Edge (Shirley and Horsfield 1945, p. 300). The latter structure lies mainly beyond the southern margin of the district though part of its gently dipping northern limb is included. To the north of this tract and apparently in part complementary to the Longstone Edge anticline, lies the basin of Wardlow Mires and farther east, at Stoney Middleton, a small plunging syncline at the limestone margin lies along the same line.

In addition to the broad folding characteristic of the limestone, there are two examples of sharp, local folds. The first is at Monk's Dale (Fig. 14), where a sharp anticline, traceable over about half a mile and of trend W.30°N., is flanked on the north-east side by two volcanic vents (see p. 300). This association suggests a penecontemporaneous origin for the folding and vulcanicity. The second sharp fold is an anticline of trend N.20°W. at Water Grove; this again is a strictly local structure, being of similar dimensions to that at Monk's Dale. Both structures are approximately parallel to local faulting and may well correspond to faults at depth; in the case of the Monk's Dale fold the fracture would also appear to have controlled the emplacement of the vents.

**Faults.** Two main sets of faults are recognizable within the Carboniferous Limestone outcrop:

(a) A dominant series of faults trending from east to east-north-east. Many of these are mineralized (see Fig. 21) especially in the eastern part of the outcrop, but in the west smaller and mainly unmineralized faults occur also. The most important of this set of faults are as follows. The Dirtlow Rake–Coalpithole Vein system in the north; this has a southerly downthrow of some 200 ft on Bradwell Moor and a small southerly downthrow at the limestone margin at Perryfoot. Moss Rake shows a small northerly downthrow at the western end, while else-where the throw is apparently small but variable in direction. The Hucklow Edge Vein system extends for nearly seven miles in an east–west direction through Great Hucklow; the vein shows a northerly throw, proved to be 80 ft at Milldam Mine, and a small variable hade which is in places reversed (see p. 315). The eastern half of this structure is a fault in the Millstone Grit at surface.

A feature of many of the veins (in addition to the demonstrable faults including those noted) is the striking lateral persistence of many of these, such as Dirty Rake, Earl Rake, much of Shuttle Rake and Horsestones Rake. This lateral persistence, coupled with the frequent presence of horizontal slickensides, suggests that many of these structures are wrench-faults though the horizontal movement is probably small. In some cases, however, veins of some importance are not faults; White Rake (Wardlow) for example crosses a knoll-reef without any displacement, either lateral or vertical.

(b) The second set of faults, of north-westerly trend, is developed particularly in the Peak Forest area and at Tideswell and throws of up to 50 ft or so are common in these. A corresponding feature in the mineralized area is a series o cross-veins of limited lateral extent and showing little or no displacement in most cases. However, the only major vein of this trend, the Old Edge Vein, shows a small north-easterly downthrow.

**Peripheral area of block.** In many places the limestone margin shows dips of up to 30°. Between the western end of the reef-belt at Perryfoot and the southern margin of the district these are considered to be of tectonic origin, reflecting the margin of the Derbyshire Block. In the reef-belt on the northern margin of the block the dips have been shown to be depositional (see p. 17) and deep-seated faulting of intra-Viséan age has been postulated to account for the establishment of the apron-reef. In the north-east, between Bradwell and Eyam, dips at the limestone margin are mainly of the order of 5° to 10° and limestone of shelf facies has been proved underground in Ladywash Mine at the relatively shallow depth of 699 ft; these facts suggest that the margin of the block here lies to the north-east, beneath the Namurian.

## STRUCTURE OF THE UPPER CARBONIFEROUS ROCKS

The Upper Carboniferous rocks of the district occur in two structurally distinct areas which are described separately below.

*Western area.* Most of this area, lying west of Charlesworth, Hayfield and Chapel en le Frith, is characterized by strong north–south folding, and by common faulting. This tract forms part of the larger belt of north–south structures which lies on the western limb of the Pennine uplift.

**Folds.** The Goyt Trough (see Fig. 23) is the best-known structure in the district and is a symmetrical syncline trending northwards from the southern margin of the district near Goyt's Bridge to the neighbourhood of Rowarth where it flattens out and disappears. Dips on the limbs reach a maximum of 45° in the southernmost part of the structure and tend to decrease gradually northwards. The structure shows a gentle plunge from both south and north towards a point between Buxworth and Birch Vale. The occurrence of contorted bands in shales in the Goyt Trough is interpreted as due to inter-stratal movement during folding, and has been discussed on p. 161. Between Taxal and the southern margin of the area a small faulted anticline almost coincides with the main synclinal axis. Boreholes along the line of the Fernilee dam show that this anticline dies out at depth. This fold is clearly of tectonic origin though it appears foreign to the general structural pattern; it is interpreted as a flexure formed by incompetent folding in the axial area of the syncline.

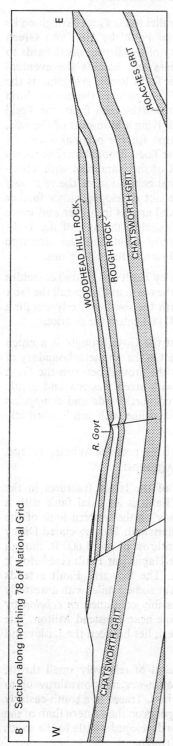

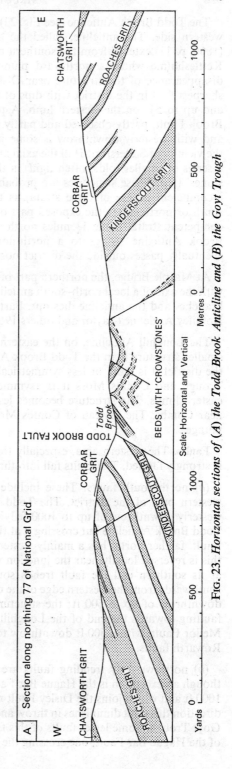

FIG. 23. *Horizontal sections of (A) the Todd Brook Anticline and (B) the Goyt Trough*

The Todd Brook Anticline (see Fig. 23) lies parallel to the Goyt Trough on its western side. The anticline, called the 'Anticlinal Fault' by Hull and Green (1866, p. 11) extends from the southern boundary of the district northwards to Kettleshulme where a northward plunge appears and heralds the eventual disappearance of the structure near Disley. The Todd Brook Anticline is the sharpest fold in the district with dips of up to 70° in parts of the eastern limb and up to 54° on the western limb. A persistent north–south fault, the Todd Brook Fault, partly observed and partly inferred, lying a little west of the axis, and with westerly downthrow is some justification for the original name. In the vicinity of Saltersford Hall the axial area of the Todd Brook Anticline shows a series of smaller folds, often tight, in that part of the Namurian with 'crow-stones' ($E_2$). These structures are probably a local expression of the relatively incompetent nature of these strata, as they are not developed either farther south or north where the exposed part of the axial area is in higher and more competent strata. Some 1½ miles north of Kettleshulme the axis of the Todd Brook Anticline swings to a north-north-westerly trend and the structure eventually passes outside the district north of Disley, before dying out.

At Marple Bridge, the northern part of the Disley Fault (see below) coincides with the axis of a local north–south anticline. To the north of Compstall the fault branches and the anticline dies out. Farther north the south-westerly trending Romiley Anticline (Taylor and others 1963, p. 84) just enters the district.

The Longhill Anticline, on the eastern limb of the Goyt Trough, is a much smaller structure than the Todd Brook Anticline. On the southern boundary of the district it is more or less symmetrical but in the ground between the Goyt Trough and Combs Moss it is asymmetrical with steep western and gentle eastern limbs. The structure becomes less marked northwards and disappears near Combs. The plateau of Combs Moss is a syncline with gentle northerly plunge.

**Faults.** The western area, especially that part lying north of Whaley Bridge, is strongly faulted. The faults fall into three main groups:

(a) north–south faults. These include many of the larger fractures in the western part of the district. The Todd Brook Fault is an axial fault with a westerly downthrow of up to 1000 ft lying mainly on the western limb of the Todd Brook Anticline but crossing it at the northern end. The associated Disley Fault, farther north, has a mainly westerly downthrow of up to 600 ft, though this is reversed locally near the junction with the Hague Bar Fault (see below); at its southern end the fault trends south-east. The Rowarth Fault extends southwards from the western edge of the district for some 6 miles with a westerly downthrow of up to 600 ft; the structure is possibly continued *en échelon* by faulting towards the end of the Longhill Anticline near Tunstead Milton. The Mellor Fault, with a 500 ft downthrow to the west, lies between the Disley and Rowarth faults.

(b) north-westerly trending faults are in general of relatively small throw, though an exception is the Hague Bar Fault of south-westerly downthrow up to 1000 ft and which joins the Disley Fault near Strines. Traced in a south-easterly direction this fault diminishes in throw and disappears on the western limb of the Goyt Trough. Some lesser faults of this trend are developed a little to the south of the Hague Bar Fault, one crossing the Goyt Trough.

(c) east–west faults are generally of small throw, the largest being one at Chinley with a 300 ft northerly downthrow; this fault splits westwards and crosses the Goyt Trough. Another prominent fault of this group follows the Sett valley at Birch Vale and has a southerly throw of up to about 150 ft.

*Central and Eastern Areas.* The Kinder Scout–Bamford area is characterized by very gentle structures. Faulting is rare and throws of the faults are small. The main structural features are two shallow anticlines with associated synclines.

The Edale Anticline is of general east–west trend and extends from west of Barber Booth to near Bamford. The structure is slightly asymmetrical (perhaps due to the proximity of the block), with dips on the southern limb of up to 12° and on the northern up to 8°. The possibility of an oil accumulation in this structure resulted in the drilling of the Edale Borehole in the axial area (Hudson and Cotton 1945). This found no oil but a small quantity of methane was encountered (ibid., p. 32). The Alport Anticline is a shallower structure and extends for a distance of over 7 miles from the Alport valley to Derwent Moors. The trend is east-south-easterly and dips on the flanks are up to about 7°. The elevated area of Kinder Scout forms the broad and very shallow Kinderscout Syncline, between the Edale and Alport anticlines, and the Fagney Syncline lies north-north-east of the latter.

The Hope Valley Anticline is a shallow structure trending east–west parallel to the Edale Anticline from Hope to Brookfield Manor and with easterly plunge at the eastern end.

In the Bradfield and Curbar areas on the eastern margin of the district the predominant structural feature is an easterly dip towards the Yorkshire and East Midlands Coalfield. On this regional dip two gentle synclines are superimposed at Bradfield Moors, and east of Stanage; both of these structures continue eastwards into the Sheffield district (Eden and others 1957). Some faulting of general south-easterly trend is also a feature of this tract, though the throws are not large.

The Abney area shows a different pattern with low dips and some easterly faults which represent a continuation of the veins of the limestone outcrop. Low dips at the limestone-shale junction suggest that part, at least, of this tract overlies the stable block.                                                     I.P.S.

## REFERENCES

COPE, F. W. 1949. Woo Dale Borehole near Buxton, Derbyshire. *Q. Jl geol. Soc. Lond.*, **105**, iv.

EARP, J. R., MAGRAW, D., POOLE, E. G., LAND, D. H. and WHITEMAN, A. J. 1961. Geology of the country around Clitheroe and Nelson. *Mem. geol. Surv. Gt Br.*

EDEN, R. A., STEVENSON, I. P. and EDWARDS, W. 1957. Geology of the country around Sheffield. *Mem. geol. Surv. Gt Br.*

GEORGE, T. N. 1962. Devonian and Carboniferous foundations of the Variscides, in north-west Europe. *in Some aspects of the Variscan fold belt*, ed. K. Coe. Manchester.

HUDSON, R. G. S. and COTTON, G. 1945. The Carboniferous rocks of the Edale Anticline, Derbyshire. *Q. Jl geol. Soc. Lond.*, **101**, 1–36.

Hudson, R. G. S. and Mitchell, G. H. 1937. The Carboniferous geology of the Skipton Anticline. *Summ. Prog. geol. Surv. Gt Br.*, part 2, for 1935, 1–45.

————— and Turner, J. S. 1933. Early and Mid-Carboniferous earth movements in Great Britain. *Proc. Leeds phil. Soc.*, **2**, 455–66.

Hull, E. and Green, A. H. 1866. The geology of the country around Stockport, Macclesfield, Congleton and Leek. *Mem. geol. Surv. Gt Br.*

Kent, P. E. 1966. The structure of the concealed Carboniferous rocks of north-eastern England. *Proc. Yorks. geol. Soc.*, **35**, 323–52.

Shirley, J. and Horsfield, E. L. 1945. The structure and ore deposits of the Carboniferous Limestone of the Eyam district, Derbyshire. *Q. Jl geol. Soc. Lond.*, **100**, 289–308.

Simpson, S. 1962. Variscan orogenic phases. *in Some aspects of the Variscan fold belt*, ed. K. Coe. Manchester.

Stille, H. 1924. *Grundfragen der vergleichenden Tektonik*. Berlin.

Taylor, B. J., Price, R. H. and Trotter, F. M. 1963. Geology of the country around Stockport and Knutsford. *Mem. geol. Surv. Gt Br.*

Teisseyre, H. 1959. Zu dem Problem der Diskordanz zwischen den Waldenburger Schichten und dem Kulm in der Innersudetischen Mulde. *Geologie*, **8**, 3–12.

White, P. H. N. 1948. Gravity data obtained in Great Britain by the Anglo-American Oil Company Limited. *Q. Jl geol. Soc. Lond.*, **104**, 339–64.

# Chapter IX

# PLEISTOCENE AND RECENT

## INTRODUCTION

DEPOSITS of Pleistocene and Recent age, although not present in substantial thicknesses, are widespread within the district and vary considerably in origin, composition and distribution. These variations reflect both the morphological differences within the district and the contrasting glacial histories of the western and eastern sides of the southern Pennines. The occurrence of the principal drift deposits is illustrated in Fig. 24.

In the western part of the district glacial deposits including boulder clay, sand and gravel, and in one place glacial lake deposits, are continuous, or nearly so, with the thick glacial deposits of the Lancashire and Cheshire plain; glacial drainage channels are also present in several places. By contrast, in the central and eastern parts of the district the only undoubted traces of glaciation are small patches of boulder clay and scattered erratics within or adjacent to the limestone area in the south.

Head deposits and valley-bulge structures, largely of periglacial origin, and landslips, believed to be of relatively recent age, are best developed in the valleys of the central, northern and eastern parts of the district occupied by the Millstone Grit Series. Extensive peat deposits occur on the moors in these areas. River terraces and flood-plain alluvium are present along the rivers Etherow, Goyt, Derwent and Noe, and also along some of their tributaries.

*Previous research.* The first comprehensive description of the glacial geology of the district was by Jowett and Charlesworth (1929) who explained the glacial phenomena on the western Pennine slopes as the product of a single ice advance and retreat. Doubt was cast on this interpretation by Tonks and others (1931, p. 181) because of the difficulty of accounting for the upper boulder clay in the Manchester district. Both Wills (1937) and Jones and others (1938, p. 114) invoked a re-advance of ice to solve similar problems in nearby districts. In more recent times controversy has focused on three interrelated questions: the number of ice advances responsible for the glacial deposits of Lancashire and Cheshire; the chronology of these deposits and the correlation of the largely indivisible upland deposits on the western Pennine slopes with the sequence found in the lowland areas. Most of the discussions of these problems have been concerned with evidence from ground to the west and south-west of the Chapel en le Frith district, and summaries are given by Taylor and others (1963) for the Stockport and Knutsford (98) district and by Evans and others (1968) for the Macclesfield (110) district. Only Rice (1957) and Johnson (1965b), in studies largely concerned with aspects of final deglaciation, have referred in any detail to glacial evidence in the western part of the Chapel en le Frith district.

The sparsely distributed glacial evidence from the central and eastern parts of the district is little referred to in the literature. The first regional compilation of these scattered glacial remains, largely consisting of erratics, was given by Green and others (1887, pp. 90–8). Subsequently Bemrose and Sargent (1915, p. 99) recorded "erratics of mountain limestone . . . together with an igneous rock (probably from the Lake District)" from the Derwent valley, a short distance south of Howden Dam. References to all previously known finds of glacial deposits and erratics on the Derbyshire uplands are given by Jowett and Charlesworth (1929) who inferred a northern and north-westerly origin for the deposits. Dalton (1945, 1953, 1958) has drawn attention to scattered glacial deposits and erratics in the southern part of the district. In an assessment of evidence from the district immediately to the south, Straw and Lewis (1962) suggest the possibility of recognizing two older glaciations in the region.

Rice (1957) concluded that the river terraces and alluvium in the Goyt valley indicate five successive post-glacial phases of stability and flood-plain development, but evidence for only the oldest two of these is present in the Chapel en le Frith district. Waters and Johnson (1958) recognized four terraces along the entire length of the Derwent valley, only the upper two, both of which were considered by these authors to be pre-last glaciation in age, being present in the district.

A palynological study of peat at Ringinglow, just outside the eastern boundary of the district, was made by Conway (1947) who subsequently (1954) extended her investigations to peat deposits on Bleaklow (to the north of the district) and Kinder Scout. The drainage and erosion of peat on the moors in the northern part of the district has been studied by Johnson (1958).

Some comments on the landslips in the valleys of the River Derwent and its tributaries have been made by Lounsbury (1962). More recently the history of the Charlesworth landslip has been outlined by Franks and Johnson (1964), and Johnson (1965a) has studied the factors controlling slip movement at this locality.

*Geomorphology.* Studies of the landforms in the southern Pennines have produced conflicting explanations of the morphological evolution of the region. These explanations are largely concerned with hypothetical pre-Pleistocene events and are not based upon depositional evidence; they involve consideration of denudation chronology, drainage evolution and tor formation, all of which have a bearing on the post-Carboniferous geological history of the Chapel en le Frith district.

Fearnsides (*in* Fearnsides and others 1932, p. 177) suggested that a vestigial summit peneplain is recognizable on the highest ground in the Peak District and that it is a relic of the post-Hercynian erosional phase, being in effect a continuation of the surface underlying the Permo-Triassic rocks to the east. In addition he considered that the River Derwent is a subsequent stream which has cut back along the shale outcrops of the lower Millstone Grit. The western edge of the district was included in a study of relict surfaces by Miller (1939) but the surfaces recognized are comparatively low and of little consequence to the district in general. Linton (1951) contended that the River Derwent is an original tributary of a proto-Trent initiated on an emergent Chalk surface which had suffered pre-Eocene warping. This contention was opposed by Sissons (1954) who suggested that an initial sub-aerial peneplain had been submerged by down-warping to the east and then subjected to periodic emer-

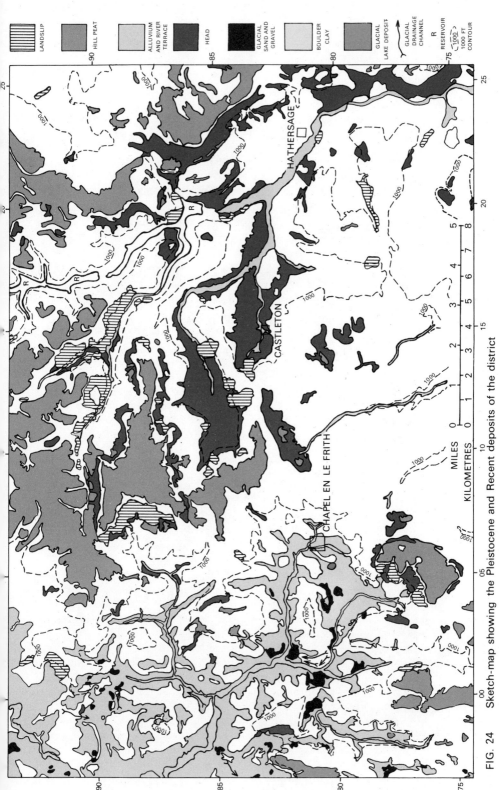

| | LANDSLIP |
| --- | --- |
| | HILL PEAT |
| | ALLUVIUM AND RIVER TERRACE |
| | HEAD |
| | GLACIAL SAND AND GRAVEL |
| | BOULDER CLAY |
| | GLACIAL LAKE DEPOSIT |
| | GLACIAL DRAINAGE CHANNEL |
| R | RESERVOIR |
| 1000 | 1000 FT CONTOUR |

HATHERSAGE

CASTLETON

CHAPEL EN LE FRITH

MILES

KILOMETRES

FIG. 24   Sketch-map showing the Pleistocene and Recent deposits of the district

gence, resulting in the formation of a series of wave-planation surfaces between 1800 ft and 500 ft O.D., which were considered to be later Tertiary in age. The drainage evolution of the Derwent and its tributaries was explained by Sissons as a sequence of consequent and subsequent streams developed as each planation surface emerged. The main planation surfaces recognized were the Holme Moss surface (1600 to 1800 ft O.D.), the Bradfield surface (1250 to 1380 ft O.D.) and the Wharncliffe surface (990 to 1070 ft O.D.). A review of the various aspects of denudation and drainage for the central and eastern parts of the district has been given by Linton (1956). Further comments on the southern part of the district and beyond have been added by Johnson (1957).

The western edge of the southern Pennines has been investigated by Johnson and Rice (1961) who suggested that the drainage in this region, including the River Goyt, may have originated as west-flowing consequent streams on a west-facing Chalk monocline, appreciably modified, after removal of the Chalk, by the structural pattern of the underlying Carboniferous rocks. These authors also recognized a widespread undulating surface at 1200 to 1400 ft O.D.

The sandstone tors in the district have been referred to in attempts to explain the origin of British tors in general. Linton (1955) postulated a two-stage process, firstly of differential decomposition of rock by deep rotting in warm, supposedly later Tertiary or interglacial climates, forming 'corestones' of unrotted rock, and secondly of exhumation of the 'corestones' by periglacial processes. Palmer and Radley (1961) did not accept the 'corestone' theory as applied to the Pennines, but advocated instead a mechanism of tor formation related to slope development and the retreat of 'free' rock faces under periglacial conditions.

G.D.G.

# GENERAL DESCRIPTION

## BOULDER CLAY

Boulder clay occurs in the western part of the district both as valley-fill and as less extensive patches on higher ground. It is a stiff till, brown in surface exposures and blue-grey at depth, which contains the familiar suite of erratics of Irish Sea type including particularly Lake District rocks from the Borrowdale Volcanic Series and the Eskdale and Ennerdale intrusions, as well as coarse granites of Criffell-Dalbeatie type.

Boulder clay is widespread in Longdendale. In the Goyt valley it also forms extensive spreads, including the deposits filling the pre-existing valley at Hague Bar and New Mills. These were described by Jowett and Charlesworth (1929, p. 324) as moraines, though this has been denied by Rice (1957). Boulder clay also extends up most of the associated minor valleys, particularly those at Rowarth, Hayfield and Chinley, and an extensive spread is present in the broad valley at Chapel en le Frith. The valley of Todd Brook shows a substantial fill of boulder clay. The higher deposits extend up to about 1000 ft O.D. on Ludworth Moor and at Fernilee, while on the western side of the Todd Brook valley they reach a height of 1200 ft O.D. at Blue Boar Farm. These patches of boulder clay and isolated erratics occurring up to heights of about 1200 ft O.D. on the western Pennine slopes bear a clear geographical relationship to the main western area of boulder clay.

Half a mile to the south-east of Chapel en le Frith the presence of a small area of boulder clay at a height of 1000 ft O.D. at Draglow lends support to the conclusion of Jowett and Charlesworth (ibid., p. 319) that the main mass of ice reached the col at Barmoor Clough.

In contrast to the above, the occurrences of both boulder clay and erratics on the limestone outcrop are very scattered and appear to result from a glaciation older than that responsible for the western boulder clay (see p. 340). The most important deposit lies south of Stoney Middleton in a hollow, which may be responsible for its preservation. Elsewhere the vestiges of glaciation are confined to scattered erratics, which are most abundant in the Great Hucklow–Eyam area and around Dove Holes.                                                I.P.S.

In the north-eastern part of the district only two occurrences of erratics are known, and both are doubtful: on Featherbed Moss, north of Kinder Scout, boulders of sandstone of a type not known *in situ* have been noted, and a cobble of igneous rock has been recorded by Bemrose and Sargent (1915, p. 99) from the Derwent valley east of Birchinlee Pasture.                          G.D.G.

GLACIAL SAND AND GRAVEL

Deposits of glacial sand and gravel are found only in the western part of the district within approximately the same geographical limits as the boulder clay. These deposits either overlie boulder clay or rest on solid rocks and are considered to have been formed during the deglaciation of the area. The only place where sand and gravel have been observed to underlie boulder clay is at Buxworth. The deposits can be grouped as follows:

(a) The Hollinsmoor–Rowarth deposits, developed along the valley sides between these two places, are associated with a glacial drainage channel at the western end. The deposits, which show evidence of strand-lines at heights of 815, 795, 775 and 715 ft O.D., are partly in the form of a delta formed in water ponded in the Rowarth valley by ice downstream at or near the junction with the Goyt valley.

(b) The Todd Brook moraine, described by Jowett and Charlesworth (1929, p. 323) blocks the pre-existing valley of Todd Brook at Pennant Nob, diverting the stream through a rock gorge on the south side. The total thickness of sand and gravel is of the order of 100 ft and the uppermost 40 ft, showing marked westerly foreset bedding, is seen in the sand pit in the upper part of the deposit (see Plate XXIV A). The steep eastern slope of Pennant Nob is considered to be an ice-contact slope. A smaller, but similar mass of sand and gravel is present in the Goyt valley near Fernilee.

(c) The Buxworth deposits, described as a possible outwash fan by Jowett and Charlesworth (ibid., p. 323–4) are interpreted here as a delta deposited from the Roosdyche channels (see below) in an area clear of ice. The maximum thickness of sand and gravel is about 100 ft and the deposits are overlain by boulder clay.

(d) Other occurrences of sand and gravel are in many cases difficult to interpret. They vary in height from about 500 to 1000 ft O.D. and the higher ones may well have been formed in earlier stages of the ice-retreat. Some small patches around the Todd Brook valley may be strand-line deposits formed at the time of the blocking of the valley by ice.

## GLACIAL LAKE DEPOSIT

A small area of laminated clay has been proved in temporary sections on the south-west side of Chapel en le Frith. The maximum observed thickness is 5½ ft. The deposit is considered to have been formed in a lake resulting from the ponding of waters in the Chapel en le Frith and Chinley valleys (see also p. 340).

## GLACIAL DRAINAGE CHANNELS

Glacial drainage channels are present in a number of places in the western part of the district. The best developed of these, the Roosdyche at Whaley Bridge, has been interpreted (Johnson 1963) as subglacial. Other channels are perhaps subaerial.                                          I.P.S.

## HEAD

Three varieties of head are recognizable in the district. They are scarp-slope deposits, valley-fill deposits and residual deposits on limestone outcrops.

Widespread deposits of head are present below scarp slopes of sandstones, the best developed examples being below 'edges' of Kinderscout Grit and Chatsworth Grit on the eastern side of the Derwent valley, and also around Kinder Scout and Combs Moss. The deposits consist of a heterogeneous matrix, varying from grey clay to rubbly sand, containing a variable proportion of angular and sub-angular rock fragments from up-slope outcrops, usually of sandstone and up to several feet in size. Where fresh exposures are seen there is little suggestion of true bedding, but there is a tendency for the coarser blocks and fragments to be concentrated on distinct horizons roughly parallel to the bedrock slope in an apparently structureless matrix. Dark brown 'iron-pan' horizons occur in places.

Scarp-slope head deposits present an irregular hummocky appearance, with low impersistent ridges normal to the slope, irregularly distributed springs, an abundance of boulders and with stream incision more marked than on head-free shale slopes. Many of these deposits merge up-slope into boulder screes or 'block-fields' below sandstone 'edges', but some pass upwards into or under landslips. In a few localities, especially on the south-eastern corner of the Kinder Scout plateau, the upper edge of the head is coincident with the base of one or more bench-like features which are suggestive of the altiplanation terraces described from Devon (Te Punga 1956) and elsewhere, and attributed to periglacial weathering. The deposits thin out laterally against spurs and slopes of increasing steepness. Downslope the matrix tends to become more clayey and there are fewer cobbles and boulders, particularly where the slope nearly levels out. Where there is an abrupt increase in slope with incised gullies, a common feature on outcrops of Shale Grit, the head is funnelled into the gully, forming 'stone rivers' similar to those described from the moors just beyond the eastern edge of the district by Eden and others (1957, p. 160). A greater tendency towards bedding is noticeable where scarp-slope head encroaches on to the margins of valley floors, and in a few places a passage into head of valley-fill type is apparent.

In the western part of the district scarp-slope head deposits appear to overlie boulder clay in a few localities. In the north-east they are overlain by peat and landslips in places, and in the Derwent valley, west and south of Hathersage, they pre-date the formation of the highest terrace. If the scarp-slope head is all of approximately the same age, and there is no evidence to the contrary, these deposits would appear to have formed towards the end of Weichselian times (see pp. 339–40) and, particularly at the higher altitudes occurring in the district, possibly continued to form until the development of a substantial vegetation cover in early post-Weichselian times.

Deposits classified as valley-fill head are most widespread in the Edale valley (where they are best developed, and up to 30 ft thick) and in the Hope Valley. They are also present as smaller spreads in the Perryfoot, Derwent, Ashop and Alport valleys. Along the margins they are similar lithologically to scarp-slope head, being ill sorted and apparently almost structureless. In the Edale valley, where exposures are most numerous, the deposit changes in two respects as it is traced towards the middle of the valley. The enclosed rock fragments become progressively more rounded and distinct stratification with localized cross-bedding and stream-layering of pebbles becomes apparent, the deposit becoming lithologically indistinguishable from river gravel. Furthermore, the basal head in the Edale valley is cemented with a tough brown ferrocrete, possibly derived from ironstone bands in the underlying rocks.

The upper margins of valley-fill head tend to follow the contours between tributary streams, but in the stream valleys the deposits extend to higher altitudes and are often continuous with 'stone rivers' and scarp-slope head on higher slopes. Valley-fill head presents a smooth concave slope levelling out towards the middle of a valley. The terraces along the rivers Derwent and Noe and their tributaries are largely contained within channels incised into valley-fill head north and west of Bamford station, but in places deposits of the highest terrace have overlapped on to the adjacent surface of head. In the Edale valley the thickest valley-fill head deposits occur along the lower courses of tributary streams and they thin towards the interfluvial areas and towards the middle of the main valley. This thickness variation suggests that the deposits consist of a number of interlinked fan-like spreads each radiating from the mouth of a tributary valley. The valley-fill head in the Hope Valley, though less well exposed, appears to have similar characteristics. The spreads in other valleys appear in places to be intermediate between valley-fill head of Edale valley type and scarp-slope deposits.

Valley-fill head appears to pre-date all terrace, alluvium, peat and landslip deposits with which it is in contact. It appears likely that the deposit formed in later Weichselian and possibly earliest post-Weichselian times, when an abundance of ground and surface water and lack of a substantial vegetation cover would be expected to facilitate this type of deposition.                    G.D.G.

Deposits of cherty loam and cherty rubble are present in a number of localities on the limestone outcrop. Lithologically the deposits appear to be similar to the Clay-with-flints found on Chalk in southern England, and it appears probable that they are residual deposits derived from limestone which, in some localities at least, have been subject to solifluxion. They have therefore been shown as head on the one-inch map.

## RIVER TERRACES

As the district only includes the upper reaches and tributaries of the rivers

Goyt and Derwent it has been found impracticable to subdivide these deposits on the one-inch map. In the Goyt valley up to three terraces are developed but their relations are complex (see p. 346). The highest observed lies about 17 ft above the floodplain between Hague Bar and New Mills. Spreads of terrace deposits are also present in the Etherow valley and in some of the larger tributaries of the Goyt.                                                      I.P.S.

Four terraces have been recognized along the rivers Noe and Derwent north-west of Hathersage Booths. The two highest terraces at 17 to 20 ft and 11 to 15 ft above river level are in part rock benches overlain by sand and gravel, often coarse and ill-sorted and up to 11 ft thick. The two lower terraces at about 8 ft and 3 to 5 ft above river level consist largely of silty clay, sand and fine gravel, with a low rock bench visible in places in the second terrace. The downstream limit of these terraces is the outcrop of Kinderscout Grit across the Derwent valley south of Hathersage. The fourth (highest) terrace is the most widely developed, with the third terrace fully developed only beyond the upstream limit of the fourth terrace. A probable explanation is that the third terrace is merely an incised meander surface within the fourth terrace, and not a separate aggradational phase. All the upper Derwent–Noe terraces post-date any scarp-slope and valley-fill head with which they are in contact. They are interpreted as resulting from alternating phases of aggradation and incision consequent upon the periodic lowering of the rock bar where the Kinderscout Grit outcrop crosses the Derwent valley south of Hathersage, and are probably Weichselian or early post-Weichselian in age.

## ALLUVIUM

Alluvial deposits largely consisting of silty clay, sand and fine gravel form floodplains along the rivers Etherow, Goyt, Derwent and their tributaries. In a number of places the alluvium terminates downstream at some feature which has reduced the river gradient and promoted deposition upstream, such as resistant rock outcrops, landslips and water levels of reservoirs.

## PEAT

The thickest and most widespread peat deposits occur on Kinder Scout and the surrounding moors, on Combs Moss and other upland areas adjacent to the Todd Brook valley and on the higher moors east of the Derwent valley. These localities are level or gently sloping, consist mainly of sandstone at outcrop and are over 1200 ft O.D. Veneers of thin peat or peat-rich weathered derivatives of the underlying rock are present in many other parts of the district above 1000 ft O.D., except on the limestone outcrop, where peat is only locally developed on residual head deposits.

Three zones can be recognized in a fully developed peat section. The basal zone, normally about 1 ft thick, is grey or grey-brown and compact, and in places contains sporadic tree 'stools' *in situ*. The middle zone of variable thickness is dark brown and fairly compact. The uppermost zone, in places constituting up to two-thirds of the total peat thickness, is red-brown, soft and spongy, and commonly has a banded appearance.

Most of the peat deposits rest directly on the weathered surface of the solid rock, which is normally a leached white rubbly sand or pale grey clay derived from sandstone or shale. Peat rests on boulder clay in the Todd Brook valley and on head in the Perryfoot valley, on Black Ashop Moor, Ashop Moor, Bamford Moor and elsewhere; it also occurs on head on the limestone outcrop at Bradwell Moor.

Much of the peat is suffering active erosion by wind and rain, and on some areas of exposed level ground or convex slopes the deposit has been reduced to residual hummocks. Detailed accounts of peat drainage and erosion have been given by Conway (1954) and Johnson (1958).

Conway has investigated the palynology of peat at Ringinglow just outside the eastern boundary of the district (1947) and subsequently of Bleaklow (just outside the northern boundary of the district) and of Kinder Scout (1954). From these investigations she concluded that peat formation commenced at the Boreal–Atlantic transition (zones VI–VIIa boundary) when pine woodland was replaced by bog vegetation on the higher moors and when damp grassy heath with open alder-birch woods developed on lower ground and in sheltered localities. Although *Sphagnum* was not initially common in the bog vegetation, it became the dominant peat-forming plant with the onset of wetter conditions, and Conway suggested that the rapid growth of thick soft saturated *Sphagnum* peat eventually built up an unstable deposit prone to bog-bursts and the initiation of rapidly eroding channels.

## BONE CAVES

Mammalian remains have been found in a number of caves and fissures in the Carboniferous Limestone.

**Dove Holes.** A cave filled with "stratified, yellowish-red clay" containing limestone and other pebbles and irregularly distributed bones and teeth was discovered during quarrying at a locality [0792 7695] a little to the east of Victoria (or Victory) Quarry, Dove Holes, and was described by Dawkins (1903). From the worn condition of some of the bones, some postulated teeth marks and the presence of *Hyaena* remains, he suggested that the fossils had accumulated in a *Hyaena* den at a higher level and had subsequently been washed deeper into the cave. The bones have been re-examined by Dr. A. Sutcliffe and Mr. H. E. P. Spencer and provisionally re-determined as follows: *Homotherium* [*Machairodus*] *crenatidens* (Fabrini), Hyaenid indet. (about the size of the present-day *Crocuta*), *Anancus* [*Mastodon*] *arvernensis* (Croizet and Jobert), *Elephas sp.* indet. and Rhinocerotid indet. Mr. Spencer also recognizes *Equus robustus* Pomel (metatarsal only; the teeth mentioned by Dawkins are not coeval with the other bones), *Dama sp.* (probably *D. nesti nesti* Forsyth Major) and *Euctenoceros sp.* The markings attributed to gnawing by hyaenas are due to some other cause. A full revision of the fauna is in preparation. It was considered by Dawkins to be of Upper Pliocene age because of similarities with Villafranchian faunas from the Red Crag of East Anglia and certain deposits in the Auvergne region of France and the Arno valley in Italy, all of which were then classified as Pliocene. Deposits containing Villafranchian faunas are now included in the Lower Pleistocene and the Dove Holes fauna is here provisionally assigned to this period.                                                       G.D.G.

**Windy Knoll.** The mammalian fauna from Windy Knoll [1266 8302] was obtained (Pennington 1875, pp. 241–4) from a fissure in the quarry, then working. The bones were found lying together in a confused mass in a loamy matrix and were considered to have been washed into a swallow hole. The fauna listed by Dawkins (1875, p. 246) comprised *"Arvicola amphibia"*, *"Bison priscus"*, *"Canis lupus"*, *"C. vulpes"*, *"Cervus tarandus"*, *"Lepus cuniculus"*, *"L. timidus"* and *"Ursus ferox"*[1]. Dawkins noted (ibid., p. 255) that none of these forms was extinct. Subsequently (1877, p. 729) he ascribed a late Pleistocene age to the deposit.

**Hartle Dale.** A small cave in this dale [1633 8039] has yielded teeth of woolly rhinoceros and bones of aurochs and mammoth (Pennington 1875, p. 241).

**Cave Dale.** The cave [1503 8267] near the entrance of the valley, described by Pennington (ibid., pp. 238–40) yielded a Recent mammalian fauna associated with one flint and was probably of Neolithic age, extending into the Bronze and Iron ages.                                                                                      I.P.S.

## LANDSLIP

Numerous and in places extensive landslips are present in the district, occurring mainly in valleys within the Millstone Grit outcrop. The landslips at Charlesworth are from the Rough Rock and those at Combs Moss involve the Chatsworth Grit, but in the central and eastern parts of the district it is the Mam Tor Beds, Shale Grit and Kinderscout Grit that have moved. Some landslips are individual with a single concave dislodgement scar. Others are compound, either from coalescing of individual slips or from the general collapse of a length of slope. Most slipped masses are well fragmented, but in a few localities, particularly in the Alport Castles landslip, and possibly in the Cowms Rocks landslips in the Ashop valley, quite substantial areas have slipped *en masse* without breaking up. In a number of places, for example on Ladybower Tor, fissures are present in the ground above and behind existing dislodgement scars, suggesting the initiation of future landslips. The character and cause of each landslip are determined by a combination of the local stratigraphy, structure, morphology and drainage. Johnson (1965a) has studied the factors controlling the Charlesworth landslips, and emphasised that the development of factors favourable to landslipping is a long-term process in which stream incision and progressive stratal weakening by groundwater are important.

The following practical subdivision into three types of landslip can be made: collapsed scarps, down-dip slides and rotational slumps. Collapsed scarps may be individual or compound; they are well fragmented and occur where steep shale slopes are present below massive sandstone scarps, particularly those of the Kinderscout Grit or Chatsworth Grit. The dislodgement scars are irregular and unless very recently collapsed they are little different from adjacent unslipped scarps. Typical examples are present around the edge of the Kinder Scout plateau. Down-dip slides are mainly compound landslips, many include areas of unbroken ground and occur where stream incision has produced steep slopes in a similar direction to the tectonic dip. Their dislodgement scars are typically wide and craggy. The best examples, involving the Shale Grit and Mam Tor beds, are those in the Ashop and Alport valleys. Rotational slumps are mainly individual landslips and are well fragmented, and most examples occur on steep slopes composed of Shale Grit and/or Mam Tor Beds which have slipped downwards and outwards in front of concave, high but relatively narrow

---

[1]Respectively: water-vole, bison, wolf, fox, reindeer, rabbit, hare, grizzly bear.

dislodgement scars. In some localities they occur below a prominent spring line; good examples are present in the southern and western slopes of the Edale valley. In some landslips movement must have been rapid. Highly polished rock surfaces have been found in places just below the exposed part of the Mam Tor dislodgement scar where the slipped debris has been excavated to a depth of about 4 ft.

Some landslips are still active, for example, those in the Ashop valley and at Mam Tor (see Plate XVI B). Others have obviously not been active for a considerable time; one example, a mile south-east of Win Hill summit, has a greatly denuded slip mass and an overgrown dislodgement scar. Some landslips certainly rest on head deposits, and others apparently overlie alluvial deposits as in the Edale, Ashop and Alport valleys; elsewhere alluvium appears to have been deposited as a result of landslips reaching rivers in these valleys (see p. 347). Franks and Johnson (1964) have shown by means of pollen analysis that the earliest traceable movement in the Charlesworth landslips was of Late Boreal (Zone VI) age.

## SUPERFICIAL STRUCTURES

**Valley-bulge structures.** The Edale valley contains numerous exposures of structures indicative of valley-bulging. Around the margins of the valley floor the highest Edale Shales and lowest Mam Tor Beds exhibit sharp, commonly nearly symmetrical folds with straight limbs. More rarely sharp asymmetrical folds are associated with small low-angle thrust dislocations. At exposures nearer the middle of the valley the lower Edale Shales, devoid of thin sandstone or ironstone bands, are more irregularly crumpled, dislocations are at higher angles but the overall intensity of deformation decreases. In the middle of the valley only a few outcrops exhibit minor folds or dislocations, but dips are relatively steeper than on the surrounding slopes, suggesting that the flanks of the Edale Anticline may have been over-steepened by valley-bulge movements.

Sharp, symmetrical, straight-limbed folds are also present in lower Mam Tor Beds and upper Edale Shales in the lower Alport valley and adjacent parts of the Ashop valley, good examples being exposed near the confluence of the rivers Alport and Ashop (see Plate XXIV B). Similar structures also occur in Mam Tor Beds in the Derwent valley and its tributary valleys, for example Abbey Brook and Ouzelden Clough. They were also seen in the floor of the Derwent valley when the level of Ladybower Reservoir was greatly reduced in the 1959 drought, and can be seen in a few outcrops in the river south of Ladybower Dam and also in the Hope valley.

Trenches excavated for the foundations of Howden and Derwent dams exposed a single large fold in each locality. These folds were seen to decrease in magnitude with depth, and strata more than 100 ft below original ground level were seen to be but slightly affected (Fearnsides and others 1932, p. 157). In the foundation trench for Ladybower Dam a large thrust-fold with a westerly-dipping axial plane was uncovered.

These structures have many similarities to those described in the Northampton region by Hollingworth and others (1944) and Hollingworth and Taylor (1951, pp. 31–5). They are believed to have resulted from the pressure exerted on predominantly argillaceous strata by superimposed beds inducing lateral

(L 501)

A. Pit in glacial sand and gravel, Pennant Nob, Whaley Bridge

PLATE XXIV

B. Valley-bulge structures in Mam Tor Beds: Rowlee Bridge

(L 498)

movement or 'squeezing out' of the argillaceous strata into adjacent load-free valley areas, producing compression folds and thrusts. Movement so caused would be possible wherever valley-deepening exposed the top of any substantial thickness of shale, but would be greatly facilitated by thawing of ground-ice following a glacial or periglacial phase.

In the Edale valley and on the southern side of the Ashop valley, near the Alport–Ashop confluence, the structures are generally truncated and where overlain by head the latter shows no clear evidence of movement. On the north side of the River Ashop, however, near Rowlee Bridge [149 892], a number of short steep ridges up to 20 ft high are present. At least one has an anticlinal core of shale. A possible explanation for these apparently later movements is the relatively large increase in load caused by the descent of large landslips on adjacent ground, as inferred by Lounsbury (1962).

No age can be assigned to these valley-bulge structures, but it seems probable that they are largely the result of glacial or periglacial episodes.

**Cryoturbation structures.** Widening of near-surface bedding planes, joints and cracks, amounting in some instances almost to fragmentation, is common in many sandstone outcrops down to depths of 5 ft, and attributable to freezing and thawing of groundwater. Examples can be seen at the top of a number of landslip-scars and in quarries such as those east of Ladybower House.

Distinct contortion of near-surface strata due to cryoturbation activity is relatively rare in the district. The largest example noted is in a quarry [2289 9268] in the Heyden Rock north of Bole Edge Plantation where the upper flaggy beds are folded and crumpled to a depth of 6 ft, the intensity of deformation decreasing sharply with depth.                                                    G.D.G.

At Combs, a sharp anticline in shales above the Roaches Grit is present in Meveril Brook [0435 7729] and another, though smaller, anticline occurs a few yards upstream; both are attributed to the effects of frost heaving.            I.P.S.

# INTERPRETATION OF GLACIAL DEPOSITS

The interpretation of the glacial chronology of the western part of the district is dependent on that of the adjacent Cheshire plain. In the east the scattered and fragmentary nature of the evidence makes correlation difficult.

*Western area.* The glacial deposits, especially the sand and gravel, commonly show relatively unmodified constructional features which are suggestive of relatively recent glaciation. The deposits of boulder clay are indivisible and there is no evidence of deposition from more than one ice sheet although minor advances and retreats of ice margins can be postulated (see p. 340). These deposits are continuous with the 'upland drifts' of the Stockport and Knutsford district to the west (Taylor and others 1963, pp. 98–9) and of the Macclesfield district to the south-west (Evans and others 1968, pp. 185–6). These 'upland drifts' were suggested by Taylor and others to be of the same age as the 'Lower Boulder Clay' of the Cheshire plain, which they considered to have been deposited during the 'Main Irish Sea Glaciation' of Wills (1937) and which was regarded as the earlier and more extensive of two major ice advances during the Weichselian glaciation. Evans and others (ibid., pp. 241–53) considered that the 'upland drift' to the south-west was deposited by the same ice sheet that laid down the 'Upper Boulder Clay' of the Cheshire plain, and that this ice was the only demonstrable ice advance into the eastern Cheshire plain during the Weichselian. They referred this ice advance to later Weichselian times commencing about 25 000 years ago. This conclusion has since been supported by radiocarbon dates from sands and gravels beneath boulder clay near Wolverhampton, which indicate that the maximum extension of an "Irish Sea Glacier to the Wolverhampton Line occurred in Late-Würm times" (Shotton 1967).          G.D.G.

At the maximum extent of the Weichselian ice sheet within the district the ice margin reached up to a height of 1000 to 1200 ft O.D. Early stages in the shrinkage of the ice sheet are marked by patches of sand and gravel at a height of 1000 ft O.D. on Ludworth Moor. Later stages in this process are difficult to place in a coherent sequence though certain features may be related to one another. (i) A tongue-like mass of ice is postulated as having remained in the Goyt valley at Whaley Bridge. This formed the sand and gravel at Pennant Nob and a smaller mass of sand and gravel at Fernilee. Subglacial channels in the ice lobe (Roosdyche and associated channels) debouched into an ice-free area at Buxworth to form a large sand and gravel delta possibly along the ice front. A small re-advance subsequently over-rode the Buxworth delta, ponding the Chinley and Chapel en le Frith valleys and leading to the formation of the laminated clay at Chapel en le Frith. (ii) Blockage of the Rowarth valley by ice downstream resulted in the formation of strand-line gravels at Rowarth. (iii) Shrinkage of the ice created a ponded mass of water at Disley which escaped, possibly by subglacial drainage, to the south of the Goyt valley. I.P.S.

*Central and eastern area.* Boulder clay on and near the Carboniferous Limestone outcrop is scarce and associated glacial features virtually absent but sufficient erratics occur to indicate the former presence of at least one ice sheet derived from the north-west. No direct conclusions can be drawn about the age of these sparse deposits, but it is unlikely that they are Weichselian: the western deposits of this age appear to be but little eroded and evidently reached a limit broadly coincident with their present distribution. The boulder clay at Stoney Middleton is more reminiscent of similar remnant patches of glacial deposits which are present in scattered localities on the eastern side of the southern Pennines, and which have been interpreted as pre-Weichselian (see for example Edwards and Trotter 1954, p. 68).          G.D.G.

## DETAILS

### BOULDER CLAY

*Etherow valley.* At Simmondley the southern part of a large low-lying spread of boulder clay in Dinting Vale on the Glossop (86) Sheet extends into the present district, passing round the solid outcrops at Charlesworth into the valley floor west of Cown Edge and thence to Chisworth.

On the southern side of Longdendale, boulder clay occurs as a thin sheet on the sloping valley side. In places on the northern side of the valley, the deposits are much thicker and a borehole [9784 9393] near Broadbottom and just outside the present district proved glacial sand and gravel to 33½ ft on 'clay' to 83½ ft.

*Hayfield, Low Leighton and Rowarth.* The valley between New Mills and Rowarth has a substantial fill of boulder clay. The eastern slope of the valley is concave, suggesting that considerable ice movement may have taken place, perhaps by overflowing from the high ground at Ludworth (see below). Fairly common exposures of boulder clay are present in the stream, for example one [0086 8875] near Rowarth, which shows 12 ft of clay with erratics. Nearby [0082 8878] a

6-ft boulder of andesite lies in the stream course. An irregular outlier of boulder clay is present at Matley Moor and east of Cown Edge where it reaches a height of about 1100 ft O.D. The flat expanse of boulder clay on Ludworth Moor lies at about 1000 ft O.D. and on the south-east side of this area the deposit slopes down to join the valley-fill at Rowarth. Mellor Moor forms another plateau-like expanse of boulder clay to the west of the New Mills–Rowarth valley.

Boulder clay forms somewhat irregular patches in the valley of the Sett between Thornset and Birch Vale, and more continuous spreads occur on the valley floors both north and south of Hayfield. The latter deposit extends up to over 1000 ft O.D. at its southern end. In the Little Hayfield valley boulder clay is restricted to the valley floor but the presence of scattered erratics on the dip-slope of the Upper Kinderscout Grit, up to the general level of the 1000-ft O.D. contour, shows the ice to have been more extensive than the boulder clay outcrop (see p. 331). An example is a 2½-ft granite boulder [0369 8866] north of Park Hall at a height of about 910 ft O.D. A capping of boulder clay is present on the western part of the Lower Coal Measures outlier at Thornset; the drift-free parts of the area show scattered erratics including a 1½-ft granite [0218 8792] at a height of about 1020 ft O.D. on the western slopes of Lantern Pike.

*Goyt valley.* In this valley boulder clay occurs as a wide spread between Bottom's Hall and New Mills and as a less continuous spread in the floor and on the sides between New Mills and Whaley Bridge. At Strawberry Hill, Hague Bar and New Mills large masses of boulder clay have obstructed the valley and these have been interpreted by Jowett and Charlesworth (1929, p. 324) as moraines which brought about the diversion of the river into rock-gorges. An alternative explanation by Rice (1957) of the diversions is that the 'moraines' are residual masses remaining after the re-excavation of the original fill of boulder clay and that the gorges are entrenched meanders.

The Strawberry Hill diversion lies partly outside the present district though the original valley as noted by Rice (1957, fig. 1) lies within it and to the east of the gorge, beneath boulder clay. The general relations

x

are similar to those at Hague Bar and here a large mass of boulder clay blocks the valley, the top lying some 60 ft above river level at Hague Bar. The relation to the Disley overflow channel is discussed on p. 344. The Goyt has been diverted through a valley cut in solid rocks between Woodend and Waterside. The boulder clay mass between Newtown and New Mills is even more extensive and has resulted in a further diversion of the Goyt with the cutting of an 80-ft gorge at New Mills. A borehole at Grove Mill [9942 8506] proved boulder clay from 10 to 49 ft on sand and gravel resting on solid rock (see pp. 366–7) at 61 ft.

Between New Mills and Whaley Bridge extensive spreads of boulder clay are present though good sections are lacking. To the south of Whaley Bridge boulder clay underlies the gravel of Horwich End and to the south-east extends round the hill on which Shallcross Manor stands. South of Taxal a small tributary valley of the Goyt is filled with boulder clay. A patch of boulder clay with scattered large erratics at Fernilee [022 785] occurs at a height of over 1000 ft O.D. Farther south, the main boulder clay outcrop extends to a point [014 780] a little north of Fernilee Reservoir and small patches are present near the dam [e.g. 016 774]. To the south scattered erratics are present in the lower part of the valley, an example being a 6-ft boulder of epidotized rhyolite (E 32833) at Bunsal Cob [0197 7558].

*Disley and Kettleshulme.* Spreads of boulder clay at Higher Disley and Disley are continuous with those already noted in the Goyt valley. From Higher Disley a tract of boulder clay extends southwards along Mather Clough and into the Todd Brook valley at Handley Fold.

The Todd Brook valley contains an extensive spread of boulder clay as far south as Green Booth. Patches are also present at Saltersford Hall and at Hooleyhey [974 753]. On the western side of the Todd Brook valley, boulder clay is present at a height of over 1200 ft O.D. at Blue Boar Farm.

*Buxworth and Chinley.* Much of the valley of Black Brook is drift-filled. Where the valley is narrow, at Buxworth, relatively thin patches of boulder clay are present on the slopes. At Whitehall and Chinley, where the valley broadens, a wide tract of boulder clay occupies its floor. The tributary valley of Otter Brook is likewise floored with boulder clay which extends up to a height of 900 ft O.D. at Chinley Head. Other valleys at Hull End and Chapel Milton contain boulder clay. A section [0587 8184] in a stream near Chapel Milton shows 4 ft of stiff brown clay with erratics, including rocks of andesite and Ennerdale granophyre type. The boulder clay at Chapel Milton extends south to join the larger spread around Chapel en le Frith and Combs. In this area many of the sandstone features show marked glacial smoothing below a height of about 1000 ft O.D. This is particularly evident in the case of the Upper Kinderscout Grit to the south of Bole Hill. The Chatsworth Grit to the west of Chinley, however, retains strong features down to a height of about 800 ft O.D.

*Chapel en le Frith and Combs.* Boulder clay forms a substantial spread in the low ground between Chapel en le Frith, Marsh Hall, Lower Crossings and Combs. To the south of the latter village patches extend for a distance of over one mile, the southernmost [041 772] being near Broadlee Farm. To the south-east of Chapel en le Frith a small patch of boulder clay is present at Draglow [069 800] at a height of over 1000 ft O.D. (see also p. 332).

*Limestone area.* Scattered erratics occur in places on the limestone outcrop. Taylor (1894, p. 519) noted the presence of "granite, andesite and flint . . . about a mile south of Dove Holes" and Dale (1900, p. 121) described a boulder of "volcanic ash containing red garnets" from "the upland beyond Fairfield". More recently Dalton (1958, p. 281) noted a concentration of erratic dolerite boulders in the Foolow–Eyam–Coombs Dale area.

During the course of the present work other erratics have been noted though only the following examples need be quoted: a 9-in cobble of andesite [0747 7650] south of Dove Holes; a 1-ft 'andesite' [1066 7855] near Peak Forest.

The only recognizable area of boulder clay on or near the limestone outcrop is south of Stoney Middleton. Its presence was established by Dalton (1945, p. 27) from a section in a trench [2279 7506 to 2313 7498]. The deposit contained erratics of biotite-granite, porphyritic plagioclase-granite, microgranite, basalt, chert, sandstone and 'grit'. In addition (ibid., p. 28) a pocket of boulder clay with erratics, some striated, including "garnetiferous ashes . . . basalt, quartzite and various grits" was noted near "Cupola Works" [210 755]. This has now been removed by quarrying though clay with erratics is still visible on the tips.       I.P.S.

*North-eastern area.* Only two traces of possible glaciation are known in this area. In the upper reaches of streams draining Featherbed Moss north of Kinder Scout, sub-rounded boulders of coarse pebbly sandstone, of Kinderscout Grit type and unlike the adjacent Shale Grit sandstones, occur on the shaly bedrock beneath peat. The nearest outcrops of Kinderscout Grit are on Bleaklow several miles to the north of the district, across Ashop Clough on Kinder Scout farther south, and in the Glossop–Chunal area several miles farther west.

The "igneous rock" reported by Bemrose and Sargent (1915, p. 99) was found, according to the records of the Derwent Valley Water Board, on or beneath the surface of the floor of the Derwent valley east of Birchinlee Pasture, an area now submerged beneath Derwent Reservoir. Thin sections (E 30261, E 30261A) of the rock, a 7-in cobble of dark grey porphyritic appearance, confirm that it is best termed a porphyrite, possibly a welded tuff. The "erratics of mountain limestone" also reported by Bemrose and Sargent (ibid., p. 99) from near Howden Dam have not been seen. Fragments of carbonate bullions from the local shales occur in some streams and angular pieces of Carboniferous Limestone can be seen in places along the line of the railway used for the construction of Howden and Derwent dams.

The "boulder clay" in the Edale valley reported by Lewis (1946) is a head deposit (see pp. 334, 345).       G.D.G.

## SAND AND GRAVEL

*Chisworth and Rowarth.* On the north-western side of the Etherow valley, a large spread of sand and gravel lies partly within the Chapel en le Frith district at Hodgefold. Sections are lacking, though a borehole on the Glossop (86) Sheet [9784 9393] proved 33½ ft of sand and gravel on boulder clay. To the north of Compstall, small patches of sand and gravel are present near Beacon Houses [972 920] and around Mortinfold [968 912] at a height of 400 to 450 ft O.D.

On the southern side of the Etherow valley isolated patches of sand and gravel are present at Lane Ends [975 902; 978 902].

On Ludworth Moor two small deposits of sand and gravel are present [006 905; 008 904] in the vicinity of Near Slack at a height of 1000 ft O.D. To the south of Ludworth Moor patches of sand and gravel occur at heights of from 760 to 825 ft O.D. with a higher patch at 900 ft O.D. [997 903]. Between Hollinsmoor and Rowarth a number of similar areas of sand and gravel are present and have a series of flat benches cut in them. These features are best seen north of Rowarth [011 894] where an elongated outcrop shows flat benches of strand-line type at 815, 795, 775 and 715 ft O.D. Similar benches can be recognized at Hollinsmoor at all but the highest level. No clear sections are present.

*Goyt valley.* To the north-east of Disley a dissected patch of sand and gravel occurs at a height of about 500 ft O.D. A section [9789 8498] in a pit shows 10 ft of clean sand with scattered small pebbles and coal fragments.

At Buxworth the large mass of sand and gravel noted by Jowett and Charlesworth (1929, p. 323) lies across the valley of the Black Brook. This shows a flat top at 670 ft O.D. The relations of the deposit to the boulder clay are clearest to the north of the Black Brook valley where it rests on boulder clay. In addition the railway-cutting at Buxworth shows that on the eastern side the sand is overlain by boulder clay with the interface sloping down to the south. A section [0186 8255] in the cutting shows 52 ft of clean brown sand with lenses of coarse gravel. The gravel bears boulders up to 1½ ft including rocks of Eskdale granite and Ennerdale granophyre type, quartz, quartzite and sandstone.

At Horwich End a mass of sand and gravel of somewhat uncertain relationships is present across the Todd Brook valley. It is in general poorly exposed but a section [0092 8076] near the River Goyt shows red-brown sandy clay with pebbles, 9 ft, on sand, 2 ft.

Near Fernilee a further mass of sand and gravel is present on the right bank of the Goyt. This occurs on the south or upstream side of the hill on which Shallcross Manor lies. The sand and gravel overlies boulder clay in the lower part of the valley but its northern edge rests on solid rock.

*Higher Disley and Lyme Park.* An elongated mass of sand and gravel is present in Lyme Park [968 836] near Horsecoppice Reservoir and a smaller patch [970 829] near Cage Hill; neither show clear sections. In the valley of Bollinhurst Brook areas of sand and gravel are present both north and south of Cocks Knoll. One of the latter at a height of over 1000 ft O.D. shows a 2-ft section [9813 8182] of clean sand and gravel.

*Todd Brook valley.* The presence of the large mass of sand and gravel at Pennant Nob was noted by Jowett and Charlesworth (1929, p. 323). Mapping shows that the deposit lies across the pre-existing valley of Todd Brook which has been diverted through a conspicuous gorge at Kishfield Bridge. The mass of sand shows steep slopes on both up- and down-valley sides and a north-westerly extension of it is present at Hawkhurst Head. The deposit is well seen in sand pits on Pennant Nob [9973 8100] (see Plate XXIV A) and these show 40 ft of clean sand and gravel with marked foreset bedding inclined to the west. Erratics include rocks of Borrowdale volcanic type, granite, quartz, quartzite, 'grit' and shale. Jowett and Charlesworth noted rocks of Criffell, Eskdale, Buttermere and Borrowdale type from this locality.

Small patches of sand and gravel are present at a height of about 900 ft O.D. near Handleyfoot, and smaller patches are also present farther south [978 791] near Charles Head, near Nab End [975 755], and at a height of over 1200 ft O.D. at Blue Boar Farm [971 765]. A sand and gravel outcrop in the Black Brook valley is more extensively developed in the Stockport district to the west.

*Chinley and Chapel en le Frith.* On the outskirts of Chinley the presence of an area of sand and gravel has been recognized [042 827] at a height of 700 ft O.D. though the deposit is little exposed. Similar deposits are present on the western side of Chapel en le Frith between Lower Crossings, Cockyard and Bradshaw Hall at heights of 700 to 750 ft O.D. These are usually only poorly exposed though a section [0390 8065] near Bradshaw Hall proved 2½ ft of sand with small pebbles on gravel 3 ft. Another area of sand and gravel is present on the northern side of the Carr Brook valley west of Tunstead Milton.

## GLACIAL LAKE DEPOSIT

*Chapel en le Frith.* An oval outcrop of laminated clay has been proved in excavations on the south-western side of Chapel en le Frith. A typical section [0528 8029] showed 5½ ft of blue-grey and brown mottled clay with silty laminae. In the absence of sections the deposit is difficult to recognize and it may be more extensively developed than shown on the map.

## GLACIAL DRAINAGE CHANNELS

*Ludworth.* The prominent dry valley [995 914] running south from the vicinity of Chisworth is a glacial drainage channel cutting the low watershed between the Etherow valley and that in which Mill Brow lies. The channel has been interpreted by Johnson (1963, p. 159) as of marginal type but is here considered to be a normal 'col' overflow channel. The intake is at about 880 ft O.D.

*Shiloh.* The channel [998 891] north of Shiloh is cut through an extensive spread of boulder clay into solid rock. The upper end of the channel lies at about 805 ft O.D. Johnson (ibid., p. 160) considers that the channel is of sub-aerial origin and this is borne out by the evident association of the eastern end of the channel with a small delta of sand and gravel (see p. 343).

*New Mills.* A small channel on the valley slopes [000 859] north of New Mills, at a height of 615 ft O.D., lies near the local upper limit of the boulder clay in which it is cut. The feature is tentatively interpreted as of marginal origin though a sub-glacial origin cannot be excluded.

*Disley.* The large channel at Disley occupies a nearly dry valley at about 600 ft O.D. This is a 'col' channel and has been considered to be of sub-glacial origin by Johnson (1963, pp. 159–60). The western end of the channel lies in the Stockport (98) district (Taylor and others 1963, p. 102).

*Whaley Bridge.* A series of well-marked glacial drainage channels have been cut in the valley sides east of Whaley Bridge. The highest and most strongly developed of these, the Roosdyche [015 807], shows a complex branching at its southern end at levels of from 780 to 720 ft O.D., the northern end lying at about 660 ft O.D. Lower channels are present below the Roosdyche: one at Horwich End [014 807] at 695 ft O.D., and another more complex group of four associated channels a little farther north [013 812]; the latter lie at about 600 to 630 ft O.D. The Roosdyche has been interpreted by Johnson (1963, pp. 155–8) as of sub-glacial origin.

## HEAD

*Glossop, Hayfield and Chinley.* Above Gnat Hole near Glossop, a terrace-like spread of head is present at and above the confluence of two streams. A section [0455 9197] shows 3 ft of clay with sandstone fragments on shale but the deposit appears to be thicker elsewhere. Farther south, similar deposits occur near the lower end of Hollingworth Clough [035 892] and in the upper part of this valley [045 897].

On Ollersett several areas of head with large sandstone boulders are present in small valleys. The lower ends of these are in contact with boulder clay which the head appears to overlie. A more extensive area of head lies on the eastward-facing slope beneath the escarpment of the Rough Rock.

*Black Hill.* A large spread of head is present below the escarpment of the Rough Rock on the western slopes of Black Hill from Higher Disley to Whaley Moor. On its western side the deposit is in contact with and apparently overlies boulder clay.

*Combs and Dove Holes.* The Chatsworth Grit escarpment of the Combs Moss outlier has given rise to substantial spreads of head in places. The most extensive lies to the north of Combs Edge and another near Dove Holes appears to have been derived both from Roaches Grit and Chatsworth Grit.

I.P.S.

*Kinder Scout.* Deposits of sandy clay and rubbly sand with sandstone boulders are present almost continuously around the plateau below the escarpment of Kinderscout Grit. On the western slopes landslips have probably concealed or destroyed more extensive deposits. The thickest and most widespread deposits occur on the northern slopes. In a number of localities, the best examples being on Black Ashop Moor, the head has flowed down into stream gullies incised into the steeper Shale Grit slopes below.

I.P.S., G.D.G.

*Alport and Ashop valleys.* Several spreads of head of the valley-fill variety are present on the lower slopes of the Alport valley and adjacent parts of the Ashop valley, occurring between and in places overlain by landslips. A section on the east bank of the Alport [1333 9183], nearly ½ mile N. of Alport Castles Farm, consists of sand 3 ft, on gravel 2 ft, on clay with angular stones 2 ft, on gravel 2 ft. A mile to the south on the east bank [1409 9030] sandy clay 4 ft, rests on fine gravel 2 ft, on coarse gravel 3 ft. A short distance downstream on the west bank [1406 9011] clay with boulders (landslip) 15 ft rests on bedded gravel (head) 8 ft.

*Edale, Perryfoot and Hope valleys.* The lower slopes of the Edale valley are uniformly covered by an even spread of valley-fill head varying from an ill-sorted angular breccia around the margins of the deposit to a well-sorted bedded gravel in the middle of the valley.

G.D.G.

A typical section near the edge of the head [1165 8444] south-east of Barber Booth showed 15 ft of a rubbly unsorted deposit with angular to subangular sandstone blocks up to 1-ft diameter in a pebbly sand matrix.

Head deposits along Grinds Brook are well exposed and show the typical lithological variation within the Edale valley. Near the northern margin of the deposit at a point [1205 8654] near Grindslow House, clay with angular sandstone boulders and cobbles 10 ft, rests on ferrocreted clayey sand with angular sandstone cobbles 6 ft. A section [1220 8620] near Grindsbrook Booth consists of clay with angular stones 12 ft, on ferrocreted clayey silt and sand with angular stones 6 ft, on black mud 6 in (which has 'intruded' upwards into cracks in the ferrocreted bed above). South of Grindsbrook Booth an exposure [1249 8565] shows bedded gravel with sub-rounded and rounded pebbles 3 ft, on grey pebbly clay 3 ft. About 60 yd to the south-west, 11 ft of gravelly head is exposed in the west bank of the brook. Some 12 ft of well-bedded sandy gravel with 'stream-layered' pebbles is visible on the west bank of the brook [1257 8536] a short distance north of the junction with the River Noe.

Similar conditions obtain in the Perryfoot valley where a sink-hole [1039 8172] showed 11 ft of sandstone rubble with angular to rounded pebbles and cobbles up to 9 in diameter.

Widespread valley-fill head is present in the Hope Valley and the adjacent part of the Derwent valley between Yorkshire Bridge and the Derwent–Noe confluence. There are few exposures, but the rubbly and pebbly sandy clay soil indicates the nature of the deposit. At a locality [1684 8480] near Fullwood Style, gravel, bedded and stream-layered, is exposed in a bluff above the back of the highest terrace. At the western end of the Hope Valley 6 ft of poorly sorted angular and rounded stones in a clay matrix was noted [1603 8303] south-west of Marsh Farm.

G.D.G., I.P.S.

*Derwent valley.* Widespread thick deposits of scarp-slope head are present in the valley, largely on the eastern side and mainly associated with Kinderscout Grit and Chatsworth Grit scarps. On the western side, small deposits are present below the Kinderscout Grit outcrops on Crook Hill, Win Hill, the northern side of Offerton Moor (where some flow into lower gullies is apparent) and on the northern side of Eyam Moor.

On the eastern side of the valley larger spreads are present below the Kinderscout Grit scarps on Howden Moors and below Low Tor, Howshaw Tor, Lost Lad Hillend [190 914], Derwent Edge, Hordron Edge, Bamford Edge and Carhead Rocks. In some of these localities and particularly below Derwent Edge, the deposit has flowed down into gullies cut into the lower, steeper slopes of the Shale Grit outcrop. The predominant lithology at all localities is a sandy clay with rubbly sand lenses and pockets, and containing cobbles, boulders and blocks of sandstone. The deposits form subdued hummocky boulder-strewn slopes with irregularly located springs.

Head of similar type forms an almost continuous deposit below the outcrops of Chatsworth Grit from Moscar Fields to Curbar. On the northern side of Bamford Moor, head derived from the Chatsworth Grit has flowed down Jarvis Clough to merge with head derived from the Kinderscout Grit below. From Hathersage Booths to Curbar, head derived from the Chatsworth Grit has flowed over the Kinderscout Grit outcrop to reach the floor of the Derwent valley in a number of places.          G.D.G.

*Limestone area.* On the eastern side of Eldon Hill a patch of cherty loam is present and has been shown on the one-inch map as head. A section [1205 8124] in a small sink-hole shows $5\frac{1}{2}$ ft of brown loam with angular to subangular chert fragments. A similar, though more extensive, deposit is present on Bradwell Moor but does not appear to exceed a few feet in thickness. Patches of cherty rubble are also present on the limestone outcrop at Burnt Heath and Lane Head, south-west of Eyam, and on the southern margin of the district near Black Harry House and south of Coombs Dale.

## RIVER TERRACES

*Etherow and Goyt valleys.* In the Etherow valley a terrace north of Woodseats lies some 10 ft above the flood-plain. Farther downstream both terrace and alluvium are discontinuous. A system of terraces is developed in the Goyt valley. The lowest downstream in the present district occurs just below the Strawberry Hill knick-point and at some 10 to 12 ft above the river. At Hague Bar, a small terrace [979 857] 9 to 12 ft above the alluvium is on a level corresponding to that of the alluvium upstream from the knick-point in the rock gorge at this locality. Between Hague Bar and New Mills, terraces are present at 5 to 8 ft above the flood-plain on the northern side [984 854] of the river and at 4, 10 and 17 ft above it on the southern side [988 852]. Upstream from the New Mills gorge two terraces are present, the lower some 4 to 5 ft above the flood-plain is not extensively developed, but the higher at 8 to 12 ft above the flood-plain is well seen at Gowhole and Bridgemont.

Isolated occurrences of a terrace at a height 6 to 12 ft above the flood-plain are present in the valley of the River Sett between Birch Vale and Hayfield. In the valley of the Black Brook a low terrace, 5 to 6 ft above river level, is seen in places.          I.P.S.

*Derwent and Noe valleys.* Four terraces have been recognised along these rivers. The highest stretches almost continuously from Carr Bottom near Bamford in the Derwent valley and Townhead north of Hope in the Noe valley to about $1\frac{1}{4}$ miles south of Hathersage where the valley narrows considerably across the Kinderscout Grit outcrop. It is the widest terrace between Hope and Bamford station, and south-east of the station it is broad and virtually uninterrupted, as lower terraces are largely absent. North and east of the station it lies 17 to 20 ft above river level, but farther south-east its height diminishes considerably, even allowing for the weir at Leadmill, near Hathersage. The upper part of the fourth terrace around Hope consists of gravel, which is exposed in a lane cutting [1683 8462] near Fulwood Style. At a point [1759 8350] east of Hope, gravel 6 ft, on shale 6 ft, is exposed. Around the confluence of the Derwent and Noe the terrace is in places a rock-cut bench with only a veneer of pebbly clay, but farther downstream 11 ft of sand and gravel are exposed [2133 8200] near Cunliffe House.

The third terrace of the Noe is nearly continuous from Carr House, north of Lose Hill, where it is 11 ft above river level, to

Townhead where it is 15 ft above river level and only a few feet below the fourth terrace; it cannot be traced beyond Townhead with certainty. In this area it is largely a rock bench with only a thin capping of gravelly clay. In the Derwent valley the third terrace forms a few impersistent benches between Yorkshire Bridge and Carr Bottom at 13 to 15 ft above river level, with gravelly deposits in places. At a point [1997 8408] near Carr Bottom, fine gravel 2 ft, on coarse gravel 3 ft, on siltstone, is exposed in the terrace, only small remnants of which occur farther downstream.

The second terrace lies approximately 8 ft above river level from near Nether Booth in the Noe valley and Yorkshire Bridge in the Derwent valley to the confluence of the two rivers. It consists largely of impersistent meander benches of silty clay and sand on sand and gravel, resting in places on a low rock bench.

The first terrace is present at 3 to 5 ft above river level, over the same stretches of the two rivers as the second terrace, and also consists of a series of impersistent meander benches of similar composition. G.D.G.

## ALLUVIUM

*Etherow and Goyt valleys.* In the north-western part of the district small spreads of alluvium are present in the Etherow valley. In the Goyt valley these deposits are more uniformly developed and up to some $\frac{1}{4}$ mile across. At Hague Bar and New Mills the flood-plain is interrupted by the rock-gorges noted above (see p. 346), which also constitute knick-points. The alluvium between Hague Bar and New Mills is on a level with the terrace developed below the Hague Bar knick-point (see also p. 346). Above New Mills a more uniform development of alluvium is present as far as Horwich End, south of which it rapidly diminishes. The tributary valleys of the Sett, Black Brook and Randal Carr Brook (the latter extending from Horwich End to Combs) bear fairly uniform spreads of alluvium.

*Perryfoot and Peak Forest.* A narrow deposit is recorded as alluvium on the limestone outcrop through Perry Dale, Peak Forest, Dam Dale, Hay Dale and Peter Dale. Sections are lacking but sandy soil with small sandstone pebbles has been seen in places [e.g. 1187 7767]. With the exception of the part overlying the perched water-table at Peak Forest this valley is dry and the deposit cannot have formed under present-day conditions (see pp. 357–8). I.P.S.

*Edale and Hope valleys.* The alluvium in the Edale and Hope valleys consists largely of clayey sand with fine gravel. Its down-stream limit in the Edale valley is the foot of the Back Tor landslip south of Nether Booth. Gravels, reported by the Derwent Valley Water Board to be up to 10 ft thick, are present along the stream running north from Bradwell, and have yielded a worn 'mammoth' tooth at an unknown depth.

*Derwent, Alport and Ashop valleys.* Narrow flood-plains graded to adjacent reservoir level occur along the lower reaches of the River Westend (north of Birchinlee Pasture), Abbey Brook and Ouzelden Clough. The river deposits in the Alport and Ashop valleys are rather coarse ill-sorted sands and gravels, and although classified as alluvium, consist of miniature suites of minor terraces at elevations of up to seven feet above river level, each suite terminating downstream at the foot of a landslip. The only exception is the flood plain in the lower Ashop valley now graded to Ladybower Reservoir, but shown on Old Series One-inch Geological Sheet 81 N.E. to have existed prior to the reservoir, when the deposit was probably graded to knick-points caused by harder outcrops of Shale Grit. In the northern bank of the Ashop [1437 8942] near Gillethey Farm several feet of landslip rest on alluvial gravel 4 ft. In the Derwent valley between Upper Padley and Curbar the alluvium includes remnants of terraces, up to 5 ft above flood-plain level in places, which are too small to be shown on the one-inch map.

## PEAT

*Kinder Scout and adjacent moors.* Large spreads of peat up to 8 ft thick are present on the nearly level moors north and west of the Ashop valley, including Leygatehead Moor, Black Moor, Coldharbour Moor, Featherbed Moss, the moors between Lady Clough and the Alport valley, Hope Woodlands, Birchinlee Pasture and Rowlee Pasture. Erosion of peat in these areas is largely limited to surface channelling directed both towards the heads of valleys and to a lesser extent towards the steep moorland edges. Only in a few localities, for example east of Cowms Rocks and southeast of Rowlee Pasture, has the peat cover been reduced to discontinuous hags. G.D.G.

Peat on the Kinder Scout plateau is variable in thickness, the maximum recorded being 12 ft near Blackden Rind. Surface channelling into the peat is widespread over much of the eastern half of the plateau and around the edges of this area has reduced the peat to residual hags which are rapidly disappearing as a result of strong wind erosion. On the western side of the plateau the maximum thickness of peat seen is 8 ft in the banks of the River Kinder [0876 8803]. Near the western scarp the deposit is undergoing strong wind erosion.

Patches of thin peat overlie head deposits on the lower slopes of Black Ashop Moor, below Blackden Edge, and in other places on the western and south-western slopes of the plateau. Farther south, peat up to about 6 ft in thickness covers Shale Grit outcrops around Brown Knoll and Colborne. Two thin patches of peat rest on head deposits in the Perryfoot valley; the westernmost of

these shows a maximum thickness of 7 ft [0954 8147].                    G.D.G., I.P.S.

*Kettleshulme and Combs Moss.* To the north-west of Kettleshulme, an elongated tract of peat, extending into the Stockport (98) district, is present on high ground from Sponds Hill to Lyme Park. In the Todd Brook valley, small and thin patches are present both on the boulder clay in the valley floor [984 779] and farther south [983 770] on solid rock. Peat is also present on the dip-slopes of both Chatsworth Grit and Rough Rock on the western limb of the Goyt Trough. The outlier of Combs Moss is almost entirely peat-covered and the maximum thickness, about 8 ft, is developed in the area lying just west of Black Edge [e.g. 0535 7621].    I.P.S.

*Eastern side of the Derwent valley.* Thin peat occurs on the Shale Grit outcrop north of Abbey Brook and elsewhere, but the largest spread east of the Derwent valley is on the dip-slope of the Kinderscout Grit north-east of Ladybower Reservoir, from Derwent Moors northwards. The peat in this area is up to 8 ft thick and forms a nearly continuous cover, being dissected to only a moderate degree by eastward-flowing surface channels. Only on parts of Holling Dale, near Back Tor and on the southern edge of Derwent Moors is there evidence of intensive peat erosion. North of Back Tor the peat continues westwards down some of the minor scarp slopes and partially overlaps on to head deposits. Widespread peat up to 10 ft thick is also present between Hallam Moors and White Path Moss.             G.D.G.

## LANDSLIP

*Charlesworth, Chinley and Combs.* A large landslip is present below the scarp of the Rough Rock at Coombes Rocks, Charlesworth. The slip has been described by Johnson (1965a) and the earliest movement dated by pollen analysis (Franks and Johnson 1964) as late Boreal (late zone VI) in age. Another smaller area of landslip is present below the Rough Rock scarp of Slack Edge, west of Simmondley. The landslip at South Head, Chinley occurs on the dip-slope of the Lower Kinderscout Grit, while at Combs, two large landslips are present below the

Chatsworth Grit scarp of Combs Edge. I.P.S.

*Kinder Scout.* On the western slope of Kinder Scout, collapse of the escarpment of the Lower Kinderscout Grit has produced one of the largest landslips in the district. In places, for example to the east of Broad Clough, its upper part is a mass of almost unbroken sandstone with a small scar at the junction with unslipped solid. In its lower part the slip degenerates into a chaotic mass of large sandstone blocks. Similar but much smaller collapsed scarp landslips are present in the upper reaches of Grinds Brook and Jaggers Clough. I.P.S., G.D.G.

*Edale valley and Mam Tor.* A number of landslips of the rotational type, mainly involving Mam Tor Beds, are present in the Edale valley, the two largest being on the southern side of the valley, one north-west of Mam Tor and the other north-west of Back Tor. G.D.G.

The Mam Tor landslip itself (see Plate XVI B) situated on the eastern side of Mam Tor, originated near the base of the Mam Tor Beds. It is one of the few active slips in the district and is well known on account of the road stability problems it raises. These result from the steepness of slope of the slipped mass, from the inherent unstable nature of the Edale Shales of which it is largely composed and from the presence of springs at the base of the Mam Tor Beds. · I.P.S.

*Ashop and Alport valleys.* Compound landslips are present along about half of the northern side of the Ashop valley, where the southerly dip has facilitated large-scale down-dip sliding. At Cowms Moor an area of ground in the middle of the slip appears either not to have moved or to have slipped without breaking up. Periodic recent movements have necessitated repeated road repairs in the Ashop valley. The most spectacular landslip scenery in the district occurs on the eastern slope of the Alport valley, where relatively large areas of ground have slipped without breaking up. The largest is some 200 yd square and has moved 50 to 100 yd.

*Derwent valley.* Small rotational slump landslips involving Mam Tor Beds and the lower part of the Shale Grit are present in a number of places in the Derwent valley. One, situated on the eastern side of Win Hill is obviously of considerable antiquity as its slipped mass is greatly denuded and its dislodgement scar is virtually overgrown. In contrast, the scarp collapse landslips involving Kinderscout Grit adjacent to Ladybower Tor are 'fresh' with steep boulderstrewn hummocky slipped masses and bare scars. Fissures opening to depths of 10 ft or more are present in the solid rocks above the scar of the more easterly landslip. G.D.G.

*Hucklow and Bretton.* The striking landslip below the western face of Hucklow Edge originates from the lowest part of the Shale Grit. In Bretton Clough extensive slipping has occurred at a similar horizon.

*Limestone outcrop.* Landslipping is not common in this area, but does occur on some slopes where masses of limestone are underlain by tuff or lava. The uppermost part of the latter tends to alter to a soft impervious clay ('toadstone clay') and the presence of a perched water-table also tends to favour slipping. Landslips of this type are, however, of very limited dimensions and none are indicated on the one-inch map. The best example is Peter's Stone [1739 7525] in Cressbrook Dale, where a mass of Upper Monsal Dale Beds has slipped on the underlying Litton Tuff. A similar but smaller slip is present above the Lower Lava at Cave Dale [148 822]. I.P.S.

## REFERENCES

BEMROSE, H. H. and SARGENT, H. C. 1915. Report of an excursion to Derbyshire. *Proc. Geol. Ass.*, **26**, 93–104.

CONWAY, VERONA M. 1947. Ringinglow bog, near Sheffield. *J. Ecol.*, **34**, 149–81.

——— 1954. Stratigraphy and pollen analysis of southern Pennine blanket bogs. *J. Ecol.*, **42**, 117–47.

DALE, E. 1900. *The scenery and geology of the Peak of Derbyshire.* London.

DALTON, A. C. 1945. Notes on some glacial features in north-east Derbyshire. *Proc. Geol. Ass.*, **56**, 26–31.

——— 1953. Glacial evidences of the Sheffield area. *NWest. Nat.*, **24**, 38–54.

——— 1958. The distribution of dolerite boulders in the glaciation of N.E. Derbyshire. *Proc. Geol. Ass.*, **68**, 278–85.

DAWKINS, W. B. 1875. The mammalia found at Windy Knoll. *Q. Jl geol. Soc. Lond.*, **31**, 246–55.

DAWKINS, W. B. 1877. The exploration of the ossiferous deposit at Windy Knoll, Castleton, Derbyshire, by Rooke Pennington, Esq., Ll.B., F.G.S. and Prof. Boyd Dawkins, M.A., F.R.S. *Q. Jl. geol. Soc. Lond.*, 33, 724–9.

———— 1903. On the discovery of an ossiferous cavern of Pliocene age at Dove Holes, Buxton (Derbyshire). *Q. Jl geol. Soc. Lond.*, 59, 105–32.

EDEN, R. A., STEVENSON, I. P. and EDWARDS, W. 1957. Geology of the country around Sheffield. *Mem. geol. Surv. Gt Br.*

EDWARDS, W. and TROTTER, F. M. 1954. The Pennines and adjacent areas. *Br. reg. Geol.*

EVANS, W. B., WILSON, A. A., TAYLOR, B. J. and PRICE, D. 1968. Geology of the country between Macclesfield, Congleton, Crewe and Middlewich. *Mem. geol. Surv. Gt Br.*

FEARNSIDES, W. G., BISAT, W. S., EDWARDS, W., LEWIS, H. P. and WILCOCKSON, W. H. 1932. The geology of the eastern part of the Peak District. *Proc. Geol. Ass.*, 43, 152–91.

FRANKS, J. W. and JOHNSON, R. H. 1964. Pollen analytical dating of a Derbyshire landslip: the Cown Edge landslides, Charlesworth. *New Phytol.*, 63, 209–16.

GREEN, A. H. FOSTER, C. LE NEVE and DAKYNS, J. R. 1887. The geology of the Carboniferous Limestone, Yoredale Rocks, and Millstone Grit of North Derbyshire. 2nd edit. with additions by A. H. Green and A. Strahan. *Mem. geol. Surv. Gt Br.*

HOLLINGWORTH, S. E. and TAYLOR, J. H. 1951. The Northampton Sand Ironstone stratigraphy, structure and reserves. *Mem. geol. Surv. Gt Br.*

———— ———— and KELLAWAY, G. A. 1944. Large-scale superficial structures in the Northampton Ironstone field. *Q. Jl geol. Soc. Lond.*, 100, 1–44.

JOHNSON, R. H. 1957. An examination of the drainage pattern of the eastern part of the Peak District of north Derbyshire. *Geogrl Stud.*, 4, 46–55.

———— 1958. Observations on the stream patterns of some peat moorlands in the southern Pennines. *Mem. Proc. Manchr lit. phil. Soc.*, 99, 110–27.

———— 1963. The Roosdyche, Whaley Bridge: a new appraisal. *E. Midld Geogr.*, 3, 155–62.

———— 1965a. A study of the Charlesworth landslides near Glossop, north Derbyshire. *Trans. Inst. Br. Geogr.*, No. 37, 111–26.

———— 1965b. The glacial geomorphology of the west Pennine slopes from Cliviger to Congleton. *in Essays in geography for Austin Miller.* J. B. Whittow and P. D. Wood (Editors), 58–93. Reading.

———— and RICE, R. J. 1961. Denudation chronology of the south-west Pennine upland. *Proc. Geol. Ass.*, 72, 21–31.

JONES, R. C. B., TONKS, L. H. and WRIGHT, W. B. 1938. Wigan district. *Mem. geol. Surv. Gt Br.*

JOWETT, A. and CHARLESWORTH, J. K. 1929. The glacial geology of the Derbyshire dome and the western slopes of the southern Pennines. *Q. Jl geol. Soc. Lond.*, 85, 307–34.

LEWIS, W. V. 1946. Stream profiles in the Vale of Edale, Derbyshire. *Proc. Geol. Ass.*, 57, 1–7.

LINTON, D. L. 1951. Midland drainage: some considerations bearing on its origin. *Advmt Sci. Lond.*, 7, 449–56.

———— 1955. The problem of tors. *Geogrl J.*, 121, 470–87.

LINTON, D. L. 1956. Geomorphology. *in Sheffield and its region*. D. L. Linton (editor), 24–43. British Association, Sheffield.

LOUNSBURY, R. W. 1962. Landslips in the Ashop valley, Derbyshire, England. (Abstract). *Spec. Pap. geol. Soc. Am.*, No. 68, 219.

MILLER, A. A. 1939. Pre-glacial erosion surfaces round the Irish Sea basin. *Proc. Yorks. geol. Soc.*, **24**, 31–59.

PALMER, J. and RADLEY, J. 1961. Gritstone tors of the English Pennines. *Z. Geomorph.*, **5**, 37–52.

PENNINGTON, R. 1875. On the bone-caves in the neighbourhood of Castleton, Derbyshire. *Q. Jl geol. Soc. Lond.*, **31**, 238–45.

RICE, R. J. 1957. Some aspects of the glacial and post-glacial history of the lower Goyt valley, Cheshire. *Proc. Geol. Ass.*, **68**, 217–27.

SHOTTON, F. W. 1967. Age of the Irish Sea Glaciation of the midlands. *Nature, Lond.*, **215**, 1366.

SISSONS, J. B. 1954. The erosion surfaces and drainage system of south-west Yorkshire. *Proc. Yorks. geol. Soc.*, **29**, 305–42.

STRAW, A. and LEWIS, G. M. 1962. Glacial drift in the area around Bakewell, Derbyshire. *E. Midld Geogr.*, **3**, 72–80.

TAYLOR, A. 1894. Note on boulder clay and other glacial deposits between Chapel-en-le-Frith and Miller's Dale. *Rep. Br. Ass. Advmt Sci.*, 519–20.

TAYLOR, B. J., PRICE, R. H. and TROTTER, F. M. 1963. Geology of the country around Stockport and Knutsford. *Mem. geol. Surv. Gt Br.*

TE PUNGA, M. 1956. Altiplanation terraces in southern England. *Biul. peryglac.*, No. 4, 331–8.

TONKS, L. H., JONES, R. C. B., LLOYD, W. and SHERLOCK, R. L. 1931. The geology of Manchester and the south-east Lancashire coalfield. *Mem. geol. Surv. Gt Br.*

WATERS, R. S. and JOHNSON, R. H. 1958. The terraces of the Derbyshire Derwent. *E. Midld Geogr.*, **2**, 3–15.

WILLS, L. J. 1937. The Pleistocene history of the west midlands. *Rep. Br. Ass. Advmt Sci.*, 71–94.

# Chapter X

# MINERAL PRODUCTS AND WATER SUPPLY

---

## COAL

THE WORKING of coal has in the past been widespread though in most cases on a small scale owing to the thinness or poor quality of the seams. In general only one seam has been worked at any one place, though at Shallcross and Whaley Bridge collieries three seams have been worked. The most extensive workings have been either in conjunction with the underlying fireclay or for coal for lime burning. In the latter case coal was formerly transported in small quantities to the quarries and the limestone burnt *in situ*; later, limestone from the Dove Holes area was transported to Buxworth by tramway and burnt with coal mined in the immediate vicinity. At present only one small working for coal and clay is in operation, though a larger mine was closed as recently as 1964. Considerable unworked areas of coal remain.

Workings in the lowest coals are very scattered. The Ringinglow Coal is workable on the eastern side of the district and has been got by day-holes on Hallam Moors; on the western side it appears only to have been worked by bell-pits at Combs Moss. The Simmondley Coal has not been worked in the east but in the western outcrops it has been mined at the type locality, at Errwood, and, more recently (1953), by adit at Charles Head.

The Six-Inch Mine is developed as a workable seam only in the north-west of the district. Though thin, and in places pyritous, there have been a few small workings (for example at Mill Brow) in conjunction with the underlying clay (see also p. 245).

The Yard Coal is the most widely worked seam. At Whaley Bridge it is thick and dirty and has been much worked for lime burning under the name 'Kiln Coal'. Where thinner, the quality of the seam improves. Large areas have been won in the axial area of the Goyt Trough between Thornset and Buxworth. The Yard seatearth has not been worked.

The Ganister Coal is usually thin and often absent. Locally, however, it has been worked in conjunction with the underlying ganister in the north-west of the district at Chew Wood.

The White Ash is the second most widely worked seam. It has been mined in conjunction with its seatearth at Furness Vale. Substantial areas were also worked at Shallcross and Whaley Bridge.

Workings in the Red Ash Coal are quite numerous though usually of small extent, probably due to the sulphurous character of the seam. Larger areas have been got at Whaley Bridge and Shallcross where it probably served for lime burning in addition to the Yard.

The Big Smut is the highest coal worked in the district. Its variable nature has discouraged working in most places, though it has been and is worked on Ludworth Moor together with the underlying fireclay.

## REFRACTORY MATERIALS
### FIRECLAY

The seatearth underlying the Simmondley Coal has been exploited on a limited scale at Errwood and at Charles Head. That underlying the Six-Inch Mine is apparently silty and unworkable in the area of the Goyt Trough where the coal is not developed. To the north of Mellor, however, it constitutes a workable fireclay and has been got together with the overlying coal. On the eastern margin of the district the equivalent Pot Clay is continuous with the important outcrops of the Sheffield area (Eden and others 1957, pp. 169–70) and is worked in a large opencast on Hallam Moors, it is used for the manufacture of steel-works refractories including furnace nozzles and stoppers. The seatearth of the White Ash was mined until 1964 at Furness Vale and used for the manufacture of fire bricks and fire backs. Analyses of the fireclay have been published by Ennos and Scott (1924, p. 25). The Big Smut seat is in places a workable fireclay and has been dug opencast on Ludworth Moor; it is at present being worked by adit in this area.

### GANISTER

Despite the occurrence of ganister at several horizons in the area there is no evidence of it having been worked other than on a small scale at Glossop Vale Mine (see p. 284), Chisworth. The principal horizons at which ganisters occur are beneath the Ringinglow Coal and beneath the Ganister Coal (see p. 284). Siliceous refractories of 'crowstone' type are present in the axial area of the Todd Brook Anticline. Workings of these were carried out on Fox Hill [981 765] and also south-west of Saltersford Hall [981 759].

## SANDSTONE

Most of the more massive sandstones of the Millstone Grit and Coal Measures have furnished building stones and some of the coarser Millstone Grit sand-stones have in the past been used for millstones. Less massive and more flaggy beds have furnished rockery stones and rough paving stone and at one locality the Rough Rock Flags were worked for roofing flags.

The Shale Grit has been worked where it is most massive. At Kinder Bank [050 880], where there is a 100-ft face nearly free from partings, the stone was worked for the construction of Kinder Reservoir and associated installations. The Shale Grit has also furnished building stone, though on a smaller scale, around Chapel en le Frith, particularly at Breck Edge [082 822] and to a lesser extent at Barmoor Clough.

Working of the Kinderscout Grit appears to have been restricted to some extent by the inaccessibility of many of the outcrops, though the western and more easily reached Upper Kinderscout Grit has been worked at a number of places between Chunal and New Smithy. At Chunal [035 912] this bed is at present being worked for the production of ornamental fireplaces. At Chinley Moor Quarry [049 852] it is worked for building stone, while in a quarry [049 844] near the Lamb Inn, Chinley, it furnishes rockery stone. The Lower Kinderscout Grit has been formerly worked in the large quarries [043 865] near

Hayfield which show a 160-ft face of massive sandstone. On the western slope of Kinder Scout, near Cluther Rocks [074 878] millstones were made *in situ* from large landslipped blocks of Lower Kinderscout Grit.                              I.P.S.

In the eastern outcrops the Lower Kinderscout Grit was worked in large quarries near Ladybower House [208 868] and more recently in Stokehall Quarry [236 769] near Grindleford, which was producing about forty pulp-stones a year in the early 1950s mainly for export to Norway and Sweden. The pulpstones were up to 5 ft in diameter. This quarry also produced building and walling stone and gravestones.

The Heyden Rock has been quarried at Thornseat [230 926].                         G.D.G.

The more massive parts of the Roaches Grit have been worked for building stone at Combs [038 773], near Ridge Hall [055 793] and Longhill [044 746].

The finer, northern facies of the Chatsworth Grit has been worked for aggregate at Birch Quarry [030 869], Birch Vale, but now furnishes dressed stone and walling stone. The coarse facies has been worked in large quarries at Buxworth [027 818] and in smaller quarries elsewhere on the flanks of the Goyt Trough. In the eastern outcrops it has been much quarried around Padley, for example, in Yarncliff Quarry [255 793], Bole Hill [249 795] and at Grey Millstone Quarries on Millstone Edge [248 806]. At Millstone Edge millstones were produced, apparently from large loose blocks below the escarpment. The Bole Hill quarry furnished some $1\frac{1}{4}$ million tons of masonry for the Howden and Derwent dams, opened in 1912 and 1916 respectively.

The Rough Rock has been quarried extensively on Chinley Churn at Cracken Edge Quarry [037 837] which communicated with Maynestone Road by a long incline. The quarry provided stone for the construction of Chinley station and embankment walls. The Rough Rock Flags have been mined in several places on Cracken Edge, north of the quarry, for roofing flags.

The Woodhead Hill Rock has been worked in a number of places; these include Loads Quarry, Ludworth and Rowarth Quarry [018 886] near Aspenshaw Hall, which is at present being worked for rockery stone. Other old quarries at this horizon are at Low Leighton [006 852] and at Walker Brow on the western limb of the Goyt Trough [001 803], south of Todd Brook Reservoir. The 'roof-rock' of the Yard Coal is worked on a large scale in Arden Quarry [022 861], Birch Vale, and used for building stone and concrete aggregate.

The Milnrow Sandstone has been worked in a few places in the axial area of the Goyt Trough. The largest of these is Furness Vale Quarry [009 832] which furnishes building stone and aggregate.

# LIMESTONE

Most horizons in the Carboniferous Limestone have in the past been worked for building and walling stone. For the most part the quarries were small and served local requirements. More recently, with the large-scale exploitation of the limestone for agricultural lime and crushed stone, for use in the metallurgical, chemical and glass industries, and for cement, quarrying has become centred in a relatively small number of large quarries, and such concentration is now encouraged by the difficulties of obtaining planning permission for the development of new sites in the Peak District National Park.

The pure and uniform Chee Tor Rock, which averages over 98 per cent $CaCO_3$, with certain beds exceeding 99 per cent purity, has long furnished lime for the chemical industry, and Jackson (1965, p. 261) has linked the development of the lime industry at Buxton with the establishment of the alkali industry by Brunner and Mond in 1872 at Northwich.

The major lime producer in the area is the Tunstead Quarry of Imperial Chemical Industries Limited, which with a working face over a mile in length, one of the largest in Europe, lies across the southern boundary of the present district. Nearly the whole thickness of the Chee Tor Rock is worked together with the uppermost part of the Woo Dale Beds. After blasting the rock is crushed and then washed in a scrubbing plant before being burned in either vertical or rotary kilns (Jackson 1965, pp. 271–4). The burnt products are lime and hydrated lime for the chemical and metallurgical industries, building and agriculture. The unburnt products include limestone of all sizes from blocks weighing several tons down to finely powdered stone for industrial and agricultural use. Very large tonnages of limestone are sold for concrete aggregate and road construction. A by-product of the Tunstead Works is cement, kiln feed being produced from the slurry waste from the washing of stone for lime-burning, together with additional limestone and small quantities of other additives. The mixture is burnt in a rotary kiln.

Messrs. Staveley Lime Products Limited produce lime from Chee Tor Rock from Holderness Quarry and from the Bee Low Limestones from Bee Low Quarry. In addition to lime the firm produces aggregate and ground limestone and limestone dust for agriculture.

Eldon Hill Quarry, in the Bee Low Limestones, is worked for tarmacadam, aggregate and fluxing limestone. The remaining limestone quarries in the district are in the uppermost part of the Monsal Dale Beds. Outlands Head Quarry produces aggregate from beds high in the Monsal Dale Beds. Furness Quarry produces roadstones and aggregate from the uppermost Monsal Dale Beds and the basal Eyam Limestones. Eyam Quarry produces limestone and macadam from the uppermost part of the Monsal Dale Beds and Dalton Quarry produces aggregates from a similar horizon.

The Hope Cement Works of Messrs. Earle obtains its raw materials from two sources: limestone from the large quarry in the uppermost part of the Bee Low Limestones and the Monsal Dale Beds, and shale from the quarries at The Folly in the lowest part of the Mam Tor Beds, here containing much less sandstone than at the type area. The problem of chert in the limestone quarry[1] is dealt with by blending cherty and chert-free limestone so that the silica content is adjusted to 5 per cent before the stone is passed to the works. When the shale component is more siliceous, limestone with a lesser silica content is supplied. A critical impurity in the limestone is fluorite from veins and scrins and it is necessary to limit the total fluorine content to 0·2 per cent in the limestone component by screening. The limestone is crushed in a 54-in gyratory crusher and then in two cone crushers. The ground limestone and clay slip from the shale quarry are finely ground in the correct proportions to produce slurry for burning. This is carried out in rotary kilns, fired by pulverized coal to produce a clinker which, after the addition of 6 per cent gypsum to control the rate of setting of the cement, is again ground to produce the finished product.

---

[1]From details kindly supplied by Messrs. G. and T. Earle Limited.

## FLUORSPAR

The main producer of fluorspar in the district is the Laporte Industries Limited subsidiary Glebe Mines, whose Cavendish Mill produces acid-grade fluorspar (97 per cent $CaF_2$) by flotation. Acid-grade spar is used mainly for the production of hydrofluoric acid. Ore for the mill is produced from Ladywash Mine, from openworks on Longstone Edge and from Sallet Hole Mine (One-inch Sheet 111) and from private producers or 'tributors' working other veins and 'hillocks'. Metallurgical spar (70 to 80 per cent $CaF_2$) is also produced for fluxing. Other uses of fluorspar have been described by Dunham (1952, pp. 2–5). Metallurgical spar is also produced from the plant at Nether Water Mine which functions as a washery, treating spar obtained elsewhere in the area.

## LEAD

Despite its former importance, there is no working primarily for lead in the district. A lead concentrate is produced at Cavendish Mill as a by-product of the flotation of fluorspar. Further discussion of lead occurrences is given on pp. 305–17.

## BARYTES

Barytes, commercial baryte, has been produced in recent years by the working of old tips or 'hillocks' in places, for example, on Eldon Hill and near Rowter, Castleton. The traditional demand for barytes in the paint industry has subsided with the growing importance of synthetic paints, though its use as a filler in the paper industry continues. Recently the North Sea Gas drilling programme has resulted in an increased demand for finely powdered barytes for use in heavy drilling muds. Barytes of this type is, like galena, produced as a by-product of fluorspar flotation at Cavendish Mill.

## CALCITE

Calcite is worked on Moss Rake [151 802] as a source of pebble-dash for building purposes. When finely ground it also finds a market as a neutral filler, for example, in the cosmetic industry. The vein of coarse rhombic calcite in Hay Dale (see pp. 316–7) has also been worked.

## DOLERITE

Messrs. Tarmac work a thick dolerite sill in Waterswallows Quarry. The rock is very fresh (see p. 294) and furnishes 'granite' chippings, aggregate and tarred and bituminous macadam.

## SAND AND GRAVEL

Many of the deposits of sand and gravel have been worked on a small scale in the past. At the present time working of these deposits is confined to a pit in the large deposit at Pennant Nob, Whaley Bridge, which produces both building sand and gravel.                                                     I.P.S.

# WATER SUPPLY

The annual rainfall over the Chapel en le Frith district varies from about 35 in near Disley in the west and 42 inches in Bradfield Dale in the north-east, to slightly over 60 in on the Kinder Scout plateau.

On the outcrop of the Carboniferous Limestone drainage is largely underground but the Millstone Grit and Lower Coal Measures areas possess well-defined surface drainage systems, notably the rivers Goyt and Derwent and their tributaries, and in the extreme north-east, tributaries of the River Don. Substantial amounts of water percolate into the sandstones, particularly those of coarser texture, giving rise to perched water-tables with spring lines at the junctions with the underlying shales.

The rivers draining the **Millstone Grit** and **Lower Coal Measures** areas, totalling about 184 square miles, have been extensively impounded, an estimated 93 square miles being included in reservoir catchments. These reservoirs supply not only local requirements but also the large population centres to the west, east and south-east of the district. The principal water-supplying authorities are the Stockport and District Water Board, the Derwent Valley Water Board, the North Derbyshire Water Board and Sheffield Corporation Waterworks.

Stockport and District Water Board, which supplies Disley, New Mills, Birch Vale and Whaley Bridge within the present district, maintains five reservoirs in the western part of the district. These include Kinder Reservoir [057 882], the recently constructed Errwood Reservoir [015 755], Fernilee Reservoir [014 770] and Horsecoppice [970 836] and Bollinhurst [974 835] reservoirs, at Disley. Lamaload Reservoir [970 752], in the south-western corner of the district, supplies Macclesfield District Water Board. The British Waterways Board use Combs Reservoir [038 796] and Todd Brook Reservoir [005 809] for canal replenishment and indirectly to supply canal-side industries in the Macclesfield, Congleton and Manchester areas.

The Derwent Valley Water Board supplies 47·7 million gallons per day to Sheffield, Nottingham, Leicester, to the North Derbyshire Water Board and to the South Derbyshire Water Board (which supplies Derby). These supplies are derived from Howden, Derwent and Ladybower reservoirs in the upper Derwent valley. Some water from the River Noe in the Edale valley, outside the natural catchment, has been diverted by aqueduct and tunnel into Ladybower Reservoir. The North Derbyshire Water Board supplies most of the villages in the eastern half of the district with water from the Derwent Valley Water Board's reservoirs.

Sheffield Corporation Waterworks draws supplies from the Strines and Dale Dike reservoirs in the north-east of the district.

In the western part of the district some industrial water supplies are drawn directly from rivers. For example, the Birch Vale Print Works of the Calico Printers Association take approximately one million gallons per day from the River Sett between Hayfield and Birch Vale, 90 per cent of this water being returned to the river after settling and treatment.                              G.D.G.

Over most of the **Carboniferous Limestone outcrop**, run-off of surface water is absent and rainfall, less a proportion which evaporates, penetrates almost completely underground. In the Castleton area, Ford (1967, p. 372) has shown that the level of the main water-table in the marginal areas of the limestone is dependent on the height of the limestone-shale boundary and that this effect

Y

persists into the limestone outcrop for some distance. These observations are in accordance with the presence of springs in Middleton Dale near the limestone edge. In the southern part of the limestone outcrop the water table is controlled by the River Wye.

As in most limestone regions widespread underground movements of water take place through fissures, and Ford (ibid., p. 372) states that dye tests have shown that drainage from the Perryfoot valley (Giant's Hole and Coalpithole Mine) emerges at the Russett Well, Castleton, after passing beneath or south of the limestone col of Windy Knoll. Certain areas are de-watered by soughs or drainage levels, the most important of these being Stoke Sough, which drains Ladywash Mine, and Moorwood Sough which de-waters Glebe Mine. Most areas on the limestone outcrop now possess piped supplies from the North Derbyshire Water Board.

Igneous rocks, particularly lavas, give rise to perched water-tables within the limestone. Some of these have furnished supplies: Wormhill Moor [1068 7590] supplies some 30 000 gallons per day from a spring above the Lower Lava to Chapel en le Frith U.D.C. Tideswell was formerly supplied by a spring [143 773] from the Lower Lava at Wall Cliff, but now receives piped supplies. The original yield of the spring is stated by Stephens (1929, p. 65) to have been at least 20 000 gallons per day during drought.

Borehole supplies of water are little used. However, Eldon Hill Quarry obtains an average of 1380 gallons per day from a borehole [1128 8156] to 605 ft in the lower part of the Chee Tor Rock and the upper part of the Woo Dale Beds; the lowest beds are probably tuff and the supply may come from perched groundwater on this. At Dove Holes a borehole [0738 7635] to 350 ft yields up to 1200 gallons per hour from perched groundwater overlying the Lower Lava; this is used as an emergency supply.

Supplies from shafts, though of potential importance (at least for industrial supplies) are not used, with the exception of Watergrove Shaft ([1882 7577] from which 30 000 gallons per hour is abstracted to supply Messrs. Glebe Mines' Cavendish Mill.

The resurgences of Peak's Hole [1487 8262] and Bradwell [174 809] are the most important supplies of 'limestone' water in the area. These together furnish 4·98 million gallons per day from filter stations on diversions a little distance downstream; the water is pumped to Ladybower Reservoir to mix with the relatively acid water from the Millstone Grit outcrop.

Springs at the limestone-shale boundary include the well-known Ebbing and Flowing Well [0845 7972] at Barmoor Clough, though this has now ceased to flow. The former ebbing and flowing action of this spring may be ascribed to the effects of a natural siphon.                                    I.P.S.

On the **Millstone Grit** and **Coal Measures outcrops**, spring supplies are now less important than formerly, having been largely replaced by reservoirs and boreholes. Former spring supplies have been listed by Stephens (1929).

The North Derbyshire Water Board still maintains spring supplies to a number of towns and villages, mainly in the western part of the district. Hayfield is supplied with at least 102 000 gallons per day from springs from the Shale Grit, near Tunstead Clough. A partial analysis of water from these springs in 1923 was as follows:

| | Milligrammes per litre (converted) |
|---|---|
| Total solid matter .. .. .. .. | 60·0 |
| Free and saline ammonia .. .. .. | 0·005 |
| Albuminoid ammonia .. .. .. | 0·01 |
| Nitrogen as nitrates and nitrites .. .. | 0·0 |
| Chlorine as chlorides .. .. .. | 10·5 |
| Oxygen absorbed in 4 hours at 80°F. .. | 0·03 |
| Temporary hardness .. .. .. | 1·3 |
| Permanent hardness .. .. .. | 22·1 |
| Total hardness .. .. | 23·4 |

Chapel en le Frith and Chinley are supplied from a number of springs from the Shale Grit at Shireoaks and The Roych, which are collected in Shireoaks Reservoir [0637 8299]; the yield is some 50 000 gallons per day. Water from the nearby Roych Clough can also be used if required. In addition, springs below the Chatsworth Grit near Pyegreave, Combs, have a constant yield of 20 000 gallons per day.

A spring [1147 8677] from the Shale Grit on the eastern slopes of Grindslow Knoll supplies 17 000 gallons per day to Edale. Springs [1454 8808] on Crookstone Hill, from landslipped Kinderscout Grit, and Brockett Booth [1480 8470] from sandstones in the Mam Tor Beds supply a total of 14 000 gallons per day to Castleton (which also receives supplies from limestone sources) and Hope (which receives its main supply from the Derwent Valley Water Board). Both these springs have given much higher yields in the past.

Macclesfield District Water Board supplies Kettleshulme with up to 10 000 gallons per day from Sponds Adit [9732 7937], an old Ringinglow Coal working. A partial analysis in August 1967 gave the following data:

| | Milligrammes per litre |
|---|---|
| Temporary hardness .. .. .. | 134·0 |
| Permanent hardness.. .. .. .. | 70·0 |
| Iron .. .. .. .. .. .. | 0·06 |
| Manganese .. .. .. .. .. | <0·025 |
| pH .. .. .. .. .. .. | 7·3 |

The more important borehole supplies are for industrial use though at Fernilee boreholes have been made in recent years to augment reservoir supplies. A borehole at Turn Lea Mills [0328 9299] near Glossop, drilled in 1928 to 386 ft through the Lower Kinderscout Grit yielded 40 000 gallons per hour originally, but by 1933 was producing less than 10 000 gallons per hour. No yields are available for the more recent Charlestown Borehole [0297 9331] nearby (see pp. 363–4).

The Forge Works, Chinley, draws supplies from a number of boreholes into the Kinderscout Grit. Borehole No. 1 [0447 8207] drilled in 1923 to an unknown depth, yielded 10 000 gallons per hour. Borehole No. 2 [0478 8202] drilled in 1937 to 730 ft through the Kinderscout Grit into the Shale Grit yielded on test

5670 and 6011 gallons per hour from pumping depths of 150 ft and 186 to 277 ft respectively, the water coming apparently from thin sandstones just below the Kinderscout Grit.

A borehole [0355 8202] at Whitehall Works, Chinley, was drilled in 1960 to below the Roaches Grit at 124 ft, to a final depth of 508 ft. There was an original artesian flow, while on test the borehole yielded up to 12 160 gallons per hour with a final pumping level 117 ft below surface.

A borehole, disused since about 1887, at the former Bennett's Print Works, penetrated the Chatsworth Grit from 159 ft to the bottom of the borehole at 202 ft and is said to have overflowed. These works are now the Birch Vale Print Works of the Calico Printers Association which is supplied directly by river water.

Garrison Bleach Works, Birch Vale, has obtained supplies from two boreholes. Borehole No. 1 [0141 8687] drilled in 1904, probably passed through the Chatsworth Grit from 137 ft to the bottom of the borehole at 188 ft. Rest water level was originally at the surface but in 1948 was 51 ft below surface with a pumping level 78 ft below surface when yielding 10 000 gallons per hour. Abstraction in 1948 was 36 million gallons. Borehole No. 2 [0151 8682] drilled in 1936–7, passed through the Chatsworth Grit from 144 to 200 ft and reached a final depth of 556 ft. Original test pumping of 23 260 and 27 850 gallons per hour depressed the water level to 46 ft and 54 ft respectively, the original rest water level not being recorded. In 1948 rest water level was 48 ft below surface, pumping level 80 ft below surface when yielding 15 000 gallons per hour and abstraction during the year was 54 million gallons. Partial analyses (in milligrammes per litre) for the two boreholes in 1937 are given below:

|  | Borehole No. 1 | Borehole No. 2 |
|---|---|---|
| Non-volatile (minerals) solids | 200·20 | 185·90 |
| Volatile and organic solids | 14·30 | 42·90 |
| Total solids | 214·50 | 228·80 |
| Temporary hardness | 55·34 | 44·47 |
| Permanent hardness | 61·06 | 75·90 |
| Total hardness | 116·40 | 120·26 |
| $SiO_2$ | 13·01 | 14·01 |
| Iron and alumina | 2·43 | 20·02 |
| Ca as CaO | 33·61 | 44·90 |
| Mg as MgO | 18·73 | 18·73 |
| Sulphates as $SO_3$ | 21·31 | 24·02 |
| Chlorides as Cl | 17·30 | 19·59 |
| Alkali as $NaHCO_3$ | 87·23 | 108·68 |

A borehole [9942 8506] at Grove Mill, New Mills (Stephens 1929, pp. 108–9), drilled in 1918–19 through the Chatsworth Grit between 197 and 240 ft to a final depth of 262 ft, is said to have overflowed, had a pH of 7·1 to 7·2, but to have been unsuitable for drinking.

A borehole [9805 8528] (see p. 364) at Disley Paper Mills, passed through the Yard 'roof-rock', the Woodhead Hill Rock and the Rough Rock to a final depth of 403 ft. An initial artesian supply of 4000 gallons per hour was increased by pumping to 10 000 gallons per hour with depression of the water level to 105 ft below surface.

Two boreholes at Fernilee are used by Stockport and District Water Board to compensate for seasonal fluctuations in Fernilee Reservoir. Borehole No. 1 [0124 7823] passed through the Chatsworth Grit from 349 ft to the bottom of the borehole at 476 ft and overflowed initially, but pumping for 25 days in July 1965 at up to 28 000 gallons per hour depressed the water level to 269 ft below surface. Borehole No. 2a[1] [0119 7863] was 326 ft deep and passed through the Woodhead Hill Rock and the Rough Rock. Pumping for 25 days in July 1965 at up to 38 000 gallons per hour depressed the water level from 41 to 100 ft below surface. The Board are licensed to abstract up to 1·75 million gallons per day to a limit of 250 million gallons per year from these two boreholes.      G.D.G.

## REFERENCES

ANON. 1920. Refractory materials: fireclays. *Mem. geol. Surv. spec. Rep. Miner. Resour. Gt Br.*, **14**.

DUNHAM, K. C. 1952. Fluorspar. 4th edit. *Mem. geol. Surv. spec. Rep. Miner. Resour. Gt Br.*, **4**.

EDEN, R. A., STEVENSON, I. P. and EDWARDS, W. 1957. Geology of the country around Sheffield. *Mem. geol. Surv. Gt Br.*

ENNOS, F. R. and SCOTT, A. 1924. Refractory materials: fireclays. Analyses and physical tests. *Mem. geol. Surv. spec. Rep. Miner. Resour. Gt Br.*, **28**.

FORD, T. D. 1967. The underground drainage systems of the Castleton area, Derbyshire, and their evolution. *Cave Sci.*, (39), **5**, 369–96.

JACKSON, R. G. 1965. The limestone quarrying operations of Imperial Chemical Industries Limited, Great Britain. *in* Opencast mining, quarrying and alluvial mining: Symposium. *Instn Min. Metall.*

STEPHENS, J. V. 1929. Wells and springs of Derbyshire. *Mem. geol. Surv. Gt Br.*

[1]Drilled 15 ft south of borehole No. 2 (see pp. 365–6).

# Appendix I

# SECTIONS OF BOREHOLES AND SHAFTS

## Alport Borehole

Ht. above O.D. about 930 ft. 6-in SK 19 S.W. Site: 115 yd S.6°E. of junction of River Alport and Swint Clough brook. Grid Ref. 1360 9105. Drilled in 1939–41 for Steel Bros. & Co. Ltd. Details published by Hudson and Cotton (1943; 1945a).

| | Thickness ft in | Depth ft in | | Thickness ft in | Depth ft in |
|---|---|---|---|---|---|
| DRIFT .. .. | 20 0 | 20 0 | Limestone .. .. | 47 0 | 1384 0 |
| | | | Mudstone, with thin limestones .. .. | 50 0 | 1434 0 |
| MILLSTONE GRIT SERIES | | | Limestone, with tuff bands in lower part | 56 0 | 1490 0 |
| Mudstone, with siltstone beds, calcareous beds and bullions (see pp. 195–7) | 1079 0 | 1099 0 | Mudstone, with thin limestones .. .. | 52 0 | 1542 0 |
| | | | Limestone (see also pp. 115–9) .. .. | 1013 0 | 2555 0 |
| CARBONIFEROUS LIMESTONE SERIES Mudstone, calcareous, with thin limestones | 338 0 | 1337 0 | | | |

## Broad Low Borehole

Ht. above O.D. 952 ft. 6-in SK 17 N.E. Site: 810 yd W.25°S. of Camphill Farm. Grid Ref. 1753 7821. Drilled in 1951. Information from examination of cuttings by Professor F. W. Shotton.

Landslip and Edale Shales to 25 ft, Eyam Limestones to 133 ft (see p. 103), Monsal Dale Beds to 390 ft (see p. 94).

## Bradwell Moor (BM) No. 32B Borehole

Ht. above O.D. about 1325 ft. 6-in SK 18 S.W. Site 675 yd W.30°S. of Bradwellmoor Barn. Grid Ref. 1475 8121. Drilled in 1958 for Messrs. G. and T. Earle Ltd. Cores examined by R. A. Eden.

| | Thickness ft in | Depth ft in | | Thickness ft in | Depth ft in |
|---|---|---|---|---|---|
| Soil .. .. .. | 4 10 | 4 10 | Limestone, grey to dark grey, fine- to medium-grained; *Lithostrotion junceum* .. | 7 10 | 11 |
| MONSAL DALE BEDS Little core; some chert and fine-grained pale limestone .. .. | 5 6 | 10 4 | Chert .. .. .. | 2 11 | 1 |

362

| | Thick-ness ft in | Depth ft in |
|---|---|---|
| Limestone, pale grey fine- to medium-grained; *Lonsdaleia duplicata* Band 6 in at 15 ft with *Dibunophyllum bipartitum konincki* and *L. duplicata duplicata* .. | 8 11 | 20 0 |
| Limestone, grey finely crinoidal, coarser below .. .. .. | 8 0 | 28 0 |
| Limestone, grey coarsely crinoidal; a little irregular chert 39 ft 3 in to 40 ft 3 in .. | 14 0 | 42 0 |
| Chert .. .. .. | 4 | 42 4 |
| Limestone, grey to dark grey, with some dark argillaceous partings; 4 in nodule chert at 46 ft 3 in .. .. | 5 8 | 48 0 |
| Limestone, grey to dark grey with some crinoid debris, some dark argillaceous | | |

| | Thick-ness ft in | Depth ft in |
|---|---|---|
| partings and scattered chert nodules.. | 6 9 | 54 9 |
| Limestone, pale grey cherty; some crinoid debris .. .. | 5 3 | 60 0 |
| Limestone, grey to dark grey cherty .. .. | 6 4 | 66 4 |
| Limestone, grey to pale grey cherty .. .. | 23 8 | 90 0 |
| Limestone, pale grey with a little coarse crinoid debris .. | 4 0 | 94 0 |
| Limestone, grey with 2 in chert at 95 ft 8 in | 11 6 | 105 6 |
| Limestone, medium to dark grey rather fine-grained; 1 in chert at 105 ft 11 in | 2 6 | 108 0 |
| Limestone, grey .. | 9 0 | 117 0 |
| Limestone, dark grey fine-grained; a few small shells .. | 3 0 | 120 0 |

## Castleton Borehole

Ht. above O.D. about 731 ft. 6-in SK 18 S.W. Site: 590 yd S.14°W. of Dunscar Farm. Grid Ref. 1410 8293. Drilled in 1952 for Derwent Valley Water Board. Cores examined by Professor F. W. Shotton and W. H. C. Ramsbottom.

| | Thick-ness ft in | Depth ft in |
|---|---|---|
| DRIFT .. .. | 15 0 | 15 0 |
| MILLSTONE GRIT SERIES | | |
| Mudstone with silty and calcareous beds (see pp. 199–200) .. | 84 7 | 99 7 |

| | Thick-ness ft in | Depth ft in |
|---|---|---|
| CARBONIFEROUS LIMESTONE SERIES | | |
| Mudstone .. .. | 14 11 | 114 6 |
| Limestone, including Beach Beds below 139 ft 3 in (see pp. 111–2) .. .. | 89 0 | 203 6 |

## Charlestown Borehole

Ht. above O.D. about 525 ft. 6-in SK 09 S.W. Site: 780 yd N.15°W. of Lees Hall. Grid Ref. 0297 9331. Drilled 1962–3 for Messrs. Olive, Partington & Co. Core examined by I. P. Stevenson.

# APPENDIX I

| | Thickness ft in | Depth ft in | | Thickness ft in | Depth ft in |
|---|---|---|---|---|---|
| DRIFT            about | 50  8 | 50  8 | Mudstone with BUTTER-LY MARINE BAND (see pp. 219–20) .. | 87  3 | 200  6 |
| MILLSTONE GRIT SERIES | | | LOWER  KINDERSCOUT GRIT (see p. 219) .. | 74  6 | 275  0 |
| UPPER  KINDERSCOUT GRIT (see p. 221) .. | 62  7 | 113  3 | Mudstone and siltstone with thin sandstone bands (see p. 216) .. | 576  6 | 851  6 |

## Chisworth Colliery, Inkerman Shaft

Ht. above O.D. 890 ft. 6-in SJ 99 S.E. Site: 340 yd N.42°W. of Intakes Farm. Grid Ref. 9922 9158. Section amended after Hull and Green (1866, p. 24).

| | Thickness ft in | Depth ft in | | Thickness ft in | Depth ft in |
|---|---|---|---|---|---|
| LOWER COAL MEASURES | | | GANISTER COAL (see p. 284)        1 in to | 6 | 121  10 |
| Mudstone (see p. 288) | 36  0 | 36  0 | Ganister   ..        .. | 2  0 | 123  10 |
| RED ASH COAL (see p. 286)      1 ft 2 in to | 1  4 | 37  4 | Mudstone   ..        .. | 60  0 | 183  10 |
| | | | Sandstone   ..        .. | 5  0 | 188  10 |
| Sandstone (see p. 285) | 44  0 | 81  4 | Mudstone   ..        .. | 55  0 | 243  10 |
| | | | YARD COAL   ..        .. | 3  0 | 246  10 |
| Mudstone   ..        .. | 40  0 | 121  4 | Seatearth   ..        .. | 4  0 | 250  10 |

## Disley Paper Mills Borehole

Ht. above O.D. about 375 ft. 6-in SJ 98 N.E. Site: 825 yd S.38°E. of Woodend. Grid Ref. 9805 8528. Drilled in 1954 for Disley Paper Mills Ltd.

| | Thickness ft in | Depth ft in | | Thickness ft in | Depth ft in |
|---|---|---|---|---|---|
| DRIFT   ..   .. | 6  0 | 6  0 | WOODHEAD HILL ROCK (see p. 279)   .. | 54  0 | 220  0 |
| LOWER COAL MEASURES | | | Mudstone   ..   .. | 99  0 | 319  0 |
| Sandstone   ..   .. | 89  0 | 95  0 | 'Fireclay and strains of coal' (see p. 246)   .. | 5  0 | 324  0 |
| Mudstone   ..   .. | 20  0 | 115  0 | 'Fakes'   ..   .. | 8  6 | 332  6 |
| YARD COAL 36 in   .. | 3  0 | 118  0 | ROUGH ROCK (see p. 244) ..   ..   .. | 67  0 | 399  6 |
| Mudstone (see p. 279) | 48  0 | 166  0 | Mudstone   ..   .. | 3  6 | 403  0 |

## Edale Borehole

Ht. above O.D. about 850 ft. 6-in SK 18 S.W. Site: 665 yd E.11°S. of Highfield. Grid Ref. 1078 8493. Drilled in 1938, for Steel Bros. and Co. Ltd. Details published by Hudson and Cotton (1945b).

| | Thick-ness | | Depth | | | | Thick-ness | | Depth | |
|---|---|---|---|---|---|---|---|---|---|---|
| | ft | in | ft | in | | | ft | in | ft | in |
| **MILLSTONE GRIT SERIES** | | | | | Limestone, cherty at top .. .. .. | | 47 | 0 | 419 | 0 |
| Mudstone, with silt-stone beds, and with calcareous beds and pebbles of limestone near base (see pp. 186, 188) .. .. | 325 | 0 | 325 | 0 | Limestone with mud-stone partings .. | | 74 | 0 | 493 | 0 |
| **CARBONIFEROUS LIMESTONE SERIES** | | | | | Limestone with some chert, tuffaceous in places to 617 ft, non-cherty below (see pp. 113–5) .. .. | | 264 | 0 | 757 | 0 |
| Mudstone, with thin limestones .. .. | 47 | 0 | 372 | 0 | | | | | | |

### Eldon Hill Quarry Borehole

Ht. above O.D. about 1230 ft. 6-in SK 18 S.W. Site: 1190 yd S.11°W. of Peakshill. Grid Ref. 1128 8156. Drilled in 1958 for Messrs. Eldon Hill Quarries Ltd. Information from borer's log and examination of cutting samples by Dr. D. C. Knill.

Bee Low Limestones to 280 ft (see p. 58), Woo Dale Beds to 605 ft (see p. 42) with igneous rock (?tuff) from 505 ft to bottom of borehole.

### Fernilee No. 1 Borehole

Ht. above O.D. 653 ft. 6-in SK 07 N.W. Site: 310 yd E.1½°S. of Normanswood. Grid Ref. 0124 7823. Drilled in 1960 for Stockport Corporation Water Works. Cores examined by I. P. Stevenson.

| | Thick-ness | | Depth | | | | Thick-ness | | Depth | |
|---|---|---|---|---|---|---|---|---|---|---|
| | ft | in | ft | in | | | ft | in | ft | in |
| DRIFT .. .. | 17 | 6 | 17 | 6 | SIMMONDLEY COAL (see p. 237) .. .. | | 2 | 0 | 323 | 0 |
| **MILLSTONE GRIT SERIES** | | | | | Mudstone, siltstone and sandstone (see p. 237) .. .. | | 26 | 0 | 349 | 0 |
| Mudstone with *G. cumbriense* Marine Band at 260 ft, *G. cancellatum* Marine Band at 288 ft 9 in (see p. 240) .. .. .. | 303 | 6 | 321 | 0 | RINGINGLOW COAL (see p. 237) .. .. | | | 3 | 349 | 3 |
| | | | | | CHATSWORTH GRIT (see p. 235) .. .. | | 126 | 9 | 476 | 0 |

### Fernilee No. 2 Borehole[1]

Ht. above O.D. 650 ft. 6-in SK 07 N.W. Site: 490 yd N.32°E. of Normanswood. Grid Ref. 0119 7863. Drilled in 1960 for Stockport Corporation Water Works. Cores examined by I. P. Stevenson.

[1]Fernilee No. 2a Borehole (see p. 361) was drilled to a depth of 326 ft, at a location 15 ft south of Borehole No. 2.

| | Thick-ness | | Depth | | | Thick-ness | | Depth | |
|---|---|---|---|---|---|---|---|---|---|
| | ft | in | ft | in | | ft | in | ft | in |
| DRIFT .. .. | 28 | 0 | 28 | 0 | MILLSTONE GRIT SERIES | | | | |
| LOWER COAL MEASURES | | | | | Seatearth (see p. 246) | 1 | 9 | 193 | 6 |
| WOODHEAD HILL ROCK | 87 | 1 | 115 | 1 | ROUGH ROCK (see p. 245) .. .. .. | 110 | 3 | 303 | 9 |
| Mudstone with *G. subcrenatum* Marine Band at 191 ft 4 in (see p. 275) .. | 76 | 8 | 191 | 9 | Mudstone with thin beds of siltstone and sandstone (see p. 240) | 220 | 9 | 524 | 6 |

### Garrison Bleach Works No. 2 Borehole

Ht. above O.D. about 525 ft 6-in SK 08 N.W. Site: 845 yd S.39°W. of Wethercotes. Grid Ref. 0151 8682. Drilled in 1936 for Messrs. J. J. Hadfield Ltd.

| | | Thick-ness | | Depth | | | Thick-ness | | Depth | |
|---|---|---|---|---|---|---|---|---|---|---|
| | | ft | in | ft | in | | ft | in | ft | in |
| DRIFT | about | 46 | 0 | 46 | 0 | p. 236) .. .. | 2 | 6 | 136 | 0 |
| | | | | | | Mudstone (see p. 236) | 8 | 3 | 144 | 3 |
| MILLSTONE GRIT SERIES | | | | | | CHATSWORTH GRIT (see p. 234) .. .. | 55 | 9 | 200 | 0 |
| Mudstone .. .. | | 87 | 6 | 133 | 6 | Mudstone with thin sandstone beds .. | 356 | 4 | 556 | 4 |
| ?RINGINGLOW COAL (see | | | | | | | | | | |

### Great Hucklow Borehole
#### (Derwent Valley Water Board No. 7)

Ht. above O.D. 992 ft. 6-in SK 17 N.E. Site: 1025 yd E.12°N. of Grundy House. Grid Ref. 1777 7762. Drilled in 1951 for Derwent Valley Water Board. Information from examination of cuttings by Professor F. W. Shotton.

Eyam Limestones to 146 ft, Monsal Dale Beds to 405 ft 6 in with Cressbrook Dale Lava below 399 ft (see p. 94).

### Great Rocks Dale Borehole T/B/21

Ht. above O.D. 1173 ft. 6-in SK 07 NE. Site: 350 yd N.38°E. of Great Rocks Lees. Grid Ref. 1034 7547. Drilled in 1966 for Messrs. I.C.I. Ltd. Cores examined by P. F. Dagger and N. Aitkenhead.

Miller's Dale Beds to 60 ft 3 in, Lower Miller's Dale Lava to 158 ft 5 in, Chee Tor Rock to 186 ft (see p. 52).

### Grove Mill Borehole

Ht. above O.D. about 395 ft 6-in SJ 98 N.E. Site: 1085 yd S.10°W. of Eaves Knoll. Grid Ref. 9942 8506. Drilled in 1918–9.

| | Thickness ft in | Depth ft in | | Thickness ft in | Depth ft in |
|---|---|---|---|---|---|
| DRIFT .. .. | 61 0 | 61 0 | RINGINGLOW COAL (see p. 238) .. .. | 1 0 | 195 0 |
| MILLSTONE GRIT SERIES | | | Seatearth .. .. | 1 6 | 196 6 |
| Mudstone .. .. | 99 0 | 160 0 | CHATSWORTH GRIT (see p. 235) .. .. | 43 6 | 240 0 |
| Sandstone .. .. | 32 0 | 192 0 | Mudstone .. .. | 22 0 | 262 0 |
| Mudstone .. .. | 2 0 | 194 0 | | | |

## Hope Cement Works Borehole

Ht. above O.D. 575 ft. 6-in SK 18 S.E. Site: Salter Barn—now part of Earle's Cement Works. Grid Ref. 1678 8228. Drilled in 1932 for Messrs. G. and T. Earle Ltd. Details published by Fearnsides and Templeman (1932).

| | Thickness ft in | Depth ft in | | Thickness ft in | Depth ft in |
|---|---|---|---|---|---|
| DRIFT .. .. | 16 0 | 16 0 | CARBONIFEROUS LIMESTONE SERIES | | |
| MILLSTONE GRIT SERIES | | | Mudstone .. .. | 17 0 | 297 0 |
| Mudstone with calcareous beds (see pp. 200–1) .. .. | 264 0 | 280 0 | Limestone .. .. | 53 0 | 350 0 |
| | | | Lava, with thin tuffs at top and base .. | 105 6 | 455 6 |
| | | | Limestone .. .. | 15 0 | 470 6 |
| | | | Lava and tuff (see also pp. 112–3) .. .. | 29 0 | 499 6 |

## Hucklow Edge No. 1 Borehole

Ht. above O.D. 1029 ft. 6-in SK 17 N.E. Site: 1050 yd N.32°W. of Waterfall Farm. Grid Ref. 1941 7781. Drilled in 1965–6 by Glebe Mines Ltd. Cores examined by I. P. Stevenson to 185 ft 8 in; record below that depth provided by Glebe Mines Ltd.

| | Thickness ft in | Depth ft in |
|---|---|---|
| MILLSTONE GRIT SERIES (see pp. 206–7) | | |
| *Reticuloceras circumplicatile* Zone ($R_{1a}$) | | |
| Mudstone, weathered | 19 8 | 19 8 |
| Mudstone, dark grey, shaly; *Dunbarella sp.*, *Homoceras sp.*, *Reticuloceras* cf. *pulchellum*, *R.* cf. *todmordenense*, mollusc spat and *Cypridina sp.* from 19 ft 9 in to 28 ft 9½ in .. .. | 10 10 | 30 6 |
| Mudstone, grey shaly; orthocone nautiloid, mollusc spat and *Cypridina sp.* at 31 ft 6 in; *Caneyella sp.*, *Homoceras henkei*, *Homoceratoides varicatus*, *Reticuloceras sp.* and mollusc spat from 31ft 11 in to 33 ft 10 in; *Caneyella sp.* and mollusc spat from 34 ft 3 in to 34 ft 9 in .. | 4 4 | 34 10 |

| | Thick-ness | | Depth | |
|---|---|---|---|---|
| | ft | in | ft | in |

Mudstone, hard dark grey calcareous .. — 4 — 35 2

Mudstone, grey; *Caneyella sp.*, *Dunbarella sp.*, *Homoceras henkei* and *H. magistrorum* .. .. — 3 2 — 38 4

Upper *Homoceras* zones (H₂)

Mudstone, grey; mollusc spat at 38 ft 11 in, *Caneyella sp.*, *Dunbarella sp.*, *Homoceras* cf. *eostriolatum*, *Homoceratoides prereticulatus* and mollusc spat from 39 ft 1 in to 39 ft 8 in .. — 1 11 — 40 3

Mudstone, hard, dark grey calcareous .. — 3 — 40 6

Mudstone, grey; *Aviculopecten sp.*, *Caneyella semisulcata* and mollusc spat from 40 ft 6 in to 41 ft 1 in; *Posidoniella sp.* and *Dimorphoceras sp.* from 41 ft 3 in to 41 ft 6 in; *Aviculopecten sp.*, *C. semisulcata*, mollusc spat and acanthodian scales from 42 ft 2 in to 45 ft 9 in; *Homoceras undulatum* and mollusc spat from 45 ft 10 in to 46 ft 3 in; *C. semisulcata* and mollusc spat from 46 ft 4 in to 47 ft 11 in; *Dimorphoceras sp.*, *Homoceras smithii*, *Hudsonoceras proteus* and mollusc spat from 48 ft 11 in to base .. — 9 6 — 50 0

Lower *Homoceras* zones (H₁)

Mudstone, silty from 50 ft 11 in to 51 ft 4 in, grey platy and

calcareous at 51 ft 6 in, with a 1½ in hard calcareous bed at 70 ft 4 in and a thin pyrite lens at 70 ft 9 in; *Caneyella sp.*, *Promytilus sp.*, *Dimorphoceras sp.* and mollusc spat from 50 ft 9 in to 62 ft 4 in; fucoid markings at 54 ft 3 in; *Aviculopecten sp.*, *Caneyella sp.*, *Homoceras sp.* and mollusc spat from 62 ft 5 in to 64 ft; *Caneyella sp.*, *Posidoniella sp.* and spat from 64 ft 9 in to 65 ft 8 in; *Aviculopecten sp.*, *Dimorphoceras sp.*, *Homoceras subglobosum* and mollusc spat from 67 ft 1 in to base .. .. — 27 1 — 77 1

*Nuculoceras nuculum* (E₂c) and *Cravenoceratoides nitidus* (E₂b) zones

Mudstone, grey, platy in places; 4-in hard calcareous bed at 88 ft 2 in, 2-in bed with sphaerosiderite at 92 ft 2 in, 6-in finegrained grey limestone at 98 ft 10 in, 1-in with sphaerosiderite at 108 ft 5 in; *C. semisulcata*, *Posidoniella sp.* and mollusc spat from 77 ft 8 in to 83 ft 10 in; *Posidonia corrugata* at 86 ft 6 in; *N. nuculum* and mollusc spat at 87 ft 1 in; *P. corrugata* from 87 ft 5 in to 97 ft 9 in; *P. corrugata*, *Dimorphoceras sp.*, *N. nu-*

| | Thick-ness | | Depth | |
|---|---|---|---|---|
| | ft | in | ft | in |

*culum* and spat from 97 ft 10 in to 99 ft 2 in; *P. corrugata* and *Coleolus sp.* from 99 ft 3 in to 105 ft 8 in; *P. corrugata, Dimorphoceras sp.* and spat from 105 ft 10 in to 106 ft 9 in; palaeoniscid scales from 107 ft 4 in to 107 ft 7 in; *P. corrugata, Posidoniella variabilis, Dimorphoceras sp.* and mollusc spat from 107 ft 8 in to base .. .. 31 3 108 4

Mudstone, grey pyritous .. .. .. 7 108 11

Mudstone; 5-in argillaceous limestone at 112 ft, 3-in with sphaerosiderite at 113 ft 3 in, traces of pyrite at 114 ft 9 in, 1-in grey argillaceous limestone at 119 ft 2 in; *P. corrugata, Posidoniella variabilis, Dimorphoceras sp.* and mollusc spat to 109 ft 2 in; smooth spiriferoid fragments, *Actinopteria regularis, P. corrugata*, coiled nautiloid indet., *Cravenoceratoides fragilis* and *nuculum* from 109 ft 3 in to 111 ft 9 in; fish fragments at 111 ft 10 in; *A. regularis, P. corrugata, P.* cf. *variabilis* and mollusc spat from 115 ft 9 in to base .. .. .. 10 3 119 2

Mudstone, soft grey pyritous .. .. 2 8 121 10

Mudstone, ¾-in with sphaerosiderite at 122 ft 8½ in, 2-in hard

fine-grained siderite at 123 ft, 1-in with sphaerosiderite at 124 ft 9 in, 5-in hard calcareous at 132 ft 4 in, 9-in hard calcareous at 135 ft 3 in and 1 ft 7 in hard calcareous bed at base. *Actinopteria regularis, P. corrugata, P.* cf. *variabilis*, indet. goniatite at 122 ft, mollusc spat, to 130 ft 6 in; *P. corrugata, Cravenoceratoides sp., Eumorphoceras sp.* and orthocone nautiloid from 131 ft 1 in to 131 ft 5 in; productoid fragments, rhynchonelloid fragment, *Rugosochonetes sp.*, and goniatites indet. from 131 ft 6 in to 132 ft 3 in; *A. regularis, P. corrugata*, orthocone nautiloid and palaeoniscid scale from 132 ft 4 in to 134 ft 10 in; *P. corrugata, Anthracoceras sp., Dimorphoceras sp.*, orthocone nautiloid and mollusc spat from 136 ft 3 in to 149 ft 3 in; *P. corrugata, P. corrugata elongata, Cravenoceratoides edalensis* and *Dimorphoceras sp.* from 149 ft 6 in to 150 ft 4 in; *P. corrugata, Anthracoceras* or *Dimorphoceras sp., Cravenoceras* cf. *subplicatum* and spat from 150 ft 4½ in to base .. 29 7 151 5

*Eumorphoceras bisulcatum* (E₂ₐ) Zone

| | Thickness | | Depth | |
|---|---|---|---|---|
| | ft | in | ft | in |

Mudstone, grey, calcareous in places and with a 2½-in fine-grained limestone at 164 ft 2½ in; *L. longirostris, P. corrugata*, goniatites indet. (at 155 ft 9 in) and mollusc spat from 154 ft to 159 ft 10 in; crinoid columnals, *P. corrugata, Obliquipecten sp., Cravenoceras sp., Dimorphoceras sp., Eumorphoceras bisulcatum* and *Kazakhoceras scaliger* from 160 ft to 165 ft; *P. corrugata*, pectinoid indet., and palaeoniscid scale from 166 ft 1 in to 169 ft 8 in; *P. corrugata, Streblochondria sp., Cravenoceras sp.* and *Eumorphoceras sp.* from 170 ft to base .. .. | 20 | 4 | 171 | 9 |

*Cravenoceras malhamense* (E₁c) Zone

Mudstone, grey; palaeoniscid scales from 171 ft 10 in to 172 ft 1 in; *Caneyella membranacea, P. corrugata* and fish fragments from 172 ft 6 in to base .. .. | 1 | 4 | 173 | 1 |

*Eumorphoceras pseudobilingue* (E₁b) Zone

Mudstone, grey; *P. corrugata, Posidonia trapezoedra* and fish fragments from 173 ft 5 in to 176 ft 9 in; *P. corrugata* and *E.* cf. *pseudobilingue C* from 176 ft 10 in to base .. .. .. | 4 | 0 | 177 | 1 |

*Cravenoceras leion* (E₁a) Zone

Mudstone, grey, with a 6-in fine-grained limestone at 178 ft; *P. corrugata* from 177 ft 4 in to 178 ft 6 in; *P. corrugata, C. leion* and palaeoniscid scale from 178 ft 7 in to base .. | 2 | 1 | 179 | 2 |

CARBONIFEROUS LIMESTONE SERIES

Mudstone (in Eyam Group); some crinoid and fish debris | 5 | 7 | 184 | 9 |

EYAM LIMESTONES

Limestone, buff to pale grey .. .. .. | 2 | 5 | 187 | 2 |

Limestone, crinoidal, compound coral at 189 ft .. .. | 2 | 10 | 190 | 0 |

Limestone, pale grey crinoidal and with fine shell debris; compound coral at 194 ft 8 in .. .. | 6 | 11 | 196 | 11 |

Limestone, buff with some crinoid debris to 204 ft 1 in; grey below .. .. | 13 | 1 | 210 | 0 |

Limestone, grey to pale grey cherty, crinoidal below 212 ft 3 in .. | 6 | 0 | 216 | 0 |

Limestone, pale grey crinoidal; compound coral 218 ft 6 in .. | 13 | 11 | 229 | 11 |

Limestone, grey finely banded .. | 11 | 5 | 241 | 4 |

Limestone, dark argillaceous .. .. | | 8 | 242 | 0 |

Limestone, grey .. | 4 | 8 | 246 | 8 |

Limestone, dark cherty crinoidal .. .. | 9 | 11 | 256 | 7 |

Mudstone, dark .. | | 2 | 256 | 9 |

MONSAL DALE BEDS

Limestone, grey to buff | 19 | 9 | 276 | 6 |

Limestone, pale grey cherty with stylolites | 5 | 6 | 282 | 0 |

Limestone, pale grey .. | 1 | 10 | 283 | 10 |

| | Thickness ft in | Depth ft in |
|---|---|---|
| Limestone, pale grey cherty; coarser grained above 286 ft 4 in .. .. .. | 6 0 | 289 10 |
| Limestone, pale grey cherty crinoidal .. | 4 5 | 294 3 |
| Limestone, pale grey with scattered fine crinoid debris .. | 8 9 | 303 0 |
| Limestone, buff to pale grey cherty .. .. | 42 0 | 345 0 |
| Limestone, pale grey with stylolites, coral colony at 372 ft .. | 46 2 | 391 2 |
| Limestone, pale grey cherty with scattered crinoid debris; compound coral at 393 ft .. .. .. | 10 4 | 401 6 |
| Limestone, variegated pale grey and buff with scattered crinoid debris and common stylolites .. .. | 2 6 | 404 0 |
| Limestone, pale grey | 30 9 | 434 9 |
| Limestone, pale grey coarsely crinoidal, with corals including Lonsdaleia sp. .. | 9 | 435 6 |
| Limestone, pale grey with scattered crinoid debris; 3 ft with productoids at 448 ft .. | 12 7 | 448 1 |
| Limestone, pale grey finely banded with occasional large shells .. .. | 14 1 | 462 2 |
| Horizon of Cressbrook Dale Lava: grey calcitized tuff with vesicular lava 3 ft 4 in at 467 ft 4 in .. .. | 11 11 | 474 1 |

## Hucklow Edge No. 2 Borehole

Ht. above O.D. 1076 ft. 6-in SK 27 N.W. Site: 265 yd N.36°E. of Shepherd's Flat. Grid Ref. 2025 7760. Drilled in 1966 by Glebe Mines Ltd. Cores examined by I. P. Stevenson.

| | Thickness ft in | Depth ft in |
|---|---|---|
| MILLSTONE GRIT SERIES (see p. 207) | | |
| No core .. .. | 207 9 | 207 9 |
| | | |
| Cravenoceratoides nitidus ($E_{2b}$) Zone | | |
| Mudstone, grey, with hard calcareous beds and a fine-grained argillaceous limestone 1 ft at 219 ft; Posidonia corrugata, Cravenoceras? from 210 ft 8½ in to 211 ft 3½ in; crinoid debris, productoid fragments, Rugosochonetes sp., smooth spiriferoids, Dunbarella? and Cravenoceras? from 211 ft 4 in to 212 ft 6 in; P. corrugata, Catastroboceras sp., orthocone fragments, Cravenoceras sp., dimorphoceratid fragments, goniatite ghosts, mollusc spat and fish fragments 212 ft 7 in to 234 ft ft 10 in; Cravenoceratoides edalensis from 235 ft 5 in to 236 ft; P. corrugata, P. corrugata gigantea and mollusc spat 236 ft 1 in to base .. | 35 10 | 243 7 |

| | Thick-ness | | Depth | | | Thick-ness | | Depth | |
|---|---|---|---|---|---|---|---|---|---|
| | ft | in | ft | in | | ft | in | ft | in |

*Eumorphoceras bisul-
catum* (E$_{2a}$) Zone
Mudstone, grey, largely
hard and calcareous,
with thin fine-grained
limestones; crinoid
debris, *Dunbarella
sp.,P. corrugata, Cra-
venoceras sp., Dimor-
phoceras sp., E. bisul-
catum* [juv.] and *Ka-
zakhoceras scaliger*
from 247 ft 2 in to
251 ft 11 in; *P. cor-
rugata* and palaeo-
niscid scales from
252 ft 1 in to 265 ft
3 in (abundant bone
and shell fragments
at 261 ft 11½ in)   ..   21  11     265  6

CARBONIFEROUS
LIMESTONE SERIES

EYAM LIMESTONES (see
   p. 101)    ..    ..    79  0    344  6
MONSAL DALE BEDS (see
   pp. 93–4) with Cress-
   brook Dale Lava
   below 550 ft 2 in  ..  266  2    610  8

## Ladywash Shaft

Ht. above O.D. 1254 ft. 6-in SK 27 N.W. Site: 215 yd S.22°W. of Ladywash
Farm. Grid Ref. 2189 7754. Date of sinking unknown.

Made ground to 29 ft. MILLSTONE GRIT SERIES: mudstone to 93 ft,
Shale Grit to 312 ft (see p. 215), mudstone to 699 ft (see p. 207) CARBONI-
FEROUS LIMESTONE SERIES: limestone to 776 ft [floor of crosscut at
728 ft].

## Ladywash Shaft, underground borehole near

Ht. above O.D. 455 ft. 6-in SK 27 N.W. Site: In 280's (crosscut) level, 60 yd
W.26°N. of Ladywash Shaft. Grid Ref. 2184 7756. Drilled in 1939 for English
Lead Mines Exploration Ltd. Record published by Schnellmann and Willson
1947.

MONSAL DALE BEDS: limestone to 112 ft, Cressbrook Dale Lava to 355 ft,
limestone to 494 ft. BEE LOW LIMESTONES: limestone to 566 ft (see p. 95).

## Lane Head Farm Borehole

Ht. above O.D. about 995 ft. 6-in SK 29 S.W. Site: 310 yd N.13°E. of Hallfield.
Grid Ref. 2349 9179. Drilled in 1936.

|  | Thick-ness | | Depth | |
|---|---|---|---|---|
|  | ft | in | ft | in |
| DRIFT .. .. | 5 | 0 | 5 | 0 |

MILLSTONE GRIT
SERIES

| HEYDEN ROCK (see p. | | | | |
|---|---|---|---|---|
| 232) .. .. .. | 82 | 6 | 87 | 6 |
| Mudstone (see p. 232) | 62 | 6 | 150 | 0 |
| KINDERSCOUT GRIT (see | | | | |
| p. 224) .. .. | 41 | 6 | 191 | 6 |

### Little Hucklow Borehole
#### (Derwent Valley Water Board No. 3)

Ht. above O.D. 841 ft. 6-in SK 17 N.E. Site: 495 yd S.40°W. of Nether Water Farm. Grid Ref. 1684 7863. Drilled in 1951. Record from examination of cutting-samples by Professor F. W. Shotton.

Eyam Limestones to 34 ft (see p. 103), Monsal Dale Beds to 270 ft.

### Litton Dale Borehole
#### (Derwent Valley Water Board No. 2)

Ht. above O.D. 930 ft. 6-in SK 17 S.E. Site: 505 yd W.38°S. of Sterndale House. Grid Ref. 1599 7498. Drilled in 1951. Information from examination of cutting-samples by Professor F. W. Shotton.

Monsal Dale Beds to 55 ft, Miller's Dale Beds to 152 ft 6 in (see p. 55), Lower Miller's Dale Lava to 231 ft (see p. 53), Chee Tor Rock to 385 ft, with uncorrelated igneous rock below 340 ft (see p. 50).

### Littonfields Borehole
#### (Derwent Valley Water Board No. 1)

Ht. above O.D. 827 ft. 6-in SK 17 N.E. Site: 225 yd N.9°W. of Littonfields. Grid Ref. 1751 7595. Drilled in 1951. Record from examination of cores by I. P. Stevenson to 280 ft; below from examination of cutting-samples by Professor F. W. Shotton.

Upper Monsal Dale Beds to 249 ft (see pp. 84–5), Lower Monsal Dale Beds to 480 ft (see p. 69) with Litton Tuff from 269 to 280 ft and Cressbrook Dale Lava from 311 to 420 ft.

### Michill Bank No. AF5 Borehole

Ht. above O.D. 1114 ft. 6-in SK 18 S.E. Site: 360 yd N.4°E. of Bradwellmoor Barn. Grid Ref. 1530 8189. Drilled in 1961 for Messrs. G. and T. Earle Ltd. Cores examined by D. V. Frost and I. P. Stevenson.

z

| Description | Thickness ft in | Depth ft in |
|---|---|---|
| Soil and yellow-brown clay .. .. .. | 3 6 | 3 6 |
| MONSAL DALE BEDS | | |
| Limestone, dark fine-grained with scattered shells; 1-in argillaceous at 12 ft: *Gigantoproductus sp.* at 6 ft 7 in .. .. | 10 6 | 14 0 |
| Chert, medium to dark grey, banded .. | 1 0 | 15 0 |
| Limestone, dark grey, fine-grained with a few shells at 16 ft 6 in .. .. .. | 3 3 | 18 3 |
| Chert, dark .. .. | 9 | 19 0 |
| Limestone, grey to dark grey, fine-grained with occasional shells | 4 10 | 23 10 |
| Upper *Girvanella* Band; chert, grey-brown with *Girvanella sp.* .. | 10 | 24 8 |
| Limestone, grey to dark grey, fine-grained .. | 2 4 | 27 0 |
| Limestone, grey-brown; *Gigantoproductus sp.* | 1 4 | 28 4 |
| Limestone, grey to dark grey, fine-grained with some paler banding and variegation below 35 ft; small chert nodules below 36 ft 3 in; *Bellerophon sp.* at 34 ft 6 in; *Gigantoproductus sp.* at 36 ft; *Syringopora sp.* at 36 ft 8 in .. .. .. | 8 6 | 36 10 |
| Limestone, grey-brown fine-grained, finely oolitic in places .. | 1 10 | 38 8 |
| Limestone, pale grey-brown fine-grained cherty .. .. | 8 4 | 47 0 |
| Limestone, grey brown crinoidal with some large columnals, dark variegation in places; some chert above 51 ft 6 in .. | 9 9 | 56 9 |
| Limestone, dark grey fine-grained, some paler bands below 64 ft .. .. | 9 0 | 65 9 |
| Limestone, pale grey fine-grained with some microstylolites | 7 6 | 73 3 |
| Limestone, dark grey finely crinoidal in places; slightly paler in lower part .. | 20 9 | 94 0 |
| Limestone, dark grey-brown with *Girvanella sp.* .. .. | 1 7 | 95 7 |
| Limestone, grey-brown finely granular .. | 1 5 | 97 0 |
| Limestone, grey-brown coarse (calcirudite) with much rounded shell debris and occasional limestone pebbles up to 2 in .. | 9 0 | 106 0 |
| Limestone, grey to grey-brown fine-grained to finely granular; a little fine crinoid debris above 116 ft .. .. | 13 3 | 119 3 |
| Limestone, grey-brown with some small shells and crinoid debris; very fine-grained to 121 ft 9 in, finely granular below; *Lithostrotion portlocki* at 129 ft 6 in .. .. .. | 33 1 | 152 4 |
| Limestone, grey-brown fine-grained oolitic with bryozoan fragments .. .. | 5 8 | 158 0 |
| Limestone, dark grey; with common gigantoproductoids and large crinoid columnals to 161 ft 9 in, sporadic fine crinoid debris below; *Diphyphyllum lateseptatum* at 162 ft .. .. | 9 9 | 167 9 |

|  | Thick-ness | Depth |
|---|---|---|
|  | ft in | ft in |
| Lower *Girvanella* Band: limestone, dark grey finely crystalline with *Girvanella sp.*, finely crinoidal with small calcite-filled cavities | 4 9 | 172 6 |
| BEE LOW LIMESTONES |  |  |
| Limestone, pale grey finely granular .. | 4 6 | 177 0 |
| Limestone, grey-brown, fine-grained finely crinoidal with many calcite-filled vughs | 30 8 | 207 8 |
| Limestone, grey to dark grey .. .. .. | 6 3 | 213 11 |
| Limestone, grey to grey-brown, fine-grained; finely crinoidal below 233 ft | 21 7 | 235 6 |
| Limestone, pale grey-brown fine-grained, |  |  |

|  | Thick-ness | Depth |
|---|---|---|
|  | ft in | ft in |
| finely crinoidal with bryozoa in places | 12 10 | 248 4 |
| Limestone, grey to grey-brown, darker below 288 ft and with many stylolites .. | 56 10½ | 305 2½ |
| Limestone, grey pyritous and tuffaceous | 4 0 | 309 2½ |
| PINDALE TUFF: tuff, grey-green with disseminated pyrite, sporadic crinoid debris and small limestone lapilli to 313 ft 4½ in; mottled grey-green calcareous to 316 ft 6 in; grey-green and brownish pyritous to 321 ft; no core (recorded as tuff) to 354 ft 9 in | 45 6½ | 354 9 |
| Limestone, grey-brown finely crystalline .. | 4 3 | 359 0 |

## Middleton Dale Borehole
### (Derwent Valley Water Board No. 4)

Ht. above O.D. 777 ft. 6-in SK 27 N.W. Site: 555 yd W.38°N. of Farnsley Farm. Grid Ref. 2052 7600. Drilled in 1951. Information from examination of cuttings by Professor F. W. Shotton.

Monsal Dale Beds to 257 ft (see p. 96) with Cressbrook Dale Lava below 93 ft 6 in.

## Nether Cotes No. 8 Borehole

Ht. above O.D. 705 ft. 6-in SK 18 S.E. Site: 1020 yd W.41°N. of Bradwell church. Grid Ref. 1679 8170. Drilled in 1960–1 for Messrs. G. and T. Earle Ltd. Cores examined by D. V. Frost.

|  | Thick-ness | Depth |
|---|---|---|
|  | ft in | ft in |
| Soil, etc. .. .. | 5 3 | 5 3 |
| MONSAL DALE BEDS |  |  |
| Limestone, grey-brown fine-grained with *Pugilis pugilis*, *Rugosochonetes sp.* and *Spirifer sp.* .. .. | 1 0 | 6 3 |
| *No core* .. .. | 9 | 7 0 |

|  | Thick-ness | Depth |
|---|---|---|
|  | ft in | ft in |
| Limestone, pale grey-brown crinoidal shelly; ¼-in fluorite vein at 9 ft 6 in; *Antiquatonia sp.*, '*Brachythyris*' *planicostata*, *Gigantoproductus sp.*, *Krotovia sp.*, *Spirifer bisulcatus* .. .. | 2 8 | 9 8 |

| Description | Thickness ft | in | Depth ft | in |
|---|---|---|---|---|
| Limestone, dark grey fine-grained, finely crinoidal; *Gigantoproductus sp.* and *Spirifer sp.* .. .. | 3 | 3 | 12 | 11 |
| No core .. .. | | 9 | 13 | 8 |
| Limestone, grey-brown fine to medium grained, finely crinoidal; *Productus productus hispidus, Spirifer bisulcatus* .. | 3 | 10 | 17 | 6 |
| No core .. .. | 1 | 3 | 18 | 9 |
| Limestone, grey medium to coarsely crinoidal with some shell fragments; thin vertical vein of purple fluorite .. | 5 | 1 | 23 | 10 |
| No core .. .. | 1 | 2 | 25 | 0 |
| Limestone, dark grey fine-grained; crinoidal below 27 ft; some fluorite veinlets | 4 | 0 | 29 | 0 |
| Limestone, grey fine-grained, crinoidal in part, with some chert, stylolites in places; vertical baryte vein to 31 ft 7 in | 8 | 4 | 37 | 4 |
| Limestone, dark grey fine-grained finely crinoidal, some stylolites and dark bituminous partings; more coarsely crinoidal and with shell fragments below 40 ft 2 in; thin fluorite veinlet at 44 ft; *Lithostrotion junceum* at 40 ft 2 in .. | 7 | 8 | 45 | 0 |
| Limestone, grey-brown, part oolitic, partly bituminous towards base; vertical vein of yellow fluorite .. | 4 | 4 | 49 | 4 |
| Limestone, dark grey-brown fine-grained finely crinoidal; chert 1 in at 54 ft 1 in and 6 in at 56 ft 6 in; *L.* | | | | |

| Description | Thickness ft | in | Depth ft | in |
|---|---|---|---|---|
| *junceum* at 56 ft 3 in | 6 | 11 | 56 | 3 |
| Limestone, grey to dark grey finely crinoidal, more coarsely crinoidal below 65 ft .. | 12 | 5 | 68 | 8 |
| Lower *Girvanella* Band; dark grey fine-grained crinoidal limestone with shell fragments, some encrusted with *Girvanella sp.* .. .. | 3 | 11 | 72 | 7 |
| BEE LOW LIMESTONES | | | | |
| Limestone, pale grey-brown oolitic and crinoidal .. .. | 6 | 10 | 79 | 5 |
| No core .. .. | 1 | 7 | 81 | 0 |
| Limestone, pale grey fine-grained; with stylolites above 91 ft 3 in; *Spirifer sp.* at 83 ft .. .. | 33 | 5 | 114 | 5 |
| Limestone, pale grey with stylolites and sporadic crinoid debris .. .. .. | 5 | 6 | 119 | 11 |
| Limestone, grey finely crinoidal with shell debris and coral fragments .. .. | 2 | 9 | 122 | 8 |
| No core .. .. | 2 | 7 | 125 | 3 |
| Limestone, grey fine-grained; fluorite vein 141 ft to 145 ft; ½ in crinoid columnal 149 ft; *Caninia sp.* 125 ft 6 in; *Lithostrotion pauciradiale* 156 ft 3 in .. .. | 32 | 7 | 157 | 10 |
| Limestone, grey crinoidal with small tuff fragments, some more coarsely crinoidal bands .. .. | 9 | 2 | 167 | 0 |
| Limestone, tuffaceous | 1 | 0 | 168 | 0 |
| PINDALE TUFF: tuff, calcareous and pyritous to 169 ft 2 in; pyritous and with limestone lapilli below: *Bollandoceras sp.* at 168 ft 7 in .. .. | 7 | 2 | 175 | 2 |

| | Thickness ft in | Depth ft in | | Thickness ft in | Depth ft in |
|---|---|---|---|---|---|
| Limestone, grey tuffaceous .. .. | 10 | 176 0 | grained .. .. | 3 | 176 6 |
| Tuff, fine-grained pale green .. .. | 3 | 176 3 | Tuff, grey-green with lapilli .. .. | 2 2 | 178 8 |
| Limestone, grey fine- | | | Limestone, pale grey-brown oolitic .. | 4 4 | 183 0 |

## Pindale No. 2 Borehole

Ht. above O.D. 805 ft. 6-in SK 18 S.E. Site: 350 yd S.45°W. of Pindale Farm. Grid Ref. 1602 8231. Details published by Eden and others (1964, pp. 117–8).

Bee Low Group: limestone to 67 ft, Pindale Tuff to 100 ft.

## Shaw Farm Borehole

Ht. above O.D. about 815 ft. 6-in SJ 98 N.E. Site: 800 yd E.31°S. of Capstone. Grid Ref. 9905 8661. Drilled in 1966 for Calico Printers Association. Cuttings and cores examined by I. P. Stevenson.

| | Thickness ft in | Depth ft in | | Thickness ft in | Depth ft in |
|---|---|---|---|---|---|
| **COAL MEASURES** | | | ty; a 10-in sandstone | | |
| *No core*, mainly shale | 74 5 | 74 5 | at 390 ft .. .. | 8 9 | 392 5 |
| WOODHEAD HILL ROCK (see p. 277) .. .. | 90 7 | 165 0 | Sandstone, grey hard massive fine to medium grained .. | 5 7 | 398 0 |
| Mudstone with *Gastrioceras subcrenatum* Marine Band 1 ft 8 in at base (see p. 276) | 84 10 | 249 10 | *No core* .. .. | 2 0 | 400 0 |
| | | | Mudstone, with silty and finely micaceous bands .. .. | 11 | 400 11 |
| **MILLSTONE GRIT SERIES** | | | Sandstone, pale grey fine to medium-grained with irregular dark partings .. | 6 9 | 407 8 |
| SIX-INCH MINE (see p. 246) | | | | | |
| Coal 11½ in .. | | | Mudstone, with thin siltstone bands .. | 1 6 | 409 2 |
| Dirt 7 in .. | | | | | |
| Coal 5½ in .. | 2 0 | 251 10 | Sandstone, flaggy bioturbated, with dark micaceous partings | 7 | 409 9 |
| Seatearth .. .. | 3 2 | 255 0 | | | |
| Mudstone .. .. | 11 3 | 266 3 | Sandstone and silty mudstone, finely banded .. .. | 7 | 410 4 |
| ROUGH ROCK (for details see p. 241) .. | 115 3 | 381 6 | Sandstone, finely micaceous with some dark micaceous and argillaceous partings .. | 1 5 | 411 9 |
| Mudstone, banded silty and finely micaceous .. .. | 10 | 382 4 | Mudstone, banded silty .. .. .. | 1 3 | 413 0 |
| Sandstone, with fine dark laminations and some cross bedding; dark argillaceous partings below .. | 1 4 | 383 8 | | | |
| Mudstone, banded sil- | | | | | |

| | Thick-ness | | Depth | | | Thick-ness | | Depth | |
|---|---|---|---|---|---|---|---|---|---|
| | ft | in | ft | in | | ft | in | ft | in |
| Sandstone, top 1-ft finely banded, more massive below .. | 5 | 7 | 418 | 7 | Mudstone, banded silty with some thin silt-stone bands .. | 7 | 4 | 440 | 0 |
| Mudstone, silty finely banded with silt-stone streaks .. | 4 | 0 | 422 | 7 | Sandstone, fine-grained finely banded with dark micaceous part-ings .. .. | 1 | 4 | 441 | 4 |
| Sandstone, grey mas-sive fine to medium-grained; large ?flute cast at base .. | 4 | 5 | 427 | 0 | Mudstone, silty mica-ceous, with inter-banded siltstone be-low 445 ft 10 in .. | 6 | 2 | 447 | 6 |
| Mudstone, finely ban-ded silty micaceous | 4 | 2 | 431 | 2 | Sandstone, fine-grained flaggy with dark mi-caceous partings .. | 3 | 1 | 450 | 7 |
| Sandstone, hard flaggy, with dark partings at top and base .. | 1 | 6 | 432 | 8 | Mudstone, silty mica-ceous .. .. | 3 | 2 | 453 | 9 |

## The Folly No. S5 Borehole

Ht. above O.D. about 695 ft. 6-in SK 18 S.E. Site: 560 yd S. 5°W. of Hope church. Grid Ref. 1717 8295. Drilled in 1962 for Messrs. G. and T. Earle Ltd. Cores examined by D. V. Frost.

Mam Tor Beds (see p. 210) to 82 ft 9 in, Edale Shales (see p. 201) to 160 ft.

## Wardlow Mires No. 1 Borehole

Ht. above O.D. 805 ft. 6-in SK 17 N.E. Site: 850 yd S.32°W. of Brosterfield. Grid Ref. 1850 7553. Drilled in 1966–7 by the Institute of Geological Sciences. Cores examined by I. P. Stevenson.

Edale Shales to 67 ft 9 in (see pp. 208–9), $P_2$ shale to 82 ft 10 in (see p. 105), Eyam Limestones to 113 ft 6 in (see p. 102), Monsal Dale Beds to 632 ft 2 in with Litton Tuff 2 ft 6½ in at 452 ft and Cressbrook Dale Lava 106 ft 8 in at 614 ft 6 in (see pp. 91, 93).

## Wardlow Mires No. 2 Borehole

Ht. above O.D. 804 ft. 6-in SK 17 N.E. Site: 800 yd W.26°S. of Brosterfield. Grid Ref. 1825 7586. Drilled in 1967 by the Institute of Geological Sciences. Cores examined by I. P. Stevenson.

| | Thick-ness | | Depth | | |
|---|---|---|---|---|---|
| | ft | in | ft | in | |
| MILLSTONE GRIT SERIES | | | | | Mudstone, grey, shaly in places, including calcareous beds with fossils, pyrite on joints and some phosphatic nodules; |
| No core .. .. | 9 | 0 | 9 | 0 | |
| Eumorphoceras bisul-catum ($E_{2a}$) Zone | | | | | *Posidonia corrugata* from 9 ft to 10 ft 7 in; *P. corrugata, Streb-* |

|  | Thickness ft in | Depth ft in |
|---|---|---|

lopteria *sp.* and *Dimorphoceras* (s.l.) *sp.* from 10 ft 8 in to 11 ft 8 in; *P. corrugata, Coleolus sp., Cravenoceras sp., Dimorphoceras* (s.l.) *sp.* and *Eumorphoceras sp.* [juv.] from 11 ft 11½ in to base .. | 5 7 | 14 7

*Cravenoceras malhamense* ($E_{1c}$) Zone
Mudstone, grey, including calcareous beds and pyrite on joints; *P. corrugata* from 14 ft 7 in to 14 ft 10½ in; *P. corrugata* and *Cravenoceras sp.* from 14 ft 11 in to base .. | 11½ | 15 6½

*Eumorphoceras pseudobilingue* ($E_{1b}$) Zone
Mudstone, grey with black beds, shaly in places, including calcareous beds which become abundant below 19 ft, sporadic pyrite and phosphatic nodules; *P. corrugata* at 15 ft 7½ in; abundant palaeoniscid scales at 15 ft 10½ in; ostracods (cf. *Geisina*) at 16 ft 4 in; *P. corrugata* and palaeoniscid scales from 16 ft 6 in to 18 ft 2½ in; *P. corrugata, Dimorphoceras* (s.l.) *sp.* and *E. pseudobilingue* from 18 ft 7 in to 19 ft 6 in; *P. corrugata* and palaeoniscid scales from 20 ft 4 in to 20 ft 6 in; *P. corrugata, P. trapezoedra, 'Cyrtoceras' rugosum, Dimorpho-*

|  | Thickness ft in | Depth ft in |
|---|---|---|

*ceras* (s.l.) *sp.* and *Eumorphoceras?* [with nodose venter] from 20 ft 9 in to base .. | 9 4½ | 24 11

*Cravenoceras leion* ($E_{1a}$) Zone
Mudstone, grey with pyrite and phosphatic nodules; palaeoniscid scales at 26 ft 8½ in .. .. | 6 1 | 31 0

Mudstone, grey with phosphatic nodules abundant in places; *Lingula sp.* at 42 ft 10 in; fish debris from 33 ft 7 in to base .. .. | 20 0 | 51 0

Mudstone, grey mainly calcareous but with thin non-calcareous beds; dark grey argillaceous limestone 6½ in at 54 ft 11 in and 2½ in at 64 ft 5½ in; crinoid stem fragment, *Caneyella membranacea, P. corrugata, Dimorphoceras* (s.l.) *sp.* and palaeoniscid scales from 51 ft 11 in to to 59 ft 1 in; *Globosochonetes?, Martinia sp.*, productoid indet., *Obliquipecten?, P. corrugata, Coleolus namurcensis,* orthocone nautiloid, *C. leion, Dimorphoceras* (s.l.) *sp.* and trilobite pygidium from 60 ft 4 in to base .. .. .. | 14 4 | 65 4

## CARBONIFEROUS LIMESTONE SERIES

### EYAM GROUP

Mudstone, dark grey, calcareous in places with a few small crinoid columnals; calcareous nodules up to ¼ in below 73 ft 0½ in; 5 in hard calcareous band at 75 ft 2 in; *Martinia sp.*, *Caneyella membranacea*, *P. corrugata*, goniatites indet., platformed conodont and fish tooth 65 ft 4 in to 68 ft 2 in; *Sudeticeras sp.* [strongly crenulate] 68 ft 2¼ in; *Martinia sp.* and *C. membranacea* from 68 ft 2¼ in to 70 ft 4 in; *P. corrugata* at 70 ft 6 in; crinoid columnal, *Globosochonetes* cf. *subminimus*, *Lingula sp.*, *Martinia sp.*, *Plicochonetes sp.* [juv.], *Rugosochonetes sp.*, *C.* cf. *membranacea*, *P. corrugata*, pleurotomarian gastropod, *Sudeticeras sp.* between 70 ft 11 in and 72 ft 6¼ in; *Paraconularia sp.* [juv.], *Martinia sp.* and *Streblochondria sp.* between 72 ft 6¼ in and 72 ft 11½ in; *Dimorphoceras sp.* 73 ft 8½ in; crinoid columnal, chonetoid [juv.], *Martinia sp.*, productoid fragments, *Rugosochonetes sp.*, *Spirifer sp.* [bisulcatus gp.], *Straparollus sp.*, *C. membranacea*, *P. corrugata*, *Girtyoceras sp.*, *Neoglyphioceras sp.*, orthocone nautiloids, *Sudeticeras sp.*, *S.* aff. *ordinatum*, trilobite glabella and fish fragments from 74 ft 1½ in to 76 ft 7 in; *P. corrugata* 76 ft 9 in;

| | Thickness | | Depth | |
|---|---|---|---|---|
| | ft | in | ft | in |
| *P. corrugata* and *Dimorphoceras* (s.l.) *sp.* 77 ft; *Martinia sp.* and *P. corrugata* from 77 ft 0½ in to 77 ft 6 in .. .. | 12 | 3½ | 77 | 7½ |
| Limestone, dark fine-grained argillaceous; *Plicochonetes sp.* and smooth spiriferoids | 1 | 6 | 79 | 1½ |
| Limestone, grey-brown granular finely crinoidal; 2 in pale brown silicified band at 82 ft 5 in; ¾ in vugh with calcite 82 ft 10 in; *Antiquatonia sp.*, *Avonia sp.*, *Buxtonia sp.*, chonetoids, *Dielasma sp.*, *Gigantoproductus sp.*, *Linoproductus sp.*, smooth spiriferoid, *Spirifer (Fusella) trigonalis* .. .. | 4 | 5½ | 83 | 7 |
| Mudstone dark; *Lingula sp.* .. .. | | 1 | 83 | 8 |
| Limestone, grey-brown, finely crinoidal .. | | 5½ | 84 | 1½ |
| Limestone (knoll-reef), pale brown coarsely crinoidal (some columnals up to 1 in) with some pale grey interstitial limestone and pebbles of same lithology; many microstylolites in top 1 in–1½ in; *Dibunophyllum bipartitum konincki*, *Antiquatonia* cf. *hindi*, *A.* cf. *insculpta*, *Buxtonia sp.*, *Eomarginifera sp.*, *Productus sp.*, *Rugosochonetes sp.*, *Schizophoria sp.*, smooth spiriferoid, *Spirifer bisulcatus*, *S. (F.) trigonalis*, trilobite pygidium fragment .. | 2 | 9 | 86 | 10½ |

| | Thickness ft in | Depth ft in |
|---|---|---|
| Mudstone, dark platy | 1 | 86 11½ |
| Limestone, dark with some fine shell debris | ½ | 87 0 |
| Mudstone, dark more calcareous below, smooth spiriferoid | 3½ | 87 3½ |
| Limestone, dark with burrows .. .. | 1 | 87 4½ |
| Mudstone, dark; *Orbiculoidea sp.*, *Rugosochonetes sp.*, smooth spiriferoid, ostracods | 5 | 87 9½ |
| Limestone, dark; irregularly banded with argillaceous streaks above 88 ft; chonetoids, productoids, smooth spiriferoid, trilobite pygidium fragment, ostracods | 4½ | 88 2 |
| Mudstone, dark calcareous, productoid .. | 1 | 88 3 |
| Limestone, dark .. | 1 | 88 4 |
| Mudstone, dark; ½-in pebble dark fine-grained limestone at base; *Buxtonia sp.*, *Eomarginifera sp.*, *Lingula sp.*, *Orbiculoidea sp.*, *Plicochonetes sp.*, *Rugosochonetes sp.*, smooth spiriferoid, *Spirifer sp.*, orthocone nautiloid, ostracods, *Weberides barkei* (Woodward), fish scale .. .. | 1 8 | 90 0 |
| Limestone, grey to dark grey, fine-grained to finely granular .. | 2½ | 90 2½ |
| Limestone, dark finely crinoidal; ostracods 90 ft 3 in .. .. | 5 | 90 7½ |
| Limestone grey to grey-brown, some crinoid debris; shelly below 91 ft 10 in; *Gigantoproductus sp.*, *Leptagonia?*, smooth spiriferoid, *S. (F.) trigonalis* .. .. | 1 9 | 92 4½ |

| | Thickness ft in | Depth ft in |
|---|---|---|
| Limestone, dark argillaceous .. .. | 1½ | 92 6 |
| Mudstone, dark .. | 2½ | 92 8½ |
| Limestone, dark with some fine organic debris; ¾ in mudstone parting 92 ft 10 in.. .. .. | 1 0½ | 93 9 |
| Limestone, grey to dark grey fine-grained; *Cyathaxonia sp.* [juv.], *Eomarginifera?*, *Plicochonetes sp.* [juv.] and smooth spiriferoid .. .. | 1 1 | 94 10 |
| Limestone, grey finely crinoidal .. .. | 3 | 95 1 |
| Limestone, dark fine-grained cherty; zaphrentoid, *Orbiculoidea sp.*, productoid, smooth spiriferoid, *S. (F.) trigonalis* .. .. .. | 2 8 | 97 9 |
| Mudstone, dark, and argillaceous limestone, interbanded | 1 | 97 10 |
| Limestone, dark fine-grained cherty; productoid and smooth spiriferoid at base .. | 11 | 98 9 |
| Limestone, pale grey, fine-grained, cherty, some calcite veinlets | 5 | 99 2 |
| Limestone, dark fine-grained with some chert; burrows at 99 ft 8 in; thin mudstone parting 100 ft 4 in; orthotetoid and productoid 100 ft 9 in | 3 2 | 102 4 |
| Limestone, dark fine-grained; thin parting dark mudstone 104 ft 3 in; *Productus sp.* 102 ft 10 in; *Spirifer bisulcatus* 104 ft 7 in; ostracod 104 ft 11 in; foraminifera, clisiophylloids, smooth spiriferoid, *Bellerophon?* and *Nati-* | | |

| | Thickness | | Depth | |
|---|---|---|---|---|
| | ft | in | ft | in |
| copsis? below 105 ft 3 in .. .. .. | 4 | 10½ | 107 | 2½ |
| Mudstone, dark hard calcareous; productoids, smooth spiriferoid and fish scale | 0 | ¼ | 107 | 3 |
| Mudstone, dark platy carbonaceous .. | 0 | ½ | 107 | 3½ |
| Mudstone, grey-brown; small rounded clay fragments in clay matrix; 1-in limestone pebble at 107 ft 6 in .. .. | 4 | | 107 | 7½ |
| Mudstone, grey very finely pyritous .. | 1½ | | 107 | 9 |
| MONSAL DALE GROUP | | | | |
| Limestone, pale grey-brown to buff; foraminifera below 109 ft; with many stylolites 111 ft–113 ft 5 in; 1 in stylolite 113 ft 10 in; scattered shell and crinoid debris below 111 ft 4 in; Gigantoproductus sp., Spirifer bisulcatus and ostracod below 111 ft 5 in .. | 6 | 8 | 114 | 5 |
| Limestone, buff finely granular; fairly common microstylolites above 115 ft 7 in; ½ in stylolite 115 ft 7½ in; Gigantoproductus sp. .. .. | 2 | 1 | 116 | 6 |
| Limestone, grey-white very fine-grained with microstylolites and common foraminifera; with irregular cracks below 121 ft 7 in .. .. | 5 | 11 | 122 | 5 |
| Limestone, pale grey 'nodular' with numerous irregular mud-filled cracks | 8 | | 123 | 1 |
| Limestone, pale grey fine-grained, a little fine crinoid debris; some fine irregular cracks above 124 ft 5 in; thin argillaceous parting 124 ft 5 in; some microstylolites between 125 ft 3 in and 130 ft 4 in; 1¼-in stylolite at 130 ft 10½ in; some foraminifera below 133 ft; Lithostrotion portlocki at 129 ft 3 in .. | 10 | 5 | 133 | 6 |

## Whaley Bridge Colliery (Winding Pit)

Ht. above O.D. about 575 ft. 6-in SK 08 S.W. Site: 850 yd W.16°S. of Mosley Hall. Grid Ref. 0122 8117. Date of sinking unknown. From graphic section on Abandonment Plan 1721.

| | Thickness | | Depth | |
|---|---|---|---|---|
| | ft | in | ft | in |
| LOWER COAL MEASURES | | | | |
| Bind and shale .. | 51 | 0 | 51 | 0 |
| Coal .. .. | | 9 | 51 | 9 |
| Fireclay .. .. | 6 | 0 | 57 | 9 |
| Bind .. .. | 18 | 0 | 75 | 9 |
| Coal (recorded as "a few inches") .. say | | 3 | 76 | 0 |
| Stone .. .. | 6 | 0 | 82 | 0 |
| Bind .. .. | 12 | 0 | 94 | 0 |
| Stone .. .. | 18 | 0 | 112 | 0 |
| Shale .. .. | 90 | 0 | 202 | 0 |
| RED ASH COAL .. | 1 | 6 | 203 | 6 |
| Stone .. .. | 24 | 0 | 227 | 6 |
| Shale .. .. | 36 | 0 | 263 | 6 |
| WHITE ASH COAL .. | 1 | 4 | 264 | 10 |
| Stone .. .. | 18 | 0 | 282 | 10 |

| | | Thick-ness | | Depth | |
|---|---|---|---|---|---|
| | | ft | in | ft | in |
| Shale | .. .. | 12 | 0 | 294 | 10 |
| GANISTER COAL | .. | | 9 | **295** | **7** |
| Stone | .. .. | 12 | 0 | 307 | 7 |
| Shale | .. .. | 18 | 0 | 325 | 7 |
| Stone | .. .. | 21 | 0 | 346 | 7 |
| Shale | .. .. | 39 | 0 | 385 | 7 |
| YARD COAL | .. | 4 | 6 | **390** | **1** |
| Gritstone rock | .. | — | — | — | — |

(See also pp. 284–6, 288 and Fig. 20.)

### Whitehall Works Borehole

Ht. above O.D. 595 ft. 6-in SK 08 S.W. Site: 460 yd N. of Eccles House. Grid Ref. 0355 8202. Drilled in 1960 for Messrs. Bernard Wardle Ltd. Cores examined by I. P. Stevenson.

| | | Thick-ness | | Depth | |
|---|---|---|---|---|---|
| | | ft | in | ft | in |
| DRIFT | .. .. | 14 | 0 | 14 | 0 |
| MILLSTONE GRIT SERIES | | | | | |
| ROACHES GRIT | .. | 110 | 0 | 124 | 0 |
| Mudstone with *R. bilingue* Marine Band at 403 ft 8 in, *R. bilingue* early form Marine Band at 440 ft 3 in, *R. gracile* Marine Band at 467 ft 6 in; thin beds of siltstone in lower part (see pp. 227–8, 221) .. .. .. | | 384 | 0 | 508 | 0 |

# *Appendix II*

# FOSSIL-LOCALITIES

# CARBONIFEROUS LIMESTONE SERIES

Locations refer to the base of each section; where sections are extensive the position of the top is given in the text.

National Grid references are given within square brackets. The position of individual lettered beds in the sections will be found in the text. In boreholes only the faunal horizons have been lettered. Geological Survey registered numbers of specimens are given without brackets, e.g. PT 7819–93.

For faunal lists see pp. 392–407.

## SHELF PROVINCE
### WOO DALE BEDS
*Locality*

1. Dale Head, main dale [1234 7662] 1020 yd W.15°N. of Wheston Hall. Beds *a–e*. PT 7875–93.

2. Dale Head, tributary valley [1248 7653] 850 yd W.11°N. of Wheston Hall. Beds *a–e*. PT 7930–59.

3. Section at junction of Hay Dale and Dam Dale [1187 7781] 535 yd E.30°S. of Loosehill Farm. Beds *a–p*. RS 444–612.

4. Hernstone Lane Head [1220 7876] 285 yd E.32°S. of Laneside Farm. Beds *a–d*. PT 7743–97.

5. Peak Forest [1170 7809] 275 yd E.8°N. of Loosehill Farm. IPS 1054–5.

6. Peak Forest [1198 7928] 390 yd N.4°E. of Laneside Farm. IPS 1059–63.

7. Peak Forest [1202 7946] 590 yd N.5°E. of Laneside Farm. IPS 1020–2.

8. Near Hartle Plantation [1040 8020] 505 yd W.37°S. of Nether Barn. Beds *a–d*. RS 729–52.

9. Perry Dale [1071 8069] 240 yd N.16°W. of Nether Barn. IPS 1016–9.

10. Conies Dale [1201 8019] 300 yd E.35°N. of Conies Farm. PT 7415–20.

### CHEE TOR ROCK

11. Great Rocks Dale [1011 7512] 100 yd S.25°W. of Great Rocks Lees. PT 8023–5.

12. Duchy Quarry [0938 7681] 280 yd E.6°S. of Peak Forest Station. Beds *a–p*. RK 6372–6, RS 235–74, RS 613–6.

13. Upperend Quarry [0833 7707] 1500 yd S.40°E. of Dove Holes station. Beds *a–t*. RS 853–79.

14. Laughman Tor [1015 7799] 980 yd W.8°N. of Kempshill Farm. Beds *a–b*. RS 709–28.

15. Middle Hill [1079 7698] 1020 yd S.16°W. of Kempshill Farm. RS 1162–9.

16. Middle Hill [1129 7706] 935 yd S.14½°E. of Kempshill Farm. Beds *a–e*. RS 1139–61.

17. Middle Hill [1059 7721] 870 yd S.34°W. of Kempshill Farm. RS 1170–7.

18. East of Hargatewall [1231 7561] 710 yd N.38°E. of Hargate Hall. Beds *a–d*. PT 8042–59.

2. Dale Head, tributary valley, see p. 384. Beds *f–g*. PT 7894–929.

1. Dale Head, main dale, see p. 384. Beds *f–i*. PT 7820–74.

19. Peter Dale [1222 7626] 250 yd south of Dale Head. PT 7798–818.

20. Monk's Dale [1304 7539] 740 yd W.12°N. of Monksdale House. PT 7985–94.

21. Monk's Dale [1347 7521] 250 yd W.10°S. of Monksdale House. PT 7995–8000.

22. Between Wheston and Dale Head [1270 7637] 600 yd. W. of Wheston Hall. PT 1719–35.

23. Wheston [1273 7630] 560 yd W.8°S. of Wheston Hall. PT 7883–4.

24. Wheston [1286 7623] 460 yd W.21°S. of Wheston Hall. PT 7972–82.

25. Wheston [1295 7618] 390 yd W.34°S. of Wheston Hall. PT 7960–71.

3. Section at junction of Hay Dale and Dam Dale, see p. 384. Beds *q–s*. RS 349–443.

26. East side of Hay Dale [1225 7721] 790 yd N.3°E. of Dale Head. PT 1681–95.

4. Hernstone Lane Head, see p. 384. Beds *e–i*. PT 7724–95.

27. Tideswell Moor [1294 7805] 530 yd W.5°S. of Potluck House. PT 1556–71.

28. Wall Cliff [1424 7738] 640 yd N.26°W. of Highfield House. Beds *a–i*. PT 2256–67.

29. Oxlow Rake [1260 8002] 450 yd N.35°W. of The Cop. PT 1601–5.

### MILLER'S DALE BEDS

30. Victory Quarry [0774 7693] 440 yd N.45°W. of Peak House. Beds *a–j*. RS 330–48.

31. Dove Holes Dale Quarry [0782 7748] 800 yd S.30°E. of Dove Holes station. Beds *a–f*. RS 753–75.

32. Holderness Quarry [0842 7817] 650 yd S.32°E. of Near Ridgeclose. Beds *a–e*. RS 1120–38.

33. Old Peak Forest railway section [0799 7828] 430 yd S.16°W. of Near Ridgeclose. Bed *a*. RS 1070–7.

34. Roadside quarry north-west of Hargatewall [1120 7598] 1215 yd N.37°W. of Hargate Hall. PT 8026–41.

35. Smalldale [0977 7700] 670 yd N.17°W. of Rock Houses. Beds *a–d*. RS 175–234.

36. Near Crossgate, Tideswell [1402 7574] 1180 yd E.10°N. of Cherryslack. IPS 1071–3.

28. Wall Cliff, see above. Beds *j–p*. PT 2226–55.

### BEE LOW LIMESTONES

37. East side of Bee Low [0970 7909] 245 yd N.8°E. of Lodesbarn. Beds *a–e*. PT 8060–110.

38. Bee Low Quarry [0922 7915] 570 yd W.32°N. of Lodesbarn. Beds *a–l*. PT 1401–6.

39. Gautries Hill [1005 8087] 890 yd W.28°N. of Nether Barn. Beds *a–c*. PT 7421–47.

40. Eldon Hill Quarry [1128 8134] 1435 yd S.10°W. of Peakshill farm. Beds *a–f*. PT 1606–62.

41. Eldon Hill [1124 8077] 1000 yd N.36°W. of Conies Farm. Beds *a–j*. PT 7385–414.

42. Old quarry near Rowter Farm [1299 8194] 265 yd W.38°S. of Rowter Farm. PT 7448–53.

43. Quarry on south-west side of Snels Low [1125 8163] 1680 yd W.34°S. of Oxlow House. Beds *a–b*. RS 880–929.

44. Middle Hill [1208 8236] 500 yd W.17°S. of Oxlow House. Beds *a–b*. RS 776–852.

45. Cave Dale [1475 8215] 860 yd S.20°W. of Castleton church. Beds *a–h*. PT 4176.

46. Cave Dale [1416 8178] 390 yd S.14°E. Hurdlow Barn. Beds *a–d*. PT 7454–64.

47. Pindale Quarry [1594 8231] 385 yd S.34°W. of Pindale Farm. Beds *a–f*. LL 883–950.

## LOWER MONSAL DALE BEDS

30. Victory Quarry, see p. 385. Beds *k–y*. IPS 1979, RK 6335–44, RS 273–313.

33. Old Peak Forest railway, see p. 385. Beds *b–f*. RS 1045–69.

48. Knoll-reef, old Peak Forest railway section. [0797 7866] 145 yd W.7°S. of Near Ridgeclose. RK 6345–71.

49. Old Peak Forest railway section. [0796 7870] about 40 yd N. of locality 48. RS 1079–84.

50. Barmoor Clough [0789 7949] 600 yd W.31°S. of Bennetston Hall. Beds *a–e*. PJ 754–88.

51. Barmoor Clough Quarry [0881 7988] 465 yd E.9°N. of Bennetston Hall. Beds *a–i*. RS 1–47.

52. Near Barmoor Clough Quarry [0890 7981] 570 yd E.1°N. of Bennetston Hall. Beds *a–c*. IPS 1031–5.

53. Bole Hill [1059 7552] 590 yd N.44°E. of Great Rocks Lees. Beds *a–f*. PT 8001–22.

54. Litton Frith, Cressbrook Dale [1716 7437] 1150 yd W.14°S. of Bull's Head Inn, Wardlow. Beds *a–k*. RK 9057–9.

55. Near Peter's Stone, Cressbrook Dale [1725 7490] 950 yd S.21°W. of Litton-fields. Beds *a–p*. RK 8845–51.

56. Litton Dale [1584 7468] 860 yd S.42°W. of Sterndale House. RAE 2060–61.

57. Tideswell [1425 7610] 130 yd N.39°E. of Crossgate. Beds *a–d*. IPS 1064–70.

58. Wheston [1383 7669] 700 yd E.28°N. of Wheston Hall. Beds *a–b*. PT 1736–64.

59. Brook Bottom [1473 7638] 560 yd S.26°E. of Highfield House. Beds *a–k*. PT 2201–25.

60. Brook Bottom [1450 7703] 180 yd N.4°W. of Highfield House. Beds *a–b*. RK 9060–1.

61. North wall of White Rake [1474 7814] 400 yd E.17°N. of Whiterake Farm. Beds *a–d*. IPS 1977.

62. Between Maiden Rake and Shuttle Rake [1436 7860] 290 yd W.21°N. of Bushyheath House. Beds a–d. PT 1572–82.

63. The Holmes [1408 7918] 20 yd N. of The Holmes. Beds a–d. PT 1583–8.

64. Near The Holmes [1411 7920] 90 yd E.30°N. of The Holmes. PT 1589–1600.

65. Bradwell Moor [1336 8060] 1525 yd N.32°E. of Brocktor. IPS 1978.

66. Knoll-reef north of Dirtlow Rake [1440 8149] 770 yd S.27°E. of Hurdlow Barn. PT 1673–80.

67. Borehole AF4 [1476 8188] 795 yd E.20°S. of Hurdlow Barn. Horizons a–b. DF 466–85.

47. Pindale Quarry, see p. 386. Beds g–s. IPS 1976.

68. Upper Jack Bank Quarry [1620 8208] 480 yd S.6°W. of Pindale Farm. Beds a–b. LL 1023–6.

69. By old explosives store south of Upper Jack Bank Quarry [1620 8202] 555 yd S.6°W. of Pindale Farm. Beds a–d. RK 9054.

70. Earle's Quarry, composite section in middle and upper bench. Beds a–g.

   c. [1608 8209] 525 yd S.19½°W. of Pindale Farm. LL 1242–311.

   d. [1608 8210] 525 yd S.21°W. of Pindale Farm. LL 1161–93.

   g. [1591 8207] 630 yd S.36°W. of Pindale Farm. LL 1116–7.

71. Earle's Quarry, lower bench (now tipped over) [1622 8191] 690 yd S.2°W. of Pindale Farm. Beds a–l. LL 1029–32.

72. Earle's Quarry, east side [1643 8189] 730 yd S.16°E. of Pindale Farm. Beds a–j. RK 9055–6.

### UPPER MONSAL DALE BEDS

73. Furness Quarry [2085 7596] 300 yd N.14°W. of Farnsley Farm. Beds a–x. PT 2546–8, 2552–65.

74. Shining Cliff [2184 7582] 675 yd N.18°W. of Highfields Farm. Beds a–u. PT 2305–52, RK 8880–906.

75. Castle Rock [2247 7569] 685 yd N.43°E. of Highfields Farm. Beds a–z. PT 2456–503, RK 8855–79.

76. Stoney Middleton [2279 7560] 900 yd E.27°N. of Highfields Farm. Beds a–h. RK 8827–54.

77. Coombs Dale [2239 7429] 1035 yd S.20°E. of Highfields Farm. Beds a–f. PT 8188–250.

78. Coombs Dale [2222 7436] 920 yd S.10°E. of Highfields Farm. Beds a–m. PT 8251–309.

79. Section by Cavendish Mill [2054 7534] 510 yd W.44°S. of Farnsley Farm. PT 8351–4.

80. Roadside exposure at head of Middleton Dale [1980 7616] 820 yd N.26°E. of Castlegate Farm. PT 2436–40.

81. Near Peter's Stone, Cressbrook Dale [1731 7537] 460 yd S.35°W. of Litton-fields. RK 8999–9000.

82. Cressbrook Dale west of Peter's Stone [1715 7526] 670 yd S.41°W. of Litton-fields. RK 8996–8.

55. Near Peter's Stone, Cressbrook Dale, see p. 386. Beds q–v. RK 8930–44.

83. North-east part of Cressbrook Dale [1736 7538] 225 yd S.31°W. of Litton-fields. Beds a–z. RAE 2069–89, RK 8907–29, 8952–5.

84. Crag on east side of Cressbrook Dale [1737 7494] 925 yd W.20°N. of Bull's Head Inn, Wardlow. IPS 1975.

85. Littonfields Borehole [1752 7595] 225 yd N.9°W. of Littonfields. Horizons a–c. IPS 1961–74.

86. Shuttle Rake [1512 7940] 390 yd S.15°E. of Berrystall Lodge. Beds a–c. PT 2566–88.

87. Small valley south-east of Little Hucklow [1650 7813] 615 yd N.12°E. of Poynton House. Beds a–l. RK 9001–16.

88. Small valley joining Intake Dale [1626 7986] 510 yd N.18°W. of Intake Farm. Beds a–i. PT 7709–19, RK 9035–44.

89. Earl Rake [1619 8007] 450 yd S.22°W. of Hartlemoor Farm. PT 7720–3.

90. Hartle Dale [1616 8046] 190 yd W. of Hartlemoor Farm. Beds a–e. RK 9030–4.

91. Open workings in Dirtlow Rake [1449 8155] 785 yd S.36°E. of Hurdlow Barn. Beds a–f. PT 1663–72.

92. North of Dirtlow Rake [1484 8186] 900 yd E.19°S. of Hurdlow Barn. PT 2589–92.

93. Bradwell Moor [1424 8045] 1830 yd S.6°E. of Hurdlow Barn. PT 1407–24.

94. Bradwell Moor [1471 8114] 1280 yd S.33°E. of Hurdlow Barn. Beds a–b. PT 1696–1718, LL 979–1006.

95. Bradwell Moor [1525 8123] 335 yd S.6½°W. of Bradwellmoor Barn. LL 1007–9.

96. Bird Mine [1563 8118] 545 yd S.45°E. of Bradwellmoor Barn. Beds a–c. PJ 1069–85.

97. Earle's Quarry, upper bench [1592 8206] 600 yd S.33°W. of Pindale Farm. Beds a–j. LL 1078–101.

98. Earle's Quarry, "main knoll" [1576 8180] 590 yd E.27°N. of Bradwellmoor Barn. LL 1039–77, LL 1118–20, 1483–1522.

99. Southern end of Bradwell Dale [1714 8039] 840 yd S.27°W. of Bradwell church. Beds a–f. PT 4029–85.

100. Northern end of Bradwell Dale [1741 8077] 355 yd S.13°W. of Bradwell church. Beds a–k. PT 3965–86, RK 9016–29.

## BOREHOLES IN CONCEALED D₂ STRATA

101. Wardlow Mires No. 1 Borehole [1850 7553] 850 yd S.32°W. of Brosterfield. GN 2960–4501.

102. Wardlow Mires No. 2 Borehole [1825 7586] 800 yd W.26°S. of Brosterfield. AG 465–545.

103. Hucklow Edge No. 2 Borehole [2025 7760] 265 yd N.36°E. of Shepherd's Flat. IPS 2524, 2326–89.

104. Ladywash Mine, underground [2182 7743] at junction of branch to Ladywash Shaft with main crosscut. Sy 1786–91.

## EYAM LIMESTONES

105. Calver Peak Limekilns [2358 7466] 1050 yd S.25°W. of Knouchley. Beds a–d. PT 2087–121.

106. Flat reef, Calver Peak Limekilns [2365 7459] 105 yd S.E. of locality 105. PT 2122–40.

107. Coombsdale Quarry [2330 7478] 1130 yd S.43°W. of Knouchley. Beds *a–f*. PT 2282–304.

108. Coombsdale Quarry [2333 7481] 45 yd north-east of locality 107. PT 2268–81.

77. Coombs Dale, see p. 387. Bed *g*. PT 8161–87.

109. High Fields [2159 7490] 575 yd W.31°S. of Highfields Farm. PT 8111–60.

76. Stoney Middleton, see p. 387. Beds *o–s*. IPS 2027–32.

75. Castle Rock, see p. 387. Beds *aa–dd*. PT 2441–55.

73. Furness Quarry, see p. 387. Beds *y–z*. PT 2549–51.

110. Eyam Dale [2199 7621] 325 yd E.43°S. of Eyam church. Beds *a–j*. PT 3858–87.

111. Ladywash Shaft inset [2189 7754] 215 yd S.22°W. of Ladywash Farm. Sy 1792–6.

112. Knoll-reef, Linen Dale [1984 7688] 165 yd S.39°W. of Waterfall Farm. PT 2365–422.

113. Northern flank of knoll, Linen Dale [1989 7689] 50 yd E.5°N. of locality 112. Beds *a–e*. PT 2353–64.

114. Linen Dale [1988 7693] 100 yd S.39°W. of Waterfall Farm. Beds *a–g*. PT 2423–35.

115. Linen Dale [1988 7704] 60 yd W.38°N. of Waterfall Farm. Beds *a–f*. PT 4241–70.

103. Hucklow Edge No. 2 Borehole, see p. 388. IPS 2279–323, 2325.

116. Knoll-reef outlier, Middleton Moor [1953 7503] 450 yd S.30°E. of Castlegate Farm, PT 8310–50

117. Water Grove [1911 7583] 450 yd S.30°E. of Brosterfield. Beds *a–h*. RAE 2110–4.

101. Wardlow Mires No. 1 Borehole, see p. 388. GN 2740–959.

102. Wardlow Mires No. 2 Borehole, see p. 388. AG 292–464.

118. Silly Dale [1818 7697] 890 yd E.43½°N. of Stanley Lodge. Beds *a–e*. IPS 2033–8.

119. Knoll-reef outlier south of Intake Dale [1641 7956] 150 yd north of Intake Farm. PJ 914–85.

120. Knoll-reef outlier south of Intake Dale [1646 7965] 260 yd N.10°E. of Intake Farm. PJ 986–1068.

99. Southern end of Bradwell Dale, see p. 388. Beds *g–h*. PT 3987–4028.

100. Northern end of Bradwell Dale, see p. 388. Bed *l*. PT 3888–964.

121. North of Brook House [0690 7534] 330 yd N.13°W. of Brook House. Beds *a–h*. PJ 830–913.

122. Section with knoll-reef near Brook House [0673 7484] 350 yd S.43°W. of Brook House. Beds *a–d*. PJ 1086–1225.

123. Tip from Watergrove Mine [1882 7577]. Horizon *a*. PJ 804–829.

### SHALES IN EYAM GROUP

123. Tip from Watergrove Mine. Horizon *b*. PJ 797–803.

124. Temporary section at Wardlow Mires [1892 7540] 570 yd W.5°S. of Castlegate. Z1 4091.

101. Wardlow Mires No. 1 Borehole, see p. 388. GN 2563–739.

102. Wardlow Mires No. 2 Borehole, see p. 388. AG 197–291.

## MARGINAL PROVINCE

### APRON-REEF

125. Perry Dale [0998 8121] 180 yd S.25°E. of Whitelee. IPS 1023–30.

126. Perryfoot [0989 8126] 115 yd S.8°W. of Whitelee. Beds *a–b*. RS 48–171.

127. Valley between Middle Hill and Snels Low [between 1159 8205 and 1154 8224] PT 1425–555.

128. West side of Peaks Hill [1170 8267] 170 yd E.15°N. of Peakshill farm. Bed *a*. IPS 1051.

129. Near West side of Peaks Hill [1170 8262] 180 yd E.5°S. of E. corner of Peakshill farm. RS 982–999.

130. Peaks Hill [1176 8267] 240 yd E.10°N. of E. corner of Peakshill farm. RS 930–81.

131. Peaks Hill [1179 8266] 285 yd E.5°N. of E. corner of Peakshill farm. RS 1000–7.

132. Near Giant's Hole [1185 8261] 350 yd E.4°S. of Peakshill farm. RS 1008–33.

133. Near Giant's Hole [1196 8271] 470 yd E.11°N. of Peakshill farm. RS 1034–44.

134. East of Giant's Hole [1236 8268] 255 yd N.43°W. of Oxlow House. IPS 1041–7.

135. Odin Fissure [1342 8346] 730 yd E.13°S. of summit of Mam Tor. RS 3630–864.

136. Treak Cliff, see inset diagram, Plate XI.

    *a.* [1339 8301] PJ 1969–91.          *m.* [1349 8305] PJ 1597–628.
    *b.* [1341 8305] PJ 1919–68.          *n.* [1350 8305] PJ 1578–96.
    *c.* [1342 8304] PJ 1831–918.         *o.* [1351 8305] PJ 1544–77.
    *d.* [1341 8318] PJ 2184–215.         *p.* [1352 8310] PJ 1496–518.
    *e.* [1343 8314] PJ 2221–71.          *q.* [1353 8306] PJ 1539–43.
    *f.* [1343 8308] PJ 1992–2131.        *r.* [1355 8308] PJ 1519–38.
    *g.* [1344 8304] PJ 1752–830.         *s.* [1356 8309] PJ 1459–95.
    *h.* [1344 8298] PJ 2132–66.          *t.* [1364 8303] PJ 1405–58.
    *i.* [1345 8294] PJ 2167–83.          *u.* [1363 8309] PJ 1399–404.
    *j.* [1345 8304] PJ 1723–51.          *v.* [1368 8306] PJ 1286–398.
    *k.* [1347 8304] PJ 1692–722.         *w.* [1369 8306] PJ 1268–85.
    *l.* [1348 8304] PJ 1629–91.

137. Winnats [1357 8262] 760 yd W.44°N. of Hurdlow Barn. PJ 2272–351.

138. Cow Low Nick [1420 8241] 330 yd N.22°E. of Hurdlow Barn. LZ 2336–7, 2498, 2519–27, 4816–39, 4861–902.

139. Castleton [1488 8277] 200 yd W.45°S. of Castleton church. PT 4184–240.

140. Above entrance to Pindale Quarry [1605 8231] 320 yd S.43°W. of Pindale Farm. LL 951–78.

141. Lower Jack Bank Quarry [1616 8234] 210 yd S.24°W. of Pindale Farm. Beds *a–d*. LL 721–61.

142. Upper Jack Bank Quarry [1616 8220] 370 yd S.15°W. of Pindale Farm. LL 583–627.

143. Nunlow Quarry [1651 8226] 400 yd S.44°E. of Pindale Farm. Beds *a–c*. LL 824–82, 1312–83.

144. Mich Low [1699 8180] 950 yd N.34°W. of Bradwell church. PT 2141–200.

145. Near Bath Cottage [1740 8187] 130 yd S.5°W. of The Bath, Bradwell. PT 7640–89.

146. Near Bath Cottage [1730 8184] 220 yd S.36°W. of The Bath, Bradwell. PT 7690–708.

147. Near New Bath Hotel [1749 8185] 170 yd S.29°E. of The Bath, Bradwell. PT 7466–639.

### POST-REEF DEPOSITS OF THE REEF BELT

148. Exposure in Castleton village [1492 8282] 140 yd W.40°S. of Castleton church. PT 4177–83.

128. West side of Peaks Hill, see p. 390. Bed *b*. 5 yd north of *a*. IPS 1048–50.

149. Castleton Borehole [1410 8293] 590 yd S.14°W. of Dunscar Farm. Bi 6793–7490.

150. Hope Cement Works Borehole [1678 8228] Salter Barn, Hope. Site approximate. Ba 812–25, 847–9.

# BASIN PROVINCE

151. Edale Borehole [1078 8493] 665 yd E.11°S. of Highfield. Bi 9952–10000.

152. Alport Borehole [1360 9105] 115 yd S.6°E. of junction of Swint Clough brook and R. Alport. 2f 4703–78, 4817–67, 2h 509, 626–32, 645, 2572–906, Bc 3679–819, Bg 1412–928.

## Appendix III

# FOSSILS COLLECTED FROM THE CARBONIFEROUS LIMESTONE OF THE SHELF PROVINCE

The name of each fossil is followed by the locality numbers (see Appendix II) and bed letters at which the fossil has been recorded. The use of '?', 'cf.' or 'aff.' before a locality number in these lists respectively indicates doubt as to the identification of, similarity to, or departure from the genus or species named.

WOO DALE BEDS

*Koninckopora inflata* (de Koninck). 6
*K. sp.* 3*i*, ?3*p*, ?10

Foraminifera. 2*c*, 2*d*, 3*n*, 3*p*, 4*a*, 4*c*, 4*d*, 10

*Caninia sp.* ?1*a*, 2*d*, 3*l*
*Carcinophyllum vaughani* Salée. 3*c*, cf. 3*c*, 4*a*, 8*b*
*C.?* 3*f*
*Chaetetes septosus* (Fleming). 8*b*
*Dibunophyllum bourtonense* φ Vaughan. 2*d*, 3*n*, 3*p*
*D. sp.* 4*d*
*Diphyphyllum sp.* 3*c*
*Heterophyllia sp.* 4*a*
*Koninckophyllum* θ Vaughan [1905, pl. 23, fig. 4]. 3*n*
*K. sp. dianthoides* (McCoy) group. 3*n*
*K. sp.* ?2*a*, 2*c*, 3*n*, 4*a*
*Lithostrotion arachnoideum* (McCoy). 3*l*
*L.* cf. *aranea* (McCoy). 2*c*
*L. junceum* (Fleming). 4*d*
*L. martini* Milne Edwards and Haime. 1*a*, 2*c*, 2*d*, 2*e*, ?3*f*, 3*i*, 3*l*, 3*n*, 3*p*, cf. 3*p*, aff. 3*p*, 4*a*, 4*c*, aff. 4*d*, 7, 8*b*, aff. 10, 10
*L.* aff. *martini* [tending to cerioid]. 3*n*
*L. pauciradiale* (McCoy). 2*c*, 2*d*, 3*n*, 3*p*, 4*d*
*L.* aff. *portlocki* (Bronn). 8*b*
*L.* cf. *sociale* (Phillips). 3*f*, 3*p*, 4*d*
*L. sp.* [fragment of cerioid form]. 3*a*
*Palaeosmilia murchisoni* Milne Edwards and Haime. 2*a*, 2*d*, 4*c*, 4*d*
*Syringopora* cf. *distans* (Fischer). 1*e*, 3*n*
*S.* cf. *geniculata* Phillips. 2*b*, 4*d*
*S.* cf. *ramulosa* Goldfuss. 1*a*, 3*n*, 4*c*
*S.* cf. *reticulata* Goldfuss. 3*a*, 3*c*, 4*d*
*S. sp.* 2*d*, 3*p*, 8*b*

Bryozoa. 3*a*, 3*f*, 3*l*, 3*n*, 3*p*, 4*c*, 4*d*, 8*b*, 10
*Fenestella spp.* 3*f*, 4*a*

392

*Antiquatonia?* 2e
Athyrid. 2a, 3n
*Athyris expansa* (Phillips). 2d, 3f, ?3n, cf. 3p
*A. sp.* 4d
*Davidsonina carbonaria* (McCoy). 3c
*Daviesiella sp.* 3c, 3d
*Dielasma hastatum* (J. de C. Sowerby). 1a, 4c
*Echinoconchus sp.* 1a, 3a, 3f, 3i, ?3l, 4c, 4d, 6
*Gigantoproductus sp.* aff. *gigantoides* (Paeckelmann). 3f
*G. sp.* cf. *semiglobosus* (Paeckelmann). 3a, 3f, 3n, 3p
*G. sp.* 3a, 3b, 3e, 3f, 3l, 3n, 4a, 4c, 4d, 8b
*Linoproductus corrugatohemisphericus* (Vaughan). 1a, aff. 3a, 3c, cf. 6, 8b, aff. 8b
*L.* cf. *corrugatus* (McCoy). 2c, 3a, 3i, 6
*L. sp. hemisphaericus* (J. Sowerby) group. 1e, 2d, 3a, 3c, 3e, 3f, 3i, ?3l, ?6, 8b, 9, ?10
*L. sp.* 3d, 4d
*Martinia sp.* 8b
*Megachonetes sp. papilionaceus* (Phillips) group. 1a, 2a, 2d, 3a, 3b, 3f, 3n, 4a, 4c, 4d,
   5, 6, 8b, 9
*M. sp.* 3l
*Orbiculoidea sp.* 4c
Orthotetoids. 3a, 3b, 3c, 3f, 3p, 5, 10
*Plicatifera* cf. *plicatilis* (J. de C. Sowerby). 3i
*P. sp.* [juv.]. 3a
*Productus sp.* 3a
*Pustula pustulosa?* (Phillips). 3i
*P. sp.* 4c
*Schizophoria* cf. *connivens* (Phillips). 3i
*S. sp.* 1e, 2d, 3a, 3f
*Schuchertella sp.* 8b
*Semiplanus sp. nov.* [of Ramsbottom *in* Fowler 1966, p. 76]. 3f, 3l
Smooth spiriferoids. 1a, 2d, 3a, 3b, 3c, 3e, 3f, 3i, 3l, 4a, 4c, 4d, 5, 6, 8b
*Streptorhynchus senilis* (Phillips). 4c


*Baylea concentrica?* (Phillips). 3n
*Bellerophon sp.* 2a, ?4a, 8b
*Eoptychia sp.* ?3l, 4c
*Meekospira?* 3p
*Naticopsis sp.* 3n, 8b
*Straparella sp.* [juv.]. 3p
*Straparollus* (*Euomphalus*) *sp.* 3p
*S.* (*S.*) *dionysii* de Montfort. 3l
*S.* (*S.*) *sp.* 3n
*S. sp.* 4a


*Edmondia sulcata* (Phillips). 3l
Pectinoids. 4a
*Sulcatopinna flabelliformis* (Martin). 4a


*Vestinautilus?* 3n
Nautiloid fragment. 3f


Ostracods. 1a, 3i, 3l, 3n, 3p, 4a

### CHEE TOR ROCK

Algal nodules.   13*o*, 13*s*
'*Girvanella*' nodules.   12*d*
*Koninckopora inflata* (de Koninck).   3*q*, 13*s*
*K. sp.*   1*g*, 2*f*, 11, 16*b*, 19

Foraminifera.   1*g*, 1*h*, 1*i*, 2*f*, 2*g*, 3*q*, 3*r*, 4*e*, 4*g*, 13*o*, 14*a*, 16*b*, 17, 19, 23, 24, 25, 26, 27, 28*b*, 28*i*

*Aulina?*   16*b*
*Caninia benburbensis* Lewis.   21
*C.* cf. *densa* Lewis [Hudson and Cotton 1945a, p. 306].   3*q*, 16*b*
*C. sp.   ? C.* cf. *densa*.   15
*C..sp. subibicina* McCoy group.   27
*C. sp.*   2*f*, ?12*j*, 18*d*
*Carcinophyllum vaughani* Salée.   1*i*, 19
*C.?* 2*f*
*Chaetetes depressus* (Fleming).   3*r*, 12*d*, 12*j*, 17
*Clisiophyllum keyserlingi* McCoy.   16*b*
*C. sp.*   3*q*, 3*r*, 14*b*, 15, 19, 24
Clisiophylloid.   1*g*, 2*g*, 12*m*
*Dibunophyllum bourtonense* Garwood and Goodyear.   3*r*, 12*j*, 16*b*
*D. bourtonense* φ Vaughan.   1*h*, 2*f*, 3*q*, 13*m*, 13*o*, 14*b*, 16*b*, 19, 21, 28*i*
*D. sp.* 4*e*, 14*a*, 25
*Hexaphyllia sp.*   4*e*, 16*b*, 19, 28*i*
*Koninckophyllum* θ Vaughan [1905, pl. 23, fig. 4].   2*f*, 3*q*, 16*b*
*K. sp.* [compound].   2*g*, 3*r*
*K. sp.* [juv.].   12*d*, 19
*Lithostrotion aranea* (McCoy).   ?16*d*, cf. 17
*L. decipiens?* (McCoy).   25
*L. junceum* (Fleming).   ?2*f*, 3*q*, 3*s*, 4*e*, 16*b*, 28*d*
*L.* aff. *maccoyanum* Milne Edwards and Haime.   17
*L. martini* Milne Edwards and Haime.   1*g*, 1*i*, 2*f*, aff. 2*f*, 2*g*, 3*q*, aff. 3*q*, 3*r*, 4*e*, aff. 4*e*, aff. 4*g*, 12*f*, 12*j*, 13*o*, aff. 14*a*, aff. 14*b*, 15, 16*b*, aff. 16*b*, aff. 17, 18*a*, 18*b*, 19. 20, 23, 24, 26, aff. 26, 28*b*
*L. martini* [tending to cerioid].   18*d*, 25
*L. pauciradiale* (McCoy).   2*f*, 3*q*, 14*a*, 14*b*, 16*b*, 16*d*, 19, 24
*L. portlocki* (Bronn).   3*r*, 22
*L.* cf. *sociale* (Phillips).   3*q*, 16*b*, 19
*L. sp.* [fragment of cerioid form].   14*b*
*L. sp.*   27
*Palaeosmilia murchisoni* Milne Edwards and Haime.   2*f*, 3*q*, 3*r*, 4*e*, 12*f*, 12*j*, 13*m*, ?13*s*, 14*b*, 16*b*, 17, 19, 23, ?24, 25, 26, 27, 29
*Syringopora* cf. *distans* (Fischer).   12*j*, 17, 19, 22
*S.* cf. *geniculata* Phillips.   3*q*, 4*e*, 4*g*, 12*j*, 13*o*, 14*b*, 18*b*, 18*d*, 25
*S.* aff. *geniculata*.   12*j*
*S. sp.*   2*f*, 16*b*, 20, 28*f*

Bryozoa.   3*q*, 3*r*, 4*e*, 13*o*, 26

*Antiquatonia* cf. *insculpta* (Muir-Wood).   2*f*
*Athyris expansa* (Phillips).   1*i*, cf. 1*i*, cf. 2*f*, cf. 3*r*, 14*b*, 18*b*, 19
*A. sp.*   20, 22, 25, ?28*b*
*Davidsonina septosa* (Phillips).   12*d*, 12*j*, 12*m*, 13*m*, 13*o*, 13*s*, 15, 24, 27
*D. septosa transversa* (Jackson).   13*o*, 13*s*

*Delepinea comoides* (J. Sowerby). 12*j*, aff. 15, aff. 26, aff. 27
*D. sp.* 12*f*
*Dielasma hastatum* (J. de C. Sowerby). 1*h*, 2*f*, 3*q*
*Echinoconchus eximius* (I. Thomas). 3*r*
*E. sp.* 1*h*, 3*q*
*Gigantoproductus dentifer* (Prentice). 3*r*, 22
*G. edelburgensis* (Phillips). 13*s*
*G. sp. edelburgensis* group. 12*j*, 18*b*, 18*d*
*G. sp. maximus* (McCoy) group. 1*h*, 4*e*, 16*b*
*G. sp.* cf. *sarytscheffi* (Paeckelmann). 3*q*
*G.? sp. nov.* [wrinkled concentric ornament]. 1*h*, 2*f*, 3*q*, 19
*G. sp.* 1*i*, 12*f*, 12*m*, 14*b*, 25, 29
*Linoproductus* aff. *corrugatus* (McCoy). 15
*L. sp. hemisphaericus* (J. Sowerby) group. 1*g*, 1*h*, 2*f*, 3*q*, 3*r*, 13*o*, 13*s*, 18*b*, 19, 20, 22, 24, 25, 26, 27, 28*b*
*L. sp.* 1*i*, 12*j*, 12*m*, 14*b*, 29
*Megachonetes sp. papilionaceus* (Phillips) group. 1*f*, 1*h*, 1*i*, 2*f*, 2*g*, 3*q*, 4*g*, 12*j*, 13*m*, 13*o*, 14*a*, 14*b*, 15, 19, 20, 22, 26
*M. sp. papilionaceus* group, cf. *hemisphaericus* (Semenov). 3*r*
*M. sp. papilionaceus* group, cf. *zimmermanni* (Paeckelmann). 3*r*
*M. sp.* 13*s*, 16*b*, 16*d*, 24
Orthotetoids [with intercostal striae]. 19, 20, 29
Orthotetoids. 1*g*, 1*h*, 1*i*, 2*f*, 14*a*, 18*d*
*Phricodothyris?* 26
*Plicochonetes?* 3*r*
Productoids. 14*a*, 17
*Pugilis sp.* ?14*b*, 16*b*
*Pustula pustulosa* (Phillips). 1*h*
*P. sp. rugata* (Phillips) group. 19
*P. sp.* 2*f*, 3*q*
*Reticularia sp.* [juv.]. 3*r*
*Rugosochonetes sp.* 4*e*
*Schellwienella sp.* 3*q*
*Schizophoria sp.* 1*g*, 3*r*, 11, 27
*Schuchertella sp.* 25
Smooth spiriferoids. 1*g*, 1*h*, 1*i*, 2*f*, 3*q*, 3*r*, 4*e*, 12*j*, 13*o*, 13*s*, 14*b*, 18*b*, 19, 20, 21, 24, 26, 28*i*

*Baylea concentrica* (Phillips). 3*r*
*Bellerophon costatus* J. de C. Sowerby. 13*o*
*B. sp.* 4*e*, 13*m*
*Eoptychia sp.* [juv.]. 21
*Meekospira sp.* 3*r*
*Naticopsis sp.* 1*g*, 1*h*, 1*i*
*Straparella fallax* (de Koninck). 3*q*, 3*r*, 19
*S.?* 11
*Straparollus* (*Euomphalus*) *sp.* 3*r*, 12*j*, 27
*S.* (*Straparollus*) *dionysii* de Montfort. 3*q*
*S.* (*S.*) *sp.* 1*g*, 2*f*, 19
*S. sp.* 1*h*, ?11, 13*o*, 16*b*

*Aviculopecten sp.* [juv.]. 1*i*

Ostracods. 2*f*, 3*q*

MILLER'S DALE BEDS

Foraminifera.   28*m*, 31*a*, 32*c*, 34, 35*c*
*Saccamminopsis sp.*   ?36

*Aulina furcata* Smith.   28*k*, 28*m*
*Caninia benburbensis* Lewis.   ?28*m*, 34
C. cf. *densa* Lewis [of Hudson and Cotton 1945a, p. 306]. 35*d*
*C.?*   35*b*
*Chaetetes depressus* (Fleming).   31*a*, 32*a*, 35*c*
*C. radians* Fischer.   35*d*
*Clisiophyllum* cf. *keyserlingi* McCoy.   28*m*
*C.?*   35*c*
*Dibunophyllum bourtonense* Garwood and Goodyear.   35*c*
*D. bourtonense* φ Vaughan.   31*a*, 32*c*, 35*b*
*Diphyphyllum sp.*   35*d*
*Hexaphyllia sp.*   28*k*
*Lithostrotion junceum* (Fleming).   28*m*, 30*i*
*L.* cf. *maccoyanum* Milne Edwards and Haime.   28*k*
*Lithostrotion martini* Milne Edwards and Haime.   28*k*, 28*o*, cf. 28*o*, cf. 32*c*, 34, 35*b*,
   35*c*
*L. martini* [tending to cerioid].   30*i*, 31*a*, 35*c*
*L. pauciradiale* (McCoy).   28*k*, 28*m*, 28*o*, 35*b*, 35*c*
*L. portlocki* (Bronn).   35*d*
*Palaeosmilia murchisoni* Milne Edwards and Haime.   30*i*, 31*a*, 34, 35*b*, 35*c*
*Syringopora* cf. *distans* (Fischer).   32*e*, 35*c*, 35*d*
*S.* cf. *geniculata* Phillips.   30*i*, 32*a*, 32*c*
*S.* cf. *ramulosa* Goldfuss.   31*e*

Bryozoa.   31*a*
*Fenestella sp.*   28*m*

*Antiquatonia sp.*   31*e*
*Athyris* cf. *expansa* (Phillips).   35*d*
*Brachythyris integricosta* (Phillips).   33*a*
*B. sp.*   30*i*, ?35*d*
*Buxtonia?*   31*e*
*Davidsonina septosa* (Phillips).   31*a*, 32*c*, 35*d*
*D. septosa* towards *transversa* (Jackson).   35*d*
*Delepinea* aff. *comoides* (J. Sowerby).   35*d*
*Dielasma hastatum* (J. de C. Sowerby).   28*m*
*Eomarginifera sp. lobata* (J. Sowerby) group.   28*m*
*Gigantoproductus crassiventer* (Prentice).   33*a*
*G. edelburgensis* (Phillips).   32*e*
*G. sp. maximus* (McCoy) group.   28*k*, 28*o*
*G. sp.* aff. *sarytscheffi* (Paeckelmann).   32*c*
*G. sp.*   30*i*, 35*d*
*Linoproductus sp. hemisphaericus* (J. Sowerby) group.   30*i*, 31*a*, 31*e*, 32*c*, 33*a*, 35*c*, 35*d*
*L. sp.*   34
*Megachonetes sp. papilionaceus* (Phillips) group.   34
*M. sp.*   35*d*
Orthotetoids.   34
*Phricodothyris?*   35*d*
*Plicochonetes sp. buchianus* (de Koninck) group.   35*d*
*Productus sp.*   30*i*, 31*e*, ?35*d*

*Pugnax pugnus* (Martin). 28*m*
*Schizophoria resupinata* (Martin). 30*i*
*S. sp.* 34
Smooth spiriferoids. 28*k*, 28*m*, 28*o*, 30*a*, 30*i*, 31*e*, 33*a*, 34, 35*d*
*Spirifer sp. bisulcatus* J. de C. Sowerby group. 34

*Bellerophon costatus?* J. de C. Sowerby. 35*d*
*B. sp.* 30*i*, 34
*Meekospira?* 34
*Naticopsis sp.* 34
*Straparollus* (*Euomphalus*) *sp.* 34, 35*d*
*S. sp.* 32*c*

*Weberides sp.* 28*m*

## BEE LOW LIMESTONES

Algae. 41*d*, 41*f*, 41*h*
'*Girvanella*' nodules. 47*f*
*Koninckopora inflata* (de Koninck). 41*b*, 41*h*, 44*b*
*K. sp.* 37*a*, 37*b*, 40*b*, 41*d*, 41*f*, 43*a*, 44*a*

Foraminifera. 37*a*, 37*b*, 37*c*, 37*d*, 37*e*, 38*a*, 38*b*, 39*b*, 41*f*, 41*h*, 41*j*, 42, 44*a*, 44*b*
*Saccamminopsis sp.* 46*b*

*Aulophyllum fungites* (Fleming). 37*d*
*Caninia* cf. *buxtonensis* Lewis. 47*f*
*C.* cf. *densa* Lewis [of Hudson and Cotton 1945a, p. 306]. 47*f*
*C. sp. subibicina* McCoy group. 44*a*
*C. sp.* 39*c*, 40*a*
*Chaetetes depressus* (Fleming) [meandrine form of Smith 1934, p. 333]. 47*f*
*C. radians* Fischer. 41*f*
*Dibunophyllum bourtonense* Garwood and Goodyear. 40*b*
*D. bourtonense* φ Vaughan. 37*a*, 37*b*, 37*c*, 47*f*
*D.?* 37*e*
*Heterophyllia sp.* 39*b*
*Hexaphyllia sp.* [juv.] 37*b*
*Koninckophyllum* cf. θ Vaughan [1905, pl. 23, fig. 4]. 37*b*
*Lithostrotion arachnoideum* (McCoy). 37*c*
*L.* cf. *aranea* (McCoy). 45*b*, 47*f*
*L.* cf. *decipiens* (McCoy). 47*f*
*L. junceum* (Fleming). 37*b*, 37*c*, 40*a*, 42, 43*a*, 46*b*
*L. martini* Milne Edwards and Haime. 37*a*, cf. 37*a*, 37*b*, cf. 37*b*, aff. 37*b*, 37*c*, cf. 37*d*,
   37*e*, 38*a*, 39*b*, 39*c*, 40*a*, 40*b*, 41*b*, 41*f*, 41*j*, 44*a*
*L. pauciradiale* (McCoy). 37*a*, 37*b*, 37*d*, 41*j*, 44*b*
*L. portlocki* (Bronn). 40*b*, 47*f*, cf. 47*f*
*L.* cf. *sociale* (Phillips). 37*b*
*Palaeosmilia murchisoni* Milne Edwards and Haime. 37*b*, 37*c*, 38*a*, 38*c*, 41*j*, 43*a*, 47*f*
*Rotiphyllum sp.* 37*e*
*Syringopora* cf. *distans* (Fischer). 37*c*, 41*b*, 41*j*
*S.* cf. *geniculata* Phillips. 38*a*, 39*c*, 47*f*
*S.* cf. *reticulata* Goldfuss. 41*j*
*S. sp.* 37*b*, 40*a*

*Spirorbis sp.* 44*b*

Bryozoa. 37*e*, 40*b*, 41*h*, 41*j*, 44*a*, 46*b*
*Fenestella sp.* 40*b*, 42, 43*a*, 44*b*
*Penniretepora sp.* 43*a*

*Acanthoplecta mesoloba* (Phillips). 43*a*, 44*b*
*Alitaria panderi* (Muir-Wood and Cooper). 43*a*
*Antiquatonia antiquata* (J. Sowerby). 43*a*, 44*b*
*A.* cf. *hindi* (Muir-Wood). 43*a*
*A. hindi wettonensis* (Muir-Wood). 43*a*
*A.* cf. *insculpta* (Muir-Wood). 39*b*, 40*a*, 40*b*
*A. sp.* 47*f*
*Athyris expansa* (Phillips). 37*b*, 40*a*, cf. 40*a*, 40*b*, 41*d*, 41*j*, cf. 44*a*
*A. obtusa* (McCoy). cf. 40*a*, 44*b*
*A.?* 39*c*
*Avonia davidsoni* (Jarosz). 43*a*
*Brachythyris integricosta* (Phillips). 44*b*
*B. ovalis* (Phillips). 44*b*
'*B.*' *planicostata* McCoy. cf. 40*b*, 44*b*
*Buxtonia sp.* 43*a*
Chonetoid. 47*f*
*Davidsonina septosa* (Phillips). 40*b*
*Dielasma hastatum* (J. de C. Sowerby). 39*b*, 40*a*, 43*a*, 44*b*
*D. sp.* 40*b*
*Echinoconchus punctatus* (J. Sowerby). 40*b*, 43*a*, 44*b*
*E. sp.* 39*b*, 46*b*
*Eomarginifera* aff. *setosa* (Phillips). 43*a*
*E. sp.* 44*a*, 44*b*
*Fluctuaria* cf. *tortilis* (McCoy). 40*a*, 44*b*
*Georgethyris obtusa* (J. Sowerby). 44*b*
*Gigantoproductus dentifer* (Prentice). 37*a*
*G. edelburgensis* (Phillips). 40*a*, 44*b*
*G. superbus* (Sarycheva). 40*a*, cf. 40*a*
*G. sp.* cf. *moderatus* (Schwetzow). 40*a*
*G.? sp. nov.* [wrinkled concentric ornament]. 37*a*
*G. sp.* 37*b*, 40*a*, 43*a*
*Krotovia laxispina* (Phillips). 43*a*
*K. sp.* [juv.]. 44*a*, 44*b*
*Linoproductus sp. hemisphaericus* (J. Sowerby) group. 37*a*, 37*b*, ?39*b*, ?40*b*, 41*h*
*L. sp.* 41*d*, ?43*a*, 44*b*
*Martinia sp.* 40*b*, 46*b*
*Megachonetes sp. papilionaceus* (Phillips) group. 37*a*, 37*b*, 37*c*, 39*b*, 41*d*, 41*j*
*M. sp.* 37*a*, 44*b*
Orthotetoid [with intercostal striae]. 47*f*
Orthotetoids. 40*a*, 41*d*, 41*h*, 44*b*
*Ovatia sp. nov.* 43*a*
*Overtonia fimbriata* (J. de C. Sowerby). 40*a*
*Phricodothyris* cf. *insolita* George. 44*b*
*P. paricosta* George. 43*a*
*P. sp.* 39*b*, 46*b*
*Pleuropugnoides pleurodon* (Phillips). 43*a*
*P. sp.* 44*b*
*Plicatifera plicatilis* (J. de C. Sowerby). 43*a*, cf. 44*a*
*P.? sp.* [juv.]. 44*b*

*Plicochonetes sp. buchianus* (de Koninck) group. 42
Productoid. 37*a*, 38*c*, 40*c*
*Productus productus* (Martin). 39*b*
*P. productus* aff. *hispidus* Muir-Wood [2 rows of spine-bases on flanks]. 44*b*
*Pugnax cordiformis* (J. de C. Sowerby). 43*a*
*P. sp.* 44*b*
*Pustula sp. rugata* (Phillips) group. 39*b*, 40*a*, 41*d*
*P. sp.* 41*j*, 43*a*, 44*a*, 44*b*
*Reticularia sp.* ?43*a*, 44*a*
*Rugosochonetes sp.* 46*b*
*Schellwienella sp.* 40*b*
*Schizophoria resupinata* (Martin). 40*b*, 44*b*, cf. 44*b*, 46*b*
*Schuchertella sp.* [with intercostal striae]. 40*a*
Smooth spiriferoid. 37*a*, 37*b*, 37*c*, 39*b*, 39*c*, 40*a*, 40*b*, 41*j*, 42, 43*a*, 46*b*, 47*f*
*Spirifer duplicicosta* Phillips. 40*b*
*S. sp. bisulcatus* J. de C. Sowerby group. 44*a*
*S. sp.* 39*b*, 43*a*
*S. (Fusella)* cf. *grandicostatus* McCoy. 44*b*
*S. (F.) trigonalis* (Martin). 44*b*
*Spiriferellina octoplicata* (J. de C. Sowerby) D North. 44*b*
*Tylothyris subconica* North. 44*b*
*T. ?* 40*b*

*Bellerophon costatus* J. de C. Sowerby. 40*b*
*B. sp.* 44*b*
*Meekospira sp.* 44*b*
*Mourlonia sp.* 44*b*
*Naticopsis ampliata* (Phillips). 40*a*
*Straparella fallax* (de Koninck). 40*b*
*Straparollus (Euomphalus) acutus* (J. Sowerby). 44*b*
*S. (Straparollus) dionysii* de Montfort. 40*b*
*S. (S.) sp.* 39*b*
*S. sp.* 40*c*, 41*j*

*Acanthopecten stellaris* (Phillips). 40*a*
*Aviculopecten planoradiatus* McCoy. 43*a*
*Obliquipecten laevis* Hind. 40*a*
*Parallelodon bistriatus* (Portlock) [juv.]. 44*b*
Pectinoid. 43*a*, 44*a*, 44*b*
*Posidoniella vetusta* (J. de C. Sowerby). 43*a*

Nautiloid. 40*b*
Orthocone nautiloid. 39*c*, 41*j*

*Bollandoceras sp. nov.* 39*c*

*Griffithides longispinus* Portlock. 40*b*
*Weberides sp.* 43*a*, 44*b*
Trilobite. 42, 44*a*

Ostracods. 37*b*, 40*b*, 43*a*, 44*a*, 44*b*

## LOWER MONSAL DALE BEDS

*Girvanella staminea* Garwood. 68*a*

*Girvanella sp.* 68*a*, 71*j*
'*Girvanella*' nodules. 33*b*, 47*h*, ?67*a*, 67*b*

Foraminifera. 30*t*, 51*f*, 57*d*, 58*b*, 63*d*
*Draffania sp.* 71*j*
*Saccamminopsis fusulinaformis* (McCoy). 33*d*
*S. sp.* 30*m*, 51*f*, 54*a*, 54*g*, 54*k*, 60*a*, 61*c*, 69*c*, 72*a*, 72*e*

*Caninia benburbensis* Lewis. 30*t*, 63*d*, 71*j*
*C. juddi* (Thomson). 58*b*
*C. sp. subibicina* McCoy group. 52*b*
*C. sp.* 53*d*
*Chaetetes septosus* (Fleming). 62*d*, 70*g*
*C. sp.* 70*d*
*Cyathaxonia rushiana* Vaughan. 52*c*
*Dibunophyllum bipartitum* (McCoy). 30*t*
*D. bipartitum konincki* (Milne Edwards and Haime). 70*g*
*D. sp.* 30*o*, 58*b*
*Diphyphyllum furcatum* Thomson. 51*d*
*D. lateseptatum* McCoy. 30*t*, 49, 51*f*, 70*g*
*Lithostrotion decipiens* (McCoy). 53*d*, 66
*L. junceum* (Fleming). 30*t*, 33*d*, 56, 57*d*, 58*b*, 62*d*, 66
*L. maccoyanum* Milne Edwards and Haime. 62*d*, 65
*L. martini* Milne Edwards and Haime. 30*t*, 33*d*, cf. 33*d*, 48, 49, 58*b*
*L. pauciradiale* (McCoy). 51*f*, 71*j*
*L. portlocki* (Bronn). 33*e*, 51*d*
*L. ?* 30*o*
*Lonsdaleia floriformis* (Martin) *laticlavia* Smith. 51*f*
*Palaeosmilia murchisoni* Milne Edwards and Haime. 33*d*, 63*d*
*Rotiphyllum costatum* (McCoy). 51*h*
*R. sp.* 51*f*
*Syringopora* cf. *geniculata* Phillips. 30*t*, 53*d*
*S. sp.* 30*y*
*Zaphrentites enniskilleni* (Milne Edwards and Haime). 51*h*
*Z. ?* 52*c*

Bryozoa. 57*c*, 58*b*
*Fenestella sp.* 70*d*

*Alitaria panderi* (Muir-Wood and Cooper). 70*c*
*Antiquatonia antiquata* (J. Sowerby). 70*c*, ?70*d*, aff. 70*d*
*A. hindi wettonensis* (Muir-Wood). 30*s*, 48
*A. insculpta* (Muir-Wood). 30*o*, 58*b*
*A. sulcata* (J. Sowerby). 70*c*, 70*d*
*A. sp.* 52*c*
*Athyris expansa* (Phillips). 53*a*
*A. sp.* 53*d*
*Avonia?* 66
*Brachythyris integricosta* (Phillips). 59*g*
*B.* aff. *ovalis* (Phillips). 70*c*
*B.* aff. *pinguis* (J. Sowerby). 70*c*
'*B.*' *planicostata* McCoy. 30*s*, 33*d*
*B. sp.* 51*h*, 58*b*
*Buxtonia sp.* 51*h*, 58*b*, 70*c*
*Dictyoclostus* cf. *pinguis* (Muir-Wood). 70*c*

*Dielasma hastatum* (J. de C. Sowerby).   64
*D. radiatum* de Koninck.   70c
*D. sp.*   30t, 30x, 58b
*Echinoconchus* cf. *punctatus* (J. Sowerby).   70d
*E. sp.* [juv.].   58b
*Eomarginifera* aff. *derbiensis* (Muir-Wood).   70c
*E. sp. ?lobata* (J. Sowerby) group.   30s
*E. tissingtonensis* (Sibly) *cambriensis* (Dunham and Stubblefield).   66
*E. spp. nov.*   70c
*E. sp.*   58b, 70d
*Gigantoproductus sp.* cf. *bisati* (Paeckelmann).   59i, 59j
*G. sp.* aff. *bisati.*   70c
*G. crassiventer* (Prentice).   30o, 48, 49, 51h
*G. dentifer* (Prentice).   30s, 30t, 30x, 48, 51h, ?58b, 62d
*G. edelburgensis* (Phillips).   30s, 30t, 30x, ?33d, 51h, 53d, 58b, 59g, 59j, 59k, cf. **70c**
*G. giganteus* (J. Sowerby).   51d
*G. giganteus* cf. *crassus* (Fleming).   49
*G. sp.* cf. *moderatus* (Schwetzow).   58b, 59g
*G. sp.* cf. *semiglobosus* (Paeckelmann).   51h, 55c
*G. sp.* [latissimoid].   30s, 70c
*G. sp.*   33e, 55e, 57d, 64, 70d
*Krotovia* cf. *spinulosa* (J. Sowerby).   70c
*K. sp.* [juv.].   58b
*Linoproductus sp. hemisphaericus* (J. Sowerby) group.   30s, 30t, 30x, 53d, 59k, 63d, 64
*L. ?*   53e
*Martinia* cf. *glabra* (J. Sowerby).   50a, 70c
*M. sp.*   30t
*Megachonetes sp. papilionaceus* (Phillips) group.   58b, 62d
*M.* cf. *volvus* (McCoy).   70c
*M. sp.*   59k
Orthotetoids.   30o, 30s, 51h, 58b, 59g
*Plicatifera ?*   70c
*Productus productus* (Martin).   64
*P. productus hispidus* Muir-Wood.   51h, 58b, 59j, 70d, 70c
*Productus sp.*   30o, 51i
Productoids.   53a, 70g
*Pugilis pugilis* (Phillips).   59k
*Pugnax acuminatus* (J. Sowerby) *platylobus* (J. de C. Sowerby).   70d
*Reticularia ?*   66
*Rugosochonetes sp.*   50a
*Schizophoria resupinata ?* (Martin).   70d
*S. resupinata* cf. *gigantea* Demanet.   59k
*S. sp.*   30s, ?70c
*Semiplanus sp. latissimus* (J. Sowerby) group.   33d
Smooth spiriferoids.   30o, 30s, 30x, 33d, 51f, 51h, 52b, 52c, 53a, 53d, 59k, 70c, 70d
*Spirifer sp. bisulcatus* J. de C. Sowerby group.   48, 59k, 70d
*S.* cf. *duplicicosta* Phillips.   70c
*S.* cf. *humerosus* Phillips.   70c
*S. (Fusella)* aff. *convolutus* Phillips.   70c
*S. (F.) sp.*   70d
*S. sp.*   51h, 58b, 66
Spiriferoid.   30o
*Striatifera striata* (Phillips).   53e, 55d, 62d

*Bellerophon sp.*   51*h*, 58*b*, 59*i*, 62*d*
*Naticopsis sp.* [juv.].   62*d*
*Straparollus (Euomphalus) pentangulatus* (J. Sowerby).   62*d*

*Conocardium?*   30*o*

*Rineceras sp.*   50*a*
Orthocone nautiloid.   50*a*
*Girtyoceras sp.*   50*a*

Trilobites.   50*a*, 70*c*

Ostracods.   58*b*

*Cladodus sp.*   52*c*

UPPER MONSAL DALE BEDS

*Aphralysia sp.*   97*a*
*Girvanella staminea* Garwood.   97*a*
'*Girvanella*' nodules.   81, 82, 91*b*, 92

Foraminifera.   74*k*, 75*p*, 76*a*, 76*g*, 77*e*, 78*c*, 80, 83*a*, 83*o*, 87*i*, 88*i*, 90*d*, 92, 94*a*, 96*b*, 99*e*, 99*f*

*Aulina?*   100*j*
*Aulophyllum fungites* (Fleming).   75*e*, 76*b*, 87*i*, 100*j*
*A. fungites cumbriense* Smith.   77*e*
*Caninia benburbensis* Lewis.   ?74*k*, 83*a*
*C. juddi* (Thomson).   ?83*q*, 87*i*
*C. juddi cambrensis* Lewis.   78*c*
*C. sp. subibicina* McCoy group.   73*r*, 99*c*
*C. sp.*   ?76*g*, 77*e*, 96*b*
*Carcinophyllum sp.*   75*j*, 83*a*
*Chaetetes depressus* (Fleming).   96*b*, 97*h*, 98, 99*c*
*C. radians* Fischer.   99*c*
*C. sp.*   94*a*
*C. sp.* [encrusting].   74*q*
*Clisiophyllum* cf. *delicatum* Smyth.   98
*C. keyserlingi* McCoy.   99*f*, 100*j*
*C. sp.*   83*a*, 87*i*, 94*a*
Clisiophylloid.   99*e*
*Dibunophyllum bipartitum bipartitum* (McCoy).   55*v*, 74*q*, 77*c*, 78*c*, 80, 83*a*, 88*b*, 100*j*
*D. bipartitum craigianum* (Thomson).   80
*D. bipartitum konincki* (Milne Edwards and Haime).   74*q*, 76*b*, 78*c*, 83*a*
*D. bipartitum.*   76*g*, 87*i*
*D. sp.*   75*x*, 85*c*
*Diphyphyllum fasciculatum* (Fleming).   76*b*, 83*o*
*D. furcatum* (Thomson).   94*a*
*D. lateseptatum* McCoy.   55*v*, 73*p*, 73*r*, 74*k*, 75*e*, 77*c*, 77*e*, 78*c*, 79, 83*a*, 83*u*, 83*y*, 96*b*, 97*a*, 98, 100*j*
*D.* cf. *parricida* (McCoy).   77*e*
*D. sp.*   76*a*, 76*g*, ?83*l*, 99*c*
*Koninckophyllum* aff. *meathopense* Garwood.   98
*K. sp.*   75*x*, 96*b*, 100*j*
*Lithostrotion decipiens* (McCoy).   99*c*, 99*f*, 100*j*

*Lithostrotion junceum* (Fleming). 55*u*, 73*p*, 73*r*, 74*k*, 74*q*, 75*p*, 76*a*, 76*b*, 76*g*, 76*h*, 77*c*, 77*e*, 78*c*, 79, 83*q*, 83*v*, 85*c*, 87*i*, 90*d*, 98, 99*c*, 99*e*, 100*j*
*L. martini* Milne Edwards and Haime. 55*u*, 75*p*, 78*c*, 83*a*, cf. 83*a*, 83*e*
*L. pauciradiale* (McCoy). 55*v*, 73*p*, 73*r*, 74*k*, 74*q*, 79, 83*a*, 83*l*, 83*o*, 83*u*, 83*v*, 90*d*, 98
*L. sp.* cf. *pauciradiale* [tending to cerioid]. 74*j*, 78*c*
*L. portlocki* (Bronn). 55*v*, 73*p*, 73*r*, 74*k*, 75*g*, cf. 75*p*, 76*a*, 76*g*, 77*c*, 83*f*, 83*l*, 83*o*, 83*q*, 83*u*, 83*y*, 85*b*, 88*i*, 89, 94*a*, 100*j*
*Lonsdaleia duplicata duplicata* (Martin). 73*p*, 73*u*, 83*r*, 88*b*, 94*a*, 95
*L. duplicata.* 96*b*, 99*e*
*L. floriformis floriformis* (Martin). 55*r*, 74*j*, 74*k*, 75*h*, 83*l*, 83*m*, 83*o*, 88*b*
*L. floriformis laticlavia* Smith. 55*u*
*Nemistium edmondsi* Smith. 83*e*, 83*q*, 94*a*
*Orionastraea* aff. *indivisa* Hudson [cf. Hudson 1926, pl. 8, fig. 3]. 76*g*
*O. placenta* (McCoy). 73*r*, 76*a*, 77*c*, 77*e*, 88*i*, 99*c*
*Palaeosmilia murchisoni* (Milne Edwards and Haime). 55*r*, 77*c*, 83*a*, 83*e*, 97*h*, 98
*P. regia* (Phillips). 74*q*, 83*q*, 83*u*, 96*b*, 98
*P. sp.* 77*e*, 100*j*
*Rotiphyllum costatum* (McCoy). 74*t*
*R. sp.* [juv.] 99*d*
*Syringopora* cf. *catenata* (Martin). 75*e*, 75*g*, 75*j*, 75*p*, 78*l*, 83*n*
*S.* cf. *distans* (Fischer). 74*k*, 75*l*, 75*p*, 83*o*, 93
*S.* cf. *geniculata* Phillips. 83*a*, 83*q*
*S.* cf. *gigantea* Thomson. 77*e*
*S.* cf. *ramulosa* Goldfuss. 73*u*, 74*k*, 83*l*, 83*o*
*S.* cf. *reticulata* Goldfuss. 74*j*
*S. sp.* 55*v*, 74*e*, 76*a*, 76*b*, 76*g*, 76*h*, 78*g*, 83*e*, 100*j*
*Zaphrentites derbiensis* (Lewis). 77*c*
Zaphrentoid. 74*m*, 75*x*, 98

'*Serpula*' *subcincta* Portlock. 99*c*

Bryozoa. 75*c*, 94*a*
*Fenestella* spp. 77*e*, 96*b*, 97*h*, 98
*Fistulipora incrustans* (Phillips). 76*a*

*Alitaria panderi* (Muir-Wood and Cooper). 77*c*
*Antiquatonia antiquata* (J. Sowerby). 77*c*, aff. 97*h*, aff. 98, cf. 99*d*
*A. hindi* (Muir-Wood). 75*v*, cf. 75*x*, 77*d*, 91*e*, 98, 99*c*
*A. hindi wettonensis* (Muir-Wood). 77*c*
*A. insculpta* (Muir-Wood). 77*d*, 77*e*, ?85*a*, 96*b*, cf. 96*b*, 98, 99*a*, 99*c*, 99*d*, cf. 99*d*, 100*k*
*A. sulcata* (J. Sowerby). 91*e*
*A. sp.* ?74*c*, ?76*g*, 86*b*, ?94*a*
*Avonia thomasi* (Paeckelmann). 77*d*, 99*d*
*A. youngiana* (Davidson). 86*b*, 99*d*
*A. sp.* 74*r*, ?98
*Brachythyris integricosta* (Phillips). 77*e*, 86*b*, 99*d*
'*B.*' *planicostata* McCoy. 83*a*, 83*f*, 85*a*, 99*c*
*B. paucicostata* (McCoy). 97*h*, 98
*B. sp.* 77*c*, 94*a*, 100*f*
*Buxtonia sp.* 98, 99*a*, 99*c*
Chonetoid. 75*x*, 94*a*
*Derbyia sp.* 78*g*
*Dictyoclostus sp.* 74*c*, ?99*e*
*Dielasma sp.* 74*q*, 85*a*, 94*a*, 98

*Echinoconchus punctatus* (J. Sowerby). 77*c*, 98
*E. sp.* 75*x*, 75*z*, 77*d*, 83*a*, 86*b*
*Eomarginifera lobata* (J. Sowerby) cf. *flexa* (Muir-Wood). 98
*E. lobata laqueata* (Muir-Wood). aff. 77*d*, aff. 86*b*, aff. 91*e*, aff. 97*h*, aff. 98, 99*d*
*E. tissingtonensis* (Sibly) *cambriensis* (Dunham and Stubblefield). 91*e*
*E. sp.* 75*x*
*Georgethyris* cf. *lobata* (Muir-Wood). 74*r*
*Gigantoproductus sp.* cf. *applanatus* (Paeckelmann). 99*e*
*G. sp.* cf. *bisati* (Paeckelmann). 100*f*
*G. crassiventer* (Prentice). ?74*l*, 74*m*, 75*l*, 78*a*, 78*g*, 78*i*, 78*l*, 86*b*, 93, 100*b*
*G. dentifer* (Prentice). ?74*l*, 74*m*, 93, cf. 97*h*
*G. edelburgensis* (Phillips). 74*c*, ?74*f*, 74*q*, 74*t*, 75*c*, 77*c*, ?77*d*, ?77*e*, 78*b*, 78*d*, ?78*g*, 78*i*, 78*l*, 86*b*, 91*e*, ?94*a*, ?96*b*, ?99*e*, ?100*b*, 100*f*
*G. giganteus* (J. Sowerby) cf. *crassus* (Fleming). 93
*G. sp.* cf. *moderatus* (Schwetzow). 74*f*
*G. sp.* aff. *moderatus*. 91*e*
*G. sp.* cf. *semiglobosus* (Paeckelmann). 78*a*, 86*b*
*G. sp.* 55*r*, 73*z*, 74*q*, 75*l*, 75*z*, 78*c*, 83*a*, 83*e*, 83*f*, 83*i*, 83*m*, 83*o*, 83*q*, 87*i*, 88*b*, 95, 99*c*
*Krotovia spinulosa* (J. Sowerby). 75*x*, cf. 98, 99*c*
*K. sp.* 77*c*, 77*e*
*Linoproductus sp. hemisphaericus* (J. Sowerby) group. 78*a*
*L. sp.* 83*q*
*Martinia sp.* 75*v*, 94*a*, 97*h*
*Megachonetes sp.* 55*r*, 83*a*
Orthotetoids [with intercostal striae]. 78*g*, 78*i*, 86*b*, 96*b*, 97*h*
Orthotetoids. 76*a*, 86*b*, 98, 99*c*
*Overtonia fimbriata* (J. de C. Sowerby). 75*x*
*O. sp.* 86*b*
*Phricodothyris sp.* 86*b*
*Plicatifera sp.* 74*r*, 77*c*
Productoids. 80, 83*l*, 90*d*
*Productus productus* (Martin). 74*r*, 74*t*, 78*i*
*P. productus hispidus* Muir-Wood. 74*r*, ?77*e*, 86*b*, ?93, 96*b*, ?97*h*
*P. sp.* 83*m*
*Pugilis pugilis* (Phillips). 88*i*, 96*b*
*Pustula spp. nov.* 98
*Reticularia sp.* 77*c*, 77*e*, ?83*a*
*Rhipidomella michelini* (Léveillé). 98
*Schizophoria resupinata* (Martin). 98
*S. sp.* ?74*c*, 83*v*, 86*b*, 88*i*, 92, ?97*h*
*Semiplanus sp. latissimus* (J. Sowerby) group. 75*v*, 75*x*, 98, 99*a*, 99*d*
Smooth spiriferoids. 73*r*, 74*m*, 74*q*, 74*r*, 74*t*, 75*l*, 75*v*, 75*x*, 75*z*, 76*b*, 77*c*, 77*d*, 78*a*, 78*c*, 78*d*, 78*g*, 78*i*, 83*a*, 83*l*, 83*v*, 86*b*, 88*i*, 94*a*, 98, 99*a*, 99*c*, 99*d*, 99*e*, 100*f*
*Spirifer sp. bisulcatus* J. de C. Sowerby group. 74*c*, 74*q*, 74*v*, 78*c*, 78*d*, 91*e*, 93
*S.* cf. *duplicicosta* Phillips. 98
*S.* cf. *furcatus* McCoy. 86*b*
*S. sp.* 75*z*, 83*f*, 92, 99*e*
*S. (Fusella) grandicostatus* McCoy. cf. 74*q*, ?77*e*
*S. (F.) triangularis* (J. de C. Sowerby). 77*c*, 97*h*, 98

*Bellerophon sp.* 88*i*, 94*a*, 98
cf. *Phanerotinus nudus* J. de C. Sowerby. 98
*Naticopsis* cf. *ampliata* (Phillips). 98
*N. ? sp.* [juv.]. 86*b*
*Straparollus (Euomphalus)* cf. *catillus* (Martin). 98

*S.* (*Straparollus*) *dionysii* de Montfort.  98

*Acanthopecten* cf. *stellaris* (Phillips).  98
*Aviculopecten sp.*  99c
*Conocardium rostratum* (Martin).  86b
*Edmondia* cf. *sulcata* (Phillips).  98
*E. sp.*  99d
Pectinoid.  75v, 78c

Nautiloid.  98
*Bollandites sp.*  98

*Cummingella sp.*  75x, 98
*Griffithides sp.*  94a
*Weberides sp.*  99d
Trilobite fragments.  76h, 86b

Ostracods.  97h

*Ctenoptychius sp.*  77c
*Petalodus acuminatus* (Agassiz).  86b
*Psephodus?*  84
Fish fragments.  92

## CONCEALED D₂ STRATA

The faunas from the Wardlow Mires No. 1 (loc. 101), No. 2 (loc. 102) and Hucklow Edge No. 2 (loc. 103) boreholes have not yet been studied in detail and none of the fossils from these boreholes are listed here. Some of the important fossils are given in the text (see pp. 91–3).

*Diphyphyllum gracile* McCoy.  104
*Lithostrotion junceum* (Fleming).  104
*L. portlocki* (Bronn).  104
*Orionastrea placenta* (McCoy).  104

## EYAM LIMESTONES

Foraminifera.  105c, 122b

*Aulophyllum fungites* (Fleming) *cumbriense* Smith.  121a, 121f, 121g
*A. fungites pachyendothecum* (Thomson).  115f
*Caninia sp. nov.* aff. *buxtonensis* Lewis.  121g
*C. sp.*  121a, 121f, ?122b
*Chaetetes depressus* (Fleming).  112, 116
Clisiophylloid.  110c
*Cyathaxonia cornu* Michelin/*rushiana* Vaughan group.  121h
*C.?*  121a
*Dibunophyllum bipartitum bipartitum* (McCoy).  110i, 121a, 121f, 121g, 122b
*D. bipartitum craigianum* (Thomson).  76s, 110i, 115f, 121f
*D. bipartitum konincki* (Milne Edwards and Haime).  76s, 110i, 115f, 117c, 121a, 121f, 121g
*Diphyphyllum fasciculatum* (Fleming).  121a, 121f
*Heterophyllia spp.*  121f
*Hexaphyllia spp.*  121f

BB

*Koninckophyllum interruptum* Thomson and Nicholson. 121*g*
*K. magnificum* Thomson and Nicholson. 121*g*
*K. sp.* 115*f*, 121*a*, 121*f*
*Lithostrotion junceum* (Fleming). 105*c*, 117*b*, 117*c*, 118*d*
*L. sp.* 121*f*
*Lonsdaleia floriformis floriformis* (Martin). 105*c*, 114*e*, 117*b*
*Palaeosmilia* cf. *murchisoni* Milne Edwards and Haime. 117*c*
*P. regia* (Phillips). 105*c*
*Rotiphyllum* cf. *costatum* (McCoy). 123*a*
*R. densum* (Carruthers). 123*a*
*R.* aff. *rushianum* (Vaughan). 121*h*
Zaphrentoids. 105*c*

Bryozoa. 100*l*, 109, 112, 115*e*, 122*d*
*Fenestella spp.* 77*g*, 99*g*, 100*l*, 105*c*, 106, 109, 112, 115*e*, 116, 119, 120, 122*a*
*Fistulipora incrustans* (Phillips). 112, 122*a*

*Acanthoplecta mesoloba* (Phillips). 100*l*, 119, 120, 122*a*
*Alitaria panderi* (Muir-Wood and Cooper). 99*g*, 106, 112, 113*c*, 119, 120, 121*a*, 121*e*, 121*f*, 122*d*
*Antiquatonia antiquata* (J. Sowerby). 77*g*, cf. 99*g*, 120, 122*a*, 122*d*
*A. hindi* (Muir-Wood). 100*l*, 109, 112, cf. 113*a*, 116, 119, 120, 122*a*, 122*d*
*A. hindi wettonensis* (Muir-Wood). 75*dd*, 100*l*, 105*b*, 112, 116
*A. insculpta* (Muir-Wood). 99*g*, 100*l*, 106, cf. 107*d*, cf. 107*e*, 109, cf. 110*j*, 112, cf. 112, cf. 113*c*, cf. 114*e*, 116, cf. 116, 119, 120, 122*a*, 122*b*, 122*d*
*A. sulcata* (J. Sowerby). cf. 112, 116
*A. sp.* 105*c*, 121*e*
*Athyris sp.* 114*e*
*Avonia davidsoni* (Jarosz). 109, aff. 111, 119, aff. 119, 120, 122*a*, aff. 122*a*
*A. thomasi* (Paeckelmann). cf. 108, 109, 112, 113*c*, 115*e*, 116, 119, 120, 121*e*, 122*a*, 122*d*
*A. youngiana* (Davidson). 105*b*, 106, 108, 109, 112, 113*c*, 116, 119, 120, 121*a*, 121*e*, 122*a*, 122*d*
*A. sp.* [juv.]. ?107*d*, 112
*Brachythyris integricosta* (Phillips). 100*l*, 119, 120
'*B.*' *planicostata* McCoy. 75*dd*, 119, 122*b*
*B. sp.* 105*c*, 109, ?116, 122*d*
*Buxtonia spp.* 75*dd*, 77*g*, 105*a*, 105*c*, 109, 110*j*, 113*c*, 114*e*, 116, 119, 120, 121*e*, 122*d*
Chonetoid. ?107*d*, 113*c*, 115*c*, 121*f*
cf. *Dictyoclostus pinguis* (Muir-Wood). 112
*D. sp.* 108, 122*d*
*Dielasma hastatum* (J. de C. Sowerby). 109, 119, 120, 122*a*
*D. sp.* 100*l*, 110*i*, 113*c*, 114*e*, 115*e*
*Echinoconchus eximius* (I. Thomas). 122*a*
*E. punctatus* (J. Sowerby). 77*g*, 100*l*, 106, 109, 112, 116, 119, 120, 122*a*
*E. sp.* 99*g*. 115*e*, 121*a*
*Eomarginifera lobata* (J. Sowerby) *laqueata* (Muir-Wood). 77*g*, 99*g*, 121*e*
*E. lobata* aff. *laqueata.* 105*b*, 105*c*, 106, 107*e*, 108, 109, 112, 115*e*, 116, 119, 120, 121*e*, 122*a*
*E. sp.* ?*lobata* group. 75*dd*
*E. sp.* 100*l*, 105*a*, 107*d*, 110*j*, 113*a*, 114*e*
*Gigantoproductus edelburgensis* (Phillips). 105*c*, 108, ?110*c*, 113*a*
*G. giganteus* (J. Sowerby). 110*c*, 115*c*
*G. giganteus crassus* (Fleming). cf. 110*b*, 121*a*
*G. giganteus* cf. *inflatus* (Sarycheva). 108

*Gigantoproductus sp* cf. *gigantoides* (Paeckelmann).   113*a*, 114*a*
*G. sp.*   73*z*, 75*aa*, 105*b*, 109, 114*e*, 115*e*, 122*b*
*Girtyella? sacculus* (J. de C. Sowerby).   109, 112, 116, 119, 120, 122*a*
*Krotovia spinulosa* (J. Sowerby).   77*g*, 99*g*, 100*l*, 106, 112, 121*e*, 122*a*, 122*d*
*K.* cf. *spinulosa* (Phillips *non* J. Sowerby).   99*g*, 100*l*
*K. sp.*   105*c*
*Lingula sp.*   123*a*
*Linoproductus sp.*   112, 120
*Martinia sp.*   ?77*g*, 107*d*, 111, 122*d*, 123*a*
*Megachonetes sp.*   99*g*
*Orbiculoidea* cf. *nitida* Phillips.   123*a*
*O. sp.*   107*d*
Orthotetoids [with intercostal striae].   116, 122*a*
Orthotetoids.   99*g*, 107*d*, 109, 110*i*, 113*c*, 114*e*, 116, 119, 121*e*, 122*a*, 122*d*
*Phricodothyris sp.*   106, 110*i*, 116, 119, 120, 121*f*, 122*a*
*Pleuropugnoides pleurodon* (Phillips).   100*l*, 106, 116, 120
*P. sp.*   ?105*c*
*Plicatifera sp.*   123*a*
*P. sp.* [juv.].   77*g*, 99*g*, 116
*Plicochonetes buchianus* (de Koninck).   cf. 106, 119, 120
*P. sp.*   112, 123*a*
*Productus productus* (Martin).   107*d*, 108, ?112, 119, 122*d*
*P. productus hispidus* (Muir-Wood).   75*dd*, 77*g*, 110*b*, ?111, 121*e*, 122*h*
*P. sp.*   121*f*
Productoids.   99*h*, 100*l*
*Pugilis pugilis* (Phillips).   105*c*, 121*a*
*Pugnax pugnus* (Martin) [small form].   100*l*, 109, 113*c*, 116, 119, 120, 122*a*, 122*d*
*P. sp.*   75*dd*, ?110*j*, 112
*Pustula sp.*   121*e*
*Reticularia sp.*   99*g*, 109, 121*a*
*Rugosochonetes celticus* Muir-Wood.   123*a*
*R. sp.*   112, 121*a*, 121*e*
*Schellwienella sp.*   121*e*
*Schizophoria connivens* Phillips.   121*e*
*S. resupinata* (Martin).   109, 116, 119, 120
*S. sp.*   105*a*, 105*c*, 107*d*, 108, 110*i*, 112, 113*c*, 114*e*, 115*e*, 121*a*, 121*f*, 122*a*
*Semiplanus sp. latissimus* (J. Sowerby) group.   77*g*
Smooth spiriferoids.   75*dd*, 77*g*, 99*g*, 99*h*, 100*l*, 105*b*, 105*c*, 107*d*, 108, 109, 110*i*, 110*j*,
    112, 113*c*, 114*e*, 115*c*, 115*e*, 116, 117*c*, 119, 120, 121*a*, 121*e*, 121*f*, 121*g*, 122*a*, 122*b*,
    122*d*
*Spirifer sp. bisulcatus* J. de C. Sowerby group.   75*dd*, 100*l*, 105*b*, 105*c*, 107*d*, 110*i*,
    115*e*, 123*a*
*S. duplicicosta* Phillips.   109, 110*i*, 110*j*, 116, 122*d*
*S. sp.*   112, 114*e*, 121*a*
*S.* (*Fusella*) cf. *grandicostatus* McCoy.   113*c*, 114*e*
*S.* (*F.*) *triangularis* (J. de C. Sowerby).   109, cf. 112, 119, 120, 121*e*, 122*a*
*S.* (*F.*) *trigonalis* (Martin).   105*c*, 107*d*, 110*i*, 110*j*, 112, 113*c*, 114*e*, 121*a*, 121*e*, 121*f*,
    122*b*
*Spiriferellina octoplicata* (J. de C. Sowerby) D North.   112
*S. perplicata* D (North).   112, 116
*S. sp.*   120


*Bellerophon sp.*   99*h*, 107*d*
*Naticopsis* cf. *elongata* (Phillips).   112
*N. sp.* [juv.].   115*e*, 121*e*, 122*b*

*Platyceras?*  121*g*
*Straparollus sp.* [juv.].  115*e*
Turreted gastropod.  105*c*

*Acanthopecten stellaris* (Phillips).  116
*Actinopteria persulcata* (McCoy).  99*g*, 122*a*
*Aviculopecten perradiatus* de Koninck.  122*a*
*A. plicatus* (J. de C. Sowerby).  122*a*
*A. sp.*  ?99*g*, 100*l*
*A. planoradiatus* McCoy.  112, 116
*A. sp.*  77*g*, 107*e*
*Cardiomorpha orbicularis* McCoy.  109, 116
*C. sp.*  122*a*
*Conocardium rostratum* (Martin).  108
*C.?*  107*d*
*Edmondia maccoyii* Hind.  110*l*
*E. sp.*  115*e*
*Leiopteria hirundo* de Koninck.  100*l*
*L. laminosa* (Phillips).  120
*L. lunulata* (Phillips).  116
*L. sp.*  99*g*, ?109, 112, 122*a*
*Limipecten dissimilis* (Fleming).  120, ?122*a*
*Palaeolima simplex* (Phillips).  122*a*
*Parallelodon bistriatus* (Portlock).  122*a*
*P. sp.*  99*g*, 100*l*
Pectinoid.  123*a*
*Posidonia corrugata* (Etheridge jun.).  123*a*
*Promytilus lingualis?* (Phillips).  122*a*
*Pterinopectinella sp.*  75*dd*, 106, 109, 122*a*
*Pteronites sp.*  122*a*
*Sanguinolites tricostatus* (Portlock).  122*a*
*Streblopteria laevigata* (McCoy).  99*g*, ?122*a*
*S. sp.* [juv.].  77*g*, 109, 116

Orthocone nautiloid.  107*d*, 122*a*
Nautiloid.  122*a*

*Sudeticeras sp.*  122*a*

*Griffithides sp.*  77*g*, 109
*Weberides sp.*  123*a*

*Cyclus radialis* (Phillips).  cf. 112, ?119

Ostracods.  77*g*, 106, 107*d*, 112, 115*e*, 116, 122*a*

*Petalodus acuminatus* (Agassiz).  99*h*

### SHALES OF P₂ AGE

Chonetoid.  124
*Crurithyris sp.*  124
*Leiorhynchus sp.*  123*b*

Pectinoid.  123*b*
*Posidonia corrugata* (Etheridge jun.).  123*b*, 124

*Coleolus* or orthocone?   123*b*
*Reticycloceras sp.*   123*b*

*Sudeticeras sp.*   123*b*

*Weberides* cf. *mailleuxi* (Demanet).   124

# Appendix IV

# LIST OF GEOLOGICAL SURVEY PHOTOGRAPHS

---

Copies of these photographs are deposited for reference in the library of the Geological Survey and Museum, South Kensington, London, S.W.7, and in the library of the Geological Survey's Northern England Office, Ring Road Halton, Leeds, 15. Prints and lantern slides may be supplied at a fixed tariff.

All these photographs belong to Series L, unless otherwise specified. The National Grid references are those of the viewpoints.

## CARBONIFEROUS LIMESTONE

| | |
|---|---|
| 163 | Dam Dale. Lowest exposed strata in Woo Dale Beds [118 778]. |
| 164–5 | Junction of Hay Dale and Dam Dale. Woo Dale Beds overlain by Chee Tor Rock [118 772]. |
| 166–7 | Marvel Stones, a limestone pavement in Chee Tor Rock [104 778]. |
| 168 | Duchy Quarry, Smalldale. Chee Tor Rock [093 768]. |
| 169–70 | Potholed surface in Chee Tor Rock, Smalldale [092 770]. |
| 171 | Victory Quarry, Dove Holes. Lower Monsal Dale Beds overlying Miller's Dale Beds [076 770]. |
| 172 | Holderness Quarry, Dove Holes. Miller's Dale Beds with Dove Holes Tuff [083 782]. |
| 173 | Windy Knoll Quarry. Back reef in Bee Low Limestones, with elaterite deposit and Neptunian dyke [126 830]. |
| 174 | Windy Knoll Quarry. Detail of Neptunian dyke [126 830]. |
| 175 | Barmoor Clough Quarry. Cherty limestones in Lower Monsal Dale Beds [087 797]. |
| A9356–7 | Pindale Quarry. Monsal Dale Beds overlying Bee Low Limestones. Lateral passage into apron-reef [158 823]. |
| 178 | Litton Edge. Feature formed by Litton Tuff [160 756]. |
| A9110 | Earle's Quarry, Hope. View of Upper Bench (1958) showing knoll-reef near base of Upper Monsal Dale Beds [158 818]. |
| 179 | Bradwell Dale, southern end. Monsal Dale Beds overlain by massive flat reef in Eyam Limestones [171 804]. |
| 180 | Bradwell Dale, northern end. Monsal Dale Beds [173 807]. |
| 181–2 | Furness Quarry, Middleton Dale. Monsal Dale Beds overlain by Eyam Limestones (dark). [210 760]. |
| 183 | Shining Cliff, Middleton Dale. Monsal Dale Beds in cliff, Eyam Limestones above and Millstone Grit in background [220 757]. |
| MLD5455 | Middleton Dale. Cliffs of Monsal Dale Beds between Shining Cliff and Stoney Middleton [220 757–227 755]. |

184      Coombs Dale Quarry. Wedge-bedding in Eyam Limestones adjacent to knoll-reef [233 747].

185      as 184 [233 748].

186–7    Linen Dale. Knoll-reef in Eyam Limestones [198 768].

188      Grind Low. Features formed by knoll-reefs in Eyam Limestones [179 774].

189      Coarsely crinoidal flat reef in Eyam Limestones, near Hazlebadge Hall [173 803].

190      Siggate, Castleton. Frontal slope of apron-reef [157 824].

191      Pin Dale, seen from The Folly. Frontal slope of apron-reef [165 828].

192      Mich Low. Feature formed by $P_1$ apron-reef limestones at the eastern end of the reef-belt [171 817].

193      Middle Hill and Peak's Hill, seen from near Windy Knoll. Features of $B_2$ apron-reef limestones [125 830].

194–5    Winnats, Castleton. Gorge in $D_1$–$B_2$ limestones at inner margin of apron-reef [133 828].

196      As above [132 827].

197      Winnats, from Speedwell Mine. As above. [139 827].

198      Snelslow. Hill at the inner edge of the apron-reef [111 815].

199      Quarry near entrance to Cave Dale, Castleton. Outward-dipping fore-reef limestones [150 826].

200–1    Peveril Castle and gorge, Castleton. Gorge cut in massive $B_2$ apron-reef [148 827].

202      Perry Dale. Algal reef, forming crags, at inner margin of apron-reef [103 811].

203      Treak Cliff, Castleton. Stromatolitic limestone in algal reef [1343 8301].

204      Treak Cliff, Castleton. Fibrous calcite in algal reef [1343 8305].

205–6    Near Speedwell Mine, Castleton. Exposure of rolled-shell conglomerate in Beach Beds [1400 8270].

## MILLSTONE GRIT

207–8    River Noe, Barber Booth. Exposure of Edale Shales ($E_{1b}$) with thin limestones [1084 8482].

209      Old quarry, near Saltersford Hall. Working in 'crowstones' of $E_2$ age. [981 759].

210      River Noe, north of Fulwood Holmes. Edale Shales ($H_1$) with bullions [1662 8537].

211–3    Mam Tor landslip-scar. Alternating sandstones and mudstones of Mam Tor Beds [131 835].

214      Mam Tor from near Odin Mine. Mam Tor, with landslip in foreground [134 835].

215      Back Tor, from near Odin Mine. Features formed by Mam Tor Beds; Edale Shales in valley bottom [133 834].

216      Do.   [133 836].

217      Ashop Clough. Exposure of Mam Tor Beds [1080 9067].

218      Eyam Edge and Bretton. Scarp of Shale Grit [196 779].

219        Upper North Grain, near Snake road. Exposure of mudstone with thin
           sandstones in Shale Grit [1005 9295].

220        Quarry S.E. of Hayfield. Lower Kinderscout Grit [041 867].

221        Pym Chair and Woolpacks from Jacob's Ladder. Crags of Lower Kinder-
           scout Grit with Shale Grit in foreground [086 859].

222        Woolpacks. Closer view [086 859].

223        Kinder Downfall. Escarpment of Lower Kinderscout Grit [077 886].

224        Do.   [078 888].

225        Do.   [080 888].

226        Strines Dyke, at junction with Hollin Dale Brook. Mudstones, with
           *Reticuloceras gracile* Marine Band, overlying Kinderscout Grit [2213
           9098].

227        Rake End. Exposure of Chatsworth Grit on eastern limb of Goyt Trough
           [025 758].

228        Stanage Edge. Escarpment of Chatsworth Grit [233 848].

229–30     Do.   with head deposit and millstones in foreground [227 844].

231        Chinley Churn from Eccles Pike. Features of Chatsworth Grit, Rough
           Rock and Woodhead Hill Rock [035 812].

232        South Head and Kinder Scout from Eccles Pike. Features of Upper and
           Lower Kinderscout Grit and Chatsworth Grit [035 812].

233        Chinley Churn from the south-east. Features of Chatsworth Grit, Rough
           Rock and Woodhead Hill Rock [045 814].

234        Chinley Churn and Lantern Pike. Features of Chatsworth Grit, Rough
           Rock and Woodhead Hill Rock [032 902].

## COAL MEASURES

235        Stream east of Fernilee. Contorted band in mudstone overlying *Gastrio-
           ceras subcrenatum* Marine Band [019 785].

236        Left bank of River Goyt, Fernilee. Steeply-dipping mudstones and thin
           sandstones underlying Woodhead Hill Rock [011 785].

237        Thornset Outlier. Escarpment of Woodhead Hill Rock crossing axis of
           Goyt Trough [008 865].

238        Arden Quarry, Birch Vale. Workings in sandstone overlying Yard Coal
           [022 861].

## INTRUSIVE IGNEOUS

239        Waterswallows Quarry. Dolerite, showing columnar structure, in fore-
           ground [085 749].

240–1      Waterswallows Quarry. View of eastern face of quarry showing horizontal
           columnar structure in dolerite [085 750].

242        Waterswallows Quarry. General view looking south [085 752].

484–5      Near Damside Farm, Peak Forest. Marmorized limestone, with nodules of
           secondary silica, above Peak Forest Sill [116 788].

## EPIGENETIC MINERAL DEPOSITS

486        Dirtlow Rake. Old open workings for lead [154 821].

487–8     Odin Vein. Old open workings for lead [134 834].

489       White Rake. Workings for fluorite in vein [145 781].

490–2     Near Hazelbadge Hall, Bradwell. Workings for fluorite in pipe-vein [173 803].

493       Smalldale, south of Earle's Quarries. Workings for fluorite in pipe-vein [160 815].

494       Do.   Close-up view [160 815].

495–6     Treak Cliff, Castleton. Workings for 'Blue John' [135 830].

497       Bradwell Moor. Block of silica rock on south side of Dirtlow Rake [154 820].

### PLEISTOCENE AND RECENT

498       Near Rowlee Bridge, Ashop Valley. Valley-bulging in Edale Shales [1475 8915].

A1        Section in foundations for Howden Dam (1904). Valley-bulging in Mam Tor Beds [170 924].

A2        Section in foundations for Derwent Dam (1904). Valley-bulging in Mam Tor Beds [173 898].

499       Glacial erratic, near Bunsal Cob. Boulder of epidotized rhyolite [0197 7558].

500–1     Gravel pit, Pennant Nob. Working in glacial sand and gravel [997 810].

511       Roosdyche, Whaley Bridge. Glacial drainage channel [015 811].

A4706     Ludworth Intakes. Glacial drainage channel [994 912].

A4707     Do.   [993 910].

512       Hollins, near Rowarth. Delta of glacial sand and gravel [006 891].

513       New Mills. Gorge cut by diversion of R. Goyt through a spur of solid rock [000 852].

514       Bull Pit, Perryfoot. Sink-hole in Carboniferous Limestone [106 814].

515       Near Tor Top, Perryfoot. Sink-hole in Carboniferous Limestone [098 813].

516       Entrance to Windy Knoll Cave, Castleton. A Pleistocene bone-cave [126 830].

517–8     Peter's Stone, Cressbrook Dale. Landslipping in beds above Litton Tuff [174 754].

519       Near Alport Bridge. Landslip resting on alluvium. Edale Shales at base of section [1437 8945].

520       East side of Alport valley. Landslip below scarp of Shale Grit [136 908].

521       Do.   [134 911].

A4703–4   Coombes Rocks. Landslip below escarpment of Rough Rock [017 922].

A4705     Coombes Rocks. Closer view of Rough Rock escarpment and landslip [018 920].

### ECONOMIC SUBJECTS

522       Eldon Hill Quarry. Working for roadstone in $D_1$ limestone [113 818].

523–4     Earle's Cement Works, Hope [159 824].

525       Dolly Pit, Buxworth. View of engine-house and tip [021 830].

## GENERAL VIEWS

# INDEX

415

Cown Edge Colliery, 236, 246
Cowper Stone, 239
Cracken Edge, 354
—— —— Quarry, 236, 354
Cracoe, 9, 152
Cravenian Stage, 13
*Cravenoceras*, 166–7, 188–90, 196–7, 200, 204, 208; *C. cowlingense*, 157, 167, 187, 189, 196, 207, 209; *C. darwenense*, 203; *C. holmesi*, 168, 190, 200; *C. leion*, 157–8, 186, 196, 200–1, 207, 209; *C. malhamense*, 157, 159, 166–7, 187–8, 196, 199–201, 207; *C. nititoides*, 168, 200; *C. subplicatum*, 168, 207
—— *cowlingense* Subzone, 159, 167
—— *leion* Zone, 159, 163, 165–6, 186, 196–7, 199–200, 207, 209
—— *malhamense* Zone, 159, 166–7, 188–9, 196, 199–200, 207, 209
*Cravenoceratoides*, 190, 204; *Ct. bisati*, 168, 190; *Ct. edalensis*, 168, 187, 189–90, 196, 207; *Ct. fragilis*, 169, 191, 196; *Ct. nitidus*, 168, 200, 207; *Ct. nititoides*, 159, 168–9, 187, 190, 196, 207
—— *edalensis* Subzone, 159
—— *nitidus* Subzone, 159
—— *nititoides* Subzone, 159
—— *nitidus* Zone, 159, 167–8, 189–91, 196, 200, 206–7
Crawshaw, 291
—— Sandstone, 259, 269, 272, 291
—— ——, beds below, 272, 291
Cressbrook Dale, 10, 27–31, 66, 69, 75, 83–5, 93, 136–8, 316, 349
—— —— Lava, 27–8, 67–9, 81, 92–5, 122–3, 315
Crinoid debris, 159, 167–8, 171, 189–90, 200, 202–3, 207
—— in reef-limestone, 150
Crook Hill, 213, 217, 225, 345
Crookstone Hill, 359
—— Knoll, 218
Cross Low Vein, 306, 308, 310, 315–6
Cross-bedding, 25, 181, 219
Crosscuts, 95, 99, 311–2, 315–6
Crossdale Head, 29
Crossgate, 55, 62, 69
Cross-measures drifts, 279, 284, 286, 291
Crowden Bank, 254
—— Brook, 169, 189, 191–5, 199, 210, 214, 216, 218, 256–7
—— Tower, 175, 217–8, 256
'Crowstones', 157, 160, 164–5, 169, 201, 205, 247–54, 307, 317, 325–6, 353
cc

'Crowstones', petrography, 251
Crozzle, see contorted beds
*Crurithyris*, 105, 111
Crustacea, 134, 150, 167
Cryoturbation structures, 339
*Cryptophyllum*, 118
*Ctenopetalus lobatus*, 110
Cucklet Dale, 100
Cunliffe House, 346
"Cupola Works", 342
Curbar, 174, 216, 225, 233–4, 327, 346–7
—— Edge, 181, 236
Current-bedding, 15, 19, 176, 178, 180, 185, 277
*Curvirimula*, 183, 227, 273–4, 280, 283, 285; *C. belgica*, 182, 237; *C. sp. nov.*, 273
Cutthroat Bridge, 223–4
*Cyathaxonia*, 9, 119; *C. cornu/rushiana*, 104; *C. rushiana* 28, 66, 118–9
Cyclic sedimentation, 162, 164, 172, 178, 236
Cyclothems, 161–2, 178, 182, 184, 267, 269–73
*Cypridina*, 193, 208, 220, 274, 289
"*Cyrtina septosa* Band", see Upper *Davidsonina septosa* Band
'*Cyrtoceras*' *rugosum*, 208

DAGGER, P. F., 43
DAKYNS, J. R., 126, 264, 303, 318, 350
DALE, E., 342, 349
Dale Bottom, 215, 217
—— Dike Reservoir, 184, 232, 357
—— Head, 37–40, 44, 49–50, 131–2
DALTON, A. C., 330, 342, 349
Dalton Quarry, 355
Dam Dale, 21–2, 37–8, 41, 49–50, 134, 347
—— ——, section at junction with Hay Dale, see Hay Dale
*Dama*, 336
Damside Farm, 41, 293, 297
*Davidsonina carbonaria*, 21, 36, 39–40, 118, 129–30; *D. septosa*, 10, 23–5, 38, 43–8, 50–1, 54–60, 129, 131–3, 135, 296; *D. s. transversa*, 46, 133
—— *septosa* Band, Lower, 22–3, 25, 44–7, 51, 56–7, 131–3, 296
—— —— ——, Upper, 22–3, 25, 43–6, 51, 56–60, 131–3, 135–6, 294
DAVIES, W., 180, 233, 264
*Daviesiella*, 21, 39–41, 129–30
'*Daviesiella* Beds', 21
DAWKINS, W. B., 336–7, 349–50

434 INDEX

Nether Water Farm, 33
—— —— Mine, 307, 314, 356
—— —— Vein, 306, 314
New Bath Hotel, 151
—— Engine Mine, 305
—— Mills, 2, 156, 182, 184, 235, 238,
241, 243–4, 267, 269–70, 277, 279–80,
282, 284–7, 331, 335, 340–1, 344,
346–7, 360
—— Rake, 306, 311
—— Smithy, 219, 221, 353
NEWELL, N. D., 16, 36, 126
Newhouses Farm, 295
Newline Quarry, 46
Newstead, see Maynestone Road
Newtown, 244, 269–70, 282, 341
Noe, River, 2, 168–9, 188–95, 198–99,
201, 210–11, 329, 334–5, 345–7
—— Stool, 175
*Nomismoceras*, 114, 117; *N. spirorbis*,
115; *N. vittiger*, 116
Non-marine faunas, 159, 183
Non-sequence, 26, 79, 81, 90–1, 322
Normans Farm, 198, 199, 210
Normanswood, 280
North Derbyshire Water Board, 357–8
—— Lees, 217
—— Pennine Orefield, 308
—— Staffordshire, 271
Northumberland, 130
North-West Province, 130, 133
Norton Coal, 290
*Nuculoceras nuculum*, 157, 168–9, 187,
191, 196, 203–7; *N. stellarum*, 168–9,
187, 190–1, 196, 200; *N. tenuis-
triatum*, 204
—— *nuculum* Subzone, 159, 168
—— —— Zone, 159, 168–9, 189–91, 196,
200, 206
—— *stellarum* Subzone, 159, 168
Nun Low, 10
NUNAN, J. M., 295
Nunlow Limestones, 34
—— Quarry, 35, 108–9, 151, 201, 204


Oaken Clough, 211
Oakenclough, 179
Oaking Clough, 183, 239
*Obliquipecten*, 208; *O. costatus*, 166, 200
Odin Fissure, 107, 135, 149
—— Mine, 3, 107, 309–10
—— Sitch, 202
—— Vein, 305–7, 312
Offerton, 225

Offerton Hall, 215
—— Moor, 173, 217, 225, 245
Oil, borehole for, 327
Old Dam, 295
—— Edge Vein, 95, 99, 207, 305–9, 315,
321, 324, 380
—— Grove Shaft, 315
—— Ladywash Mine, 309, 316
—— Moor, 26, 28, 56
—— Tor Mine, 312
—— Wham Vein, 311
Oldfield, 242
Oldgate Nick, 235
Olefinite, 308
Oligoclase-andesine, 185
Olive and Partington's works, 211
Olivine, 120–2, 124, 294–6, 298, 303
Olivine-basalt, 20, 24, 27, 120–2
Olivine-dolerite, 293, 302
Oller Brook, 190, 193–5, 210, 214
Ollerbrook Booth, 191, 193
Ollersett, 282, 344
*Orbiculoidea*, 33, 102; *O. nitida*, 109, 177,
190, 220; *O. tornacensis*, 118
Orchard, The, 194
Ore minerals, 305, 307
Ore-bodies, form of, 308
Ores, secondary, 305, 310, 315
Orient Lodge, 323
*Orionastraea indivisa*, 80, 134, 139; *O.
placenta*, 30–1, 78, 80–2, 86, 89, 95,
129, 134, 137–9
—— Band, 10, 31, 68, 75–8, 80–3, 86–7,
89–93, 95, 134, 137–9
ORME, G. R., 19, 30–1, 125–6, 153
*Orthoceras*, 186, 188–9, 210, 229
Orthoclase, 172, 247, 251–63
Orthopyroxene, 298, 300
Orthotetoid, 41, 167, 196
Ostracods, 58, 117–8, 188, 226, 273–4, 285
Otter Brook, 227, 342
Outlands Head, 89, 313
—— —— Quarry, 355
Outwash fan, 332
Ouzelden Clough, 209, 213, 338, 347
*Ovatia sp. nov.*, 135, 149–50
Over Lee, 286–7
—— Leigh, 286
Overdale Brook, 215
Overstones Farm, 235
Overton, 280
*Overtonia fimbriata*, 58, 79, 116, 118,
149–5
Ox Low, 24, 51, 58
Oxlow House, 60, 134
—— Rake, 38, 63, 132, 295, 306–7, 311–3